Lecture Notes in Energy

Volume 47

Lecture Notes in Energy (LNE) is a series that reports on new developments in the study of energy: from science and engineering to the analysis of energy policy. The series' scope includes but is not limited to, renewable and green energy, nuclear, fossil fuels and carbon capture, energy systems, energy storage and harvesting, batteries and fuel cells, power systems, energy efficiency, energy in buildings, energy policy, as well as energy-related topics in economics, management and transportation. Books published in LNE are original and timely and bridge between advanced textbooks and the forefront of research. Readers of LNE include postgraduate students and non-specialist researchers wishing to gain an accessible introduction to a field of research as well as professionals and researchers with a need for an up-to-date reference book on a well-defined topic. The series publishes single- and multi-authored volumes as well as advanced textbooks.

Indexed in Scopus and EI Compendex The Springer Energy board welcomes your book proposal. Please get in touch with the series via Anthony Doyle, Executive Editor, Springer (anthony.doyle@springer.com)

Roberto Scipioni · Maria Elisa Gil Bardají ·
Linda Barelli · Manuel Baumann · Stefano Passerini
Editors

Hybrid Energy Storage

Case Studies for the Energy Transition

Springer

Editors
Roberto Scipioni
Department of Thermal Energy
SINTEF Energy Research
Trondheim, Norway

Maria Elisa Gil Bardají
Institute for Micro Process Engineering
Karlsruhe Institute of Technology
Karlsruhe, Germany

Linda Barelli
Department of Engineering
University of Perugia
Perugia, Italy

Manuel Baumann
Institute for Technology Assessment
and System Analysis
Karlsruhe Institute of Technology
Karlsruhe, Germany

Stefano Passerini
Helmholtz Institute Ulm
Karlsruhe Institute of Technology
Ulm, Germany

ISSN 2195-1284 ISSN 2195-1292 (electronic)
Lecture Notes in Energy
ISBN 978-3-031-97754-1 ISBN 978-3-031-97755-8 (eBook)
https://doi.org/10.1007/978-3-031-97755-8

This Springer imprint is published by the registered company Springer Nature Switzerland AG
The registered company address is: Gewerbestrasse 11, 6330 Cham, Switzerland

Preface

The urgent challenge of climate change demands for a fundamental shift towards a low-carbon economy, built on clean, secure, and resilient energy systems. As the global energy landscape rapidly transforms, energy storage is taking on an increasingly central role. Additionally, growing geopolitical tensions have made national energy security and the reduction of reliance on fossil fuel reserves more critical than ever. Energy storage systems are becoming key enablers of renewable energy integration, grid stability, electrification, and energy autonomy. However, as the deployment of renewable energy sources continue to grow, the intermittency of many of these technologies, particularly wind and solar, poses new technical challenges. Traditional storage solutions, typically centered on a single technology, often struggle to meet the diverse and evolving demands of modern energy systems. From fast response times to long-duration storage, from scalability to sustainability and cost efficiency, there is no one-size-fits-all solution—no silver bullet technology—that can address all requirements.

In this context, hybrid energy storage systems emerge as a promising solution. By combining the strengths of diverse storage technologies—electrochemical, thermal, chemical, mechanical, and superconducting magnetic—hybrid systems can offer tailored performance characteristics and synergistic advantages that individual technologies alone cannot achieve. Whether at the material, device, or system level, the hybridization of storage unlocks new possibilities for flexibility, resilience, and efficiency. This integrated approach enables optimal matching of storage capabilities to specific use cases, ranging from grid support, renewable energy smoothing, industrial continuity, heating and cooling applications, or electric mobility infrastructure.

The recognition of hybrid energy storage systems is gaining momentum among policymakers, researchers, and industry leaders alike. In Europe, the transition to a decarbonized energy system is inseparable from the broader goals of energy independence, climate neutrality, and economic stability.

Beyond policy and planning, hybrid energy storage is finding real-world relevance across sectors. Microgrids in remote regions depend on hybrid storage to boost self-sufficiency and reduce dependence on unreliable central grids. In transportation, the rise of electric vehicles demands fast, intelligent, and flexible charging infrastructure, something hybrid systems are uniquely positioned to support. Within industry, hybrid storage not only ensures reliable backup power during grid disturbances, but also enables the recovery and reuse of waste heat from industrial processes.

This book offers an in-depth exploration of hybrid energy storage systems, tracing their development, configurations, and practical applications through a series of compelling case studies. It is a journey into the heart of one of the most promising enablers of the clean energy transition. We hope it inspires new research, guides policy innovation, and encourages technological deployment in pursuit of a sustainable and resilient energy future.

Trondheim, Norway — Dr. Roberto Scipioni
Karlsruhe, Germany — Dr. Maria Elisa Gil Bardají
Perugia, Italy — Prof. Dr. Linda Barelli
Karlsruhe, Germany — Dr. Manuel Baumann
Ulm, Germany — Prof. Dr. Stefano Passerini
March 2025

Acknowledgments The editors want to thank all authors for supporting this project from the very beginning with their expertise and valuable time. In addition, the editors acknowledge financial support from the EU-Project Storage Research Infrastructure Eco-System[1] and the Joint Programme on Energy Storage within the European Energy Research Alliance.[2]

[1] https://www.storiesproject.eu (Grant Agreement Number 101036910).
[2] https://www.eera-energystorage.eu.

Overview of This Book

This book offers a comprehensive and novel perspective on the integration of various energy storage solutions to tackle the challenges and opportunities of a successful energy transition. Recognizing that no single technology can simultaneously meet all performance, sustainability, reliability, and cost requirements, this work advocates for the strategic deployment of hybrid energy storage solutions tailored to specific applications. Through a detailed series of case studies, this volume demonstrates how combining technologies can deliver balanced, high-performing solutions that are both economically feasible and environmentally responsible. The book examines five key application areas: Industry, Grid, Off-grid, Transport, and Buildings.

For each application, a curated selection of case studies is presented to illustrate the practical implementation of hybrid energy storage solutions. These case studies explore the key drivers behind system adoption and identify the technology-neutral requirements of each application. They examine how hybrid configurations meet these needs, followed by detailed technical, economic, and sustainability assessments to quantify both costs and accrued benefits. Where applicable, the studies also outline deployment timelines and highlight key implementation challenges. Taken together, these case studies offer more than a snapshot of current technological capabilities: they serve as a foundation for assessing the broader potential of hybrid energy storage in driving the energy transition.

But can these expectations be ever fulfilled? This book tries to give a first answer to this critical question by combining the perspectives of technology developers with those of system analysts, hybridizing energy storage that could contribute to short- and long-term storage in different application areas. Moreover, the perspective analysis of the potential sustainability challenges and the opportunities of these emerging technologies in a future decarbonized economy are provided. In this way, the reader gathers knowledge not only at the technological level, but also on a systemic scale.

Contents

Contributors

Youssef Achour LERMA Lab, College of Engineering and Architecture, International University of Rabat, Sala Al Jadida, Morocco

Giacomo Alessandri VGA Srl, Deruta, Italy

Davide Aloisio Istituto di Tecnologie Avanzate per L'Energia "Nicola Giordano", Consiglio Nazionale delle Ricerche, Messina, Italy

Laura Andaloro Istituto Di Tecnologie Avanzate Per L'Energia "Nicola Giordano", Consiglio Nazionale Delle Ricerche, Messina, Italy

Linda Barelli Department of Engineering, University of Perugia, Perugia, Italy

Manuel Baumann Institute for Technology Assessment and Systems Analysis (ITAS), Karlsruhe Institute of Technology (KIT), Karlsruhe, Germany

M. Rocío Bayón Thermal Energy Storage Unit, CIEMAT, Madrid, Spain

Ilenia Belviso Istituto Di Tecnologie Avanzate Per L'Energia "Nicola Giordano", Consiglio Nazionale Delle Ricerche, Messina, Italy

Asmae Berrada LERMA Lab, College of Engineering and Architecture, International University of Rabat, Sala Al Jadida, Morocco

Marcos Blanco Unidad de Accionamientos Eléctricos, CIEMAT, Madrid, Spain

Urte Brand-Daniels Institute of Networked Energy Systems, German Aerospace Center (DLR), Oldenburg, Germany

Giovanni Brunaccini Istituto Di Tecnologie Avanzate per l'Energia "Nicola Giordano", Consiglio Nazionale delle Ricerche, Messina, Italy

Luisa F. Cabeza GREiA Research Group, University of Lleida, Lleida, Spain

António Coelho INESC TEC—Instituto de Engenharia de Sistemas e Computadores, Tecnologia e Ciência, Porto, Portugal

Sergio Costa Typhoon HIL, Bajci Zilinskog bb, Novi Sad, Serbia

Giovanni De Carne Institute for Technical Physics, Karlsruhe Institute of Technology, Eggenstein-Leopoldshafen, Germany

Michele De Gennaro AIT Austrian Institute of Technology GmbH - Center for Transport Technologies, Vienna, Austria

Anisa Emrani National School of Applied Sciences, Cadi Ayyad University, Safi, Morocco

Hüseyin Ersoy Institute for Technology Assessment and Systems Analysis (ITAS), Karlsruhe Institute of Technology (KIT), Karlsruhe, Germany;
Center for Environmental and Sustainability Research (CENSE), NOVA School of Science and Technology, Caparica, Portugal

Marco Ferraro Istituto di Tecnologie Avanzate per L'Energia "Nicola Giordano", Consiglio Nazionale delle Ricerche, Messina, Italy

Domenico Ferrero Department of Energy, Politecnico di Torino, Torino, Italy

Gabriella Ferruzzi Portici Research Centre, ENEA—Italian National Agency for New Technologies, Energy and Sustainable Economic Development, Portici, Naples, Italy

Andrea Frazzica Istituto di Tecnologie Avanzate per L'Energia "Nicola Giordano", Consiglio Nazionale delle Ricerche, Messina, Italy

João Pedro Freitas LEPABE—Laboratory for Process Engineering, Environment, Biotechnology and Energy, Faculty of Engineering, University of Porto, Porto, Portugal;
ALiCE—Associate Laboratory in Chemical Engineering, Faculty of Engineering, University of Porto, Porto, Portugal

Federico Gallorini VGA Srl, Deruta, Italy

Olav Galteland Department of Thermal Energy, SINTEF Energy Research, Trondheim, Norway

Elahe Ghanaee ETSI Caminos Canales y Puertos, Universidad Politécnica de Madrid, Madrid, Spain

Clara Gouveia INESC TEC—Instituto de Engenharia de Sistemas e Computadores, Tecnologia e Ciência, Porto, Portugal

Kai Heussen Department of Wind and Energy Systems, Technical University of Denmark, DTU, Lyngby, Denmark

Mari Juel Sustainable Energy Technology, SINTEF Industry, Trondheim, Norway

Hanne Kauko Department of Thermal Energy, SINTEF Energy Research, Trondheim, Norway

Sridevi Krishnamurthi Battery Technology, SINTEF Industry, Trondheim, Norway

Helmut Kuehnelt AIT Austrian Institute of Technology GmbH - Center for Transport Technologies, Vienna, Austria

Marcos Lafoz Electric Power System Unit, CIEMAT, Madrid, Spain

Raffaele Liberatore Department of Energy Technologies and Renewable Sources, ENEA—Italian National Agency for New Technologies, Energy and Sustainable Economic Development, Rome, Italy

Michela Longo Department of Energy, Politecnico di Milano, Milano, Italy

Paolo Marocco Department of Energy, Politecnico Di Torino, Torino, Italy

Seyede Masoome Maroufi Institute for Technical Physics, Karlsruhe Institute of Technology, Eggenstein-Leopoldshafen, Germany

Elias Martinez IREC, Catalonia Institute for Energy Research, Sant Adrià del Besòs, Barcelona, Spain

Adélio Mendes LEPABE—Laboratory for Process Engineering, Environment, Biotechnology and Energy, Faculty of Engineering, University of Porto, Porto, Portugal

Antonio Morandi Dipartimento di Ingegneria dell'Energia Elettrica e dell'Informazione "Guglielmo Marconi", Bologna, Italy

Nicola Musicco Department of Mechanical and Industrial Engineering, University of Brescia, Brescia, Italy

Jorge Nájera Unidad de Accionamientos Eléctricos, CIEMAT, Madrid, Spain

Gustavo Navarro Electric Power System Unit, CIEMAT, Madrid, Spain

Valeria Palomba Istituto di Tecnologie Avanzate per L'Energia "Nicola Giordano", Consiglio Nazionale delle Ricerche, Messina, Italy

Stefano Passerini Center of Transport Technologies, Austrian Institute of Technology (AIT), Vienna, Austria

Dario Pelosi Department of Engineering, University of Perugia, Perugia, Italy

Juan Ignacio Pérez-Díaz ETSI Caminos Canales y Puertos, Universidad Politécnica de Madrid, Madrid, Spain

Andrea Rampini Carlo Rampini S.P.A., Passignano Sul Trasimeno (PG), Italy

Eduardo Rausell Electric Power System Unit, CIEMAT, Madrid, Spain

Margarita M. Rodríguez-García Thermal Energy Storage Unit, CIEMAT, Almería, Spain

M. Esther Rojas Thermal Energy Storage Unit, CIEMAT, Madrid, Spain

Jorge Salgado-Beceiro Department of Thermal Energy, SINTEF Energy Research, Trondheim, Norway

Massimo Santarelli Department of Energy, Politecnico di Torino, Torino, Italy

José Ignacio Sarasúa ETSI Caminos Canales y Puertos, Universidad Politécnica de Madrid, Madrid, Spain

Roberto Scipioni Department of Thermal Energy, SINTEF Energy Research, Trondheim, Norway

Francesco Sergi Istituto di Tecnologie Avanzate per L'Energia "Nicola Giordano", Consiglio Nazionale delle Ricerche, Messina, Italy

José Vicente Silva LEPABE—Laboratory for Process Engineering, Environment, Biotechnology and Energy, Faculty of Engineering, University of Porto, Porto, Portugal;
ALiCE—Associate Laboratory in Chemical Engineering, Faculty of Engineering, University of Porto, Porto, Portugal

Ricardo Silva INESC TEC—Instituto de Engenharia de Sistemas e Computadores, Tecnologia e Ciência, Porto, Portugal

Mattia Simonazzi Dipartimento di Ingegneria dell'Energia Elettrica e dell'Informazione "Guglielmo Marconi", Bologna, Italy

Filipe Joel Soares INESC TEC—Instituto de Engenharia de Sistemas e Computadores, Tecnologia e Ciência, Porto, Portugal

Kyrre Sundseth Sustainable Energy Technology, SINTEF Industry, Trondheim, Norway

Ragnhild Kjæstad Sæterli Department of Thermal Energy, SINTEF Energy Research, Trondheim, Norway

Seyede Zahra Tajalli Department of Wind and Energy Systems, Technical University of Denmark, DTU, Lyngby, Denmark

Davide Tommasini Department of Thermal Energy, SINTEF Energy Research, Trondheim, Norway

Juan Camilo Gomez Trillos Institute of Networked Energy Systems, German Aerospace Center (DLR), Oldenburg, Germany

Alberto Vazzola Carlo Rampini S.P.A., Passignano Sul Trasimeno (PG), Italy

Dario Zaninelli Department of Energy, Politecnico di Milano, Milano, Italy

Gabriel Zsembinszki GREiA Research Group, University of Lleida, Lleida, Spain

Industry and Power Generation

Preface

Hybrid energy storage systems have gained interest in recent years to cope with the unstable and poorly predictable generation of renewable energy sources. In this part different energy storage and generation systems are concerned due to their potential to outperform single energy storage systems in certain situations. Chapter "Industrial Application of High-Temperature Heat and Electricity Storage for Process Efficiency and Power-to-Heat-to-Power Grid Integration" explores the potential of thermal energy storage (TES), considering its coordinated management with electrochemical energy storage and renewable energy sources (RES). It covers various TES technologies, including sensible heat storage (SHS), latent heat storage (LHS), and thermochemical energy storage (TCS), each offering unique benefits and facing specific challenges. The integration of TES into industrial parks is highlighted, showing how these systems can optimize energy management and reduce reliance on external sources. District heating is also addressed by demonstrating the economic and environmental advantages of a multi-energy management strategy over single-energy approaches. Chapter "Repurposing Coal-Fired Power Plants Integrating the Use of Second-Life Batteries, Flywheels, and High Temperature Molten Salt Storage" further explores a HESS consisting of a thermal energy storage section but derived from a pre-existent coal-fired power plant. In this use case, thermal energy storage is used in place of the coal combustion chamber, coupled with second-life batteries and flywheels. This chapter is also connected with Second Part, *Power Grids and Renewable Transition*, since the system proposed and its operation control is developed to provide auxiliary services to the electric grid and power smoothing and storage for PV and wind generation. Finally, Chapter "Optimizing Renewable Power Systems: Hybrid Gravity-Battery Energy Storage System for Wind/PV Integration and Load Balancing" deals with hybrid energy storage systems combining gravity energy storage (GES) with high-power electrochemical energy storage, which have gained interest in recent years due to their potential to outperform single energy storage systems. The chapter focuses on the integration of such a hybrid Gravity/

Battery storage system with wind and PV energy generation, including an energy management system for the overall optimization. Reliability and efficiency in the utilization of renewable energy are enhanced by the integration of forecast models into the study and the hybridization of energy storage, which enhance the renewable power system's optimal operation and design accuracy.

Industrial Application of High-Temperature Heat and Electricity Storage for Process Efficiency and Power-to-Heat-to-Power Grid Integration

António Coelho, Ricardo Silva, Filipe Joel Soares, Clara Gouveia, Adélio Mendes, José Vicente Silva, and João Pedro Freitas

Abstract This chapter explores the potential of thermal energy storage (TES) systems towards the decarbonization of industry and energy networks, considering its coordinated management with electrochemical energy storage and renewable energy sources (RES). It covers various TES technologies, including sensible heat storage (SHS), latent heat storage (LHS), and thermochemical energy storage (TCS), each offering unique benefits and facing specific challenges. The integration of TES into industrial parks is highlighted, showing how these systems can optimize energy manage-ment and reduce reliance on external sources. A district heating use case

A. Coelho · R. Silva · F. J. Soares · C. Gouveia (✉)
INESC TEC—Instituto de Engenharia de Sistemas e Computadores, Tecnologia e Ciência, Porto 4200-465, Portugal
e-mail: clara.s.gouveia@inesctec.pt

A. Coelho
e-mail: antonio.m.coelho@inesctec.pt

R. Silva
e-mail: ricardo.emanuel@inesctec.pt

F. J. Soares
e-mail: filipe.j.soares@inesctec.pt

A. Mendes · J. V. Silva · J. P. Freitas
LEPABE—Laboratory for Process Engineering, Environment, Biotechnology and Energy, Faculty of Engineering, University of Porto, Porto 4200-465, Portugal
e-mail: mendes@fe.up.pt

J. V. Silva
e-mail: jvsilva@fe.up.pt

J. P. Freitas
e-mail: joaofreitas@fe.up.pt

J. V. Silva · J. P. Freitas
ALiCE—Associate Laboratory in Chemical Engineering, Faculty of Engineering, University of Porto, Porto 4200-465, Portugal

R. Scipioni et al. (eds.), *Hybrid Energy Storage*, Lecture Notes in Energy 47,
https://doi.org/10.1007/978-3-031-97755-8_1

also demonstrates the economic and environmental advantages of a multi-energy management strategy over single-energy approaches. Overall, TES technologies are presented as a promising pathway to greater energy effi-ciency and sustainability in industrial processes.

Keywords District heating · Industry decarbonization · Microgrids · Multi-energy systems · Thermal storage

1 Introduction

The industrial sector plays a major role in global greenhouse gas emissions, highlighting the urgent need for its decarbonization to halt climate change. In the first quarter of 2024, industrial emissions in the EU reached 669 thousand tonnes of CO_2 equivalents, nearly three times the emissions produced by households during the same period (Eurostat 2025). One promising approach to address this challenge is the integration of high-temperature heat and electricity storage systems, which can enhance process efficiency and support the transition to RES. These systems allow for the storage of excess renewable energy, which can be converted back into electricity or used directly for industrial heating processes, thereby improving process efficiency and supporting grid stability by balancing supply and demand (SETIS 2023; Pompei et al. 2023; Akar et al. 2021). This chapter explores the mid- to long-duration applications of these storage systems, focusing on their role in power-to-heat-to-power (P2H2P) grid integration.

The motivation for this chapter stems from the need to address the dual challenges of energy efficiency and sustainability in industrial operations. By leveraging TES technologies, industries can optimize energy use, reduce emissions, and improve resilience against energy supply fluctuations. TES technologies are being utilized in various industrial applications to supply thermal loads and incorporate RES. For example, industries are using TES to store solar thermal energy, which can be used to meet heating demands during periods of low solar irradiance. Additionally, TES systems are being employed to capture and store waste heat from industrial processes, which can then be reused, thereby reducing overall energy consumption and emissions (IRENA 2020). This is particularly relevant as industries seek to balance the increasing demand for energy with the imperative to reduce their carbon footprint (IRENA 2020; Energy.gov 2023; Sun et al. 2023).

Achieving these objectives requires the optimal management of TES, which in industrial settings often involves the integration with Combined Heat and Power (CHP) systems. CHP systems can simultaneously generate electricity and useful heat, and when combined with TES, they can store excess heat for later use, enhancing overall system efficiency. This integration allows industries to better manage energy loads, reduce peak demand, and improve energy security (Benalcazar 2021).

On the other hand, the maturity of TES technologies varies, with some solutions being well-established and widely adopted, while others are still in the development or early deployment stages. Sensible heat storage, using materials like water or molten salts, is a mature technology with numerous industrial applications. On the other hand, advanced TES technologies, such as thermochemical storage, are still being researched and developed to improve their efficiency and cost-effectiveness (Ali et al. 2024).

Multi-energy management, which involves the coordinated use of various energy sources and storage systems, offers significant advantages over single-energy management. By integrating TES with other energy systems, such as electrical grids, gas networks, and RES, industries can achieve greater flexibility, resilience, and efficiency. This holistic approach allows for better optimization of energy flows, reduced operational costs, and enhanced sustainability (Coelho 2023; Glücker et al. 2024).

Following this introduction, the chapter focuses on specific applications of TES in industrial processes, including cogeneration, biomass gasification and hydrogen production. Subsequent sections present a methodology for optimizing P2H2P systems, discuss a practical use case, and consider the broader implications and future directions of TES integration in industrial settings.

2 Thermal Energy Storage: A Brief Overview

TES is an emerging technology that stores thermal energy from diverse sources, such as industrial waste heat, concentrated solar power, or electricity generated from RES, for future heat and/or power generation applications, as illustrated in Fig. 1. These systems can provide mid-to-long-term storage duration, from weeks to months, and can present remarkable low-cost energy storage and high round-trip efficiencies by employing suitable materials and proper insulation systems. These systems can be used in space and water heating, cooling, power generation, and peak shaving applications (SINTEF 2018).

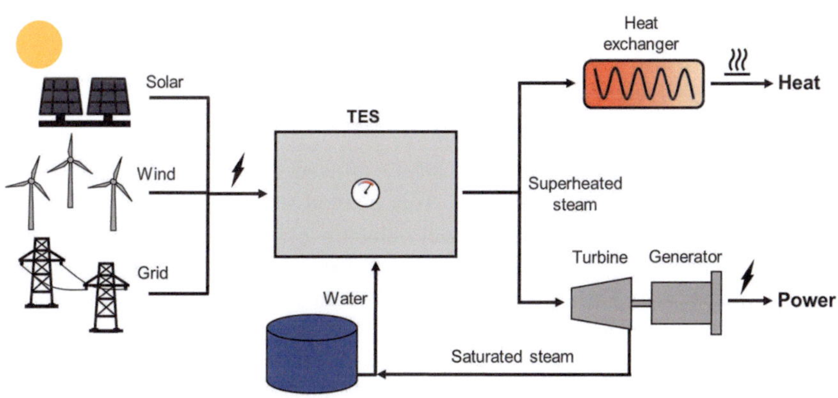

Fig. 1 Schematic integration of a thermal energy storage system

TES systems can be classified into four groups according to the operating temperature:

- Cold TES—$T < 20$ °C
- Low-temperature TES—20 °C $< T <$ 100 °C
- Mid-temperature TES—100 °C $< T <$ 300 °C
- High/very high-temperature TES—$T >$ 300 °C (high), T > 500 °C (very high).

Solutions for all temperature ranges are needed to meet industry needs. Although most global heat demand is for low-to-medium-temperature applications, several hard-to-decarbonize industry areas, like cement and steel production, require very high temperatures, typically above 1000 °C to operate (IRENA 2023).

Like any other energy storage system, TES systems can be described in terms of the following parameters:

- **Capacity**—presents the total energy stored in the system, generally in kWh.
- **Power**—defines the amount of heat transferred per second for the storage medium in charging or from the storage medium in discharging, generally in kW.
- **Efficiency**—quantifies the amount of energy retrieved compared with the energy given to the system, thus accounting for heat losses during storage, commonly in %.
- **Storage period**—defines the duration of storage, which can be given from seconds to months depending on the type of storage.
- **Charge and discharge time**—outlines how much time the system takes to charge or discharge, usually in hours.
- **Cost**—states the cost per unit of energy stored, typically in $€ \cdot kWh^{-1}$.

However, considering its applications the following specific characteristics are also relevant, its temperature range for different applications and the characteristics of its materials (e.g. density, specific heat and thermal conductivity. The next sections characterize the most relevant types of thermal energy storage technologies for industrial applications.

2.1 *Sensible Heat Storage*

Sensible Heat Storage (SHS) is extensively utilized due to the practicality and simplicity of its underlying principles. This method stores energy by varying the temperature of a storage medium without inducing a phase change. Figure 2 illustrates the typically linear relationship between the amount of heat stored and the system's temperature. Heat is absorbed and released through radiation, conduction, and/or convection processes.

The storage capacity of a material is primarily determined by its specific heat capacity, which is the amount of heat required to raise the temperature of 1 kg by 1 °C. A material with a lower specific heat capacity must reach a higher temperature to store the same amount of energy as a material with a higher specific heat capacity.

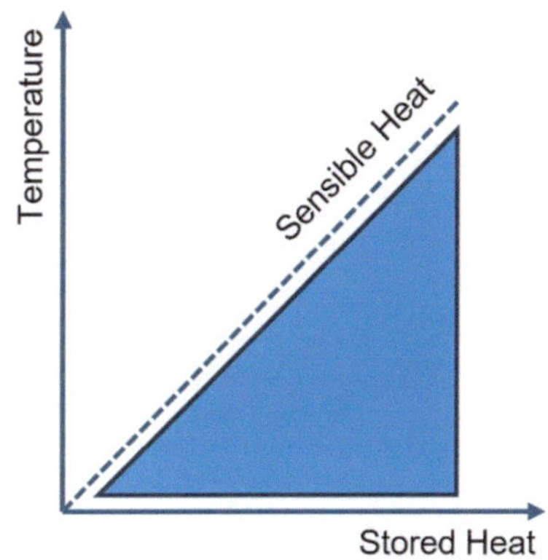

Fig. 2 Linear relationship between the amount of stored heat and the system's temperature, demonstrating how temperature increases in direct proportion to heat stored

For example, 1 kg of water ($c_p = 4.18$ kJ kg^{-1} °C^{-1}) stores approximately 200 kJ when heated from 20 to 68 °C. In contrast, 1 kg of air ($c_p = 1.005$ kJ kg^{-1} °C^{-1}) must be heated to around 199 °C to store the same heat.

SHS systems utilize the specific heat capacity to store heat in a medium by raising its temperature. The quantity of heat stored is given in (1),

$$Q_{stored,\,SHS} = \int_{T_i}^{T_f} m \cdot c_p \cdot \mathrm{d}t \tag{1}$$

where $Q_{\text{stored, SHS}}$ is the heat stored, m is the storage medium mass, c_p is the specific heat capacity, and T_i and T_f the initial and final temperatures, respectively.

Table 1 provides an overview of the most used materials for this technology, highlighting water as the most effective SHS liquid-state material due to its low cost, abundancy, and high specific heat. The primary characteristics of widely utilized solid-state thermal storage materials, such as sand-rock minerals, concrete, fire bricks, and molten salts are also detailed.

Molten salts are currently the state-of-the-art in energy storage for mid to high/very-high temperature applications such as concentrated solar power. With a storage cost ranging from 4 to 20 € kWh^{-1}, these systems can operate in the salt's thermally stable liquid phase, between 250 and 600 °C, having a storage energy density of ca. 200 kWh m^{-3} (European Commission 2023). However, TES systems based on sensible heat face significant limitations compared to other available methods. These limitations stem from the materials used, which typically have lower energy density and poor thermal conductivity. This complicates the device design, particularly in terms of efficiently extracting stored energy, as well as in impractically large storage units. Additionally, the variable outlet temperature, in the case of electricity production, can result in unused thermal capacity when the outlet temperature falls below the turbine's operational threshold.

Table 1 Liquid (highlighted in blue) and solid-state materials (in orange) utilized in TES systems (Alva et al. 2018; Bauer et al. 2021)

Material	Temperature range (°C)	Density (kg m^{-3})	Specific heat (kJ kg^{-1} °C^{-1})	Thermal conductivity (W m^{-1} K^{-1})
Water	0–100	998	4.18	0.6
Thermal oil	12–400	900	2.00	0.1
Molten salts	250–600	2160	0.85	7.0
Sand-rock minerals	200–300	1700	1.30	1.0
Reinforced concrete	200–400	2200	0.85	1.5
Silica fire bricks	200–700	1820	1.00	1.5
Magnesia fire bricks	200–1200	3000	1.15	5.0

2.2 Latent Heat Storage

Latent Heat Storage (LHS) involves phase change materials (PCMs) with high energy density, high thermal conductivity, and small volume change to provide a constant outlet temperature at the melting or vaporization point. The phase transition can be solid–solid, such as the crystalline form transitions in iron; solid–liquid, consisting of a melting-solidification process; or liquid–vapor, where vaporization occurs. Figure 3 illustrates the characteristic plateau in the temperature profile during phase change, representing the constant temperature at which latent heat is absorbed or released.

Although the liquid–vapor transition usually possesses higher energy, it involves a huge volume expansion, which precludes technological implementation. On the other hand, solid–liquid transitions encompass an average 10% volume expansion with a significantly higher energy density than sensible heat systems, making them

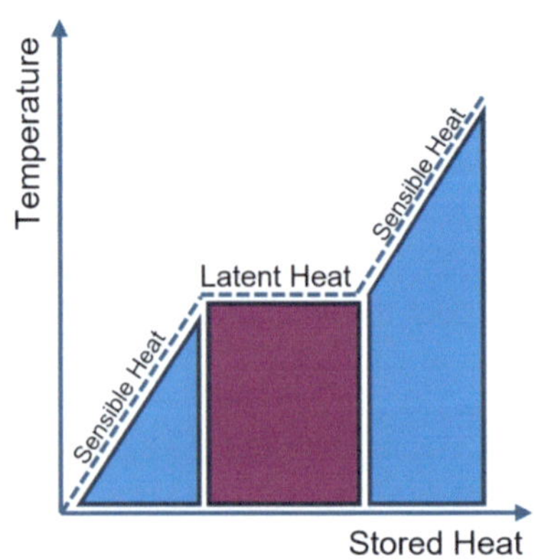

Fig. 3 Temperature profile showing the plateau during phase change, where latent heat is consistently absorbed or released

the focus of study in latent-TES systems. The heat stored for a liquid-solid phase transition can be determined in (2),

$$Q_{stored,\,LHS} = \int_{T_i}^{T_m} m \cdot c_{p,s} \cdot \mathrm{d}t + m \cdot h_f + \int_{T_m}^{T_f} m \cdot c_{p,l} \cdot \mathrm{d}t \tag{2}$$

where $c_{p,s}$ and $c_{p,l}$ represent the specific heat capacity of the solid and liquid phase, h_f the enthalpy of fusion and T_m the melting temperature.

PCMs can be organic (sugar alcohols), inorganic (salts, metal alloys, etc.), and eutectic mixture materials. Sugar alcohols appear as promising PCM candidates for mid-temperature TES applications for their high specific heat storage capacity, non-toxicity, non-flammability, and absence of corrosion to the insulation materials. However, the low thermal conductivity (ca. 0.5 W m^{-1} k^{-1}) and reduced thermal endurance with T_m over 150 °C present significant limitations to their implementation.

On the other hand, inorganic materials can withstand high temperatures with greater thermal stability. Molten salts (e.g., nitrates, carbonates, and chlorides), which are especially used in the liquid state for sensible heat TES applications, can also function as PCMs for mid- to high-temperature TES systems. Besides a high specific heat storage capacity, they are generally low-cost, thus contributing to a high specific heat storage capacity per unit cost. Nevertheless, molten salts are largely affected by their low thermal conductivity (< 1 W m^{-1} k^{-1}), high volume expansion ratio, and high levels of chemical corrosion to the shell materials.

Finally, metal alloys have been gaining attention for their superior energy density, given by metal's high densities. Thus, more compact TES systems can be achieved. In addition, metallic PCMs present high thermal conductivities and very low volume expansion ratios during phase transition. The challenges here are related to high chemical corrosion exhibited by the liquid metal alloys, which force the use of ceramics over the structural materials of the storage units.

Tables 2 and 3 present the PCMs candidates considered for integration into the mid, high, and very high-temperature TES. Organic and salt PCMs were grouped together in Table 2, while metal alloys are presented in Table 3. Metal alloys containing lead (Pb) and cadmium (Cd) were excluded for their threats to human life and environment (EPA 2000, 2011).

2.3 Thermochemical Heat Storage

Thermochemical Heat Storage (TCS) operates through two primary mechanisms: chemical reactions and sorption processes. In chemical reactions, energy is stored as the heat of reversible reactions, while sorption processes involve storing thermal energy either through adsorption (physical bonding) or absorption (dissolution of a

Table 2 Organic and molten salts PCMs for mid and high temperature TES system (Takahashi et al. 1988; Nomura and Akiyama 2017; Liu et al. 2022; Tye et al. 1976)

Material	Melting temperature (°C)	Latent heat ($kJ\ kg^{-1}$)	Density ($kg\ m^{-3}$)	Energy density ($kWh\ m^{-3}$)[a]	Thermal conductivity ($W\ m^{-1}\ K^{-1}$)
D-mannitol[b]	167	316	1520	133	0.7
Pentaerythritol[b,c]	186	285	1399	111	0.1–0.5
$AlCl_3$	193	290	2440	197	(Not found)
$NaNO_3 + KNO_3$ (60/40 wt%)	222	110.7	1712	53	0.55
$LiNO_3$	252	379	2400	253	0.51

[a]Considering only latent heat
[b]Organic material (sugar alcohol)
[c]Solid-solid transition

material). TCS technologies can accommodate a wide range of temperatures, from as low as 0 °C to as high as 900 °C, and typically provide storage durations ranging from several hours to days and potentially up to several months (EASE 2023).

This technology is recognized for its high energy density, as it stores large amounts of energy within compact volumes by converting heat into reversible chemical bonds. This capability makes it well-suited for long-term storage, with minimal energy losses over extended periods and reduced heat dissipation compared to other storage methods.

Despite its potential, thermochemical storage faces several significant challenges compared to more established technologies, namely the high cost of materials necessary for chemical reactions, which impacts the economic competitiveness of these systems. Also, the slow reaction kinetics observed in some thermochemical processes does not allow for rapid absorption and release of energy, affecting the dynamic response of the TES. Finally, the complexity of system design—requiring precise control over temperature, pressure, and chemical conditions—contributes to higher operational and maintenance costs. Integration with existing energy infrastructure is also demanding due to the need for high-pressure vessels and sophisticated heat exchangers.

3 Integration in Industrial Parks

Integrating TES systems into industrial parks plays a pivotal role in enhancing energy efficiency, reducing emissions, and facilitating the transition to RES. TES enables the storage of excess energy—such as waste heat from industrial processes or surplus electrical generation from renewable sources—for later use in heating, cooling, or power generation. This allows industrial parks to optimize energy management by

Table 3 Metals and metallic alloys PCMs for mid, high, and very high- temperature TES systems (Costa Pereira and Kenisarin 2022; Morando et al. 2014; Manasijevic et al. 2021; The Engineering ToolBox 2005; King 1988; Bilek et al. 2006; Wang et al. 2006)

Material	Melting temperature (°C)	Latent heat ($kJ\ kg^{-1}$)	Density ($kg\ m^{-3}$)	Energy density[a] ($kWh\ m^{-3}$)	Thermal conductivity ($W\ m^{-1}\ K^{-1}$)	Energy T_{amb} to T_m ($kWh\ m^{-3}$)	Coefficient of thermal expansion ($\mu m\ m\ °C^{-1}$)
In/Sn[b] (52/48 wt%)	120	24.9	7220	50	40.8 (s), 61.4 (l)	47	(Not found)
Sn/Bi[b] (43/57 wt%)	139	49.1	8560	117	19 (s), 15.5 (l)	47	(Not found)
In	156.8	28.5	7030	56	86 (s), 36.4 (l)	61	32.1
Li	178	442	530	65	84.8 (s), 44 (l)	100	46
Sn/Zn[b] (91/9 wt%)	198	65.4	7270	132	61.9 (s), 56 (l)	88	20.88
Sn/Ag[b] (96.5/3.5 wt%)	221	63	7500	131	78 (s)	109	22.5
Sn/Cu[b] (99.3/0.7 wt%)	227	62.3	7300	126	65(s)	93	(Not found)
Sn	232	60.5	7280	122	67 (s), 61 (l)	91	22
Bi/Zn (96.5/3.5 wt%)	251.4	50.3	9340	131	50	≈ 75	(Not found)
Bi	271.4	52.2	9747	141	8.2 (s)	83	13.4
Mg/Zn (46.3/53.7 wt%)	340	185	4600	236	(Not found)	≈ 294	(Not found)
Al/Zn[b] (5/95 wt%)	382	118	6623	217	110 (s)	280	27
Zn	420	100.9	6730	189	120 (s), 100 (l)	290	30.2

(continued)

Table 3 (continued)

Al/Si[b] (88/12 wt%)	577	560	2700	420	160 (s)	406	20
Mg	650	358	1738	173	156	309	24.8
Al	660	386.9	2700	290	234 (s), 210 (l)	432	23.1
Si/Mg/Ca (49/30/21 wt%)	865	305	2250	191	(Not found)	264	(Not found)
Si/Mg (56/44 wt%)	946	757	1900	400	(Not found)	386	(Not found)
Ag	961.8	104.8	10,490	305	426 (s), 356 (l)	578	18.9
Cu	1085	206	8960	513	399 (s), 340 (l)	1020	16.5
Si	1414	1800	2425	1225	149 (s)	664	2.6

[a]Considering only latent heat
[b]Eutectic mixture

balancing supply and demand, reducing reliance on external energy sources, and improving overall system flexibility.

Below are key applications where TES can be integrated into industrial processes to maximize energy use efficiency:

- **Concentrated Solar Plants (CSP)**. As already implemented with molten salts, TES systems in CSP allow the storage of the harvested solar energy during the sunlight hours. This capability enables CSP facilities to maintain continuous electricity generation and dispatchability, enhancing grid stability and reliability.
- **Cogeneration Plant Integration**, where TES allows excess electricity generation or heat to be stored for later use, enabling the dispatch of thermal energy when the cogeneration plant is offline. This approach enhances energy reliability and ensures that industrial processes continue to receive power and heat as needed, even during periods of reduced generation.
- **Biomass Gasification** process depends on the supply of superheated steam at temperatures as high as 800 °C, which is crucial for converting biomass into syngas. However, achieving such high temperatures presents considerable challenges, including increased operational costs and safety concerns related to high-temperature operations and equipment durability.
- **Peak Shaving Applications for Electrolyzers**. Alkaline (AE) and proton exchange membrane (PEM) electrolizers require a stable direct current power supply to ensure high efficiency and purity of the H_2 produced. TES can be used in peak shaving applications to ensure the constant power supply of this equipment. In this case, thermal energy in the form of superheated steam can be converted to electrical energy using steam turbines and generators, thereby increasing the operational efficiency of the electrolyzers during high-demand periods.
- **Integration with Renewable Methanol Plants**. In renewable methanol production, integration with industrial park energy systems can occur in multiple ways. For instance, thermal energy can be supplied to both gasification and distillation units to improve the efficiency of methanol production. Additionally, steam turbines may be utilized to power the entire methanol plant or specific units within the facility, reducing the reliance on external energy sources.
- **Hydrogen and Oxygen Purification via Temperature Swing Adsorption**. In hydrogen and oxygen production from electrolyzers, purification can occur through temperature swing adsorption (TSA). This technique involves the adsorption of water from the gas streams using an adsorbent bed. To regenerate the adsorbent, heat is applied at a constant temperature, creating a significant thermal demand. The required temperature for regeneration ranges from 100 to 300 °C, depending on the adsorbent material employed.

4 Power-to-Heat-to-Power Grid Integration—A Methodology to Optimize P2H2P Systems

Industrial parks typically integrate multi-energy resources and multi-energy networks, as represented in Fig. 4. Decarbonizing industrial parks needs to consider both electric resources (like photovoltaic panels (PVs), batteries, and heat pumps) and thermal resources (like CHP systems and TES). Two energy networks may also need to be considered, namely an electricity network and district heating. The industrial loads (to consume heat), the CHPs (to provide heat), and thermal storage systems (to provide heat) are connected to the district heating. The CHPs are also connected to the gas network (to consume gas) and to the electricity network (to provide electricity), while the thermal storage is connected to the electricity network (to consume electricity). The PVs, batteries, and heat pumps are only connected with the electricity network.

The industrial park energy resources (see Fig. 4) can be operated in a coordinated way as a multi-energy microgrid (MES), in order to minimize the energy-related costs, considering participation in the energy markets (electricity, gas, and carbon). Figure 5 presents the main steps involved in the optimization of the day-ahead bidding process, minimizing the energy costs of the industrial microgrid.

In this chapter, the Iberian Energy market operation was considered. It was assumed that the industrial park operator (IO) will act as a price-taker. This way, the IO submits demand bids (electricity, natural gas, and carbon) at cap-prices, and submits supply bids (electricity and secondary reserve) at floor prices during the day-ahead phase of the markets. The day-ahead phase occurs in the day before (d-1)

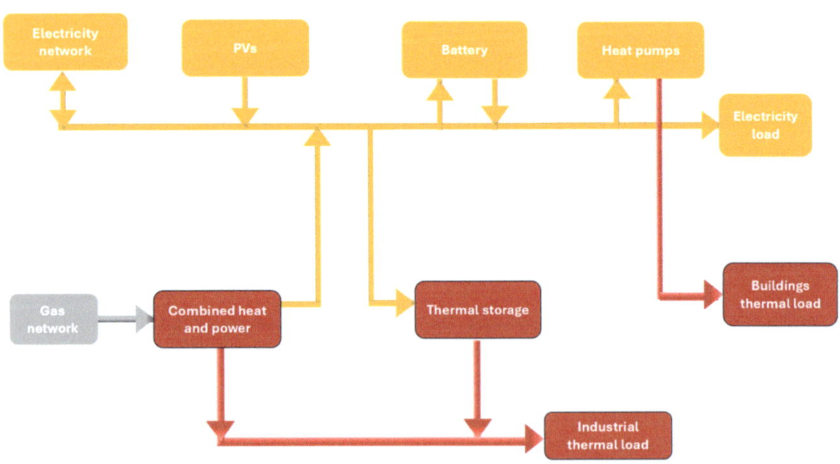

Fig. 4 Multi-energy resources

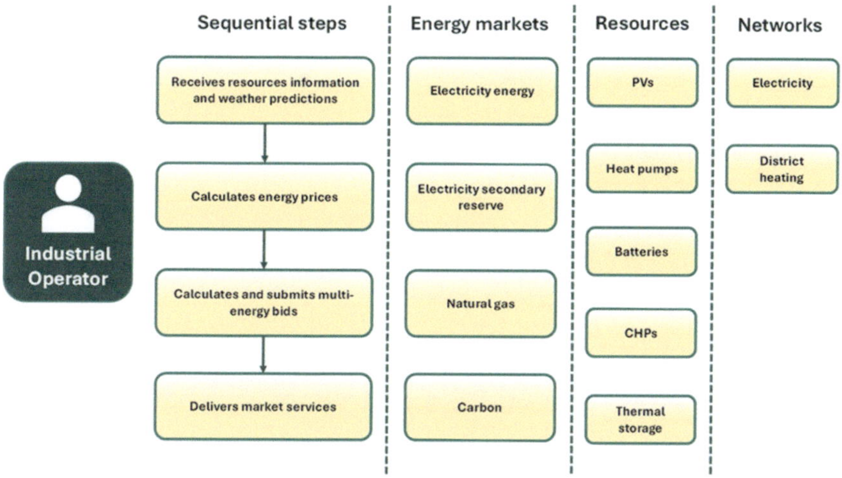

Fig. 5 Scheme of the methodology with the sequential steps, energy markets, resources, and networks considered

of the trading day (d). During the trading day (d), the IO needs to deliver the market services that were bought/sold during the day-ahead session.

The objective function of the optimization problem is to minimize the net cost of trading energy services in the multi-energy markets (electricity energy, secondary reserves, natural gas, and carbon), as in (3). The system cost can be divided into four terms: the first term is the net cost of buying and selling energy and secondary reserves,, f_t^E, the second term is the cost of buying natural gas, f_t^G, and the third cost are the costs of buying carbon allowances $f_t^C + f^{CFA}$.

$$\min \sum_{t \in T} \underbrace{f_t^E}_{\substack{\text{Buying energy} \\ \text{Selling energy} \\ \text{Selling secondary reserves}}} + \underbrace{f_t^G}_{\text{Buying natural gas}} + \underbrace{f_t^C + f^{CFA}}_{\text{Buying carbon allowances}} \tag{3}$$

$$f_t^E = \lambda_t^E E_t^E \Delta t - \lambda_t^B (U_t^E + D_t^E) + \left(\lambda_t^{D,E} \phi_t^D D_t^E - \lambda_t^{U,E} \phi_t^U U_t^E \right) \Delta t \tag{4}$$

$$f_t^G = \lambda_t^G E_t^G \Delta t + \left(\lambda_t^{U,G} \phi_t^U U_t^G - \lambda_t^{D,G} \phi_t^D D_t^G \right) \Delta t \tag{5}$$

$$f_t^C = \lambda^{CO2} \sum_{j \in J_n} \left(P_{j,t}^{CHP,E} + U_{j,t}^{CHP,E} . \phi_t^U - D_{j,t}^{CHP,E} . \phi_t^D \right) \alpha^{CO2,G} \Delta t \tag{6}$$

$$f^{CFA} = \lambda^{CO2} . \mathrm{A}^{+,CO2} \tag{7}$$

According to the rules of secondary reserve market, the upward secondary reserve band must twice as much as the downward secondary reserve band offered (Iria et al.

2019), as described in (8).

$$U_t^{DA} = 2.D_t^{DA}, \quad \forall t \in T \tag{8}$$

Constraint (9) and (10) define the carbon allowances that the IO needs to buy from the heat produced by the CHPs.

$$\mathrm{A}^{+,\mathrm{CO2}} - \mathrm{A}^{-,\mathrm{CO2}} = \sum_{t \in T} \sum_{j \in J_n} \left[\left(P_{j,t}^{CHP,H} + U_{j,t}^{CHP,H}.\phi_t^U - D_{j,t}^{CHP,H}.\phi_t^D \right).\alpha^{\mathrm{CO2},G}.\Delta t \right] - FA^{\mathrm{CO2}} \tag{9}$$

$$\mathrm{A}^{+,\mathrm{CO2}}, \mathrm{A}^{-,\mathrm{CO2}} \geq 0 \tag{10}$$

To facilitate the readability of the chapter, the rest of the formulation will not be fully presented. Instead, we provide an overview of how the bids, resources, and networks were modeled.

The electricity energy bids were calculated considering the consumption from heat pumps, batteries, thermal storage systems, and inflexible loads, and the injection from PVs, batteries, and CHPs. The natural gas bids were calculated considering the consumption from CHPs. The carbon bids were calculated considering the emissions of CO_2 from CHPs. The upward and downward bids were calculated coidering the flexibility provided by PVs, batteries, heat pumps, and CHPs.

A large set of constraints have been considered for the energy resources, which are related to their operational limits and technical characteristics of the following:

- PV—limits related with maximum and minimum power and available reserves.
- Heat pumps—limits related with maximum and minimum power, thermal characteristics of the building's thermal loads, and available reserves.
- Batteries—limits related with maximum and minimum power and capacity, and available reserves.
- Combined heat and power—limits related with maximum and minimum power and available reserves.
- Thermal storage—limits related to maximum and minimum power and capacity.

See references (Coelho 2023; Coelho et al. 2021) for a detailed description of the constraints considered.

In relation to the electricity and district heating networks, they each have their own formulation. The electricity network is modelled using a non-convex formulation of the branch flow model (Baran and Wu 1989). This model considers the branch power flows, and the limits of voltage and current. The district heating is modelled using a hydraulic and a thermal model (Cao et al. 2019). With these models it is possible to calculate the mass flows and temperatures of the pipes and nodes of the network.

To optimize the mathematical problem considering the restrictions of the multi-energy resources, markets, and networks, we decompose the problem into sub-problems using a distributed algorithm, the alternating direction method of multipliers. By doing this, we will have three main sub-problems: one considering the restrictions of the markets and resources, another one considering the restrictions of the electricity network, and another one considering the restrictions of the district heating. This way, the problem becomes easier to solve as the sub-problems are smaller.

4.1 Day-Ahead Dispatch of a MES: The Case of the Manchester Microgrid

In order to illustrate the benefits of an integrated energy dispatch in a MES, the Manchester microgrid case is describe in this section, considering real data from a university campus's multi-energy microgrid, which features both electricity and heat networks, as shown in Fig. 6.

The data for the networks is available in (Martínez Ceseña et al. 2020) and includes network parameters and inflexible load profiles for electricity, gas, and heat in the buildings. In the electricity network, the voltage limits were set between 0.9 and 1.1 p.u., with the slack bus voltage fixed at 1 p.u.. For the heat network, the mass flow limit was set at 40 kg/s, the generator supply temperature was fixed at 85 °C, each load's outlet temperature was set to 70 °C, and the ambient ground temperature was defined as 7 °C.

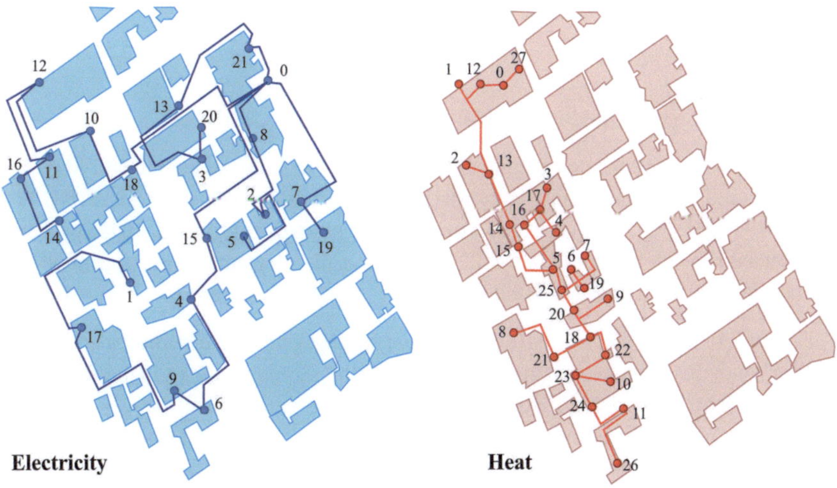

Fig. 6 Electricity and heat networks of the study case. Adapted from Martínez Ceseña et al. (2020)

Table 4 Resources' parameters

Resources	Power (kW)	Capacity (kWh)	Efficiency	COP
PV	750–1500	–	–	–
Battery	100	250	0.9	–
Heat pumps	750	–	–	3.45
CHP	5000	–	0.35 (electricity) 0.45 (heat)	–
Thermal storage	375	750	1	–

The Distributed Multi-Energy Resources (DMERs) can be connected to the electricity and heat networks. The DMERs connected to the electricity network are PV systems, energy storage systems (ESSs), heat pumps (HPs), and CHPs. The CHPs are also connected to the heat network. The district heating flexible loads are connected to the heat network. Table 4 presents the resources' parameters.

The buildings connected to the HPs and district heating system are maintained within a comfort temperature range of 19–23°C between 7 and 18 h, and 16–26 °C for the remaining hours of the day. Real outdoor temperature profiles were used in the simulations performed.

The electricity market data considers the energy prices, secondary reserve prices, upward and downward tertiary reserve prices, and ratios of upward and downward mobilizations (Fig. 7) (ENTSO-E 2025; REN 2025). The gas market data considers the gas prices (63.07 €/MWh) (MIBGAS 2025). The carbon market data considers the price of CO_2 emissions (82.3 €/permit) and number of free allowances (2.8). The data chosen is from 15 of January of 2023.

4.1.1 Results Analysis

For the analysis performed two bidding strategies were considered:

- Single-energy (S) strategy: under this strategy, separate dispatch problems are considered for the electricity network (S-ELE) and for the heat network (S-HEAT);
- Multi-energy (MULTI) strategy: where according to the optimal dispatch problem described in Sect. 1.4.1, all DMERs and both the electricity and heat networks are operated with a coordinated strategy.

Figure 8 presents electricity (energy), gas, and upward and downward secondary reserve band bids. The two bidding strategies exhibit similar patterns in the day-ahead energy market (electricity). Both demand (positive values) and supply (negative values) bids are primarily influenced by PV production and market prices. The aggregator predominantly places supply bids during forecasted PV generation periods. Additionally, higher demand bids are placed around 4 and 23 h when prices are lower.

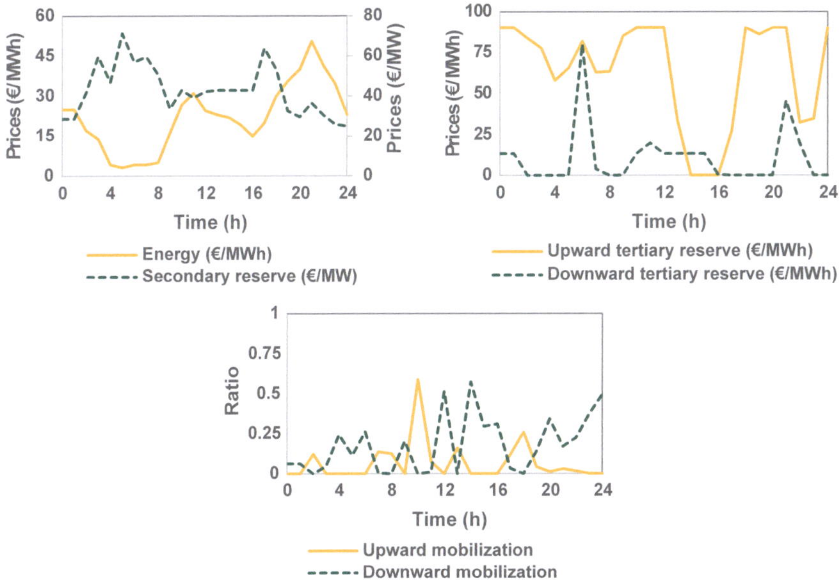

Fig. 7 Electricity prices for energy, secondary band, and tertiary reserves and mobilization ratios

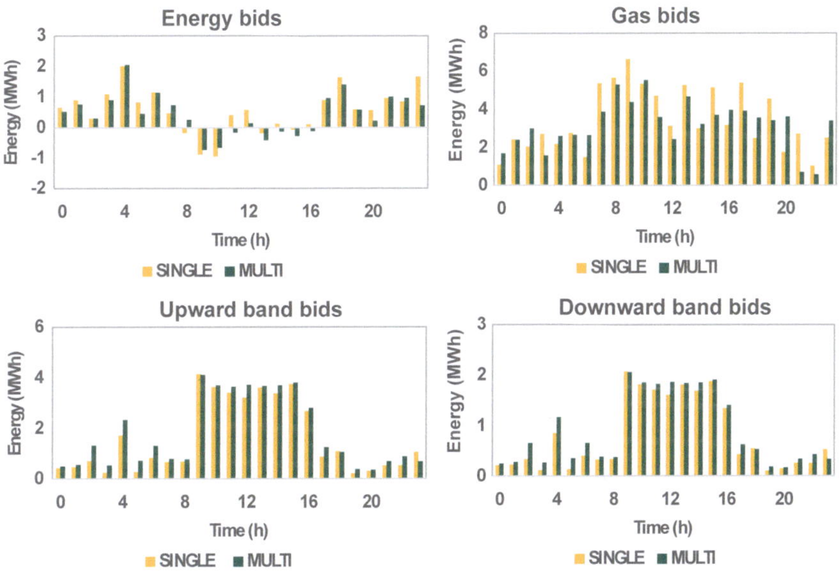

Fig. 8 Aggregator's bids

In the day-ahead gas market, the two bidding strategies also show comparable bid placement patterns. Gas bids are mainly driven by prices and heat requirements. Most of the gas is procured to fuel the CHPs, which provide heat to meet the needs of prosumers connected to the heat network. Heat load requirements are more stringent between 8 and 10 h, leading the aggregator to purchase more gas during these hours. Conversely, from midnight to 7 h and from 23 h onwards, gas bids remain relatively stable, indicating that the aggregator is primarily meeting the gas and heat load requirements expected from the CHPs during these times. This pattern is clearly seen in the SINGLE strategy, while in the MULTI strategy it diverges slightly.

The two bidding strategies also present a very similar placement of upward and downward secondary reserve band bids. It is possible to observe that they have higher values in the middle of the day, as it is the time the PV systems can provide more reserves.

Figure 9 presents the result of the optimal dispatch of PVs, batteries, CHPs, and thermal storage systems for the SINGLE and MULTI strategies. Analyzing the PV and CHP generation, we can observe that they both have a similar profile in the two strategies. Nonetheless, in the MULTI strategy the PVs produce more electricity while the CHPs produce less. In relation to the batteries, they have different profiles, but they are used throughout the day in both cases. The MULTI strategy presents higher values of charging/discharging while in the SINGLE strategy their behavior exhibits a more flatten curve (up until the end of the day). Finally, the profiles of the thermal storage system charging/discharging have some differences in each strategy. In the SINGLE strategy, they are in a cycle of charging/discharging throughout the 24 h period. On the other hand, in the MULTI strategy, they are used right at the beginning of the day to charge (as prices are lower), in the middle of the day and at the end of the day. As we can see, the thermal storage systems, which are charged by electric resistance, end up using the overproduction of electricity by the PVs.

Figure 10 presents the upward and downward reserve bids offered by each resource. In both SINGLE and MULTI strategies, the resources that provided more upward and downward reserve bids were the PVs, followed by the CHPs and batteries. It is also possible to observe that in the MULTI strategy, the PVs and CHPs provided more upward reserves than the SINGLE strategy, while batteries provided less upward reserves. In relation to the downward reserves, the PVs and batteries provided more downward reserves while the CHPs almost did not provide any reserves. This way, in the MULTI strategy, it is possible to observe a better use of flexibility available in the system and integrated between the electric (PVs and batteries) and thermal (CHPs) resources. This is because in the SINGLE strategy, the ratio between upward secondary reserves and downward secondary reserves of each resource is clearly influenced by the secondary reserve market rule stating that the upward reserve must be twice as much as the downward reserve. This shows a limit imposed on the resources which are not able to provide full potential for their flexibility. On the other hand, in the MULTI strategy, it is possible to observe a different distribution of the upward and downward reserves offered by each resource, indicating better use of the flexibility available in the system. This ends up increasing the reserve bids offered to the market and thus, the profitability of the system.

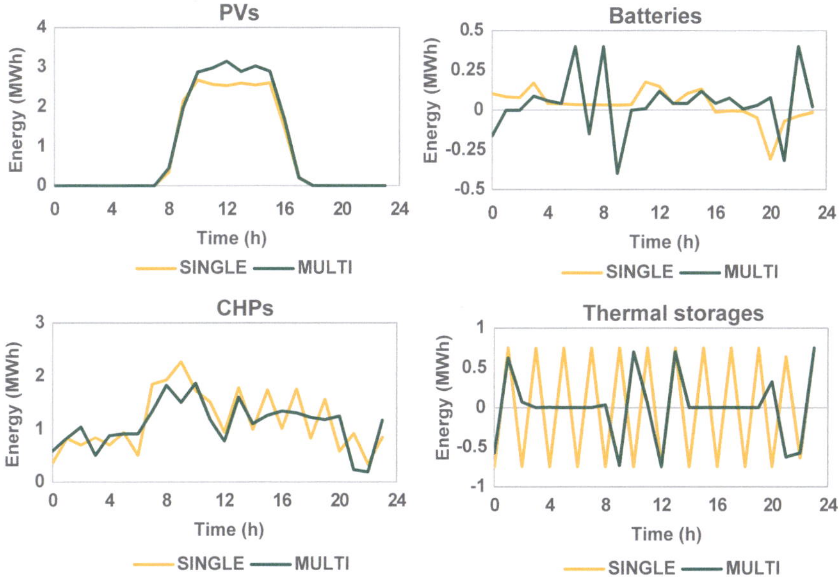

Fig. 9 Disaggregation of the electricity bids by DMER

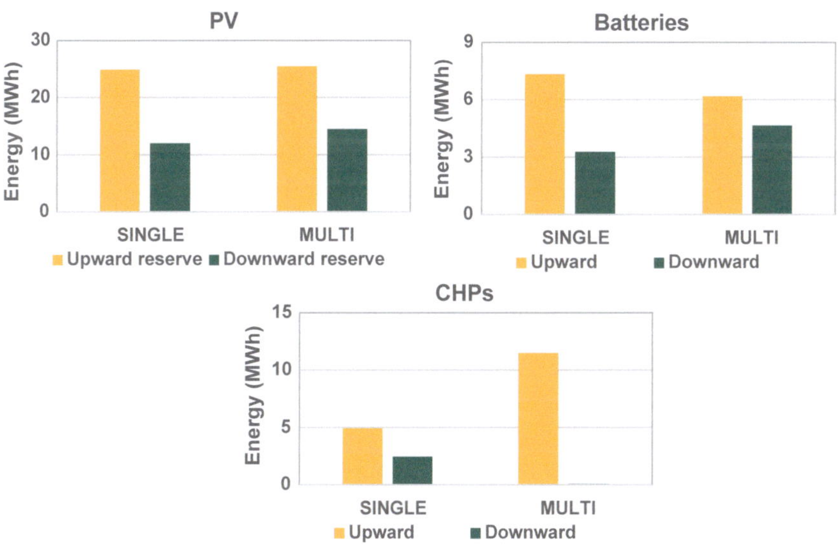

Fig. 10 Disaggregation of the secondary reserve band bids by DMER

Table 5 presents the cumulative costs obtained for the two bidding strategies (SINGLE and MULTI). The costs of the SINGLE strategy are the sum of the costs of strategy S-HEAT and S-ELE. Positive values represent costs and negative values represent income.

The results in Table 5 show that the MULTI strategy produced the most profitable outcome. The MULTI strategy outperformed the SINGLE strategy with 28% lower costs, which allows concluding that a multi-energy aggregator exploits better the flexibility of DMERs than single-energy aggregators.

The electricity energy cost is negative for S-HEAT, suggesting it might be benefiting from some form of revenue or cost reduction in electricity. Conversely, S-ELE has a positive cost, indicating higher expenses related to electricity. The SINGLE strategy has a higher electricity cost than the MULTI strategy. Gas energy costs are highest for S-HEAT, reflecting higher gas consumption from the CHPs. The gas cost for S-ELE is significantly lower, due to reduced gas use. In this case, the SINGLE strategy has higher costs than the MULTI strategy. Regarding the costs associated with secondary reserve bands and activation, the S-HEAT strategy shows lower revenues than the S-ELE strategy as it has less resources participating in secondary reserves. Comparing the SINGLE and MULTI strategies, we can observe that the MULTI strategy has higher revenues from participating in secondary reserve markets. Carbon costs are incurred for all strategies except S-ELE, as it does not consider the CHPs. The MULTI strategy has slightly lower costs than the SINGLE strategy.

In conclusion, the MULTI strategy ends up benefiting from lower costs/higher revenues of gas energy, secondary reserve participation and carbon allowances. Nonetheless, the main factor is the provision of secondary reserves which are higher in the MULTI strategy. As previously stated, this occurs due to a better use of the multi-energy flexibility available in the energy system, benefiting the aggregator.

Regarding the use of TES, the MULTI strategy demonstrated a more strategic use of thermal storage systems by charging during low-price periods and discharging during high-demand times, effectively leveraging PV overproduction. Economically, the MULTI strategy was more profitable, with 28% lower costs than the SINGLE strategy. This was achieved through better exploitation of multi-energy flexibility, leading to lower electricity and gas costs, higher revenues from secondary reserve

Table 5 Costs of each strategy

Resources	S-HEAT	S-ELE	SINGLE	MULTI
Electricity—energy (€)	– 420	687	267	199
Gas—energy (€)	5068	128	5196	4810
Electricity—secondary reserve band (€)	– 300	– 2282	– 2582	– 2969
Electricity—secondary reserve activation (€)	– 32	– 257	– 289	– 247
Carbon (€)	241	0	241	233
Total (€)	4557	– 1725	2832	2026

markets, and reduced carbon costs. Overall, the MULTI strategy's superior performance was driven by more effective use of available flexibility, enhancing reserve bids and overall profitability.

This case clearly demonstrates the significant advantages of implementing an optimal MES management strategy. It not only enhances the integration of RES, reducing reliance on carbon-intensive alternatives, but also delivers substantial economic benefits.

5 Conclusion

In this chapter, TES systems were first analyzed to highlight their potential to enhance energy efficiency, reduce greenhouse gas emissions, and support the transition to RES. The analysis covered various TES technologies, including SHS, LHS, and TCS, to assess the advantages and disadvantages of each solution for storing thermal energy from sources like industrial waste heat, concentrated solar power, and renewable electricity.

The classification of TES systems based on operating temperatures was discussed, addressing the demands of different industrial applications. Despite its simplicity and practicality, SHS faces limitations such as lower energy density and poor thermal conductivity. On the other hand, using PCMs, LHS offers higher energy density with constant outlet temperature discharge, which is crucial for power generation and heat applications requiring precise temperature control. Finally, although TCS presents the highest energy density and long-term storage capabilities, it faces issues due to high material costs, slow reaction kinetics, and low TRL state of the technology.

Furthermore, the integration of TES systems into industrial parks was examined, demonstrating their role in enhancing energy efficiency by storing excess thermal energy for later use, optimizing energy management, and reducing reliance on external energy sources. Key applications include concentrated solar plants, cogeneration plants, biomass gasification, electrolyzers for hydrogen production, renewable methanol plants, and TSA for gas purification.

The chapter also highlighted the economic and environmental benefits of a multi-energy management strategy, including TES, as demonstrated in a multi-energy microgrid use case. This strategy showed significant cost savings and increased revenues compared to single-energy strategies by leveraging the flexibility of diverse energy resources and enhancing the integration of RES.

Future research directions were identified, including exploring the role of aggregators as price makers in energy markets, incorporating stochastic approaches to model uncertainty, and integrating optimization frameworks into energy system planning tools. These advancements could further improve the efficiency, sustainability, and economic performance of TES systems in industrial applications.

Overall, TES technologies offer a promising pathway to achieving greater energy efficiency and sustainability in industrial processes, supporting the broader goals of decarbonization and renewable energy integration.

Acknowledgements This work has received funding from the European Union's Horizon Europe Research and Innovation program under grant agreement No. 101096992 (Storage Innovations for Green Energy Systems [SINNOGENES]) and was supported by national funds through FCT/MCTES (PIDDAC): LEPABE, UIDB/00511/2020 (https://doi.org/10.54499/UIDB/00511/2020) and UIDP/00511/2020 (https://doi.org/10.54499/UIDP/00511/2020) and ALiCE, LA/P/0045/2020 (https://doi.org/10.54499/LA/P/0045/2020).

References

Akar S, Kurup P, Belding S, McTigue J, Cox J, McMillan C, Boyd M, Lowder T (2021) Renewable thermal energy systems designed for industrial process solutions in multiple industries. https://research-hub.nrel.gov/en/publications/renewable-thermal-energy-systems-designed-for-industrial-process--2

Ali HM, ur Rehman T, Arıcı M, Said Z, Duraković B, Mohammed HI, Kumar R, Rathod MK, Buyukdagli O, Teggar M (2024) Advances in thermal energy storage: fundamentals and applications. Progr Energy Combust Sci 100(101109). https://doi.org/10.1016/j.pecs.2023.101109

Alva G, Lin Y, Fang G (2018) An overview of thermal energy storage systems. Energy 144:341–378. https://doi.org/10.1016/j.energy.2017.12.037

Baran M, Wu FF (1989) Optimal capacitor placement on radial distribution systems. IEEE Trans Power Delivery 4(1):725–734. https://doi.org/10.1109/61.19265

Bauer T, Odenthal C, Bonk A (2021) Molten salt storage for power generation. Chem Ing Tec 93(4):534–546. https://doi.org/10.1002/cite.202000137

Benalcazar P (2021) Optimal sizing of thermal energy storage systems for CHP plants considering specific investment costs: a case study. Energy 234:121323. https://doi.org/10.1016/j.energy.2021.121323

Bilek J, Atkinson JK, Wakeham WA (2006) Thermal conductivity of molten lead-free solders. Int J Thermophys 27(1):92–102. https://doi.org/10.1007/s10765-006-0035-4

Cao Y, Wei W, Wu L, Mei S, Shahidehpour M, Li Z (2019) Decentralized operation of interdependent power distribution network and district heating network: a market-driven approach. IEEE Trans Smart Grid 10(5):5374–5385. https://doi.org/10.1109/TSG.2018.2880909

Ceseña M, Alejandro E, Loukarakis E, Good N, Mancarella P (2020) Integrated electricity—heat-gas systems: techno-economic modeling, optimization, and application to multienergy districts. Proc IEEE 108(9):1392–1410. https://doi.org/10.1109/JPROC.2020.2989382

Coelho A, Iria J, Soares F (2021) Network-secure bidding optimization of aggregators of multi-energy systems in electricity, gas, and carbon markets. Appl Energy 301:117460. https://doi.org/10.1016/j.apenergy.2021.117460

Coelho AMF (2023) Network-secure participation of aggregators of multi-energy systems in multi-energy markets. https://hdl.handle.net/10216/151765

Costa Pereira S-C, Kenisarin M (2022) A review of metallic materials for latent heat thermal energy storage: thermophysical properties, applications, and challenges. Renew Sustain Energy Rev 154:111812. https://doi.org/10.1016/j.rser.2021.111812

EASE (2023) Thermal energy storage. https://ease-storage.eu/publication/thermal-energy-storage/

Energy.gov (2023) Storage Innovations 2030. Energy.Gov, 19 July 2023. https://www.energy.gov/oe/storage-innovations-2030

ENTSO-E (2025) ENTSO-E Transparency Platform. https://transparency.entsoe.eu/dashboard/show

EPA (2000) Cadmium compounds. https://www.epa.gov/, Jan 2000. https://19january2017snapshot.epa.gov/sites/production/files/2016-09/documents/cadmium-compounds.pdf

EPA (2011) Lead compounds. https://www.epa.gov/, Sept 2011. https://19january2021snapshot.epa.gov/sites/static/files/2016-09/documents/lead-compounds.pdf
European Commission (2023) RoHS directive—European Commission, 7 Dec 2023. https://environment.ec.europa.eu/topics/waste-and-recycling/rohs-directive_en
Eurostat (2025) Air emissions accounts for greenhouse gases by NACE Rev. 2 activity—quarterly data, 14 Feb 2025. https://ec.europa.eu/eurostat/databrowser/product/page/ENV_AC_AIGG_Q
Glücker P, Pesch T, Benigni A (2024) Optimal sizing of battery energy storage system for local multi-energy systems: the impact of the thermal vector. Appl Energy 372(C). https://ideas.repec.org//a/eee/appene/v372y2024ics0306261924011152.html
IRENA (2020) Innovation Outlook: thermal energy storage, 29 Nov 2020. https://www.irena.org/publications/2020/Nov/Innovation-outlook-Thermal-energy-storage
IRENA (2023) Power to heat and cooling: status. https://www.irena.org/Innovation-landscape-for-smart-electrification/Power-to-heat-and-cooling/Status
Iria J, Soares F, Matos M (2019) Optimal bidding strategy for an aggregator of prosumers in energy and secondary reserve markets. Appl Energy 238:1361–1372. https://doi.org/10.1016/j.apenergy.2019.01.191
King JA (1988) Materials handbook for hybrid microelectronics. In: Electronic materials and devices library. Artech House. https://books.google.pt/books?id=SR9TAAAAMAAJ
Liu X, Chen M, Xu Q, Gao K, Dang C, Li P, Luo Q et al (2022) Bamboo derived sic ceramics-phase change composites for efficient, rapid, and compact solar thermal energy storage. Solar Energy Mater Solar Cells 240:111726. https://doi.org/10.1016/j.solmat.2022.111726
Manasijevic D, Balanovic L, Markovic I, Gorgievski M, Stamenković U, Božinović K (2021) Microstructure, melting behavior and thermal conductivity of the Sn–Zn alloys. Thermochim Acta 702:178978. https://doi.org/10.1016/j.tca.2021.178978
MIBGAS (2025) Preços diários do MIBGAS. https://www.mibgas.es/pt/market-results
Morando C, Fornaro O, Garbellini O, Palacio H (2014) Thermal properties of Sn-based solder alloys. J Mater Sci Mater Electron 25(8):3440–3447. https://doi.org/10.1007/s10854-014-2036-6
Nomura T, Akiyama T (2017) High-temperature latent heat storage technology to utilize exergy of solar heat and industrial exhaust heat. Int J Energy Res 41(2):240–251. https://doi.org/10.1002/er.3611
Pompei L, Nardecchia F, Miliozzi A (2023) Current, projected performance and costs of thermal energy storage. Processes 11(3):729. https://doi.org/10.3390/pr11030729
REN (2025) Sistema de informação de mercados de energia. https://mercado.ren.pt/PT/Electr/InfoMercado/InfOp/MercOmel
SETIS (2023) Novel thermal energy storage in the European Union, 25 Oct 2023. https://setis.ec.europa.eu/novel-thermal-energy-storage-european-union_en
SINTEF (2018) Applications of thermal energy storage in the energy transition—benchmarks and developments. SINTEF. https://www.sintef.no/en/publications/publication/1646046/
Sun M, Liu T, Wang X, Liu T, Li M, Chen G, Jiang D (2023) Roles of thermal energy storage technology for carbon neutrality. Carbon Neutral 2(1):12. https://doi.org/10.1007/s43979-023-00052-w
Takahashi Y, Kamimoto M, Abe Y, Nagasaka Y, Nagashima A (1988) Heat capacity, heat of transition, and thermal conductivity of pentaerythritol and its slurry. Netsu Bussei 2(1):53–58. https://doi.org/10.2963/jjtp.2.53
The Engineering ToolBox (2005) Thermal conductivity of metals and alloys: data table & reference guide. https://www.engineeringtoolbox.com/thermal-conductivity-metals-d_858.html
Tye RP, Bourne JG, Destarlais AO (1976) Thermal energy storage material thermophysical property measurement and heat transfer impact. Dynatech R/D Co., Cambridge, MA, USA. https://www.osti.gov/biblio/7253431
Wang X, Liu J, Zhang Y, Di H, Jiang Y (2006) Experimental research on a kind of novel high temperature phase change storage heater. Energy Convers Manag 47(15):2211–2222. https://doi.org/10.1016/j.enconman.2005.12.004

Repurposing Coal-Fired Power Plants Integrating the Use of Second-Life Batteries, Flywheels, and High Temperature Molten Salt Storage

M. Esther Rojas, Marcos Lafoz, Margarita M. Rodríguez-García, Eduardo Rausell, Gustavo Navarro, and M. Rocío Bayón

Abstract The chapter explores the transformation of coal-fired power plants into sustainable energy facilities through the integration of advanced storage systems. Historically, coal-fired plants operated with inflexible schedules and slow response times, making them unsuitable for modern grids reliant on renewable energy. Repurposing these plants into Thermal Storage Power Plants (TSPP) replaces coal-fired combustion with molten salt thermal energy storage, leveraging existing infrastructure to store renewable energy for electricity generation. To enhance flexibility, second-life batteries (SLBs), repurposed from electric vehicles, are integrated. These batteries, despite reduced capacity, provide cost-effective, sustainable energy storage for medium-term demands, extending their lifecycle through optimized conditions. Flywheels are also included, offering instantaneous power output to stabilize the grid while reducing SLB degradation. The resulting Hybrid Energy Storage System (HESS) combines TSPP, SLBs, and flywheels under a centralized control scheme. Flywheels manage short-term fluctuations, SLBs address intermediate demands, and TSPPs provide long-term, stable power. This system improves efficiency, minimizes degradation, and ensures grid stability. Case studies and examples demonstrate the potential of this model, highlighting its economic and environmental benefits. This innovative approach aligns with global decarbonization goals, offering a pathway to modernize energy systems while repurposing obsolete infrastructure effectively.

Keywords Thermal storage power plant · Second-life battery · Flywheel · Sustainable power generation · Hybrid energy storage

M. E. Rojas (✉) · M. R. Bayón
Thermal Energy Storage Unit, CIEMAT, Madrid, Spain
e-mail: esther.rojas@ciemat.es

M. Lafoz · E. Rausell · G. Navarro
Electric Power System Unit, CIEMAT, Madrid, Spain

M. M. Rodríguez-García
Thermal Energy Storage Unit, CIEMAT, Almería, Spain

R. Scipioni et al. (eds.), *Hybrid Energy Storage*, Lecture Notes in Energy 47,
https://doi.org/10.1007/978-3-031-97755-8_2

1 Introduction: Requirements of the New Electric Grids

In the current climate situation, there is a global shift in the way electric power systems are structured. The traditional model, which relies on centralized generation facilities, is being replaced by a distributed approach that utilized renewable energy sources with energy storage solutions to mitigate intermittency.

Historically, thermal power plants have operated according to a fixed daily schedule. In particular, coal-fired power plants are able to provide flexibility services, including: partial load following; non-spinning reserve (meaning that they can be brought online within hours if needed); voltage support, by means of reactive power supply; inertia provision to stabilise the grid; operation at partial load, although it can increase wear and tear on the plant. However, these conventional coal-fired power plants are being shut down due to environmental reasons. Moreover, they have some flexibility limitations due to their slow ramping rate and long start-up time, which are associated with their inherent thermal processes, which are slower than other types of thermal power plants, such as natural gas, as well as hydroelectric plants.

The new paradigm of electricity generation gives rise to new considerations in relation to the operational requirements of power generation plants. These must be met in order to ensure reliability, efficiency, and stability, while also enabling integration with renewable energy sources and compliance with grid codes (Martínez-Lavín et al. 2022). Furthermore, the provision of greater flexibility to the system will result in a more economically viable power plant.

Gas turbines, small hydroelectric plants and energy storage technologies with faster response, such as batteries, flywheels and supercapacitors, are better placed to provide flexibility services. Combining a coal-fired power plant with one or more of these storage options would enhance its flexibility and security.

This chapter explores the potential for repurposing coal-fired power plants by utilising thermal energy storage in place of the coal combustion chamber. Furthermore, an additional energy storage system based on second-life batteries is integrated to provide new auxiliary services, extending the concept of repurposing. Finally, in order to implement the performance characteristics of second-life batteries, extending their lifetime, and achieve even higher system flexibility, a third energy storage technology is considered: a short-term, high-power storage unit based on flywheels.

The system is designed to be a fully integrated hybrid energy storage system, capable of meeting significant power demands through the integration of three distinct storage technologies.

2 Repurposing Coal-Fired Power Plants

Coal-fired power plants worldwide are facing challenges due to low-capacity utilisation levels and environmental concerns. An increasing number of governments have implemented plans to achieve a carbon-free future. To date, over 100 countries have joined an alliance with the objective of achieving net-zero emissions (Song et al. 2021). A number of countries, including Norway, Finland, Sweden, and Uruguay, have set a target to achieve net-zero emissions by 2030. Other countries, such as UK and Canada, anticipate reaching net zero carbon emissions by 2050, while China is targeting 2060.[1] Furthermore, coal-fired power plants have become unprofitable to utilities and uneconomical to customers (Forbes 2018). As outlined by (Song et al. 2021), countries have implemented mainly two measures to regulate the current coal-fired power plants. One approach is the coal retirement mechanism, which would allow for the elimination of existing coal-fired power plants within 10–15 years, rather than the currently expected 30–40 year to remain operational. The second measure is to replace obsolete coal-fired power plants with an integrated mix of efficiency measures, including, for example energy storage systems or renewable power generation.

Kefford et al. (2018), analysed the impact of **early retirement** of coal-fired power plants on asset owners and communities in four regions, namely, China, India, the European Union, and the United States. Notable examples in USA of Repurposed Coal Fired Power Plants are the following:

- Plant Scherer, Georgia[2]: Once one of the largest coal-fired power plants in the U. S., Plant Scherer successfully transitioned into a biomass plant.
- Fisk Generating Station, Chicago (see footnote 2): The Fisk Generating Station underwent a remarkable transformation by converting its operation to natural gas-fired power generation.
- Mount Tom Power Plant, Massachusets[3]: Currently operated by Mt. Tom Solar LLC, it was commissioned on 2018 as a PV power plant and started energy operation the same year with a peak capacity of 7.6 MW_e.

The conversion to a biomass power plant (first example) allows the power block of the former coal-fired power plant to be still used. In contrast, the transformation to a gas-fired or to a PV power plant (second and third examples) only utilises the land available at the power plant, along with the civil and power transmission system (Thomas and Akhtar 2023).

Reducing the use of fossil fuel consumption in coal-fired power plants, while increasing their peaking capability, is the approach followed by AES Gener in Chile. In 2009, they began replacing the 7 MW_e of coal-fired thermal capacity with 20 min lithium-ion battery capacity. AES Gener deployed a similar strategy at the 554 MW_e Eléctrica Angamos plant in 2012, installing 20 MW_e of 20 min battery capacity, and at the 531 MW_e Cochrane Power Station in 2016, where a comparable volume was installed. Another method for enhancing the flexibility of existing coal-fired power plants is to integrate a thermal storage system within the power block, to be

[1] https://chinaeucn.com/carbon-neutrality-china/#:~:text=In%20September%202020%2C%20president%20Xi,%E2%80%9Ccarbon%20neutral%E2%80%9D%20by%202060

[2] https://petroedgeasia.net/repurposing-and-recommissioning-coal-fired-power-plants-for-A-sustainable-energy-transition/

[3] https://database.earth/energy/power-plant/mt-tom-solar-project-hybrid

charged by the main or reheated steam and discharged to the low-pressure turbine or condenser (Wang et al. 2023; Zhang et al. 2024).

3 Thermal Storage Power Plants

An interesting option for the early retirement of coal-fired power plants is so-called Thermal Storage Plants. This reuses existing equipment such as steam turbine, heat recovery boilers or heat exchangers, but replace the coal combustion chamber by a high-temperature thermal energy storage system, mainly of the type already developed and commercially available for Concentrating Solar Thermal Power (CSTP) plants, which is charged by renewables such as PV or wind, in practice adopting the Carnot Battery concept.

Thermal storage systems for commercial CSTP plants are based on a system known as the double-tank configuration. The storage medium is a non-eutectic mixture of sodium and potassium nitrate salts (solar salt) in the liquid state, stored at high temperature in one of the tanks. When the system is discharged, the thermal energy is transferred to the power block, and the molten salts are stored in another tank, called the cold tank (Fig. 1). The limitation imposed by the maximum size of metal tanks on the amount of energy that can be stored is overcome by using several systems in parallel. Notable CSTP plants with thermal storage systems of large capacity include Noor II and Noor III, which have 200 MW_e/1200 MWhe (6 h of nominal power output without solar input) and 150 MW_e/1125 MWh_e (7, 5 h of storage capacity), respectively (SENER, 2022a, 2022b), or Atlántica Solana Generating Station, in Arizona, with 250 MW_e/1500 MWh_e (6 h at nearly nominal power, (Atlantica Sustainable Infrastructure PLC)). The current average investment cost for CSTP-TES systems is around 40 €/kWh_e (Crespo 2020).

Fig. 1 Molten salt thermal storage system at MASEN (Morocco)

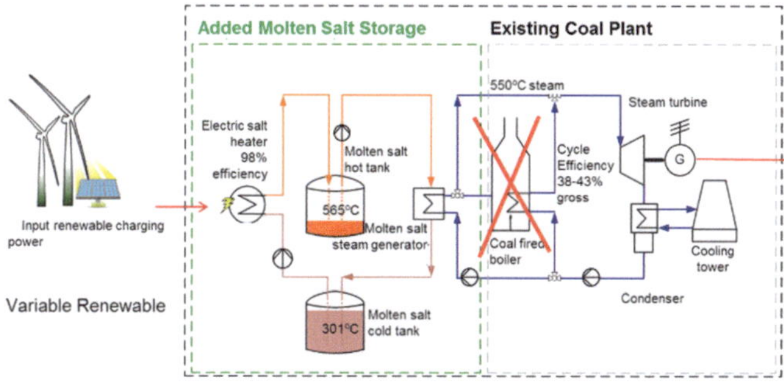

Fig. 2 Scheme of a thermal storage power plant: integration of a two-tank molten salt storage into an existing coal-fired power plant (adapted from Geyer et al. 2020)

In the retrofitted (coal-fired) power plant, the molten salt would be heated using electric resistance heaters or microwaves (Rodríguez-García et al. 2023), powered by renewable electricity. In this way, the excess or curtailed variable power available in the grid from photovoltaic and wind power plants can be stored as thermal energy. The stored thermal energy is then released by pumping the hot salt through a turbine steam generator system. In this process, heat is transferred from the salt to the turbine steam, which is then returned to the cold tank after cooling. The turbine steam is then used by the existing steam cycle of the former coal-fired power station to generate electricity (see Fig. 2). The components of a thermal storage plant are mature technologies, but the combination of technologies represents a novel approach (Geyer et al. 2020).

The concept of Thermal Storage Power plants has been developed since 2018 (Deign 2021), but not such a project has been realized yet. This may change in the near future thanks to the Alba project, which will transform the Angamos coal-fired power plant (560 MW_e), located in Mejillones (northeast Chile), into a thermal storage power plant. The Chilean authorities approved the environmental study submitted by AES Andes in December 2023 (EnerData 2023).

4 Hybrid Energy Storage with Second-Life Batteries

To enhance the flexibility of the Thermal Storage Power Plant, the potential use of batteries, such as lithium-ion, has been explored. The use of batteries in a (thermal energy storage) power plant offers the following advantages (Koike et al. 2018):

- Batteries are able to respond almost instantaneously (within milliseconds to seconds) to frequency deviations. This makes them an ideal solution for primary frequency regulation, fast ramping for load following and other types of fast responses to changes in demand or fluctuations in renewable energy generation.
- Batteries can be precisely tuned to deliver the optimal output power.

The development of electric vehicles (EV) has increased in recent years in line with EU goals to reduce the CO_2 emissions from the transport sector. Batteries are typically discarded after being used in an EV when they have reduced by of 20–30% of their nominal capacity. However, they have some remaining capacity that can be used in other applications, such as stationary energy storage with lower technical requirements. This extends the battery's useful life and contributes to environmental sustainability, enhancing the circular economy (Thakur et al. 2022) and reducing the dependence on critical materials (Wu et al. 2020).

These batteries are referred to as second-life batteries (SLB), also known as repurposed or reused batteries. They offer a promising solution to extend the lifecycle of EV batteries, which represents a significant cost factor (around 40% according to Shahjalal et al. 2021). It is anticipated that by 2025, approximately 3 million discarded EV batteries from vehicles with a total capacity of 953 GWh will still have potential for use as SLB. With regard to battery technology, lithium nickel manganese cobalt (NMC) batteries offer a more circular and environmentally sustainable option to lithium iron phosphate (LFP) batteries, with a Global Warming Potential of 4–6% and an Abiotic Depletion Potential of 13–16% for the minerals used (Picatoste et al. 2024).

The initial decision when considering a battery from an electric vehicle is to either retain those battery cells for a second-life application and extend their operation beyond the original and first intent (SLB), or send them to recycling facilities (Takahashi et al. 2023).

An SLB system comprises a set of battery cells or packs, sourced from different EVs and, each with distinct characteristics. The SLBs are characterized by measuring internal resistance increase and capacity, and repackaged for a second life. They are classified into groups of cells with similar conditions and performance. SLB systems are designed to operate beyond the so-called 'aging knee', which is the threshold where the battery is expected to start degrading rapidly. However, it is possible to halt this process by optimizing the working conditions, thereby achieving a second-life potential of 4000–6000 cycles between 75 and 50% of the State of Health (SoH) (Gao et al. 2024).

It is essential to regulate the maximum electrical current of SLBs in order to maintain the desired level of degradation while ensuring an optimal SoH. This may result in the system being sized to the maximum peak power, which would entail a higher initial investment.

5 The Additional Support of Flywheels

To extend the operational range and to improve the power performance, it is advisable to consider the further hybridisation of the storage system with a technology capable of frequently delivering ultra-fast power peaks without experiencing degradation of SLBs. Thus, in order to enhance the SLB performance, it would be beneficial to consider complementary storage technologies, such as flywheels. This approach is supported by various research studies, including those by (Lee and Wang 2008; Barelli et al. 2019; Glücker et al. 2021; Arani et al. 2020; Ayodele et al. 2020).

The use of batteries and flywheel combination in a power plant has been referenced in the case of a hydroelectric power plant (Casarin et al. 2023). Flywheels contribute to the flexibility market in two main ways:

1. they can provide high power peaks almost instantaneously (in milliseconds) to grid frequency deviations for primary frequency control and/or support secondary frequency control by providing energy over a few minutes. This makes them ideal for providing fast frequency response (FFR) services.
2. being connected to the grid through power electronic converters, they are able to provide voltage support and reactive power compensation.

In addition to the above, flywheels offer very high round-trip efficiency (85–95%), an almost unlimited number of charge-discharge cycles, a significantly lower environmental impact compared to batteries, and a sustainable manufacturing process that uses basic resources.

Similarly to the use of flywheels, the technology of supercapacitors can be combined with batteries, with a comparable operational profile than flywheels, although with some different features primarily associated with the specific application area. Some references of the use of batteries and supercapacitors operating co-ordinately for grid application are included in (Guo and Sharma 2016; Vaca et al. 2016; Akram and Khalid 2017; Kim et al. 2016).

With regard to the specific operation of the flywheel subsystem, it should be noted that it is composed of a set of parallel connected number of machines. They operate at a common DC voltage, connected to a grid-tie converter. However, they can provide independent power, depending on the system command (Torres et al. 2020).

6 The Complete Hybrid Storage System and Its Control Scheme

This section outlines the description of the hybrid energy storage system (HESS) that integrates a thermal storage power plant (TSPP) with second life batteries (SLB) and flywheels (FW) connected to the grid. The objective is to optimise plant operation for the management of both short-term power fluctuations (fast frequency response) and longer-term energy demands. By combining the rapid response of FW with the medium-term power regulation provided by SLB and the Rankine cycle run by the thermal energy storage system, this approach is set to offer enhanced efficiency, reduced SLB degradation, and improves grid stability.

The system configuration is shown in Fig. 3. It consists of a renewable energy source connected to the grid, where surplus energy from renewables is supplied to the thermal storage power plant or used to recharge SLB and FW. Flywheel and SLB are also implemented to handle short-term frequency and power fluctuations, delivering the rapid response that the thermal storage plant, due to its slower dynamics,

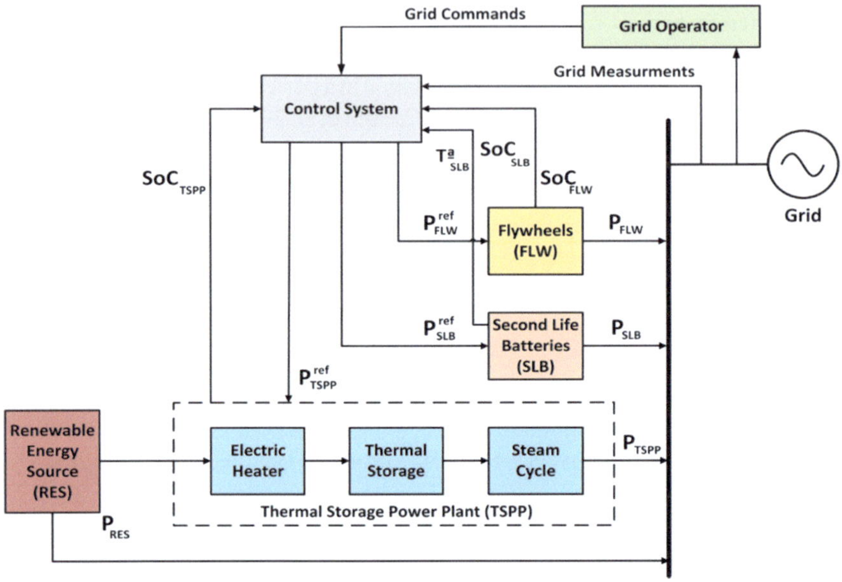

Fig. 3 Hybrid Energy Storage System (HESS) Configuration Scheme

cannot provide. A centralized control system operates the entire system, receiving grid commands and measurements, as well as monitoring the state of charge across all energy storage components.

The respective roles of each subsystem in maintaining optimal power output and grid performance under various operational conditions are described as follows:

1. **Power Request Input (Grid Commands)**: The power grid controller sends a command to adjust the power in accordance with grid requirements, such as frequency control or load following. Alternatively, the power command could aim at controlling grid variables according to measurements right at the connection point.
2. **Flywheel Operation (Fast Response)**: Due to their rapid response time, the flywheels are ideally suited to providing an immediate high-power output to smooth out short-term fluctuations. However, due to their limited capacity to store energy, they are only able to manage immediate peaks or drops in demand for short periods of time. Once the batteries or the thermal storage power plant are able to meet the required load, the flywheels should be switched off.
3. **SLB Operation (Intermediate Response)**: As the flywheels discharge, the batteries respond more slowly to prevent high power spikes and their associated degradation. The maximum batteries power output is regulated by its state of charge (SoC_{SLB}) and temperature ($T^{a}{}_{SLB}$). In the event that the batteries overheat or have a low charge, the system will reduce the discharge rate in order to extend their lifespan.

4. **Thermal Storage Power Plant Operation (Slow Response)**: Due to its slow dynamics, the Rankine cycle is the last to respond. It gradually increases power output to absorb long-term demand, ensuring a steady supply once activated. The thermal plant adjusts its output based on sustained grid demand, reducing reliance on the flywheel and batteries once stable conditions are reached.

The different operation states of the Hybrid Thermal Storage Power Plant (HTSPP) are described in the scheme of Fig. 4. The system considers a certain power P_{RE}, coming from renewable energy sources, and a certain power demand coming directly from the grid operator (TSO) or based on measurements at the connection point, P_{grid}. Based on the difference between them, it is evaluated whether there is a surplus energy from renewables or not. If the grid demand is covered by the renewables (OP1) the HTSPP does not operate.

If there is a surplus on the renewable power, not used by the grid, this power is firstly used to recharge FW (OP2) and SLB (OP3), in case they need to be charged (SoC below the limits). The rest is used to supply the electric heater in order to charge the thermal energy storage system (OP4). Once being at this point, the HTSPP is fully charged (OP5) and further excess power will not be possible to be managed.

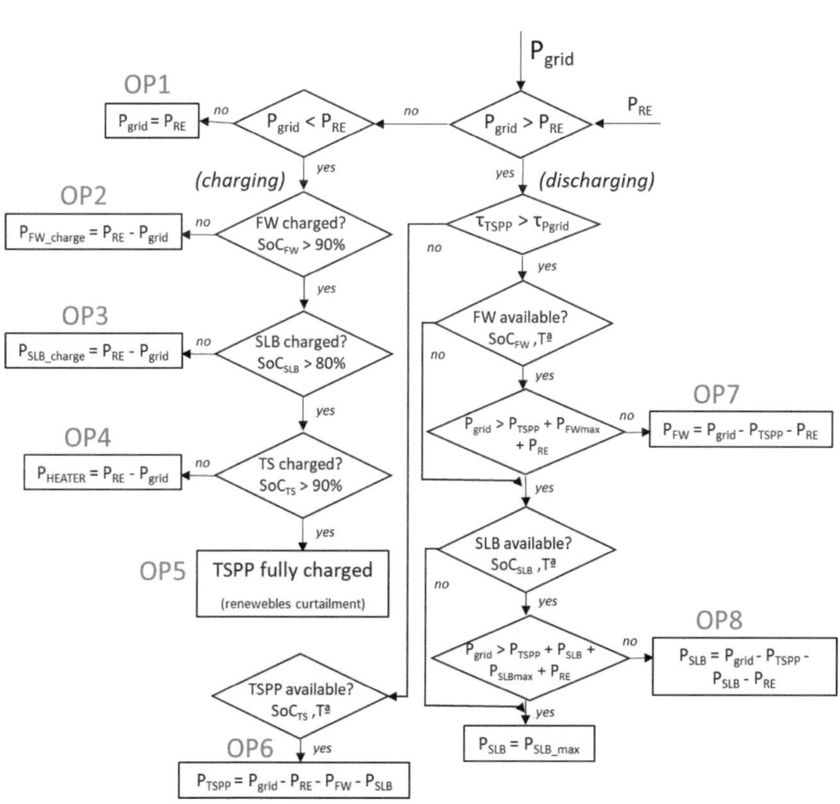

Fig. 4 Flux process diagram for the hybrid energy storage power plant

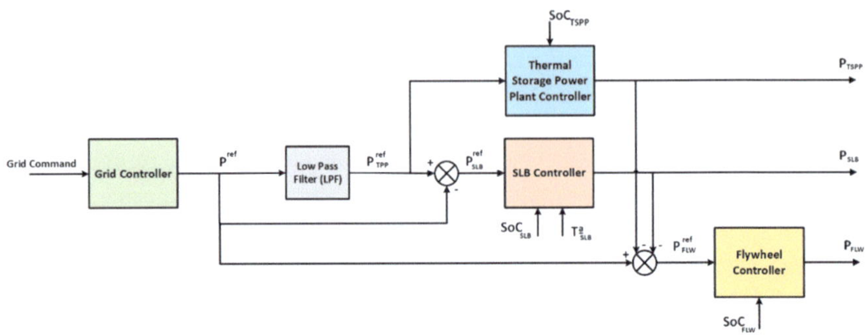

Fig. 5 Control scheme

Considering that the grid requires power beyond the renewable energy supply, the power is provided by the different subsystems of the HTSPP. In the case of rapid oscillations, if the time constant of the power requirements ($\tau_{P_{grid}}$) is lower than the time constant of the thermal power cycle (τ_{TSPP}), the FW is firstly in charge of providing the faster response, considering that the TSPP will require a longer time to supply the required power, $P_{FW} = P_{grid} - P_{TSPP} - P_{RE}$ (OP7). Once the FW reaches the minimum SoC or maximum power limitation, the SLB contribute to the power, $P_{SLB} = P_{grid} - P_{TSPP} - P_{SLB} - P_{RE}$ (OP8). The TSPP will contribute to the required power whenever the response time and the rate value allow to provide it, $P_{TSPP} = P_{grid} - P_{RE} - P_{FW} - P_{SLB}$ (OP6).

Coordinated Operation of a Thermal Power Plant with SLB and Flywheels

The coordinated control of the Hybrid Energy Storage System is illustrated in Fig. 5 and described in the following paragraphs.

In case of stable power demand stable, with minor fluctuations, the flywheel responds to handle short-term fluctuations. Meanwhile, the battery stays in standby mode if its contribution is not needed, maintaining an optimal SoC and temperature to preserve its lifespan. The thermal power plant operates at a steady output, ensuring the base load power supply.

In case of sudden increase in the power demanded by the grid (P_{grid}) FW switch on immediately to absorb the required power spike, delivering high power for some seconds until it reaches its limited energy capacity. As the energy of the flywheels decreases, the battery activates after a short period, with its power output depending on the following factors:

- Batteries SoC: If the SoC_{SLB} is high ($> 80\%$), it can discharge at full power; if it is low ($< 20\%$), it limits power output to preserve battery health.
- Batteries Temperature, T^{a}_{SLB}: If the batteries temperature is high, power output is restricted to prevent thermal degradation.

As shown in Fig. 6, the thermal power plant remains unaffected, as the duration of the spike is too short to trigger its slow ramp-up process. A similar process occurs

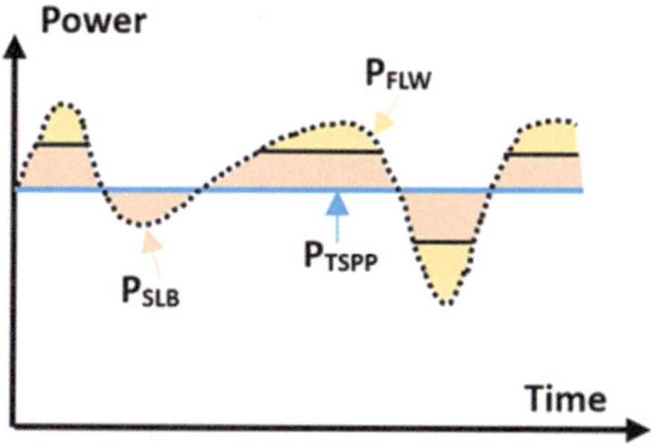

Fig. 6 Power demand in time

during a sudden decrease in the power required by the grid, but in this case, flywheels and batteries charge until they reach their SoC limit. SLB handle more of the energy from these fluctuations, while flywheels can provide faster power peaks, reducing the load on SLB to absorb high power demands, which helps minimize their degradation.

On the other hand, when the grid demand increases the power supply for an extended duration, lasting several minutes to hours, the flywheels discharge, but they soon reach its minimum state of charge limit and is kept in a standby state until operating conditions allow them to be recharged. Subsequently, batteries take over to supply intermediate-term power, with their discharge rate adjusted according to SoC_{SLB} and $T^a{}_{SLB}$.

As demand persists, the thermal power plant slowly ramps up its output, delivering stable, long-term power to ensure a sustained supply. A similar process occurs during a sudden decrease in power, but in this case, flywheels and batteries charge until the SoC is limited and the thermal power plant ramps down slowly.

In conclusion, the control system prioritizes the flywheel for fast short-term power regulation, followed by the SLB for short and medium-term operation, and finally the thermal storage for long-term, stable power supply.

6.1 How Does This Hybrid Technology Improves the KPIs?

Table 1 describes the different KPIs associated with the three energy storage technologies involved (thermal storage, SLB and flywheels), as well as the KPI improvement from the combined installation and operation of the three technologies as a hybrid thermal storage power plant (HTSPP).

The main advantages of integrating the three technologies are:

1. Temporal complementarity. Thermal plant provides energy supply for many hours, batteries support loads for minutes to hours and flywheels provide power response in the range of milliseconds for grid stabilization.
2. Cost optimization. Thermal storage and SLB have low long-term costs. Flywheels reduce the cycling stress on the SLB, extending their life cycle.

Table 1 KPI analysis associated to the HTSPP

KPI	Description	SLB values	FW values	TES values	Hybrid storage
Energy storage capacity	Maximum energy the system can store, measured in kWh	• Industrial: 50–500 kWh • Grid-scale: > 1 MWh	• Small: 5–50 kWh • Large: Up to 500 kWh	up to GW_eh	Sum of capacities of thermal storage and batteries. FW handle short-term high-power scenarios
Energy density	Energy stored per unit of weight or volume	• 200–500 kWh/m^3 (approximately 70–80% of C_N)	• 100–400 kWh/m^3 for grid applications	• 26–80 kW_eh/m^3	The low energy density of TES is compensated by high energy densities from SLB and FW
Output power	Maximum power the system can deliver	• Industrial: 50–100 kW • Grid-scale: 1–5 MW	• Industrial (e.g., UPS): 10–500 kW • Grid-scale systems: 1–5 MW	10–500 MW	Flywheels provide rapid-response power, while batteries and thermal storage handle sustained and larger-scale loads
Response time	The time taken by the system to respond to changes in load demand	• < 1 s (similar to first-life batteries)	• < 1 ms for power delivery • Extremely fast ramp-up	• Minutes to hours depending on system design	Improved significantly since FW < 1 ms batteries: secs and TES in minutes to hours
Round trip efficiency	Ratio of energy charged to energy discharged (%)	• 85–95% (depending on the cells' state of health)	• Typical flywheels: 85–95% • Advanced vacuum-sealed systems: Up to 98%	30–40% (considering both the thermal and electric cycles)	Depending on the ratio of usage of the dif. Technologies, 60–80%

(continued)

Table 1 (continued)

KPI	Description	SLB values	FW values	TES values	Hybrid storage
Cycle life and durability	The number of charge-discharge cycles before the system degrades	• 1000–5000 additional cycles (depending on usage and maintenance)	• Commercial systems: 10^5 to 10^6 cycles	• 5250–10,000 cycles (25–35 years)	FW increase the life cycle of SLB
CapEx and OpEx	CapEx: Initial investment. OpEx: operating and maintaining costs	• Installation: 50–150 €/kWh (significantly lower than new batteries)	• Installation cost: 500–1000 €/kWh • Operational cost: \$0.01–\$0.05 per kWh/cycle	Installation: 20–55 €/kWh	CAPEX is increased but OPEX is reduced when using certain patterns
Size and weight	Physical dimensions and weight of the system	• Similar to original battery size	• Compact designs: 0.1–0.5 m^3 per 100 kWh • Weight: 10–50 kg/kWh	5–15 m^3 per MWh for molten solar salts	Not significantly modified from the initial size and weight since TES is dominant
Carbon footprint ($kgCO_2eq$)	CO_2 emissions during manufacturing and operation	• 30–70% reduction compared to manufacturing new batteries	• Minimal emissions during operation. Not relevant from fabrication	• Minimal emissions during operation. Not relevant from fabrication	Not significantly modified from the initial values
Recyclability and end-of-life Impact	Recycling and disposing of system components	• Lithium batteries have not very highly recyclable active materials	• Made from recyclable materials (e.g., steel, carbon fiber)	• Recyclable materials (e.g., molten solar salts, ceramics)	SLB reduce waste; flywheels and thermal storage use recyclable materials

3. Resilience of the system. Hybrid storage provides flexibility and redundancy to the power system.

References

Akram U, Khalid M (2017) A coordinated frequency regulation framework based on hybrid battery-ultracapacitor energy storage technologies. IEEE Access 6:7310–7320. https://doi.org/10.1109/access.2017.2786283

Arani AAK, Gharehpetian GB, Abedi M (2020) A novel control method based on droop for cooperation of flywheel and battery energy storage systems in Islanded microgrids. IEEE Sys J 14(1):1080–1087. https://doi.org/10.1109/jsyst.2019.2911160

Atlantica Sustainable Infrastructure plc (no date) Solana. Available at https://www.atlantica.com/web/en/company-overview/our-assets/asset/Solana/. Accessed 28 Sept 2023

Ayodele TR, Ogunjuyigbe ASO, Oyelowo NO (2020) Hybridisation of battery/flywheel energy storage system to improve ageing of lead-acid batteries in PV-powered applications. Int J Sustain Eng 13(5):337–359. https://doi.org/10.1080/19397038.2020.1725177

Barelli L et al (2019) Flywheel hybridization to improve battery life in energy storage systems coupled to RES plants. Energy 173:937–950. https://doi.org/10.1016/j.energy.2019.02.143

Casarin S, Cavazzini G, Pérez-Díaz JI (2023) Battery and flywheel hybridization of a reversible pumped-storage hydro power plant for wear and tear reduction. J Energy Storage 71:108059. https://doi.org/10.1016/j.est.2023.108059

Crespo L (2020) The double role of CSP plants on the future electrical systems. In: WBG conference 'Concentrating Solar for Power and Heat'

Deign J (2021) Latin America's energy storage leader is getting creative. Canary Media

EnerData (2023) Chile approves AES Andes' coal-fired power plant conversion project. Available at https://www.enerdata.net/publications/daily-energy-news/chile-approves-aes-andes-coal-fired-power-plant-conversion-project.html. Accessed 17 Sept 2024

Forbes (2018) Energy innovation: policy and technology. India coal power is about to crash: 65% of existing coal costs more than new wind and solar. Forbes Newsletter

Gao W et al (2024) Comprehensive study of the aging knee and second-life potential of the Nissan Leaf e+ batteries. J Power Sources 613:234884. https://doi.org/10.1016/j.jpowsour.2024.234884

Geyer M, Trieb F, Giuliano S (2020) Repurposing of existing coal-fired power plants into thermal storage plants for renewable power in Chile. Bonn, Germany

Glücker P et al (2021) Prolongation of battery lifetime for electric buses through flywheel integration. Energies 14(4):899. https://doi.org/10.3390/en14040899

Guo NF, Sharma R (2016) Hybrid energy storage systems integrating battery and ultracapacitor for the PJM frequency regulation market. IEEE Power Energy Soc Gen Meet. https://doi.org/10.1109/pesgm.2016.7741867

Kefford BM et al (2018) The early retirement challenge for fossil fuel power plants: a multi-region analysis of asset stranding and community impact. Energy Policy 123:367–379. https://doi.org/10.1016/j.enpol.2018.08.014

Kim Y, Raghunathan V, Raghunathan A (2016) Design and management of battery-supercapacitor hybrid electrical energy storage systems for regulation services. IEEE Trans Multi-Scale Comp Syst 3(1):12–24. https://doi.org/10.1109/tmscs.2016.2627543

Koike M et al (2018) Optimal scheduling of battery storage systems and thermal power plants for supply–demand balance. Control Eng Pract 77:213–224. https://doi.org/10.1016/j.conengprac.2018.05.008

Lee D, Wang L (2008) Small-signal stability analysis of an autonomous hybrid renewable energy power generation/energy storage system part I: time-domain simulations. IEEE Trans Energy Convers 23(1):311–320. https://doi.org/10.1109/tec.2007.914309

Martínez-Lavín M et al (2022) Evaluation of the latest Spanish grid code requirements from a PV power plant perspective. Energy Rep 8:8589–8604. https://doi.org/10.1016/j.egyr.2022.06.078
Picatoste A et al (2024) Comparing the circularity and life cycle environmental performance of batteries for electric vehicles. Resour Conserv Recycl 210:107833. https://doi.org/10.1016/j.resconrec.2024.107833
Rodríguez-García MM et al (2023) Experimental and theoretical investigation on using microwaves for storing electricity in a thermal energy storage medium. AIP Conf Proc 2815(1):060003. Available at https://doi.org/10.1063/5.0148703
Sener (2022a) Planta termosolar CCP NOOR Ouarzate II. Available at https://www.group.sener/proyecto/planta-termosolar-ccp-noor-ouarzate-ii/. Accessed 28 Sept 2023
Sener (2022b) Planta termosolar de torre central NOOR III, en Ouarzazate (Marruecos). Available at https://www.group.sener/proyecto/termosolar-torre-nooro-iii-ouarzazate/#. Accessed 28 Sept 2023
Shahjalal M et al (2021) A review on second-life of Li-ion batteries: prospects, challenges, and issues. Energy 241:122881. https://doi.org/10.1016/j.energy.2021.122881
Song F et al (2021) Review of transition paths for coal-fired power plants. Global Energy Interconnection 4(4):354–370. Available at https://doi.org/10.1016/j.gloei.2021.09.007
Takahashi A, Allam A, Onori S (2023) Evaluating the feasibility of batteries for second-life applications using machine learning. iScience 26(4):106547. https://doi.org/10.1016/j.isci.2023.106547
Thakur J, De Almeida CML, Baskar AG (2022) Electric vehicle batteries for a circular economy: second life batteries as residential stationary storage. J Clean Prod 375:134066. https://doi.org/10.1016/j.jclepro.2022.134066
Thomas T, Akhtar K (2023) Repurposing coal-fired power plants: benefits and challenges, Hatch. Available at https://www.hatch.com/About-Us/Publications/Blogs/2023/03/Repurposing-coal-fired-power-plants-benefits-and-challenges. Accessed 27 Aug 2024
Torres J et al (2020) Dimensioning methodology of energy storage systems for power smoothing in a wave energy conversion plant considering efficiency maps and filtering control techniques. Energies 13(13):3380. https://doi.org/10.3390/en13133380
Vaca SM, Patsios C, Taylor P (2016) Enhancing frequency response of wind farms using hybrid energy storage systems. In: IEEE International conference on renewable Eeergy research and applications (ICRERA). https://doi.org/10.1109/icrera.2016.7884560
Wang B et al (2023) Effects of integration mode of the molten salt heat storage system and its hot storage temperature on the flexibility of a subcritical coal-fired power plant. J Energy Storage 58:106410. Available at https://doi.org/10.1016/j.est.2022.106410
Wu W et al (2020) Does energy storage provide a profitable second life for electric vehicle batteries? Energy Econ 92:105010. https://doi.org/10.1016/j.eneco.2020.105010
Zhang Q et al (2024) Dynamic characteristics and economic analysis of a coal-fired power plant integrated with molten salt thermal energy storage for improving peaking capacity. Energy 290:130132. Available at https://doi.org/10.1016/j.energy.2023.130132

Optimizing Renewable Power Systems: Hybrid Gravity-Battery Energy Storage System for Wind/PV Integration and Load Balancing

Asmae Berrada, Anisa Emrani, and Youssef Achour

Abstract Hybrid energy storage systems have gained interest in recent years due to their potential to outperform single energy storage systems in certain situations. However, optimally designing and sizing HESS to effectively replace a single energy storage technology remains a complex challenge. In this study, a HESS combin-ing gravity energy storage (GES) with high-power electrochemical energy storage is integrated into a hybrid energy system (PV/WIND) to balance supply and demand in a renewable energy power system. This chapter investigates the design and optimization of this HESS. To achieve the best possible integration between the hybrid Gravity/Battery storage system and the WT/PV system, an energy management system linked to an optimization process has been developed. Prediction models have been used to forecast solar and wind generation. The obtained results have shown that the energy cost is significantly affected by the reliability percentage of the plant. When the reliability level is reduced to 20%, the optimal energy cost is 0.09 €/kWh, compared to 0.22 €/kWh at 100%, reliability. Reliability and efficiency in the utilization of renewable energy are enhanced by the integration of forecast models into the study and the hybridization of energy storage, which enhance the renewable power system's optimal operation and design accuracy.

Keywords Hybrid energy storage · Gravity energy storage · Optimization · Energy forecast · Energy management

A. Berrada (✉) · Y. Achour
LERMA Lab, College of Engineering and Architecture, International University of Rabat, Sala Al Jadida, Morocco
e-mail: asmae.berrada@uir.ac.ma

A. Emrani
National School of Applied Sciences, Cadi Ayyad University, Safi, Morocco

R. Scipioni et al. (eds.), *Hybrid Energy Storage*, Lecture Notes in Energy 47,
https://doi.org/10.1007/978-3-031-97755-8_3

1 Introduction

Significant efforts worldwide are being made to reverse global warming and eliminate its causes, especially with fossil fuel reserves nearing depletion (Wright 2023). Additionally, significant growth in population has led to increased energy demand. For these reasons, world nations are searching for clean sustainable alternative to avoid crises (IEA 2024). Renewable energy sources present an excellent solution to these challenges. PV technology, for instance, has been the optimal energy source in space stations and satellites (Photovoltaics for Space 2023). However, they are characterized by intermittent production, unlike standard power generation methods such as coal and oil (IRENA 2019). In addition, renewable energy systems (RES) have other limitations, such as low to medium efficiency and, in some cases, high costs (IRNEA 2022a, b). The hybridization of renewable energy systems has been considered a promising solution to the problem of intermittent energy production. A photovoltaic (PV)/Wind hybrid system was studied back in 1981 to solve the energy scarcity in remote areas (Castle et al. 1981). The study demonstrated that combining PV/wind managed to significantly increase the efficiency of the system. The hybridization of RESs has been the main focus of a number of studies, either modeling and optimization or experimental setups (Sinha and Chandel 2014, 2015). This configuration allows for high-efficiency and low-emissions, which make such systems affordable and environmentally friendly (IRNEA 2022a, b). PV panels with batteries as an example are an optimal energy system for off-grid applications in farms, especially with the recent levelized costs of energy for renewable systems (Achour et al. 2023; Solar PV Utility Scale Levelised Cost of Energy Index Based on Average Annual Input Costs, 2018–2024—Charts—Data & Statistics—IEA, n.d.).

Energy production from renewables varies throughout the hour, highlighting the need for effective energy storage systems (ESS) (Douglas 2016). A common distinction between the different ESSs is to categorize them by the type of energy stored, which can be divided into electrochemical, mechanical, chemical, thermal, and electrical energy storage (Mitali et al. 2022). Promising mature technologies such as electro-chemical storage or battery energy storage have been widely utilized (Park et al. 2024). Nevertheless, batteries are accompanied by several challenges regarding their energy density, accelerated degradation, and high prices (Guo et al. 2022). Another promising technology is pumped hydro storage (PHES), which has been extensively researched. Despite its low energy storage costs and long lifespan (50 years), it requires a significant initial cost and very specific geographical conditions (Zakeri and Syri 2015; Mostafa et al. 2020). Gravity Energy Storage (GES) operates similarly to PHS. It utilizes gravity to push down the piston to generate electricity during discharge mode. When there is excess electricity, a motor lifts the piston back up during charge mode (Berrada et al. 2017; Ameur et al. 2022, 2023). Emrani et al. (2024) published a recent review regarding various energy storage technologies, applications, costs, and developments.

Efficient energy management has been the focus of many researchers for its important effect on the grid's stability. The critical sizing of each system in a hybrid installation is essential to guarantee a smooth operation and efficient energy flow (Ameur et al. 2021). Paudel et al. (2019) proposed a game-theoretic model for real-time prosumer-based community microgrid (PCM) energy management. The model was built on data input from renewable generation and load profiles. Lamnatou et al. (2022) reviewed the recent advancements in smart grids. The main considerations of this work were PVs, storage technologies, buildings, and the environmental impact. Researchers have developed several models to enhance the various aspects regarding energy management. For instance, tube model predictive control (TMPC) has emerged as a promising solution to minimize the gap between predictive and actual information. The TMPC method achieves this by decoupling the standard Model Predictive Control (MPC) approach into two distinct parts: an offline robust control subproblem and an online open-loop feedback control subproblem (Xie et al. 2024). A renewable energy system with hybrid energy storage was suggested in a recent paper (Emrani et al. 2024). The system managed to meet 100% of the load demand and achieved a low cost of energy (COE) of about 0.29 $/kg.

Gravity energy storage has demonstrated excellent potential. Nonetheless, it is still a recent technology that requires further investigation. This chapter delves into the aspects of dry gravity storage for industrial applications. While there is plenty of research that was dedicated towards the development of this technology, only a handful of papers paid attention to their hybridization with battery storage in an off-grid application. This chapter is one of the few studies that integrate hybrid GES/Battery system within a hybrid PV/Wind power plant. The study proposes the optimal configuration of hybrid energy production/storage to cut down the costs significantly. Advanced forecast algorithms have been deployed to predict energy generation. This research area, to the authors' knowledge, remains a clear gap in the literature. The aim is to address this gap with the hope of incorporating an optimal system and guaranteeing an accurate, stable energy balance for the off-grid industrial application.

Following this introductory section, this chapter is structured as follows: The modeling of the hybrid renewable energy system and the assumptions considered in this study are presented in Sect. 2. Section 3 is dedicated to discussing the obtained results. Finally, the conclusion with the key findings and future perspectives is presented in Sect. 4.

2 Hybrid Energy System Model

2.1 System Description

This chapter investigates the potential of combining gravity storage with high-power electrochemical energy storage for balancing supply and demand of renewable energy power system. This hybrid system will manage the energy load of an industrial application. Forecast models for power generation from each renewable energy source

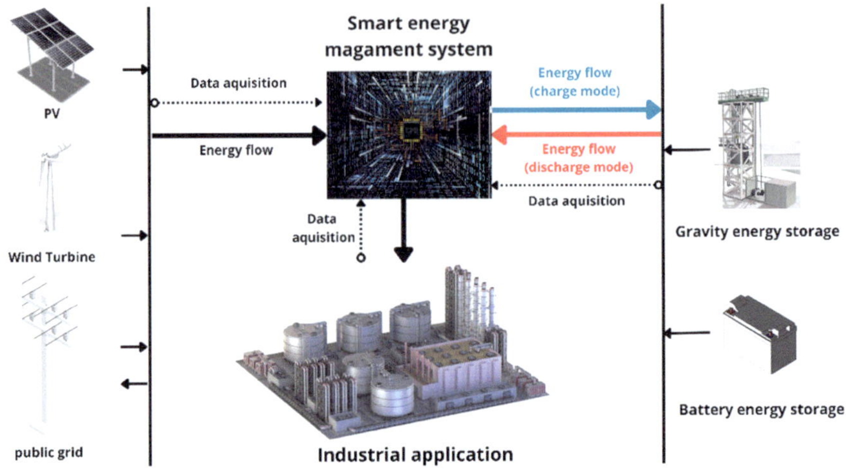

Fig. 1 Hybrid PV/Wind/GES/BAT system schematic

were developed in order achieve this objective and build an energy system that is cost-effective.

A schematic of the proposed system is depicted in Fig. 1 where the different sub-components along with energy flow are displayed.

This work explores the integration of wind (operates in AC) and PV (operates in DC) to supply energy to an industrial load (operates in AC) while coupled to a hybrid energy storage system. The aim of this study is to meet the full load demand in an off-grid set up.

Energy generated from the PV system flows into the grid through a DC-AC converter, whereas wind power feeds directly into the distribution station via the AC grid After frequency adaptation and regulation (50 Hz for Morrocco's case). Shortages of the used RE source are expected because of its intermittent supply. Thus, the hybrid GES/BAT storage system is incorporated into the energy system to improve its reliability. This is achieved by storing excess energy and releasing it when it's needed (Fig. 2).

2.2 PV Prediction Model

Mitsubishi Electric (Model: PV-MLT260HC) PV panels are utilized in this study. The rated power of these modules can reach 255 Wp under standard test conditions (STC). PV panel prices have declined in recent years. Solar panels are found in various applications, from household usage to large scale power plants. The detailed characteristics of the considered modules are presented in Table 1. PV generated power is determined by Eqs. 1–2.

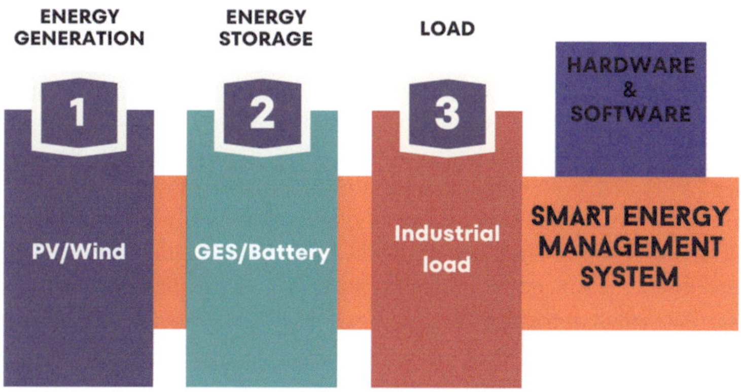

Fig. 2 Model components

$$P_{PV}(t) = E_p \frac{G_f(t)}{G_{STC}}[1 + k(T(t) - T_{STC})] \quad (1)$$

$$T(t) = (T_{air}(t) + 0.0318 \times G_f(t) \times (1 + 0.031\ T_{air}(t)) \times (1 - 0.042 \times V)) \quad (2)$$

With k being the coefficient of temperature, T(t) is the cell's temperature, $T_{air}(t)$ is the ambient temperature, T_{STC} is the reference temperature, and V is the wind speed. The forecasted PV power generation is denoted by $P_{PV}(t)$, the maximum output power under standard test conditions (STC) is denoted by E_p, and the actual solar radiation is represented by $G_f(t)$.

Table 1 WTs, PV modules, and battery characteristics

Parameter	Value		
	WT	PV	Battery
Type	3 bladed horizontal axis	m-Si (78 mm × 156 mm)	Lead acid
Rated power	30 kW	255 Wp	–
Hub height	18 m	–	–
Rotor diameter	13.8 m	–	–
Swept area	150.0 m^2	–	–
Rotor speed, max	60.0 U/min	–	–
Open circuit voltage	–	37.8 V	58.4 V
Short circuit current	–	8.89 A	–
Max power voltage	–	31.2 V	51.2 V
Current at max power	–	8.18 A	400 A
Rated capacity	–	–	20 kWh
Efficiency	–	16%	90%
Depth of discharge	–	–	80%

PV generation is governed by several key factors, mainly temperature and irradiation. Other parameters may interfere with the energy production process, such as the shading on the modules' surfaces, dust, maintenance, etc. Thus the forecasting of energy production by PV is essential. It helps achieve a smooth off-grid operation and plans energy balance ahead of time. Dark Sky Application Programming Interface (API) enables the prediction of diffuse, direct, and total irradiance. These predictions are based on the percentage of cloud cover and geographic information (latitude and longitude). The model retrieves a JSON file containing the temperature and cloud cover percentage for a specific site each day. The data is used as input to forecast solar radiation (Eqs. 3–6).

$$DNI = G_0 * 0.73^{\left(\frac{1}{\cos(Zenithangle)}\right)^{0.678}} \tag{3}$$

$$I_f = (1 - cloud\,cover\,precent) * DNI \tag{4}$$

$$D_f = 0.2 * I_f \tag{5}$$

$$G_f = I_f + D_f \tag{6}$$

With G_o, DNI I_f, D_f, and G_f being the solar constant (1366 W/m^2), direct irradiance, predicted direct irradiance, predicted diffuse irradiance, and predicted global irradiance, respectively.

2.3 Prediction Model of Wind Energy

Wind energy has been harnessed for decades, and it is considered a mature technology. It is utilized across various scales and applications (Bird et al. 2016). The primary factor for wind turbines' performance is wind speed. This latter is categorized into rated speed (v_{rate}), cut-in speed (v_{cut-in}), and average speed. The cut-in speed varies depending on the technology and scale. Wind turbines only produce electricity when the wind speed (v) exceeds the initial speed, or $v_{(cut\text{-}in)}$. The wind turbine (WT) produces a nominal power of $P_{(wind\_rate)} = 30$ kW once it achieves the nominal speed (v_{rate}).

Equation 7 is used to determine the power generated by the WTs. While the characteristics of the utilized WTs are detailed in Table 1.

$$P_w = \begin{cases} 0 & v < v_{cut-in} \\ \frac{N_{wt} \times P_{wind\_rate} \times \left(v^3 - v_{cut-in}^3\right)}{v_{rate}^3 - v_{cut-in}^3} & v_{cut\,in} < v < v_{rate} \\ N_{wt} \times P_{wind\_rate} & v_{rate} \le v \le v_{cut-out} \\ 0 & v_{cut-out} \le v \end{cases} \tag{7}$$

With N_{wt} denotes the number of WTs and P_{wind_rate} refers to the produced rated power by the WT.

3 Hybrid Gravity/Battery System Model

3.1 Gravity Energy Storage

Gravity energy storage has proven to be an interesting storage technology for medium-to-large scale. Numerous studies have been conducted to evaluate this technology (Emrani et al. 2022a, b; Berrada and Loudiyi 2016). The GES system is comprised of a motor/generator, a drum, a wire rope, and a heavy piston. For the storage phase, excess electricity pulls up the heavy piston, guided within a shaft, upwards by converting electrical energy into gravitational potential energy. In the discharge phase, the reverse process occurs. The wire rope, wound around the drum, serves as a flexible conduit to transfer the gravitational potential energy from the descending piston to the drum, which is connected to the generator. This allows for converting the stored gravitational potential energy back into electrical energy through regenerative breaking. Consequently, for the GES configuration, power is directly proportional to both weight and vertical velocity, while energy storage is fundamentally related to the product of mass and vertical displacement (Loudiyi and Berrada 2017). From a design perspective, the energy capacity of the GES system is predetermined by the product of height and weight. Nonetheless, the power output can be adjusted by varying the piston's speed. In addition, like any other storage system, losses may occur during both the charging and discharging phases. The operational capacity of the GES system can be calculated by Eq. 8.

$$E = \mu * m * g * h(m) \tag{8}$$

where E is energy capacity in joules (J), μ is storage efficiency, m is the mass of the pistons (Kg), h represents the lifting height (m), and g is the gravitational acceleration (m/s^2).

This equation can be rewritten by incorporating the piston density (ρ_p), height (h(m)) and diameter (d), as follows (Eq. 9):

$$E = \mu * \rho_p\left(\frac{1}{4}\pi d^2 h(m)\right) * g * h(m) \tag{9}$$

As stated in Eq. 9, P_{ch}, and P_{disch} are the stored and discharged power at any given moment, respectively. These two parameters depend on the system efficiency, and the piston velocity.

$$\begin{cases} P_{ch} = \frac{m*g*V}{\eta_{ch}} \\ P_{disch} = \eta_{disch} * m * g * V \end{cases} \tag{10}$$

3.2 Battery Energy Storage

Batteries are electrochemical means for electrical energy storage. They have been used for decades and have undergone extensive research and development. Their operating conditions change significantly across different technologies. The state of charge and discharge (SOC/SOD) of batteries, expressed in Eqs. 11 and 12, are typically used for assessing the performance of energy storage devices. During the charging and discharging phases of the ESS (energy storage system), energy losses can occur. Consequently, the self-discharge rate and efficiency of an ESS influence its SOC. The SOC is determined by the amount of energy generated and stored between two consecutive time points, i.e. t and t − 1, as well as the previous SOC state. Depending on the load demand at any given time, the total power generation may or may not be sufficient to supply the load. Typically, this is assessed on an hourly time step.

$$\begin{aligned} Charge : SOC(t) &= SOC(t-1) * (1-\sigma) \\ &\quad + \left(P^g(t) - \frac{P^l(t)}{\mu^*_{inv}(t)} \right) * \mu_{inv} \end{aligned} \tag{11}$$

$$\begin{aligned} Discharge : \mathrm{SOD}(t) &= SOD(t-1)*(1-\sigma) \\ &\quad + \left(\frac{P^l(t)}{\mu^*_{inv}(t)} - P^g(t) \right) \Big/ \mu_{bf} \end{aligned} \tag{12}$$

Equation 13 determines the maximum energy storage capacity of battery bank, represented as $\mathrm{E}_{\mathrm{Batt-charg,max}}$. This equation takes into account the total number of batteries (NBAT), capacity of individual battery (EBAT), and the battery capacity during a given hour (i).

$$E_{Batt-ch\arg,\max} = N_{BAT} \times E_{BAT} - E_{battery(i)} \tag{13}$$

The maximum amount of electricity discharged may be expressed mathematically as (Eq. 14)

$$E_{Batt-disch,\max} = \eta_{bat} \times \left(E_{battery(i)} - (1 - DOD) \times E_{BAT} \right) \tag{14}$$

The different characteristics of the PV modules, Wind turbines, and battery are shown in Table 1.

4 Energy Management and Optimization Strategies

After developing simulation models for the different components of the hybrid power plant, it is crucial to implement an energy management system that would control the energy flow transferred between the various plant components. To address this issue, this section first presents an optimization problem formulation that includes the intended objective and constraint function. The decision variables are also discussed in this section. Additionally, based on the previously developed models, the energy flow management algorithm under consideration is provided, which will determine energy flow behavior using real-world scenarios.

4.1 Objective Function

The hybrid PV/WT/GES/BAT system's cost of energy (COE) is minimized in the developed model as shown in Eq. 15. This objective function is achieved using a life-cycle cost analysis (LCCA). In order to determine which configuration is the most cost-effective, it assesses the techno-economic performance of each and determines the total cost of the project.

$$\boldsymbol{COE} = \frac{\sum \frac{r(1+r)^{L_i}}{(1+r)^{L_i}-1} \times (\boldsymbol{C}_{Cap-i} + \boldsymbol{C}_{O\&M-i} + \boldsymbol{C}_{REP-i} + \boldsymbol{C}_{DISP-i})}{\boldsymbol{E}_{output}} \tag{15}$$

$$\boldsymbol{C}_{Cap} = \boldsymbol{Cap}_{PV} + \boldsymbol{Cap}_{GES} + \boldsymbol{Cap}_{BAT} + \boldsymbol{Cap}_{WT} \tag{16}$$

$$\boldsymbol{C}_{O\&M-i} = \boldsymbol{O\&M}_i\% \times \boldsymbol{C}_{Cap-i} \tag{17}$$

where and L_i is the system component i's lifetime (e.g., PV, GES, batteries, wind turbines), $\boldsymbol{E}_{output}$ is the total energy output and r is the discount rate or interest rate. The operation and maintenance (O&M) cost of component i is represented by $\boldsymbol{C}_{O\&M-i}$. The capital cost of component is denoted $\boldsymbol{C}_{Cap-i}$. The cost to replace or repair component i, and its disposal cost are by represented by $\boldsymbol{C}_{REP-i}$, and $\boldsymbol{C}_{DISP-i}$, respectively.

The hybrid power plant's LCCA is calculated by taking into account the capital costs of all of its components, which include the PV, wind, GES, and battery system (See Table 2). Operating and maintenance costs are also factored in by the LCC and are expressed as a percentage of the initial expenditure. The cost of disposal, recycling, and replacement are all considered in the LCCA analysis (Berrada et al. 2021).

The objective function uses the discount rate to determine the whole life-cycle cost of the system components, which includes capital expenditures, operating and maintenance (O&M) costs, replacement costs, and disposal costs. The results are

Table 2 Hybrid system components' costs

	PV (Diab et al. 2020)	WT (Diab et al. 2020; Yang et al. 2021)	GES (Berrada et al. 2021)	Battery (Berrada and Loudiyi 2016; Emrani et al. 2022a, b)
Capital Cost	$N_{PV} \times 112$	$58563 \times N_{WT}$	$0.5 \times E_{GES}$ (Wh)	$0.13 \times E_{BAT}$ (Wh)
O&M Cost (%)	1	3	1	1
Replacement Cost	0	1350 €/kWh	0	$0.16 \times E_{BAT}$ (Wh)

Table 3 Optimal capacity configuration of the hybrid energy system for different reliability levels

REL (%)	Number of PVs	Number of WTs	GES (kWh)	Battery (kWh)	COE €/kWh
100	79	15	29.51	98.1	0.226
90	61	12	25.1	80	0.181
70	50	10	20.5	66.5	0.152
50	43	9	16.3	45.4	0.135
20	29	6	10.9	27.2	0.09

then discounted to present value. The COE is then calculated by dividing this total cost by the system's total energy production.

Equations (16) and (17) offer additional details on how each component's particular cost was calculated.

Table 3 gives insights about the system costs. It's important to note that when estimating the life cycle cost of GES, disposal and recycling charges are typically excluded due to the lack of well-established demonstration plants for the system (Berrada et al. 2021).

4.2 Model Constraint

The proposed hybrid system's primary objective is to provide a consistent energy supply at a minimum cost. As a result, a preset percentage of dependability (REL) limits the suggested target function. This is a crucial indicator of the hybrid system's effectiveness and capacity to handle the required load. REL may be calculated in detail using Eq. 18. The predetermined REL requirement is considered an input parameter in the optimization process. During the iterative optimization process, the adaptive energy management system uses the solution—that is, the design variables gained in each iteration—as input to calculate the surplus and deficit energy and provide the relevant REL value for the current solution. To assess whether the solution is acceptable or whether it has to be repeated for a more precise answer, the latter is compared to the predefined REL criterion. In an attempt to minimize the cost of energy (COE), the simulation runs continuously until the dependability requirement is satisfied.

$$REL = 100 - \frac{\sum_{i=1}^{168}\left(P_{load}(t_i) - \left(P_{pv}(t_i) + P_w(t_i) + P_{BG}(t_i) + P_{dsich}(t_i)\right)\right)}{\sum_{i=1}^{8760} P_{load}(t_i)} \quad (18)$$

4.3 Optimization Model

The current problem has to achieve two main goals which include optimizing the hybrid energy system design as well as the hybrid energy storage size (GES, and batteries). That is, the optimal capacities for effective storage must be achieved. This optimization problem is constrained by the requirement that the system's reliability must be maximized. The Fmincon solver from MATLAB's optimization toolbox is used to solve this nonlinear optimization problem. This objective function must comply with the constraints that ensure a maximized reliability level.

In order to fulfill the stated restrictions and optimize the dependability level, this objective function must comply. This optimization problem is represented mathematically in Eq. 19. Minimizing the COE (objective function) within the constraints of the problem is the objective function.

$$\begin{cases} \min\limits_{x} function(x) \\ constrainte(x) \leq 0 \\ constrainte_{eq}(x) = 0 \\ L_b \leq x \leq U_b \end{cases} \quad (19)$$

where $constrainte(x)$ and $constrainte_{eq}(x)$ and function (x) are functions that return a scalar, all determined using Eq. (15), and x is a vector of the design variables.

The interaction between the SEMS, the forecast model, and the optimization algorithm is depicted in Fig. 3. Initially, the forecast models predict the power that could be generated from the renewable energy sources. Then, the Fmincon solver is utilized to determine the optimal values of the design variables using the interior point method, with an aim to minimize the objective function while satisfying the constraints. In each iteration, the EMS determine the reliability level (REL) and evaluate the solution's adequacy.

The use of Fmincon and the interior point method can be justified by a number of considerations. Initially, Fmincon employs a state-of-the-art optimization method known for its reliability. Because of its adaptability, it can effectively tackle both minor and large optimization issues. The solver can provide helpful Hessian information to increase the effectiveness of the optimization.

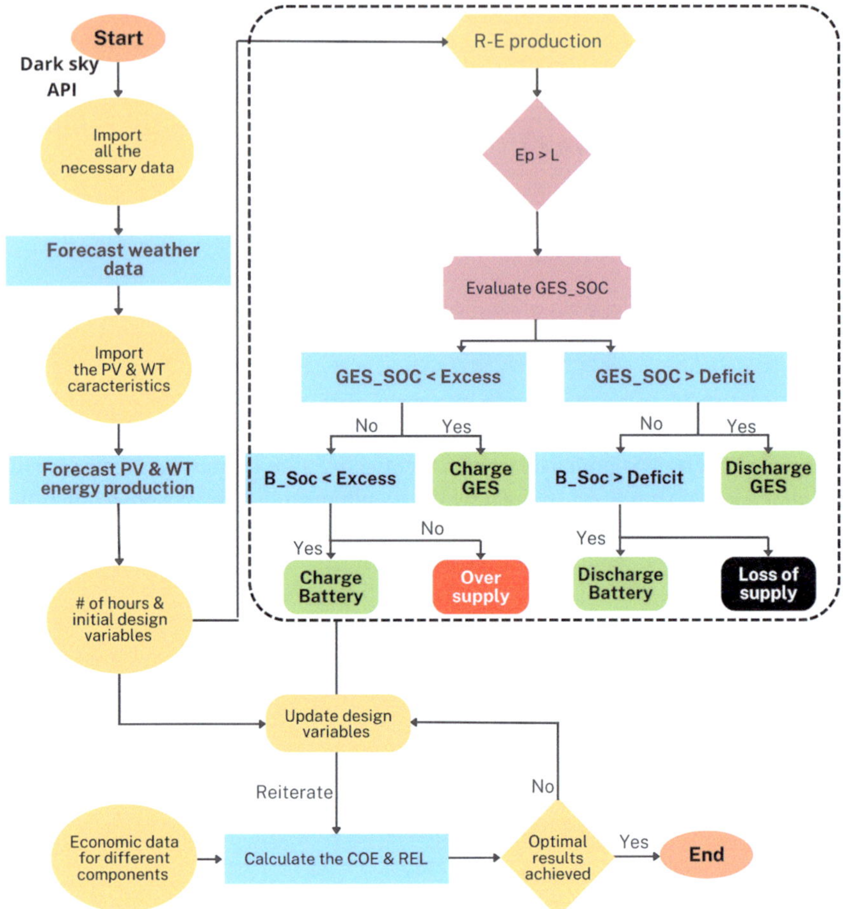

Fig. 3 SEMS and optimization Problem Flowchart

4.4 *Smart Energy Management System*

The energy flowing into and out of the energy storage system is controlled by the energy management system. The hybrid GES/BAT system, when used in an off-grid hybrid energy system, is charged and discharged based on the power plant's surplus or deficit of energy. The flowchart shown in Fig. 3 illustrates the energy management approach for the system.

The symbols SOC_{BAT} and SOC_{GES} represent the battery system and GES's respective states of charge. The energy generated from renewable sources is denoted by E_{gen} which is the sum the Energy produced by the PV system and the wind turbine. E_l stands for load demand. The energy used by the motor to raise the heavy piston is denoted by E_{GES_ch} and the energy used to charge the battery is E_{BAT_ch}. The energy produced during the discharge mode of both the energy storage system GES and battery is E_{GES_disch} and E_{BAT_disch}, respectively.

Maintaining a power balancing constraint between energy supply and demand is the goal of the proposed operational approach, which ensures that the load is supplied on an hourly basis. The power generated by WTs and PV, along with the hourly electricity released by GES and the battery system, should equal the network's total energy loss and load demand. This is achieved by accounting for both the varying load demand and the intermittent nature of PV and WT systems. Based on the projected meteorological data, the model estimates the power generation of the PV and WTs on an hourly basis.

In case the generated power exceeds the demand, GES is charged with excess power, and E_{GES_ch} is computed to guarantee that the piston position stays within its upper bound. The excess energy is used to charge the battery, when GES is fully charged. If the production renewable energy system is not able to meet the load demand, power is discharged from GES $\left(E_{GES\_disch}\right)$, taking into account the GES's state of charge (SOC) If the required energy surpasses what the GES can provide, additional energy is discharged from the battery, while considering its state of charge.

5 Results and Discussion

5.1 Power Generation Forecast for the Investigated Case Study

An industrial facility in Laayoune, Morocco, has been used in this case study. Figure 4 depicts the load demand pattern that has to be met. Over the course of a week, it shows the energy usage in hourly increments. There is a peak electrical usage of around 15.8 kW. This happens between Thursday, which is considered an operating day. Over the weekend, there is a notable decline in load demand, with high-power usage of around 10 kW on Sunday and 8 kW on Saturday.

Laayoune, Southern Morocco, is classified as a BWh climate (dry winters and very hot summers) by the Köppen-Geiger climate classification system (Beck et al. 2018). Based on this categorization, it appears to have optimal conditions for solar energy integration. This location is also well known for having a high potential for wind energy integration.

The forecast model uses the geographic coordinates of the investigated region in order to predict the weather parameters for a seven-day period and, as a result, forecast the power generated by renewable sources during that time. Figures 5 and 6 show the findings obtained for wind speed, ambient temperature, and solar irradiation, respectively.

The predicted solar irradiation is shown in Fig. 5, where it varies from 0 W/m2 in cloudy circumstances to 970 W/m^2 in clear skies. The obtained solar radiation

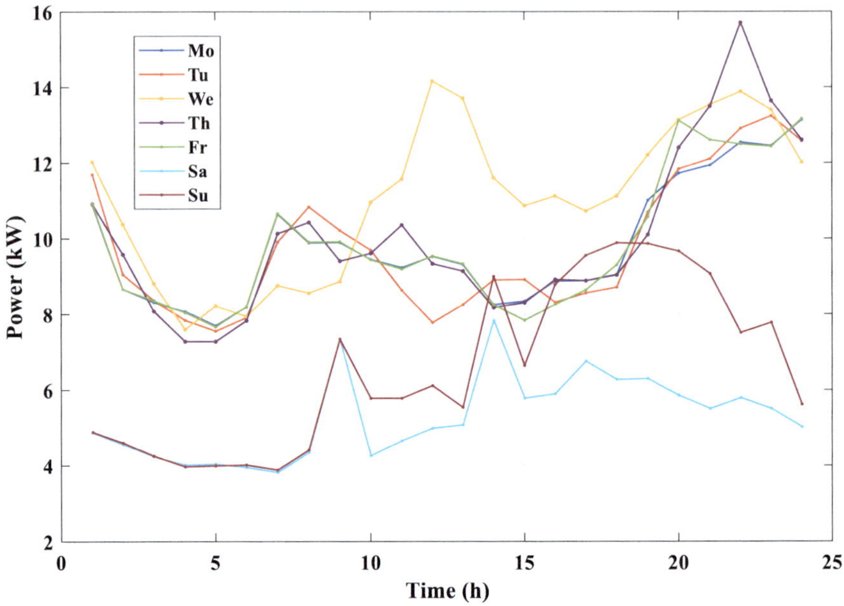

Fig. 4 Industrial load demand over the week

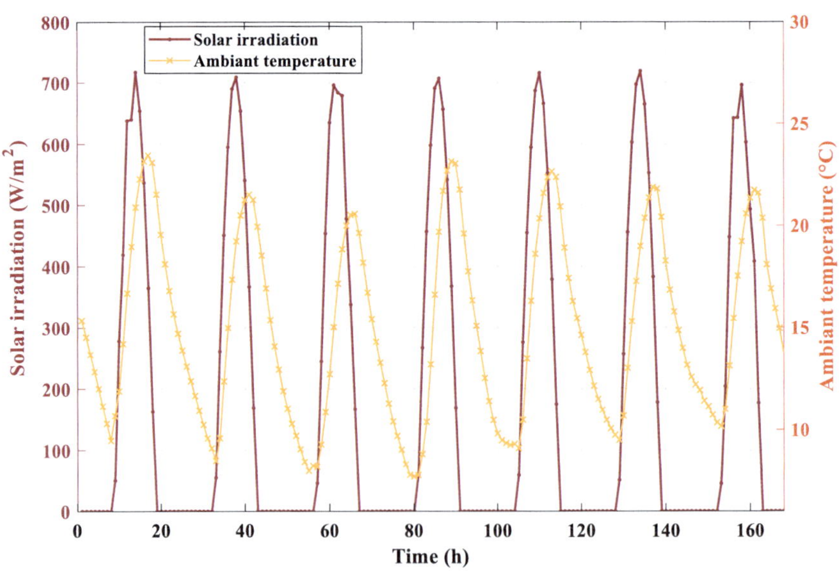

Fig. 5 Forecasted solar irradiation and ambient temperature

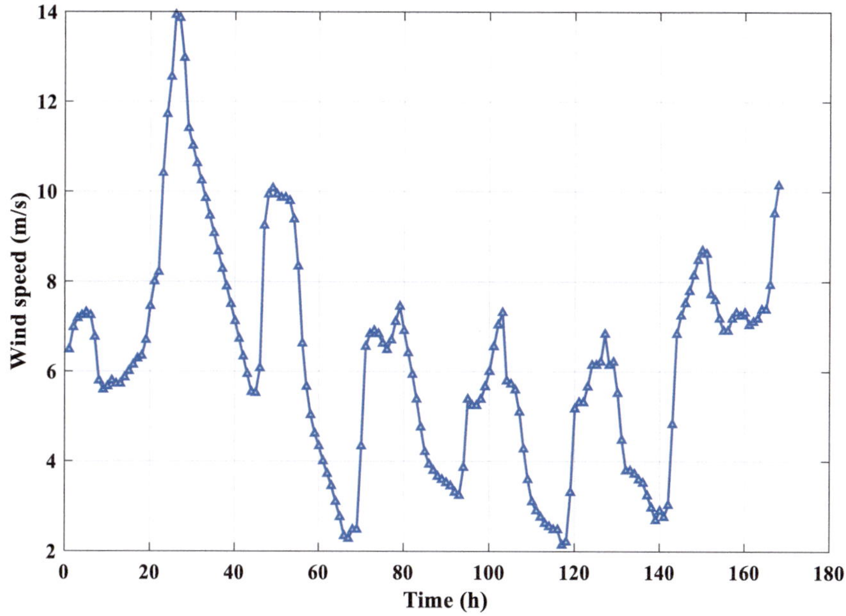

Fig. 6 Forecasted wind speed

indicates the region's potential for solar energy generation. A crucial factor in forecasting PV power generation, in addition to solar radiation, is ambient temperature. The model forecasts the ambient temperature, which ranges from 8.1 to 24.8 °C as shown in Fig. 5. These temperature predictions are crucial for estimating PV power production and evaluating the overall performance of the PV system, as temperature directly impacts the efficiency of converting solar energy into electrical power. The temperature and irradiation profiles exhibit a distinct daily trend significantly influencing the PV system's output.

To forecast wind power generation, the model first predicts the wind speed which shown in Fig. 6. The data, which ranges from 2.4 to 13.9 m/s, shows that there will be wind over the week, which is optimal for wind turbine (WT) operation. An average wind speed of approximately 6.25 m/s is obtained, indicating a potential wind resource.

The outcomes of the weather prediction were crucial in estimating the potential for producing renewable energy in the investigated region and, as a result, in directing the optimal capacities of these energy systems when coupled with a hybrid energy storage system (GES/Battery). Figure 7 shows the generation contribution of these systems (PV and WT) which has been estimated based on the expected meteorological data. It illustrates a comparison of the daily expected generation over the course of a week between one WT and one PV module. When compared to solar PV sources, WT output shows good performance. This is mostly because of the site's large potential for wind speed.

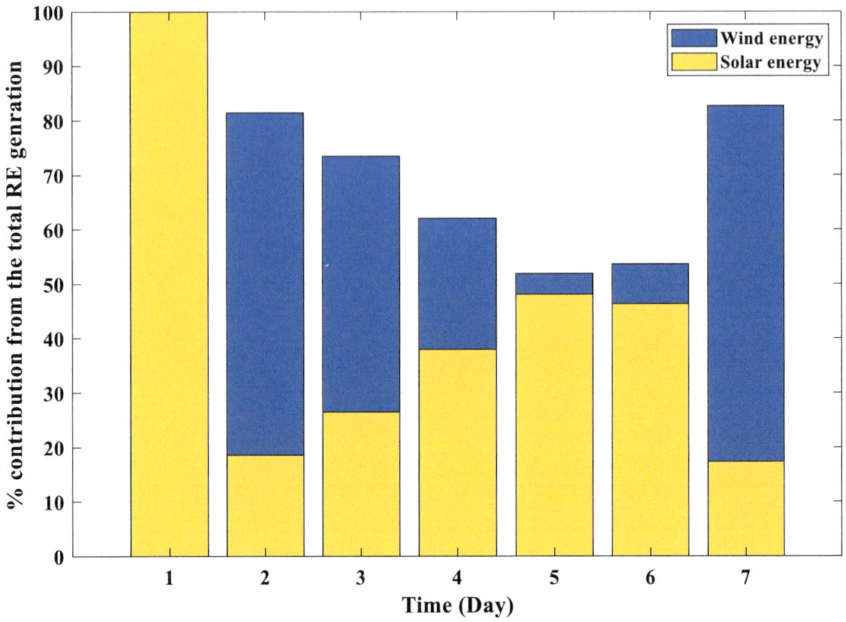

Fig. 7 Contribution of each form of RE from the total power output

5.2 *Design Optimalization of the Hybrid System*

The optimization process was used to calculate each renewable energy source's optimal capacity after power generation predictions were determined. This allowed the hybrid system to achieve the lowest possible overall cost while meeting load demand requirements with a predetermined reliability level.

Figure 8 illustrates the process of optimizing the design variables to obtain the optimum values and, hence, the lowest cost while meeting 100% of the load demand. The quantity of PV modules and WTs, as well as the capacities of batteries and GES systems, are the optimal design variables considered by the model. This latter achieves the objective function's best possible value after 110 iterations. For 100% reliability, the lowest cost obtained is equal to 0.226 €/kWh.

The obtained optimal design variables are displayed in Table 3. Additionally, Fig. 9 illustrates their variance with regard to the reliability criteria. Compared to WTs, PV panels are found to be more common. This is explained by the fact that the purchasing costs of this latter different from those of other components. As the objective function (COE) is minimized, the optimization method favors the least cost system. It is shown that lowering the reliability criterion results in a declining trend of COE, as seen in Fig. 10. This illustrates how a minor modification to the reliability objective can have a larger influence on cost results.

To meet the necessary energy demand, the model determines the daily power generation of each renewable energy system. Figure 11 illustrates the expected power

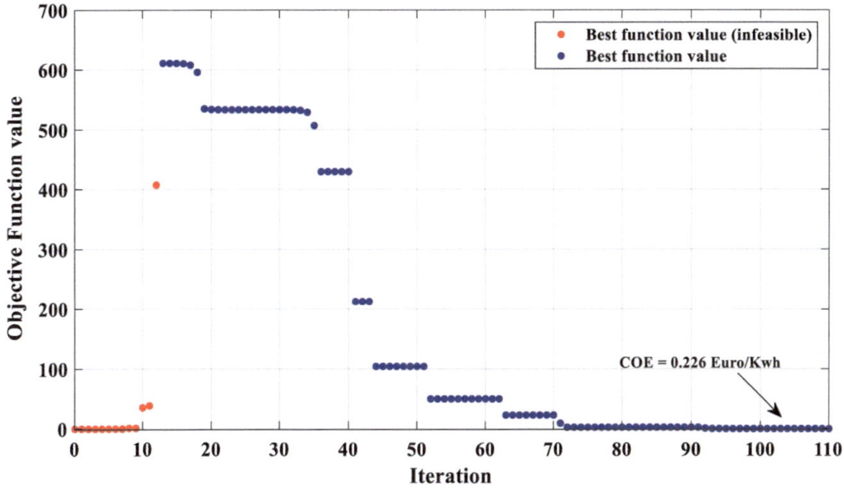

Fig. 8 Progress of optimization procedure for REL = 100%

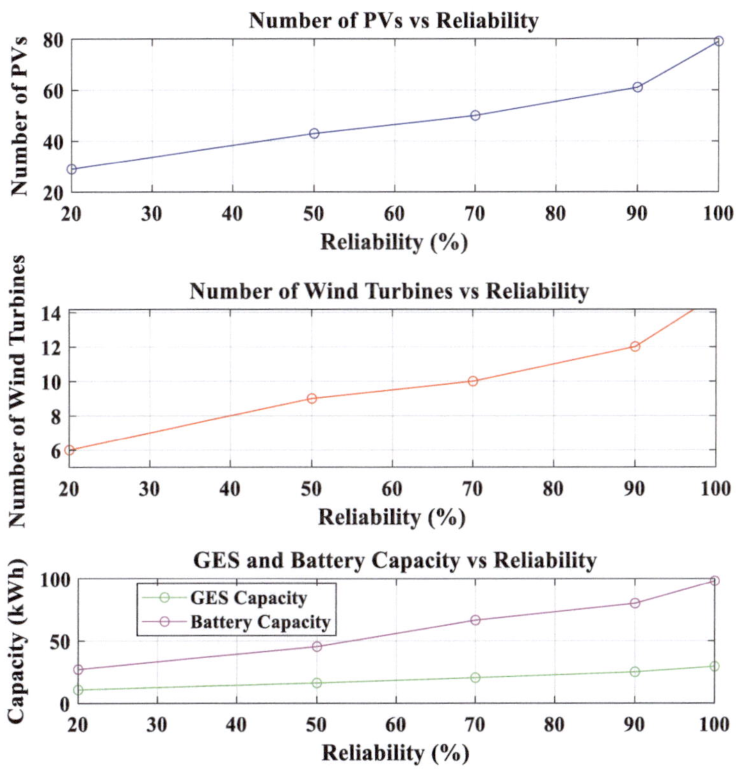

Fig. 9 Variation of design parameters with the reliability requirement

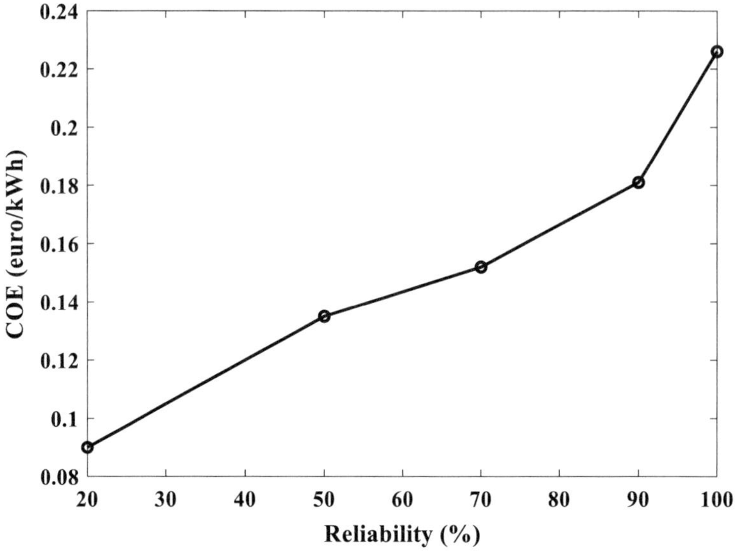

Fig. 10 Variation of COE with the reliability requirement

generation of WTs in the scenario when REL = 100%. It has a maximum power of 45 kW and follows the pattern of wind speed. WT production is in line with the pattern of varying wind speed.

The expected photovoltaic production, as shown in Fig. 12, is subject to weekly variations that closely correspond to variations in solar radiation. Its maximum output energy is expected to reach 12.8 kW. Furthermore, there are other important factors which impact the PV production besides solar irradiation. This includes temperature, shade, routine maintenance, and cable connectivity.

Understanding the operation and reliability of an off-grid hybrid energy system requires an understanding of the hourly dynamic simulation of the total ideal renewable generation together with the energy demand, as shown in Fig. 13. Based on the obtained data, it is clear that RE sources alone cannot provide the entire energy demand. This emphasizes the need for a hybrid GES/BAT storage system. As a matter of fact, there is an extraordinary excess during the first two days, which is followed by a shortage of renewable energy that cannot satisfy demand for the next two days. The seventh day shows another surplus phase, while the other days show the existence of an energy shortfall.

The intermittent nature of RE sources is offset by the hybrid energy storage system (GES/BAT). Figure 14 the fluctuation of the total generated (PV/Wind) plus discharged power (GES/BAT) versus the demand power. The optimal designed system with a REL = 100% was able to fully meet the load demand.

The operation of GES system versus the renewable production of each source while meeting the load demand is depicted in Fig. 15. This figure illustrates the contribution of GES in balancing energy supply and demand. This latter has been

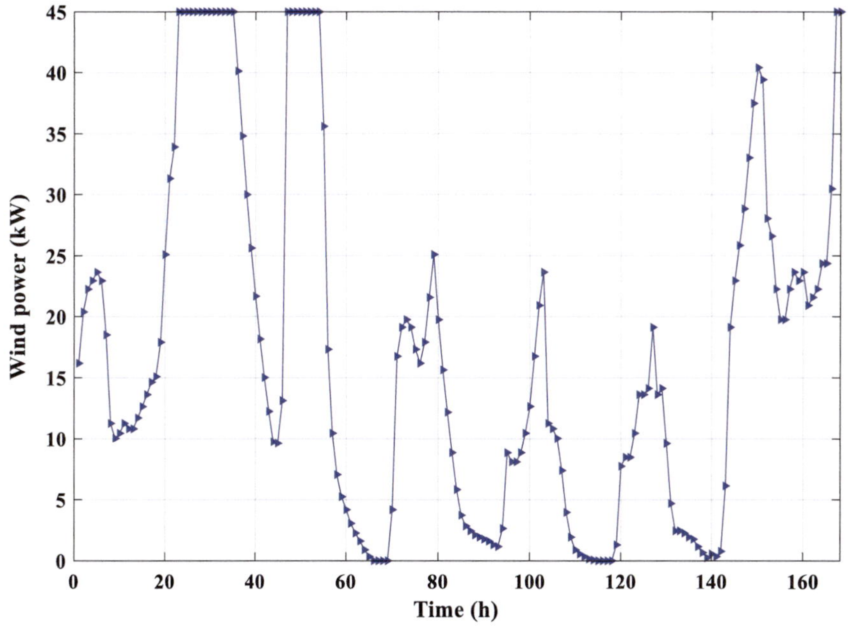

Fig. 11 Optimal Forecasted WT power generation

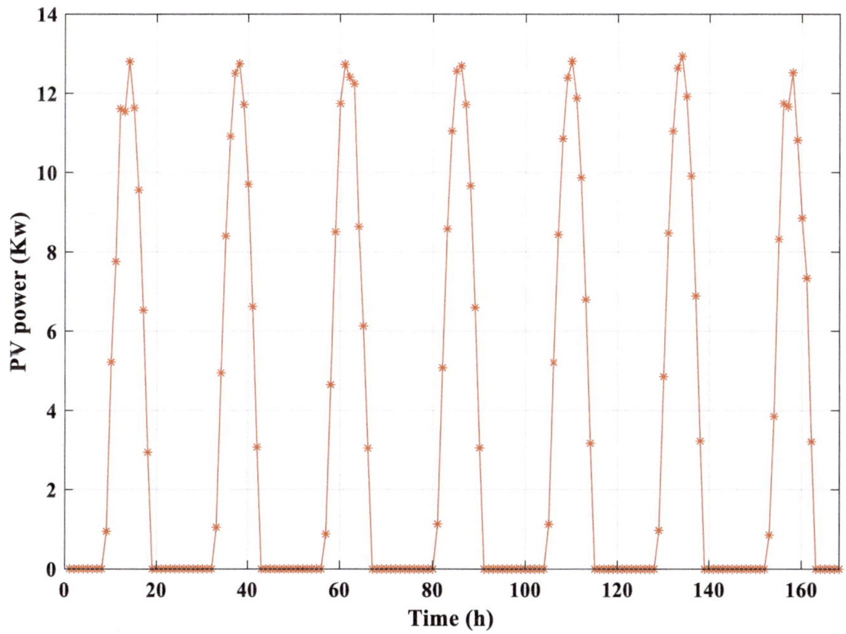

Fig. 12 Optimal forecasted PV power generation

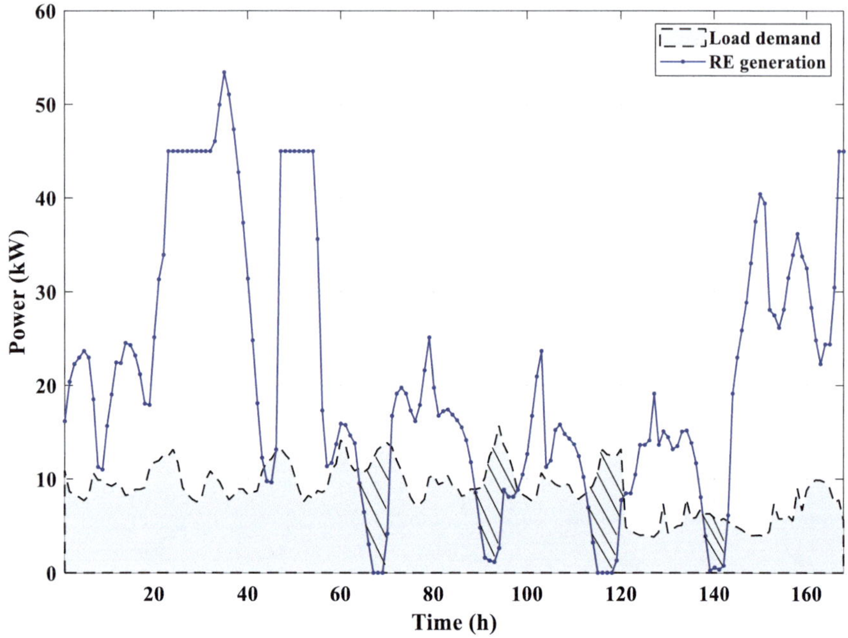

Fig. 13 Total optimal RE generation and load requirement

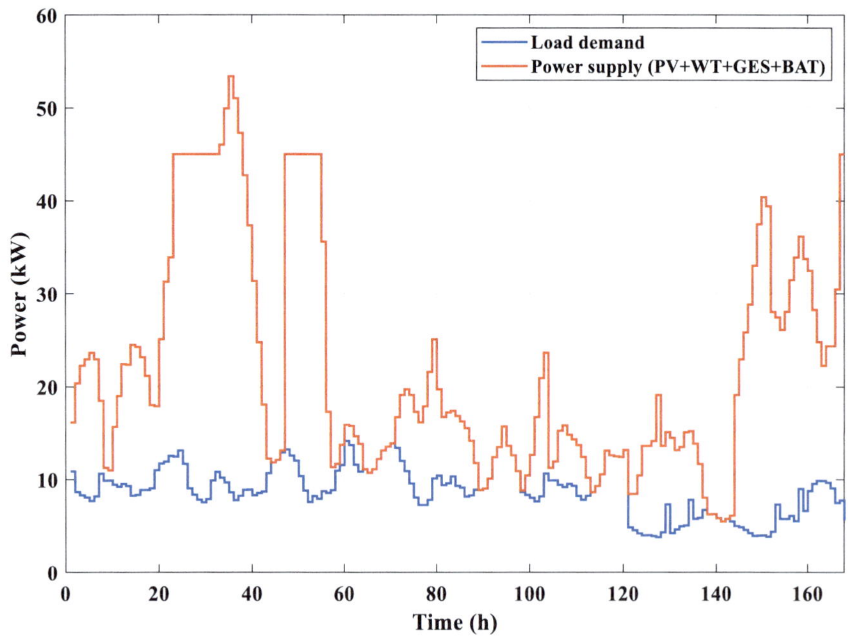

Fig. 14 Total optimal power demand and supply including hybrid storage contribution

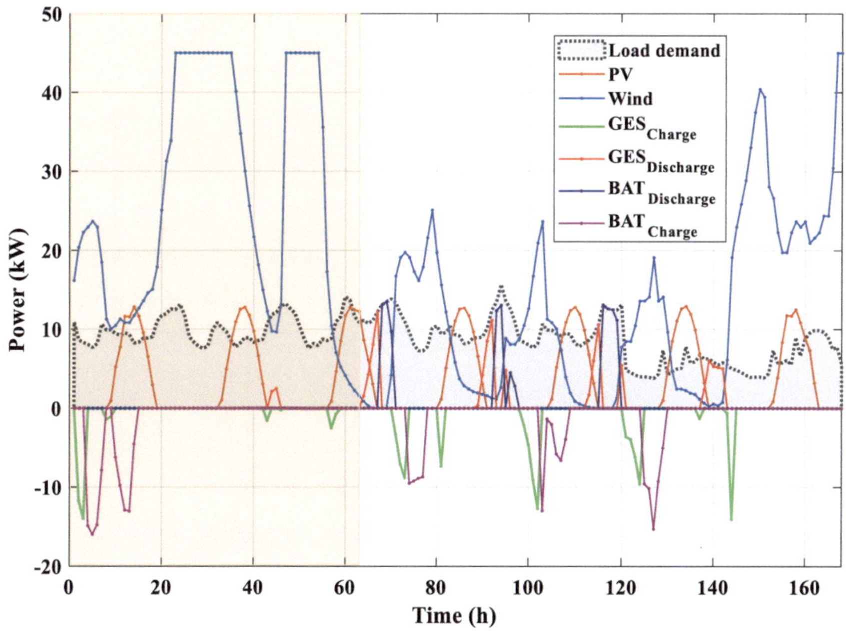

Fig. 15 Dynamic simulation of the energy flow of the different hybrid system components

also compared to the operation of the battery system. In some cases, the contribution of the hybrid energy storage system is minimal or nonexistent in the highlighted zone of Fig. 15. This is due to the high production of the renewable energy technologies which were able to fully satisfy the demand.

The energy charged and discharged from the hybrid GES/Battery system and the state of charge (SOC) of GES system over the investigation period are shown in detail in Figs. 16 and 17. The simulation highlights the role of the hybrid GES/BAT storage system and highlights how effective it is at maintaining a steady supply of energy, especially during times of peak energy demand, by balancing variable renewable energy sources.

During periods of oversupply, the hybrid GES/BAT storage system first stores energy in the GES system until it is fully charged, then switches to the battery system. This sequence is based on an analysis of the system's operations, load demand, and total renewable generation over the course of a week. GES discharges energy to satisfy the load when the generation from renewable energy systems decreases. The battery is then discharged if the GES is fully drained. This illustrates how the hybrid GES/BAT storage system can shift load by storing excess energy produced throughout the day and using it during peak hours.

Figure 17 shows that the hybrid GES/BAT system transitions to a fully charged state at around the 17th hour, or the start of the second day. Without using the stored energy, renewable energy production can already provide nearly all of the energy needed. This demonstrates how effectively the combination of an optimally designed

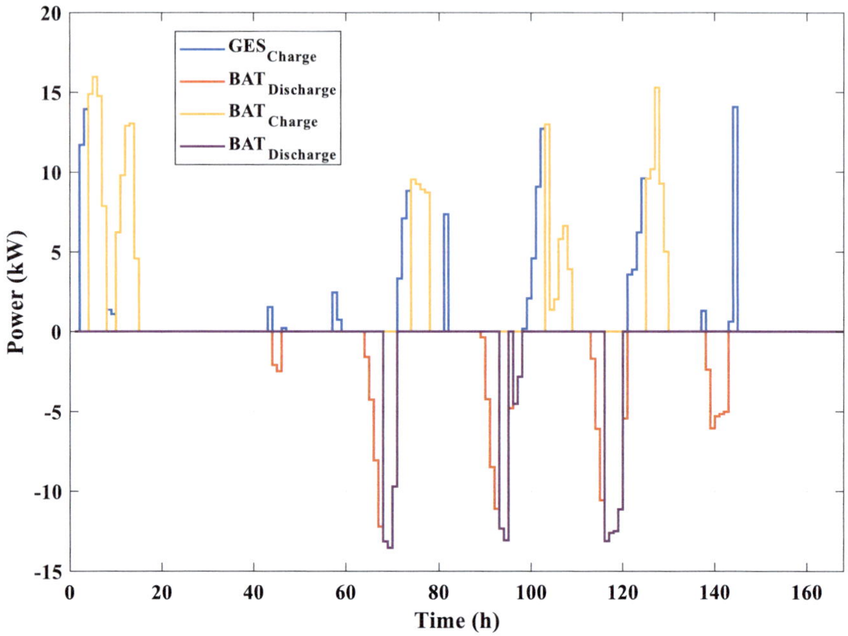

Fig. 16 Dynamic simulation of the energy charged and discharged from the hybrid GES/Battery system

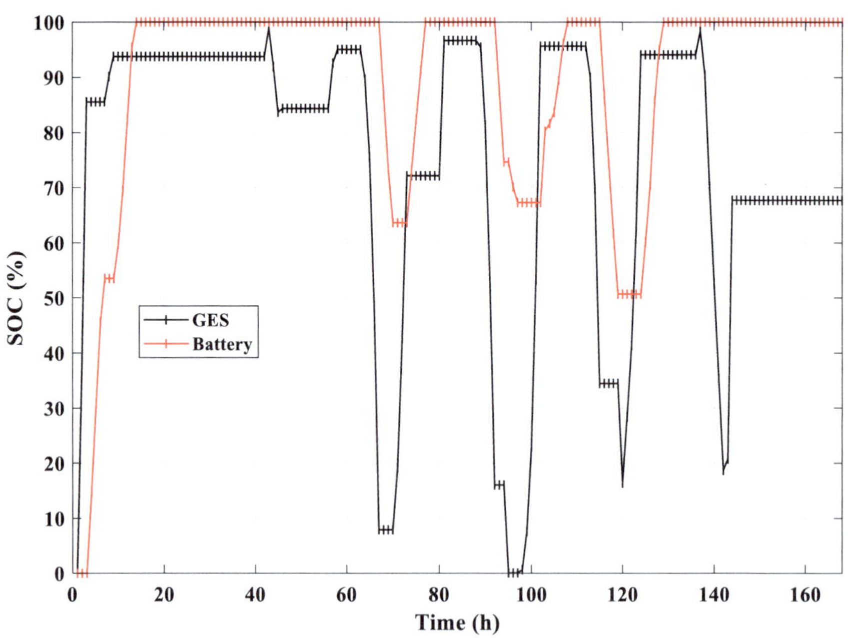

Fig. 17 SOC of the hybrid GES/Battery system

hybrid storage system can enhance the utilization of renewable resources. In addition to having positive economic effects, such an optimal hybrid energy system coupled with hybrid energy storage would reduce greenhouse gas emissions.

6 Conclusion

The chapter proposes a methodology to optimally size a hybrid renewable energy system coupled with hybrid energy storage incorporating forecast models. The hybrid energy storage is composed of a battery and a gravity energy storage system. This study's main objective was to develop a prediction model to determine the generation of renewable electricity from PV and WT systems before implementing an intelligent energy management system to efficiently balance the supply and demand for energy. To determine each energy system's optimal capacity design that will reduce energy costs and satisfy reliability requirements, an optimization process utilizing Fmincon has been implemented. In fact, reducing the total energy cost of the system allowed for the determination of the most cost-effective hybrid combination.

The optimization yields the number of WTs and PV modules as well as the capacity of GES and battery systems needed to fulfill various reliability levels while maintaining the lowest possible value of the objective function. The optimal number of wind turbines and photovoltaic modules needed to satisfy the energy demand has been found through optimization. The daily energy generation from each source has been computed using these optimal designs.

The percentage contributions of each source to the total energy production have been determined by comparing the daily energy generated by wind turbines and PV modules. The largest relative contributions of renewables to fulfilling the energy demand in the optimum system design, in the case of REL = 100%, are 18% (day 3) for PV and 92% (day 1) for WTs, respectively. Because of the wind speed's high potential in the case study under investigation, wind energy has a higher contribution than solar energy. For hybrid systems with reliabilities of 20% and 100%, respectively, the resultant COE ranges from 0.09 to 0.22 €/kWh, depending on the reliability level.

This study proposes an interesting techno-economic optimization of a hybrid GES/BAT system coupled with PV/wind energy systems. The results of this study will be helpful to ESS practitioners as they provide a route for the deployment of hybrid GES/BAT systems. These systems can balance the supply and demand of energy in hybrid RE systems by choosing innovative and appropriate hybrid storage solutions while taking cost-effectiveness and reliability related objectives into account.

Future work on hybrid energy systems with hybrid energy storage is needed. While hybrid energy production systems have been the subject of many research, the hybridization of storage systems in a single application is still largely unexplored. Additionally, the energy usage of commercial, industrial, and agricultural applications may be further improved by combining a number of various renewable energy technologies. The use of artificial intelligence techniques for energy generation and demand predictions is another future perspective. Artificial intelligence can improve failure prediction and optimize maintenance.

Nomenclature

C	Cost ($)
C_{cap}	Capital expenditures ($)
$C_{O\&M}$	Operation and maintenance expenditures ($)
d	Diameter (m)
D_f	Predicted diffused irradiance (W/m^2)
DNI	Direct irradiance (W/m^2)
DOD	Depth of discharge (%)
E	Energy (J)
$E_{battery(i)}$	Battery Capacity at a specific hour (kWh)
E_{BAT}	Battery Capacity (kWh)
G_{STC}	Solar irradiance (W/m^2)
g	Gravitational acceleration (m/s^2)
G_0	Solar constant (W/m^2)
G_f	Actual solar irradiance
h	Height (m)
I_f	Predicted direct irradiance (W/m^2)
k	Temperature coefficient
m	Mass (kg)
N_{pv}	Number of PVs
N_{WT}	Number of wind turbines
P	Power (kWh)
r	Discount rate (%)
SOC	State of charge (%)
T	Temperature (k)
V	Piston velocity (m/s)
v	Wind speed (m/s)
$v_{cut\text{-}in}$	Starting speed (m/s)
$v_{cut\text{-}out}$	Stopping speed (m/s)
V_{rated}	Rated wind speed (m/s)
x	Variable vector

Greek Symbols

σ	Discharge rate (%)
η_{bat}	Efficiency (%)
η_{b-dich}	Discharge Efficiency (%)

η_{b-ch} Discharge Efficiency (%)
ρ Density

References

Achour Y, Berrada A, Arechkik A, El Mrabet R (2023) Techno-economic assessment of hydrogen production from three different solar photovoltaic technologies. Int J Hydrogen Energy. https://doi.org/10.1016/j.ijhydene.2023.05.017

Ameur A, Berrada A, Emrani A (2022) Dynamic forecasting model of a hybrid photovoltaic/gravity energy storage system for residential applications. Energy Build 271:112325. https://doi.org/10.1016/j.enbuild.2022.112325

Ameur A, Berrada A, Emrani A (2023) Intelligent energy management system for smart home with grid-connected hybrid photovoltaic/ gravity energy storage system. J Energy Storage 72:108525. https://doi.org/10.1016/j.est.2023.108525

Ameur A, Berrada A, Loudiyi K, Adomatis R (2021) Performance and energetic modeling of hybrid PV systems coupled with battery energy storage. In: Hybrid energy system models, pp 195–238. Elsevier. https://doi.org/10.1016/B978-0-12-821403-9.00008-1

Beck HE, Zimmermann NE, McVicar TR, Vergopolan N, Berg A, Wood EF (2018) Present and future Köppen-Geiger climate classification maps at 1 Km resolution. Sci Data 5. https://doi.org/10.1038/sdata.2018.214

Berrada A, Loudiyi K, Zorkani I (2017) Toward an improvement of gravity energy storage using compressed air. Energy Procedia 134:855–864. https://doi.org/10.1016/j.egypro.2017.09.542

Berrada A, Loudiyi K (2016) Modeling and material selection for gravity storage using FEA method. In: 2016 International renewable and sustainable energy conference (IRSEC), pp 1159–64. IEEE. https://doi.org/10.1109/IRSEC.2016.7983956

Berrada A, Emrani A, Ameur A (2021) Life-cycle assessment of gravity energy storage systems for large-scale application. J Energy Storage 40. https://doi.org/10.1016/j.est.2021.102825

Bird L, Lew D, Michael Milligan E, Carlini M, Estanqueiro A, Flynn D, Gomez-Lazaro E et al (2016) Wind and solar energy curtailment: a review of international experience. Renew Sustain Energy Rev 65:577–586. https://doi.org/10.1016/j.rser.2016.06.082

Castle JA, Kallis JM, Moite SM, Marshall NA (1981) Photovoltaic specialists conference, 15th, Kissimmee, FL, 12–15 May 1981, conference record. Institute of Electrical and Electronics Engineers Inc., New York, NY

Diab AAZ, Sultan HM, Kuznetsov ON (2020) Optimal sizing of hybrid solar/wind/hydroelectric pumped storage energy system in Egypt based on different meta-heuristic techniques. Environ Sci Pollut Res https://doi.org/10.1007/s11356-019-06566-0/Published

Douglas T (2016) Dynamic modelling and simulation of a solar-PV hybrid battery and hydrogen energy storage system. J Energy Storage 7:104–114. https://doi.org/10.1016/j.est.2016.06.001

Emrani A, Berrada A (2024) A comprehensive review on techno-economic assessment of hybrid energy storage systems integrated with renewable energy. J Energy Storage 84:111010. https://doi.org/10.1016/j.est.2024.111010

Emrani A, Berrada A, Ameur A, Bakhouya M (2022a) Assessment of the round-trip efficiency of gravity energy storage system: analytical and numerical analysis of energy loss mechanisms. J Energy Storage 55:105504. https://doi.org/10.1016/j.est.2022.105504

Emrani A, Berrada A, Arechkik A, Bakhouya M (2022b) Improved techno-economic optimization of an off-grid hybrid solar/wind/gravity energy storage system based on performance indicators. J Energy Storage 49:104163. https://doi.org/10.1016/j.est.2022.104163

Emrani A, Achour Y, Sanjari MJ, Berrada A (2024) Adaptive energy management strategy for optimal integration of wind/PV system with hybrid gravity/battery energy storage using forecast models. J Energy Storage 96:112613. https://doi.org/10.1016/J.EST.2024.112613

Guo Y, Yang D, Zhang Y, Wang L, Wang K (2022) Online estimation of SOH for Lithium-ion battery based on SSA-Elman neural network. Protect Control Mod Power Syst 7(1):40. https://doi.org/10.1186/s41601-022-00261-y

IEA (2024) Clean energy market monitor—March 2024. International Energy Agency

IRENA (2019) Climate change and renewable energy: national policies and the role of communities, cities and regions (report to the G20 climate sustainability working group (CSWG)), pp 1–60

IRENA (2022a) Renewable power generation costs in 2021—chart Data. In: Renewable power generation costs in 2021

IRNEA (2022b) Renewable power generation costs in 2021. International Renewable Energy Agency

Lamnatou C, Chemisana D, Cristofari C (2022) Smart grids and smart technologies in relation to photovoltaics, storage systems, buildings and the environment. Renew Energy 185:1376–1391. https://doi.org/10.1016/J.RENENE.2021.11.019

Loudiyi K, Berrada A (2017) Experimental validation of gravity energy storage hydraulic modeling. Energy Procedia 134:845–854. https://doi.org/10.1016/j.egypro.2017.09.541

Mitali J, Dhinakaran S, Mohamad AA (2022) Energy storage systems: a review. Energy Storage Saving 1(3):166–216. https://doi.org/10.1016/j.enss.2022.07.002

Mostafa MH, Abdel Aleem SHE, Ali SG, Ali ZM, Abdelaziz AY (2020) Techno-economic assessment of energy storage systems using annualized life cycle cost of storage (LCCOS) and levelized cost of energy (LCOE) metrics. J Energy Storage 29:101345. https://doi.org/10.1016/j.est.2020.101345

Park S-W, Jung-Un Y, Lee J-W, Son S-Y (2024) A comprehensive review of battery-based power service applications considering degradation: research status and model integration. Appl Energy 374:123879. https://doi.org/10.1016/j.apenergy.2024.123879

Paudel A, Chaudhari K, Long C, Gooi HB (2019) Peer-to-peer energy trading in a prosumer-based community microgrid: a game-theoretic model. IEEE Trans Industr Electron 66(8):6087–6097. https://doi.org/10.1109/TIE.2018.2874578

Bailey SG, Hepp AF, Ferguson DC, Raffaelle RP, Durbin SM (2023) Photovoltaics for space. Elsevier. https://doi.org/10.1016/C2019-0-04344-0

Sinha S, Chandel SS (2014) Review of software tools for hybrid renewable energy systems. Renew Sustain Energy Rev 32:192–205. https://doi.org/10.1016/J.RSER.2014.01.035

Sinha S, Chandel SS (2015) Review of recent trends in optimization techniques for solar photovoltaic-wind based hybrid energy systems. Renew Sustain Energy Rev 50:755–769. https://doi.org/10.1016/J.RSER.2015.05.040

Solar PV Utility Scale Levelised Cost of Energy Index Based on Average Annual Input Costs, 2018–2024—Charts—Data & Statistics—IEA (n.d.) Accessed 28 July 2024. https://www.iea.org/data-and-statistics/charts/solar-pv-utility-scale-levelised-cost-of-energy-index-based-on-average-annual-input-costs-2018-2024

Wright VP (2023) World energy outlook. IEA report

Xie P, Wang H, Jia Y (2024) Decentralized energy management regime for prosumer community microgrids using flexible tube model predictive control. Electr Power Syst Res 236:110553. https://doi.org/10.1016/j.epsr.2024.110553

Yang J, Yang Z, Duan Y (2021) Optimal capacity and operation strategy of a solar-wind hybrid renewable energy system. Energy Convers Manag 244. https://doi.org/10.1016/j.enconman.2021.114519

Zakeri B, Syri S (2015) Electrical energy storage systems: a comparative life cycle cost analysis. Renew Sustain Energy Rev 42:569–96. https://doi.org/10.1016/j.rser.2014.10.011

Power Grid and Renewable Transition

Preface

Growing renewable generation, several challenges must be overcome to enable clean energy transition, due to the high fluctuating and intermittent character of renewable sources, specifically of solar and wind energy, and the contextual loss in grid inertia. This leads to frequent and unpredictable energy imbalances across the grid infrastructure, requiring flexible solutions to be implemented. Among potential solutions, the integration of energy storage systems (ESSs) will have a relevant impact. Several services are required to ESSs to guarantee safety and stability of future power grids, e.g. power smoothing, load levelling and shedding, voltage and frequency support, as well as the energy service. Such services usually require ESSs to operate over multiple storage timeframes, posing challenges with reference to technical features of available technologies. The hybridization of complementary technologies allows to extend the operating ranges and the lifespan of the single devices, as well as to optimize overall round-trip efficiency thanks to suitable real time power management. Costs for operation and devices' replacement are reduced for Hybrid ESSs (HESSs), while installation costs generally increase requiring detailed economic assessments. Frequency regulation is addressed in Chapter "Integration of Run-of-River/Pumped Hydro with an Energy Storage System Based on Batteries and Supercapacitors for Enabling Ancillary Services and Extending the Lifetime of Generating Equipment" which presents a HESS constituted by pumped hydroelectric system coupled with Li-ion batteries and supercapacitors. Efficiency and lifespan of all components are improved while enhancing HESS contribution to frequency regulation. Chapter "Power-Intensive Energy Storage Systems" introduces basic principles and integration of power-intensive technologies, i.e. supercapacitors and superconducting magnetic energy storage systems used for compensating fast-dynamic phenomena, such as faults or voltage dips. Their hybridization with energy-intensive ESSs is also discussed, highlighting control possibilities and benefits. Finally, Chapter "Hybrid Energy Storage Systems Coupled with Renewable Power Plants for Power Smoothing Applications" deals with HESS coupling with RES generation for power smoothing. Li-ion battery/flywheel HESSs coupled to both wind and wave energy plants are investigated. HESS benefits and economic feasibility are assessed.

Integration of Run-of-River/Pumped Hydro with an Energy Storage System Based on Batteries and Supercapacitors for Enabling Ancillary Services and Extending the Lifetime of Generating Equipment

Jorge Nájera, Marcos Blanco, Juan Ignacio Pérez-Díaz, José Ignacio Sarasúa, and Elahe Ghanaee

Abstract This chapter explores the integration of run-of-river and pumped-storage hydroelectric power plants with lithium-ion batteries and supercapacitors to enhance frequency regulation while minimizing mechanical stress on hydropower units. Hydroelectric plants play a key role in grid stability, but frequent power adjustments required for ancillary services accelerate wear and tear on their mechanical components. The proposed hybrid system follows a cascading control approach, where each energy storage element protects the next: lithium-ion batteries handle medium-term power fluctuations, reducing the need for constant hydro unit adjustments, while supercapacitors absorb rapid transients, shielding both the batteries and the hydro units from excessive cycling and mechanical fatigue. This setup improves the efficiency and lifespan of all components while enhancing the plant's ability to participate in ancillary service markets. The chapter reviews state-of-the-art solutions and presents a detailed analysis of technical requirements for hybrid hydropower-storage systems. It also discusses results from simulations and laboratory tests that validate the effectiveness of the proposed configuration. Findings indicate that integrating batteries and supercapacitors with hydropower can significantly improve frequency regulation quality, extend asset lifespan, and facilitate higher penetration of renewable energy sources, making hybrid hydropower systems a crucial component of future power grids.

Keywords Hydroelectric power plant · Lithium-ion battery · Supercapacitor · Frequency regulation · Hybrid energy storage

J. Nájera (✉) · M. Blanco
Unidad de Accionamientos Eléctricos, CIEMAT, Madrid, Spain
e-mail: jorge.najera@ciemat.es

J. I. Pérez-Díaz · J. I. Sarasúa · E. Ghanaee
ETSI Caminos Canales y Puertos, Universidad Politécnica de Madrid, Madrid, Spain

R. Scipioni et al. (eds.), *Hybrid Energy Storage*, Lecture Notes in Energy 47,
https://doi.org/10.1007/978-3-031-97755-8_4

1 Case Study Description and State-of-the-Art

Electric power systems have suffered severe changes over the last decades, mainly related to the increasing integration of non-dispatchable renewable generation units, large nondeferrable loads (e.g. fast charge of electric vehicle fleets), and other actors such as energy storage systems. This paradigm, where imbalances between power generation and consumption can happen all of a sudden and are hardly predictable, forces the electric power system (namely the transmission system operator and distribution system operator) to make use of the ancillary services at its disposal in order to keep the power quality within the safe operation range. Among those ancillary services, the ones related to the frequency control are typically provided by the power generation units and energy storage systems.

Hydroelectric power plants have historically taken a strong share in relation to frequency control ancillary services. In this sense, a hydroelectric power plant is forced to adjust its generation power when providing primary and secondary frequency control, causing wear and tear to the different mechanical parts (moving components of the hydro-mechanical governors). In the new paradigm of electric power system, the needs of providing frequency control are continuously growing and, thus, the lifespan of hydroelectric power plants is being compromised.

Several solutions have been proposed in the literature for solving the issues associated with the hydroelectric power plants' frequency control. The most relevant alternatives integrate energy storage systems with the hydroelectric power plant, with the aim of extending the lifetime of the hydropower units, enlarging the plant's active power regulation capacity, enhancing the quality of the frequency control provided by the plant, and increasing the revenue that the plant obtains from the energy and ancillary services markets.

In 2007, the Pacific Northwest National Laboratory (PNNL) started a project for the Bonneville Power Administration (BPA) aimed to address the issue of mitigating additional variability and fast ramps that occur at higher penetration of variable resources, including wind generation in the BPA and California independent system operator (CAISO) control areas, through the energy exchange between the control areas and through the use of energy storage and other generation resources. In Phase 1 of the abovementioned project, the PNNL developed a control algorithm for an aggregate regulating unit (a hydropower unit and a flywheel) to track the area control error (ACE) signal while minimizing the quadratic deviations of the flywheel State of Charge (SOC) and the power generated by the hydropower unit, with respect to a target SOC and the power generated by the hydropower unit at the best efficiency point. The analyzed flywheel had a power rating of 7.5–12.5% of that of the hydropower unit. The algorithm was tested by simulations and provided satisfactory results, which are summarized as follows: the unit provided a robust and accurate signal tracking; the flywheel contributed to operating the hydro unit near its best efficiency; the hydro unit contributed to maintaining the flywheel's SOC within a specific range. In the second phase of the project, the PNNL tested the algorithm by means of hardware-in-the-loop simulations. Findings are detailed in Makarov et al.

(2008), Lu et al. (2010) and Jin et al. (2011). For this purpose, the research team of the project used a Beacon Power's flywheel (25 kWh, 100 kW). Some years later, the control algorithm was revised and improved as presented in Jin et al. (2013). The algorithm used a SOC band control to give the flywheel priority to follow the regulation signal and thus further reduce the frequency of the hydro unit output adjustment. As long as the flywheel SOC is within a specific band, the flywheel takes over the responsibility to follow the ACE signal. The hydro unit provides frequency regulation only when the flywheel SOC is outside the band. The original control algorithm required the hydro unit to adjust its output every 4 s, whereas the improved version required less than one adjustment per hour. A patent entitled "Controller for hybrid energy storage" (US 8,754,547 B2), was developed based on this concept.

In 2014, the project's team began working on the control and sizing of a flywheel energy storage system for the frequency control of El Hierro system. A droop-based control strategy, similar to the one implemented in the Rhode hybrid test facility (Ortega and Milano 2016), as well as a nonlinear proportional control strategy which modifies the input/output power set-point signal of the flywheels as a function of the SOC, were tested by means of simulations. The frequency regulation effort was distributed between the flywheels and the pumped-storage power plant thanks to the use of different control dead bands. The results of the study were presented in Sarasúa et al. (2016) and point out that in El Hierro system a small flywheel energy storage system with an installed power capacity of 3% of that of the pumped-storage power plant contributes to notably reduce the amplitude of frequency deviations caused by wind generation variability, and to integrate more wind power in the system at a reasonable cost. The results also highlight the importance of the control strategy of the flywheel energy storage system in order to maximize the effective integration of wind generation.

Moghaddam and Chowdhury (2016) presented a method for the optimal sizing of a hybrid power plant comprising pumped-storage and a sodium sulfur battery so as to reduce the wind power imbalances in the BPA control area. The results presented in the paper showed that the storage capacity required for the hybrid power plant to compensate for the wind power forecast error is smaller than the one required when only one of the storage systems is utilized. Gevorgian et al. (2017) worked on the coordinated operation of a variable-speed run-of-river hydropower plant and a supercapacitor energy storage system for the provision of frequency regulation. The results presented in the paper demonstrated that the suggested coordination strategy between the supercapacitor energy storage system and the hydropower plant can help increase the quality of the system's frequency. Laban (2019) developed two control algorithms for the real-time energy management of a hybrid power plant comprising hydropower and Li-ion batteries. The hybrid power plant is assumed to provide frequency containment reserves for normal operation (FCR-N) in the reserve market operated by Svenska Kraftnät. Laban (2019) concluded that the hybrid power plant is able to provide the same amount of FCR-N as the hydropower plant, with a much lower wear and tear of the hydropower units.

Guezgouz et al. (2019) and Javed et al. (2020) presented two optimization algorithms for the hourly power dispatch of a hybrid power plant comprising a pumped-storage power plant and a Li-ion battery energy storage system (BESS). The hybrid power plant is assumed to be connected to an isolated power system which is exclusively fed by variable renewable generation. Both articles concluded that with a proper power dispatch algorithm the hybrid power plant helps increase the reliability of the electricity supply in the systems under study. Parastegari et al. (2015) developed an optimization model for the generation scheduling of a hybrid power plant comprising a pumped-storage power plant and a BESS. The hybrid power plant is assumed to operate in a coordinated fashion with a set of wind and solar PV power plants. The results presented by Parastegari et al. (2015) demonstrate that with the help of a proper scheduling model the hybrid power plant can contribute to increase the profit of the wind and solar PV plants. Dawn et al. (2018) presented an optimization model for the hourly power dispatch of a hybrid power plant comprising a pumped-storage power plant and a generic converter-interfaced energy storage system. The hybrid power plant is assumed to operate in a coordinated fashion with a thermal power plant and a set of wind and solar PV plants. The results obtained by Dawn et al. (2018) showed that with a proper power dispatch model, the hybrid power plant contributes to increasing the profit of the thermal and renewable power plants. Pérez-Díaz et al. (2020) discussed the benefits of the integration of flywheels in pumped-storage power plants for the enhancement of the quality of the system's frequency in a scenario of high renewable penetration. Román et al. (2019) suggested the possibility of coordinating the operation of a hydropower plant and the batteries of a fleet of electric vehicles to modulate the fluctuation of the flow released by the hydropower plant. Thus, the environmental impact of hydropeaking on the river reach downstream of the hydropower plant is mitigated. Anindito et al. (2019) compared the use of batteries and re-regulation reservoirs for the mitigation of the environmental impact of hydropeaking. Anindito et al. (2019) concluded that batteries can help mitigate the environmental impact of hydropeaking as effectively as a re-regulation reservoir, though their feasibility strongly depends on the prices of the energy and ancillary services markets.

Bucher and Schreider (2017) discussed the benefits of the coordinated operation of a pumped-storage power plant and a BESS for the provision of frequency containment reserves (FCR) in the market currently implemented in Central Europe. Bucher et al. (2018) describe the Li-ion BESS (12.5 MW/13 MWh) that the company ENGIE integrated in 2017 in the Pfreimd pumped storage system. The batteries operate in a coordinated fashion with several ternary pumped-storage units. The latter are used to charge/discharge the batteries when the SOC is lower/higher than 25%/75% so as to meet the requirements for participation in the FCR market. FORTUM company integrated a Li-ion BESS (5 MW/6.2 MWh) in the Forshuvud run-of-river hydropower plant in 2019. The batteries provide FCR-N in the reserve market operated by Fingrid. FORTUM uses the hydropower units as backup reserve when the BESS cannot provide FCR-N as required by Fingrid because of having reached the maximum or minimum SOC.

Eiper et al. (2019) described the advantages of ANDRITZ company Hydro's HyBaTec product (ANDRITZ 2024). It consists in the integration of a Li-ion BESS in a hydropower plant and its control system to extend the lifetime of the hydro units and maximize the plant's revenue. Eiper et al. (2019) illustrate some advantages of the HyBaTec product by simulating the coordinated operation of a run-of-river hydropower plant and a Li-ion BESS for the provision of FCR. The results of the simulations show that the coordinated operation allows reducing the wear and tear of the hydro units. ANDRITZ Hydro is involved in the XFLEX HYDRO H2020 project (https://xflexhydro.net/) in which the coordinated operation of a run-of-river hydropower plant and a Li-ion BESS is being demonstrated in an operational environment. The French Alternative Energies and Atomic Energy Commission (CEA) integrated in 2021 a 600 kW Li-ion BESS in EDF's Vogelgrun run-of-river hydropower plant which is equipped with 4 35.5 MW low head Kaplan turbines. ANDRITZ Hydro has developed the control system of the hybrid power plant and a digital twin of the plant for the real-time assessment of the Kaplan units' wear and tear.

Verbund company integrated in 2020 a Li-ion BESS (8 MW/14 MWh) in the Wallsee run-of-river hydropower plant, which is equipped with 6 Kaplan turbines, of 35 MW each.

As discussed above, the solutions explored in the literature integrate a BESS (Battery Energy Storage System), supercapacitors (SC), or flywheels with hydropower units to address the issues mentioned earlier. Among these energy storage technologies, BESS stands out as the most effective one in extending the lifespan of a hydroelectric power plant and enhances its regulation capacity across a wider range of events and situations, due to its higher energy density. However, it is well known that BESS units experience aging, which is influenced by cycling and environmental conditions. As a result, the lifespan of a combined hydropower and BESS plant is not only constrained by the wear and tear of the hydro unit, but also by the aging of the BESS, which can accelerate significantly in scenarios with high penetration of non-dispatchable renewable energy sources.

Therefore, it is crucial to analyze the combined integration of BESS and fast response energy storage systems (FRESS) with a hydropower plant operating in cascade. In this configuration, the BESS and FRESS protect the hydropower system from rapid power fluctuations, while the FRESS also shields the BESS from excessive aging due to frequent cycling.

2 Technology Neutral System Requirements

From a technology-neutral perspective, this case study explores the hybridization of a large-scale energy generation system, such as a hydroelectric plant (with a focus on run-of-river and pumped-storage types), with additional elements to reduce the fatigue (wear and tear) of its mechanical components when providing ancillary regulation services. Simultaneously, the integration of new energy storage systems should extend and enhance the regulation capacity of the original technology, potentially opening access to new markets for ancillary services.

Hydroelectric plants are capable of generating substantial amounts of power and energy, and they play a critical role in grid services such as frequency regulation, reactive power provision, and black start capability. However, reducing the mechanical fatigue associated with these services requires smoothing the hydro's response to power set-point changes. To achieve this, another system component must be introduced to provide or absorb the power that the hydro plant cannot manage at the point of common coupling (PCC). This additional element should be an energy storage system (ESS) with sufficient energy and power capacity to smooth the hydro's set-point curve over a desired time window, which could extend over several hours depending on the regulation service. Additionally, the ESS must have a faster response time than the hydro plant, as it needs to act more quickly than the hydro's mechanical components to protect them.

When integrating different technologies into a hybrid plant, the overall performance is constrained by the weakest component. Therefore, to reduce fatigue and extend the lifespan of the hydroelectric plant, the durability of the hybrid system will be limited by the lifespan of the ESS. Energy storage technologies capable of supporting hydro plants in providing ancillary services tend to have shorter lifespans. Devices with low aging rates typically have very low energy densities, while those with sufficient energy density are subject to significant aging. To address this, the case study examines the feasibility of hybridizing with not just one, but two different ESS types, each with complementary characteristics. In this way, the high energy density ESS can be protected by a low-aging ESS, which will also safeguard the other systems in the hybrid plant.

From a technical standpoint, the key requirements of a hybrid system comprising a hydro plant, a high energy density ESS, and a fast-responding ESS (FRESS) for reducing hydro fatigue while expanding its regulation capacity are time response and aging. From a general point of view, the technical requirements for both hybrid pumped hydro plant, and hybrid run-on-river plant are summarized in the next section (aging not included since it is highly dependent on the application and regulation services).

3 Hybrid Energy Storage Capabilities

To date, the most feasible hybrid power plant configuration combining hydro, a high-energy density ESS, and a FRESS consists of a hydro plant, Li-ion batteries, and supercapacitors. Analysing the components individually, their technical characteristics are summarized in the following tables (Tables 1 and 2):

Comparing the performance of the different elements, it is evident that both batteries and supercapacitors (see Tables 5 and 6, respectively) surpass the hydro units in terms of efficiency and response time (see Tables 3 and 4)—two critical factors for enhancing the hydro plant's regulation capacity and reducing mechanical stress.

Table 1 Hybrid pumped hydro plant requirements

Power capacity	50–500 MW
Storage capacity	> 8 h full load
Efficiency	> 80%
Reaction time (50–100%)	< 10 s
Reaction time (0–100%)	< 1 min

Table 2 Hybrid run-on-river plant requirements

Power capacity	10-100 MW
Storage capacity	> 1 h full load
Efficiency	> 80%
Reaction time (50–100%)	< 10 s
Reaction time (0–100%)	< 1 min

Table 3 Pumped hydro

Power capacity	50–500 MW
Storage capacity	> 8 h full load
Efficiency	70-75%
Reaction time (50–100%)	~ 15 s
Reaction time (0–100%)	< 2 min

Additionally, the long lifespan of supercapacitors makes them ideal for protecting both the hydro and the batteries during ramp-up events, preventing excessive aging and mechanical strain.

From the comparison of the technical requirements in Tables 1 and 2, it is clear that the inclusion of both batteries and supercapacitors alongside the hydro plant is sufficient to achieve the desired technical objectives.

Table 4 Run-on-river hydro

Power capacity	10-100 MW
Storage capacity	–
Efficiency	> 80%
Reaction time (50–100%)	~ 15 s
Reaction time (0–100%)	< 2 min

Table 5 Li-ion batteries

Power density	> 3000 W/kg
Energy density	> 280 Wh/kg
Efficiency	> 95%
Reaction time	< 1 s
Lifetime	> 5000 cycles

Table 6 Supercapacitors

Power density	> 20 kW/kg
Energy density	> 8 Wh/kg
Efficiency	> 98%
Reaction time	< 1 s
Lifetime	> 1 M cycles

4 Techno-economic and Sustainability Analysis

With the help of suitable control algorithms for the real-time energy management and optimization algorithms for the generation scheduling and redispatch, the coordinated operation of a hydropower plant (with or without pumping capability), Li-ion batteries and supercapacitors, can be beneficial from different perspectives.

For a reservoir hydropower plant, the proposed coordinated operation can contribute to the following: extend the lifetime of the hydropower units; enlarge the plant's active power regulation capacity; enhance the quality of the frequency control provided by the plant; increase the revenue the plant obtains from the energy and ancillary services markets; reduce the environmental impact in the river reach downstream of the plant.

For a run-of-river hydropower plant, the proposed coordinated operation can contribute to: extend the lifetime of the hydropower units; allow the plant to provide frequency control ancillary services and participate in the corresponding markets (or enlarge the plant's active power regulation capacity and enhance the quality of the frequency control provided by the plant); increase the revenue the plant obtains from the energy and ancillary services markets.

For a closed- or open-loop pumped storage power plant, the benefits are similar to those described for a reservoir hydropower plant. In addition, the proposed coordinated operation can contribute to increase the plant's energy storage capacity. In all these cases (reservoir, run-of-river and pumped-storage hydropower plant), the proposed coordinated operation can eventually contribute to improve the quality of the system's frequency.

Furthermore, the proposed hybrid power plant can contribute to mitigate the adverse effects of wind and solar generation variability and unpredictability by the coordination with wind and/or solar PV farms. These adverse effects are the imbalance costs and the impact on the system's frequency. To date, this impact has not been a concern in most large, interconnected power systems, but the situation might well change on the way to reach the declared renewable share generation targets.

5 Implementation

Implementing a hybrid power plant to perform the regulation tasks outlined in this case study requires a high-level control (HLC) system, which sets the working references for the various elements, specifically the hydro units, batteries, and supercapacitors. While hybrid power plant controls can cover a wide range of architectures, a feasible approach is proposed in this section. The high-level control scheme is illustrated below in Fig. 1.

According to the scheme depicted in Fig. 1, the HLC provides a power reference for the entire hybrid plant, based on the regulation service that needs to be delivered. This reference passes through a low-pass filter, establishing the power reference for the hydro. This filtering helps to reduce the wear and tear on the hydro's mechanical components, as it follows a smoother and less abrupt reference.

The state of charge (SoC) of both the batteries and supercapacitors is regulated by the hydro units. If either the batteries or supercapacitors reach high or low SoC levels, the hydro units respond by charging or discharging them, ensuring that both systems maintain a SoC that allows the hybrid power plant to remain flexible in responding to different scenarios. SoC control, represented at the top of the scheme, prevents situations where the batteries or supercapacitors would receive charging commands when near 100% SoC, or discharging commands when near 0% SoC. Such scenarios would place an excessive burden on the hydro units, as the hybrid power plant would then operate based solely on the hydro system. To avoid this, the SoC control adjusts the filtered reference power for the hydro units based on the SoC of the batteries and supercapacitors. The final output of the SoC control is the adjusted reference power, which is sent to the hydro units to ensure optimal operation.

The reference power for the lithium-ion batteries is determined by subtracting the filtered reference power from the unfiltered reference provided by the HLC. In this way, the batteries handle the high-frequency component of the unfiltered reference, helping to reduce mechanical stress on the hydro. This reference power is applied to the batteries, and under ideal conditions, the unfiltered reference power would be split between the hydro and battery systems.

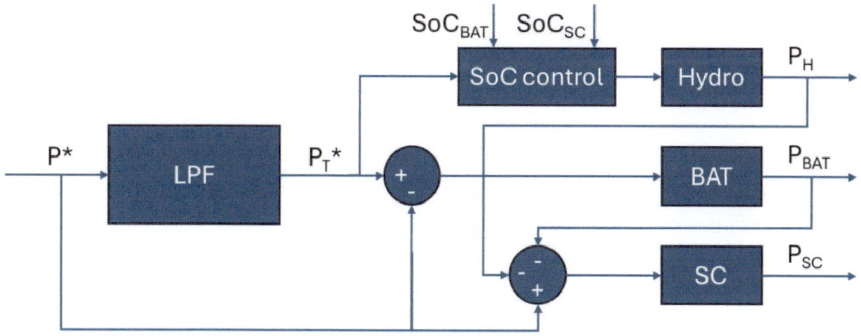

Fig. 1 High-level control scheme for the proposed hybrid power plant

In the proposed HLC, the supercapacitors manage any deviations between the unfiltered reference power and the actual power output of the hydro and battery systems. Therefore, the reference power for the supercapacitors is calculated by subtracting the real power output of the hydro and battery systems from the unfiltered power.

This HLC structure is relatively simple, with the three components of the hybrid power plant operating in a cascade arrangement. It allows for both the reduction of mechanical stress on the hydro plant and the delivery of the desired regulation, with the power reference coming from a higher-level control scheme. Additionally, the supercapacitors protect the batteries from excessive and aggressive cycling, thereby reducing the aging process associated with the batteries.

Proper tuning of the HLC parameters is essential to meet the desired objectives. The selection of parameters for the low-pass filter and SoC control depends on the specific application, the regulation tasks performed by the hybrid plant, and the sizing and characteristics of the actual devices within the plant.

For illustrative purposes, the following paragraphs describe the behaviour of the proposed control system when the hybrid power plant provides primary and secondary frequency regulation services. This example corresponds to a Hardware-In-the-Loop (HIL) system, where the hydro component is emulated on the HIL platform, while the batteries and supercapacitors are real devices. The HLC control is implemented in the laboratory testing facility of the Unidad de Accionamientos Eléctricos at CIEMAT (C/Julián Camarillo 30, 28,037 Madrid, Spain).

The laboratory testing facility includes the following components: a lithium-ion battery energy storage system (Li-BESS), a supercapacitor (SC) storage system, a DC/DC converter connected to the SCs, an inverter linking the facility to the main grid, the HIL platform, and a control platform. The general layout is shown in Fig. 2.

As shown in the layout, the installation connects to the grid on the left via a transformer, an inductive link, and a bidirectional inverter. The Li-BESS is directly connected to the DC bus, with a nominal voltage of 736 V and an energy capacity of 70kWh. The SC system is connected to the DC bus through a bidirectional DC/DC converter, with an energy capacity of 768Wh and a usable voltage range between 691

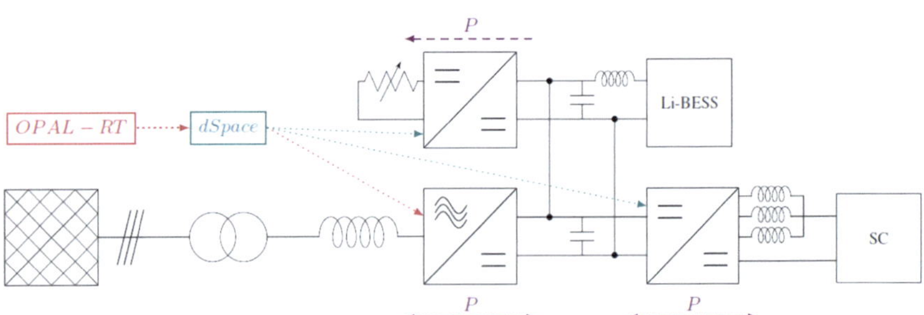

Fig. 2 Laboratory testing facility scheme

and 345 V (half of the maximum voltage, equivalent to 75% of the stored energy). Additionally, there is a controllable load emulator with a 50 kW resistor.

The laboratory setup is controlled by a dSPACE MicroAutoBox III platform (dSPACE 2024), which sets the power references for the load emulator, the power stored or delivered by the SCs (through the DC/DC converter), and the power exchanged with the grid via the inverter. The power managed by the Li-BESS is indirectly controlled as the difference between the inverter's power and the combined power of the SCs and the load emulator. The DC bus voltage is regulated by the Li-BESS since it is directly connected to the bus. Finally, the hydropower plant is emulated on an OPAL-RT 4510 HIL platform (OPAL-RT 2018), which communicates with the dSpace system. The emulated hydroelectric plant consists of two units, each with a rated power of 10.24 MWe, a nominal flow rate of 10 m^3/s, and a nominal head of 113 m.

As previously mentioned, the hybrid power plant has been tested for its ability to provide primary and secondary frequency regulation. The HLC is designed to reduce the wear and tear on the hydro system when performing correction actions. A selection of historical time-series data from real measurements of the UK power system was used to evaluate the difference in hydro valve (wicket gate movement) operation when the hydro operates alone versus when it operates alongside batteries and supercapacitors.

The behaviour of the hybrid solution, combined with the proposed control system, is illustrated through two examples: a non-aggressive frequency event (see upper graph of Fig. 3) and an aggressive frequency event (upper graph of Fig. 4). In both cases, the power reference for the plant, which dictates the correction action required, is processed by the HLC.

Figure 3 presents the results for the non-aggressive event. The upper portion of the figure shows a frequency time-series with an over-frequency event ranging between 50.06 Hz and 50.13 Hz. This results in a power reference for the hydro, prompting the plant to take corrective action for frequency regulation. The lower part of Fig. 3 compares the behavior of the hydro valve in two scenarios: one where the hydro handles the entire correction by itself (blue curve), and another where the hydro operates as part of the hybrid system with batteries and supercapacitors (red curve).

As shown in the figure, when the hydro operates without the hybrid system, the valve's movement is somewhat abrupt and exhibits high-frequency fluctuations, which contribute to mechanical stress. In contrast, when the hybrid system is implemented, the valve's operation is much smoother, reducing stress and allowing the hydro to provide correction while being protected from excessive wear.

The results for the aggressive event are shown in Fig. 4. The upper part of the figure displays an extreme over-frequency event ranging between 49.9 Hz and 50.3 Hz, which triggers a significant correction action for the hydro. The lower part of the figure compares the hydro valve's behavior in the same scenarios as before.

In this case, it is evident that operating with the hybrid power plant significantly reduces the steepness of the valve's movement, thereby lessening the mechanical stress on the system. Additionally, the overall valve operation is smoother. As in the

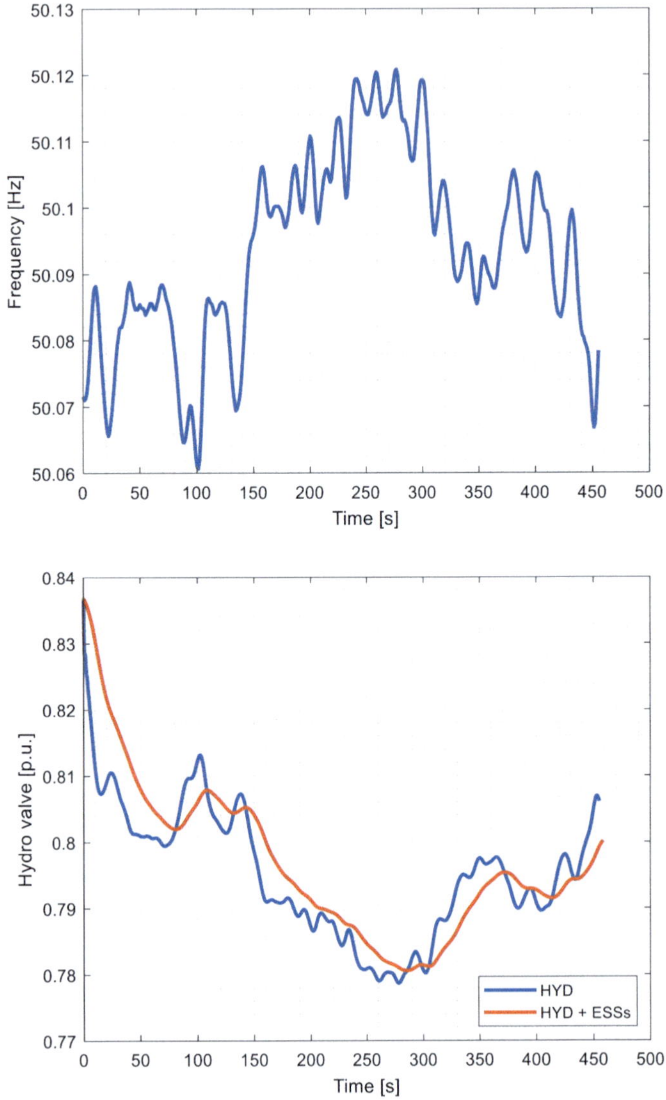

Fig. 3 Non-aggressive event. Frequency time-series and hydro valve evolution

non-aggressive event, the hydro's operation is much more stable when part of the hybrid system compared to when it operates alone.

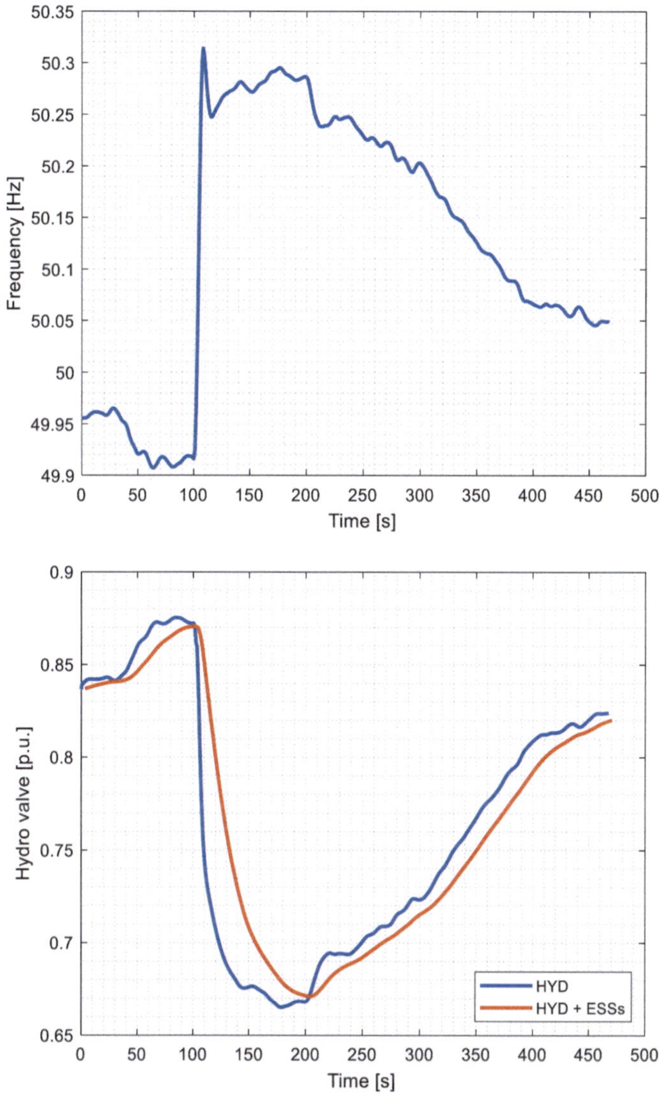

Fig. 4 Aggressive event. Frequency time-series and hydro valve evolution

References

ANDRITZ (2024) HyBaTec hybrid solutions. ANDRITZ https://www.andritz.com/products-en/hydro/products/hybrid-solutions/hybatec

Anindito Y, Haas J, Olivares M, Nowak W, Kern J (2019) A new solution to mitigate hydropeaking? Batteries versus re-regulation reservoirs. J Clean Prod 210:477–489

Bucher R, Schreider A (2017) On the pooling of hydro assets and grid-scale battery energy storage systems. Int J Hydropower Dams 5:60–64

Bucher R, Schreider A, Lehmann S (2018) Live test results of the joint operation of a 12.5 MW battery and a pumped-hydro plant. In: Proceeding 2018 HYDRO Conference, Octobe r, pp 15–17

Dawn S, Tiwari PK, Goswami AK (2018) Efficient approach for establishing the economic and operating reliability via optimal coordination of wind–PSH–solar-storage hybrid plant in highly uncertain double auction competitive power market. IET Renew Power Gener 12(10):1189–1202

dSPACE (2024) MicroAutoBox III. dSPACE https://www.dspace.com/en/pub/home/products/hw/micautob/microautobox3.cfm

Eiper T, Hell J, Kadam S (2019) Hybrid storage systems: welcome batteries in hydro powerplants. Proc Hydro

Gevorgian V, Muljadi E, Luo Y, Mohanpurkar M, Hovsapian R, Koritarov V (2017) Supercapacitor to provide ancillary services. In: 2017 IEEE energy conversion congress and exposition (ECCE). IEEE, pp 1030–1036

Guezgouz M, Jurasz J, Bekkouche B, Ma T, Javed MS, Kies A (2019) Optimal hybrid pumped hydro-battery storage scheme for off-grid renewable energy systems. Energy Convers Manage 199:112046

Javed MS, Zhong D, Ma T, Song A, Ahmed S (2020) Hybrid pumped hydro and battery storage for renewable energy based power supply system. Appl Energy 257:114026

Jin C, Lu N, Lu S, Makarov Y, Dougal RA (2011). Coordinated control algorithm for hybrid energy storage systems. In: 2011 IEEE power and energy society general meeting. IEEE, pp 1–7

Jin C, Lu N, Lu S, Makarov YV, Dougal RA (2013) A coordinating algorithm for dispatching regulation services between slow and fast power regulating resources. IEEE Trans Smart Grid 5(2):1043–1050

Laban D (2019) Hydro/battery hybrid systems for frequency regulation. Master's thesis, Universitat Politècnica de Catalunya

Lu N, Rudolph F, Loutan C, Makarov YV, Murthy S, Chowdhury S, Weimar MR, Arseneaux J (2010) The wide-area energy storage and management system–Phase II. PNNL, Richland, WA, USA, Tech Rep PNNL-19669

Makarov YV, Yang B, DeSteese JG, Lu S, Miller CH, Nyeng P, Ma J, Hammerstrom DJ, Vishwanathan, V. V. (2008). Wide-area energy storage and management system to balance intermittent resources in the Bonneville Power Administration and California ISO control areas (No. PNNL-17574). Pacific Northwest National Lab (PNNL), Richland, WA, United States

Moghaddam IN, Chowdhury B (2016). Optimal sizing of hybrid energy storage systems to mitigate wind power fluctuations. In: 2016 IEEE power and energy society general meeting (PESGM). IEEE, pp 1–5

OPAL-RT Technologies (2018) RT-LAB product sheet: BDL45–100. OPAL-RT https://www.opal-rt.com/wp-content/uploads/2018/07/OPAL_FICHES_BDL45-100_RTLAB.pdf

Ortega A, Milano F (2016) Modeling, simulation, and comparison of control techniques for energy storage systems. IEEE Trans Power Syst 32(3):2445–2454

Parastegari M, Hooshmand RA, Khodabakhshian A, Zare AH (2015) Joint operation of wind farm, photovoltaic, pump-storage and energy storage devices in energy and reserve markets. Int J Electr Power Energy Syst 64:275–284

Pérez-Díaz JI, Lafoz M, Burke F (2020) Integration of fast acting energy storage systems in existing pumped-storage power plants to enhance the system's frequency control. Wiley Interdiscip Rev Energy Environ 9(2):e367

Román A, de Jalón DG, Alonso C (2019) Could future electric vehicle energy storage be used for hydropeaking mitigation? An eight-country viability analysis. Resour Conserv Recycl 149:760–777

Sarasúa JI, Torres B, Pérez-Díaz JI, Lafoz M 2016 Control strategy and sizing of a flywheel energy storage plant for the frequency control of an isolated wind-hydro power system. In: 15th wind integration workshop. Energynautics GmbH, Darmstadt, Germany

Power-Intensive Energy Storage Systems

Giovanni De Carne, Seyede Masoome Maroufi, Antonio Morandi, and Mattia Simonazzi

Abstract The current energy system requires fast and long-term energy storage delivery. While energy-intensive energy storage systems, like batteries or hydrogen, can provide energy in the long term (e.g., hours or days), their fast dynamic power delivery is either constrained or strongly affects their lifespan. Power-intensive energy storage systems are specific technologies developed for allowing short-term high-power energy delivery. This specific characteristic is particularly useful for compensating fast-dynamic phenomena, such as faults or voltage dips, or optimizing energy-intensive technologies' efficiency and lifespan. This chapter describes the basic principle, operation, and network integration of the two most common power-intensive energy storage systems: the supercapacitor energy storage system (SCES) and the superconducting magnetic energy storage system (SMES). Their system hybridization with energy-intensive ESS is discussed, highlighting the control possibilities.

Keywords Flywheels · Batteries · Energy storage · Power system management · Fuzzy logic · Frequency control · Fluctuations · State of charge · Control systems · Vehicle dynamics · Flywheels · Hybrid energy storage · Power management · Ramp Rate (RR) · State of Charge (SoC)

G. De Carne (✉) · S. M. Maroufi
Institute for Technical Physics, Karlsruhe Institute of Technology, Eggenstein-Leopoldshafen, Germany
e-mail: giovanni.carne@kit.edu

S. M. Maroufi
e-mail: seyede.maroufi@kit.edu

A. Morandi · M. Simonazzi
Dipartimento di Ingegneria dell'Energia Elettrica e dell'Informazione "Guglielmo Marconi", Bologna, Italy

R. Scipioni et al. (eds.), *Hybrid Energy Storage*, Lecture Notes in Energy 47,
https://doi.org/10.1007/978-3-031-97755-8_5

1 Introduction to Power-Intensive Energy Storage Systems

Energy storage devices offer fast, flexible, and controllable bidirectional energy transfer, making them ideal for precise power exchanges with the grid to meet various regulation needs (Mao et al. 2020). In particular, power-intensive ESSs are essential components in modern energy infrastructure, especially as we transition towards more renewable energy sources like solar and wind. These systems are designed to manage high power demands over short durations, critical in stabilizing the grid, enhancing efficiency, and supporting the integration of variable renewable energy (Farhadi and Mohammed 2016).

Battery energy storage systems (BESSs) currently dominate the energy storage landscape due to their high energy density, scalability, flexibility, and efficiency (Barelli et al. 2022; Pelosi et al. 2023). Batteries are advancing rapidly, making their production both cheaper and more environmentally friendly. They are capable of delivering the required power during fast transients (such as during low-inertia frequency disturbances). However, as electrochemical devices, they are vulnerable to degradation, particularly from power surges and deep depths of discharges (DoDs) (Pelosi et al. 2024). As a result, frequent replacements, or, in alternative large oversizing, would be necessary when using BESSs to handle the high-power spikes.

To avoid the need for overly large BESSs in such scenarios, hybridization with short-term energy storage systems like flywheel, Supercapacitor Energy Storage (SCES), and Superconducting Magnetic Energy Storage (SMES), which feature high power-to-capacity ratios, is recommended. These devices can manage high-power spikes, reducing stress on the energy-intensive storage device (Jia et al. 2023; Mongird et al. 2020). Due to their fast response times and high efficiency, high-power energy storage systems are suitable for frequency regulation, voltage control, and oscillation damping in the power grid (Gyuk et al. 2013; Kirby and Hirst 1997). Additionally, combining complementary energy storage technologies in hybrid systems can mitigate the limitations of individual devices, greatly expanding their operational capabilities across various conditions and timeframes (Barelli et al. 2019). Hybrid storage systems combining power- and energy-intensive energy storage, allows to integrate the capability of providing energy for a relatively long period of time (e.g., hours for batteries, days for hydrogen), while extending as much as possible the energy-intensive resource lifetime, by letting the power-intensive resources address the fast- or repetitive transients.

This chapter provides an overview of two well-known power-intensive energy storage systems, such as SCES and SMES, focusing on their basic principles, grid integration, and existing applications and projects.

2 Energy and Power Comparison Among Energy Storage Systems

To better understand the potential of power-intensive energy storage systems, it is important to highlight their difference with energy-intensive systems. Figure 1 presents the specific power (per unit of mass), specific energy (per unit of energy), and Discharge Time (DT) ranges of various energy storage technologies, providing a clear description of their categorization based on performance metrics (Carne et al. 2024).

Technologies like SCES and SMES exhibit high specific power capabilities which reflects their capacity to deliver rapid bursts of energy over short durations from seconds up to a minute. Power density of SCES ranges from 500 to 7000 W/kg (Chatzivasileiadi 2012; Morandi et al. 2008). As for SMES, the power per unit mass ranges between 500 and 2000 W/kg, provided that the magnet (and the power electronics) are designed for withstanding fast charge/discharge operation. However, both technologies are constrained by relatively low specific energy values, with SCES ranging between 2.5 and 15 Wh/kg and SMES between 0.5 and 5 Wh/kg (with only the active mass of the magnet considered) (Cicéron et al. 2017; Mukherjee and Rao 2019; Zakeri and Syri 2015).

In contrast, batteries like lithium-ion are located in a more balanced area, offering a more moderate specific power and higher specific energy, making them versatile for applications and services involving longer (tens of minutes or hours) delivery of moderate (and slowly fluctuating) power (Hu et al. 2017; Milano and Manjavacas

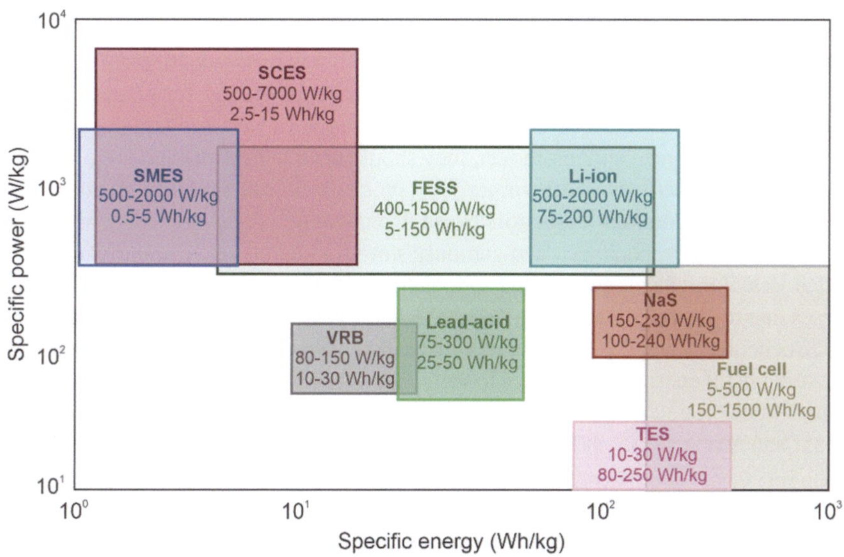

Fig. 1 Comparison between specific energy and specific power and discharge time

2019). Fuel cell storage systems, with their high specific energy but low specific power (5–500 W/Kg and 150–1500 Wh/Kg), are ideal for scenarios requiring large energy reserves for delivery or absorption over long periods (hours or days) (Farivar et al. 2022). The diagram of Fig. 1 highlights the trade-offs between power and energy capabilities, guiding the selection of appropriate storage technologies based on specific application needs.

3 Electric and Magnetic Energy Storage: Basic Principles

Supercapacitors and superconducting magnetic energy storage systems follow two different principles for storing energy. While the first ones store the energy in the form of an electric field, the latter store the energy in a magnetic field, similar to common inductors but with no joule dissipation. For capacitors, the stored electric energy $W_c\,(t)$ at the generic time instant t is expressed by the relation:

$$W_C(t) = \frac{1}{2}C_{vc}(t)^2 \tag{1}$$

where C is the capacitance and v_C the voltage across the capacitor. For SMES, like inductors, the stored electric energy $W_m(t)$ at the generic time instant t is expressed by the equation:

$$W_m(t) = \frac{1}{2}Li_L(t)^2 \tag{2}$$

where L is the inductance and i_L the current through the superconducting inductor. In order to achieve an amount of stored energy that is of interest for practical applications the capacitor must be characterized by a large capacitance and the inductor by a large inductance, and, moreover, they should be able to withstand large voltage and current, respectively. Furthermore, they should be able to operate with no or very reduced losses. Common capacitors based on conventional technologies only reach limited values of capacitance and can only store limited energy. Similarly, conventional inductors, also have limited inductance and are also characterized by Joule losses due to winding resistance. Recent technological advancements in the field of electrochemistry and superconductivity have provided capacitors and inductors that are suitable for energy storage. Due to their superior performance with respect to their conventional counterpart, this new generation of electrical components are referred to as "supercapacitors" and "superinductors" (or more commonly superconducting inductors or SMES), respectively.

By assuming a time interval from t_0 to t during which the storage device exchanges a power $P(t)$ with the grid, the stored energy $W(t)$ is:

$$W(t) = \int_0^t P(\tau)\mathrm{d}\tau \tag{3}$$

where $W(t_0)$ is the energy stored at the initial instant t_0. Based on energy conservation, by neglecting any power loss (both intrinsic in the storage device and in power converters), the capacitor voltage can be expressed as:

$$v_c(t) = \sqrt{v_c(t_0)^2 - \frac{2}{C}\int_{t_0}^t P(\tau)\mathrm{d}\tau} \tag{4}$$

and the inductor current as

$$i_L(t) = \sqrt{i_L(t_0)^2 - \frac{2}{L}\int_{t_0}^t P(\tau)\mathrm{d}\tau} \tag{5}$$

In Fig. 2 an example of the capacitor voltage or inductor current profiles during a discharging and subsequent charging phase is shown. At the initial time instant t_0 the storage device was charged, being assumed $v_C \neq 0$ and $i_L \neq 0$, then the two variables follow the trend described by (4) and (5) according to the power profile $P(t)$.

An important difference between electric and magnetic storage, and their respective power conditioning circuitry, is the stand-by operation. In fact, in order to maintain constant stored energy (and, hence, constant charge and voltage) of a capacitor, open circuit operation with zero current must be applied ($i_C = 0$). In contrast, to maintain constant stored energy (and, hence, constant magnetic flux and current) in an inductor, short circuit operation with zero voltage should be applied ($v_L = 0$).

From a practical point of view, it is simple to create nearly ideal open circuit condition through power electronic circuitry in the OFF state, while the short circuit

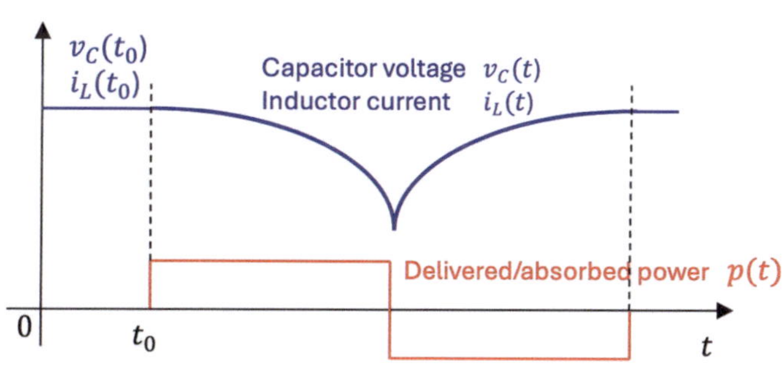

Fig. 2 Capacitor voltage or inductor current profile (blue curve) during a discharging and subsequent charging phase

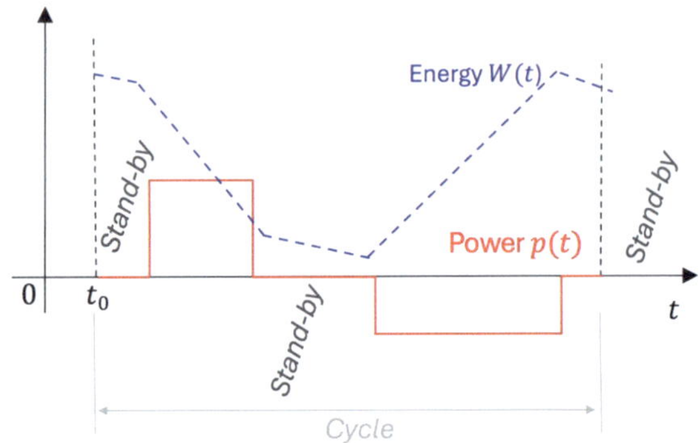

Fig. 3 Power (red curve) and energy (blue curve) profiles during a discharging and subsequent charging with stand-by phases

is more problematic since dissipation is unavoidable in power electronic components in ON state transporting a current. As a practical consequence, electric storage is suitable for applications in which long standby phases are involved, while magnetic storage is not.

Figure 3 shows a possible charge–discharge cycle for a capacitor or and inductor energy storage device. The considered cycle consists of a power delivery (positive power according to generator convention) and a power absorption (negative power) phase with the same total energy, interleaved by stand-by phases. The figure highlights that part of the energy of the device is lost in the stand-by phases due to both possible intrinsic phenomena in the storage devices and due to additional losses in the system. In order to assess an overall energy balance of practical electric or magnetic energy storage systems, all sources of power consumption must be considered. These include intrinsic dissipation in the storage device, dissipation in the power converters, dissipation during stand-by and possible auxiliary power that is required to maintain proper operating conditions.

The latter is particularly important in magnetic energy storage through superconducting inductors whereby cooling power must be constantly provided to keep the material at the proper cryogenic temperature required for superconductivity. Similarly, temperature control must also be provided for supercapacitors (and batteries) through ventilation and heat exchangers to prevent excessive temperature. However, in the former case the energy required is much higher due to the need to reach and maintain very low temperature, in the order of 77 K (– 196.15 °C) at most.

During cycling operation, in which the energy W_{del} is delivered to the grid and the energy W_{abs}, that takes all the loss mechanisms into account, is supplied to the device during the charging phase, the round-trip efficiency of the storage can be defined as:

$$\eta = \frac{W_{del}}{W_{abs}} = \frac{W_{del}}{\frac{W_{del}}{\eta_i \eta_c} + P_{aux}\Delta t + P_{standby}\Delta t_{standby}} \tag{6}$$

where η_i and η_c are the intrinsic and the converters' efficiency respectively, P_{aux} is the required auxiliary power (e.g., for cooling purposes) and $P_{standby}$ the power consumed during the stand-by phase with duration $\Delta t_{standby}$.

3.1 Grid Integration by Means of Power Electronics

In order to exchange energy with the power grid, capacitor and inductors must be interfaced through a proper power conditioning circuit able to control their voltage or current, as it is schematized in Fig. 4.

As explained above, despite being both power-intensive energy storage systems, supercapacitors and SMES work with two different principles. From the electrical point of view, supercapacitors can be assumed as voltage sources (Sect. 5.4.1 provides a detailed modeling of it). To regulate the energy flow between the SCES and the grid, a DC–DC converter is combined with a DC/AC front-end converter. The DC–DC converter performs two tasks: (i) transforms the DC voltage from the variable one of the SCES (dependent on their state of charge) to the constant one of the front-end converter DC links; (ii) regulates the current at the SCES side I_{sces}, and thus its power exchange. The DC link is then connected to a DC–AC front-end converter, that regulates the power flow between the AC and DC side, in order to keep the DC voltage V_{dc} at the nominal value. As a result, a controlled power flow is established between the SCES and the grid.

On the other side, a SMES can be assumed as a constant current source. While the front-end converter performs the same tasks as the SCES one, the DC–DC converter regulates instead the SMES voltage V_{smes} in order to extract a certain amount of energy from the SMES. In this case a controlled power flow is established between the SMES and the grid as a result.

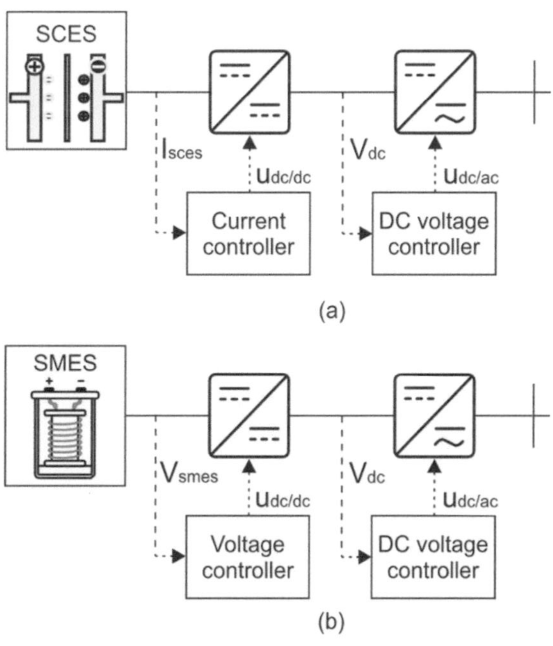

Fig. 4 Schematic representation of the grid connection architecture of **a** an electric storage device, i.e., a Supercapacitor Energy Storage (SCES) and **b** a magnetic energy storage device, i.e., a Superconducting Magnetic Energy Storage (SMES)

In the following, the SCES and SMES technologies are described in more detail, focusing on the specific modeling, services, and applications in power systems.

4 Supercapacitor Energy Storage Systems

4.1 SCES Modelling

This section introduces the most used modelling approaches for SCES (De Carne et al. 2022). The supercapacitor models consist of 4 main contributions (Fig. 5): (i) the main RC branch and (ii) its series RC elements, (iii) the parallel RC branches, and (iv) the leakage resistor branch.

Main RC Branch

The RC branch represents the basic behavior of the supercapacitors and describes the storing of electrostatic energy. It consists of a resistive element R in series with a capacitor $C(V)$ that accounts for the non-linear behavior of the capacitance, that is linearly dependent on the voltage:

$$C(v) = C_0 + k \cdot v \tag{7}$$

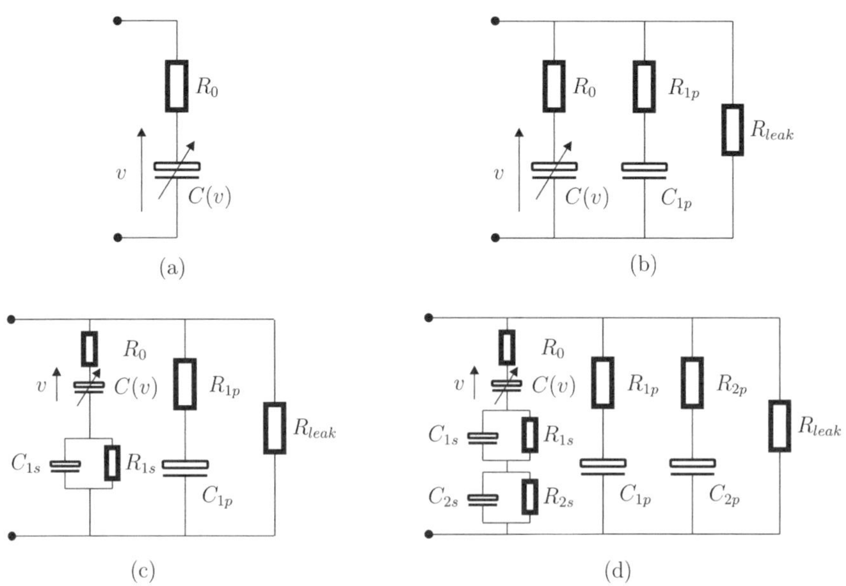

Fig. 5 Supercapacitor models proposed in literature (De Carne et al. 2022): **a** RC, **b** single parallel (1P) branch, **c** single parallel and single series (1P + 1S) branches, **d** double parallel and double series (2P + 2S) branches model

where C_0 is the supercapacitor main capacitance, and k describes the capacitance dependency on the voltage $C(v)$. This model accounts for the static behavior of the supercapacitor, without taking into account the dynamics and internal losses, that require additional modeling details.

Series RC Elements

To consider the fast voltage dynamics across the supercapacitors (e.g., caused by fast power changes), additional RC elements must be included in series to the main branch. These additional RC elements can be modeled as linear, and they target high frequency dynamics (e.g., 1 Hz and above). Despite several elements can be connected in series to represent a large dynamic spectrum, only a few RC elements (usually two) are considered in the main branch. These elements represent mostly the lower part of the high-frequency spectrum (e.g., up to 1 kHz).

Parallel RC Branches

To accurately represent the ion diffusion dynamics through the electrode power, that usually are slower dynamics in the range of seconds, additional parallel RC branches can be added. Similar to the series branches, in common practice up to two parallel branches are considered in supercapacitor models to represent the ion diffusion dynamics.

Leakage Resistor Branch

The supercapacitor cell tends to self-discharge during stand-by operations, although this phenomenon may require days or weeks. To take into account this property, a leakage resistor is usually added in the supercapacitor models.

4.2 SCES: Advantages and Disadvantages

This section describes the advantages and disadvantages in employing SCESs as energy storage solution.

- **Advantages**

SCESs offer several advantages over traditional energy storage systems, making them highly suited for applications requiring rapid power delivery and frequent charge–discharge cycles (Krpan et al. 2021). One of the most notable benefits is their high power density, which allows them to deliver bursts of energy much faster than batteries (Luo et al. 2015). This is especially advantageous in applications such as regenerative braking in electric vehicles or grid stabilization, where quick energy transfer is critical. Additionally, SCES have a long operational life due to their ability to withstand millions of charge–discharge cycles without significant degradation (Carne et al. 2024). This makes them highly reliable for applications where longevity and maintenance-free operation are crucial. Another advantage is their ability to operate efficiently across various temperatures, unlike batteries that may suffer from performance losses or may face severe safety issues in extreme conditions. Furthermore, SCES are eco-friendly as they are typically made from materials that are less hazardous than the chemicals used in traditional batteries (Saikia et al. 2020).

- **Disadvantages**

Despite their strength points, SCES also have several limitations that can restrict their usage in specific applications. A significant disadvantage is their relatively low energy density compared to batteries, meaning they cannot store as much energy in the same volume or weight (Song et al. 2014). This limits their effectiveness in applications that require long-duration energy storage, such as powering electric vehicles over extended distances (Lemian and Bode 2022). Additionally, SCES experience higher self-discharge rates than batteries, meaning they lose stored energy faster when unused. This makes them less ideal for applications requiring long-term energy retention. SCESs are also more expensive per unit of energy stored, which can increase the overall cost of energy storage solutions when used in systems requiring large amounts of stored energy (Zakeri and Syri 2015). Lastly, while they excel at providing high power output, they typically need to be paired with other storage systems, such as batteries, in hybrid configurations to efficiently handle both power and energy demands (Sutikno et al. 2022).

4.3 Application in Power System

- **Grid Stabilization**:

In modern power systems, the integration of renewable energy sources (RES), such as wind and solar, plays a crucial role in reducing reliance on fossil fuels and lowering carbon emissions (Dwyer and Teske 2018). However, these renewable sources are typically connected to the grid through power electronic converters, which isolate the energy source from the load, leading to a significant reduction in grid inertia (Nguyen et al. 2018). Grid inertia is essential for maintaining stability during power imbalances, as it helps to absorb disturbances and prevents cascading failures. When grid inertia is low, the system becomes vulnerable to undesirable effects such as load shedding (Sarojini and Kaliannan 2021). In this context, fast-acting energy storage systems like SCES can play a pivotal role. By injecting power immediately after a disturbance or the loss of a generation unit, SCES effectively emulate inertia, providing additional support to the grid. This inertia emulation can be achieved by generating power proportionally to the grid's frequency, ensuring stability. Numerous studies have highlighted the effectiveness of SCES in addressing the challenges posed by low inertia in modern power grids (Kim et al. 2019).

In addition to grid stabilization through inertia emulation, SCES are also valuable for frequency regulation, which is a critical function in balancing electricity supply and demand (Hatziargyriou et al. 2020; Wang et al. 2024). Frequency disturbances can destabilize power grids, especially those with a high share of renewable energy. SCES can absorb or deliver power almost instantaneously, making them ideal for providing short-term ancillary services within the power system (Barton and Infield 2004). Several research works propose using SCES to provide dynamic frequency support in isolated power systems, where they can effectively counteract sudden frequency deviations caused by fluctuations in generation or load (Delille et al. 2012; Tan et al. 2017).

Moreover, voltage regulation is another key area in which SCES finds applications. In power grids with significant penetration of renewable energy, voltage fluctuations can occur due to the intermittent and variable nature of generation from wind and solar sources (Ramirez and Murillo-Perez 2006). With their fast response times, power converters of SCES can be designed and controlled to rapidly inject or absorb reactive power to stabilize voltage levels during such fluctuations (Bensmaine et al. 2015). At the same time, the ability to charge and discharge faster than conventional batteries allow SCES to manage peak loads effectively, thereby adding active power control capability and contributing to grid stability. One notable industrial application is the partnership between TransnetBW and Hitachi Energy in Germany, where SCES are central to the implementation of STATCOM-GFM (Static Synchronous Compensator with Grid-Forming capabilities) systems. These systems provide enhanced voltage regulation by managing reactive power and offering short-term active power support, showcasing the role of SCES in modern grid management (TransnetBW 2024).

Thus, SCES contribute significantly to grid stability through multiple mechanisms grid inertia support, frequency regulation, and voltage stabilization making them indispensable in power systems with a growing share of renewable energy.

- **Electric Mobility**:

SCES are increasingly being integrated into transportation systems due to their unique ability to provide rapid bursts of energy combined with low weight and long cycle life. In electric vehicles (EVs), SCESs complement traditional battery systems by enabling regenerative braking, which captures energy during braking and stores it for later use (Wang et al. 2020). This process enhances energy efficiency, allowing quicker acceleration and smoother deceleration. Additionally, supercapacitors can supply high power during peak demand situations, such as when vehicles require immediate acceleration, thus improving overall performance (Horn et al. 2019). In public transport systems, like electric buses and trams, supercapacitors store energy for short trips between charging stations, allowing for quick recharges at stops and minimizing downtime (Mir et al. 2009; Zhang et al. 2017). Their lightweight design and ability to operate in various temperature ranges make them ideal for use in rail systems, where they can recover energy during braking and use it for acceleration.

Several industrial applications illustrate the effectiveness of SCES in mobility. For instance, Nidec Conversion is developing an all-electric, zero-emission passengers ferry that utilizes fast-charging supercapacitors to enhance performance and sustainability (Nidec Conversion 2024). Sunwin, a state-owned company, is implementing supercapacitors in China and the UAE in their buses to improve energy efficiency and operational capabilities (Shanghai Sunwin Bus Corporation 2024; Enterprises 2024). In Italy, Chariot Motors, in partnership with local company E-CO, is deploying buses in Rome and Turin capable of recovering over 35% of energy during braking (Bus 2024; s.r.l 2024). Furthermore, Skeleton and Siemens are collaborating on train systems that utilize supercapacitors to maximize energy recovery and operational efficiency (Mobility 2024; Technologies 2024).

5 Superconducting Magnetic Energy Storage Systems

Superconducting Magnetic Energy Storage (SMES) systems store energy in the magnetic field generated by a DC current flowing in a resistanceless superconducting coil connected to the grid through suitable power conditioning apparatuses. Conceptual SMES systems were first proposed by Ferrier in 1969 to store diurnal energy in France (Boicea 2014), while the first SMES was realized at the University of Wisconsin in 1971(Adetokun et al. 2022). The first commercial operating device was introduced in 1981 in the 500 kV Pacific Intertie corridor, connecting California to the Northwest region of the United States (Torre and Eckroad 2001). A SMES unit with 10 MW 1 s rating are used since 2008 to prevent power interruptions in a semiconducting manufacturing plant in Japan (Nagaya et al. 2012).

To maximize the current circulating in the in the SMES coil, so to maximize the stored energy, and to prevent joule losses during current circulation that would jeopardize the efficiency of the system, superconducting materials are used for the windings (Morandi et al. 2016; Schoenung et al. 1991). Energy exchange with the grid is then managed by power converters, controlled by algorithms designed to modulate the coil current according to (5) to ensure that the charging and, more importantly, the discharging of the coil occur when needed and with the promptness required. This makes SMES able to store energy with minimal losses (overall efficiency up to 95% or more) and release it almost instantaneously, resulting in a highly efficient and responsive system that can mitigate voltage deeps and brief interruptions, thereby ensuring continuity of service (Schoenung et al. 1991; Vulusala and Madichetty 2018).

5.1 *Superconducting Magnet System*

A practical SMES system consists of a lossless (or extremely low loss) superconducting magnet and a power conditioning system, required for connecting the coil to the (usually three phase) power grid. The cooling system, required for maintaining the temperature of the coil below the critical value so to ensure that it operates in the superconducting state, is an integral and essential part of the system. SMES systems based on low-temperature superconductors are an established power-intensive storage technology (Filippidis et al. 2021; Boenig and Hauer 1985; Morandi et al. 2008; Ottonello et al. 2006). In these systems the superconducting coil, made of industrially mature NbTi wires routinely used in MRI applications, operates at 4.2 K (− 268.5 °C) and is cooled by immersion in a liquid helium bath, obtained by closed cycle cooling machines. Despite complete and qualified systems have been developed, LTS-SMES technology has failed to find broad market penetration due to the high cost (capital and operational) related to liquid helium. Besides the high cost, liquid helium also implies complicated management, supply (also facing shortage concerns), maintenance and safety (overpressure and explosion risk in the case of quench), which make its use not trivial for grid or industry systems. Improvement of SMES prototypes based on first-generation HTS materials (Bi2223/Ag conductor) and cryogen-free cooling (not using liquid helium) at 20 K have been also proven (Hawley and Gower 2005). However, the high cost of practical 1G-HTS conductors (requiring a high silver content) does not offer prospects for commercial development of this SMES technology.

A drastic improvement in SMES technology, creating a breakthrough increase of both the technical and economic attractiveness, can be obtained today using modern (second generation) HTS materials ReBCO coated conductors. These materials offer much improved transport capacity and magnetic field withstand ability and, moreover, are able to operate at much increased temperature, in the range of 20–50 K and beyond for SMES applications. These characteristics open the way to more compact magnet designs, requiring a much simplified and cost-effective cooling system. The magnesium diboride (MgB2), a superconductor with intermediate electro-thermal performance compared to LTS and HTS and able to operate in the temperature range of 10–20 K, can also be considered as an alternative for SMES applications due to its excellent industrial maturity and low cost (Morandi et al. 2018). HTS-SMES prototypes based on 2G HTS tapes have already been developed and validated in the lab (Cicéron et al. 2018a, b; Hawley and Gower 2005; Ren et al. 2015).

5.2 *SMES: Advantages and Disadvantages*

- In general, SMES can have significant advantages in terms of power density and response speed, but they also have limitations in energy density and cost, which affect their adoption in various sectors. More specifically, the advantages of SMES technologies are:
- High efficiency: SMES systems can achieve round-trip efficiencies greater than 98%.
- Fast transient dynamics: SMES systems offer very small response times, making them suitable for grid stabilization and power quality applications.

- High power density: SMES can reach very high-power output capability, meaning that it can deliver large amounts of power in short periods.
- Unlimited charge–discharge cycles: SMES systems have no intrinsic degradation over time with cycling (unlike batteries and supercapacitors), offering a long service life.
- Zero carbon emissions: during operations SMES does not contribute with greenhouse emissions and does not involve hazardous materials.

SMES disadvantages can be instead summarized as:

- Cryogenic cooling: SMES systems involve superconductors whose require extremely low temperatures (typically using liquid helium or nitrogen), leading to high operational complexity and costs, including maintenance.
- High initial cost: the superconducting materials, cooling infrastructure, and supporting systems make SMES installations considerably expensive.
- Low energy density: SMES has excellent power density, but it has relatively low energy density, making it unsuitable for long-duration energy storage.

Overall, SMES systems are the best solution for high-power, short-time energy storage. While they offer exceptional efficiency and fast response times, their high costs and cooling requirements still remain significant challenges limiting their large-scale diffusion (Luongo 1996).

5.2.1 SMES Modeling

As seen in the previous sections, SCESs consist of many supercapacitor cells or small modules arranged in series or parallel in order to reach the required voltage, current, and power rating. Modeling of individual supercapacitor cells is crucial to accurately reproduce the overall behavior of system. Concerning SMES, although intense modeling efforts are required in the design phase of the coil, the interaction of the SMES with the converters and the grid can be well reproduced, at least for the typical time-scale of power exchange cycles, by representing the coil as a simple inductor. This is due to the fact that the superconducting windings are embedded in one unique cryostat with two current leads connecting it to the external environment at room temperature. Overall, the SMESs form a two-terminals component which can be modeled as an ideal inductor. Indeed, further phenomena that occur into the coils during operations (i.e., AC losses) do not significantly affect the terminals behavior of the SMES in normal conditions. Of course, these phenomena, which would require complex modeling, are duly considered in the design phase of the SMES system. It is to be pointed out that the ideal inductor model only holds in normal operating conditions, during which all the thermal and electromagnetic parameters are within the design range. If special operating conditions need to be investigated more complex models need to be introduced. For example, quench phenomena involving overheating would require distributed non-linear resistive network, whereas investigation of fast voltage transients would require distributed L-C circuits.

6 Hybrid Energy Storage Systems

Hybrid energy storage systems (HESS) represent an advanced approach to energy management. They integrate multiple energy storage technologies to harness their complementary attributes. Typically, HESS combines high power-density devices, such as SMES and SCES, with high energy–density devices, such as batteries (Liu et al. 2024; Zhu et al. 2012). With their rapid charge/discharge capabilities, high-power devices can manage short-term fluctuations and peak power demands, while high-energy devices are suited for long-duration energy storage and stable power delivery over extended periods (Etxeberria et al. 2010). The potential of HESS lies in its ability to provide both immediate power response and sustained energy delivery. For example, supercapacitors can handle high-power transients that last seconds to minutes, such as during grid frequency stabilization or rapid acceleration in electric vehicles (Jia et al. 2023; Sun et al. 2017). Meanwhile, batteries or other high-energy storage technologies deliver the required energy over a longer duration, ensuring consistent power supply even under variable load demands (Rouholamini et al. 2022). This synergy leads to increased system efficiency, reduced stress on individual storage components, and an extended lifecycle for the energy storage system as a whole (Li et al. 2021). Among implementations of HESS, in wind applications, battery/supercapacitor HESS has been introduced to improve efficiency and reduce costs by extending battery life (Sarrias-Mena et al. 2014). A SMES/battery HESS was considered to stabilize fluctuating loads in railway substations (Ise et al. 2005; Zhou et al. 2010) demonstrated that combining short- and long-term ESS improves renewable energy penetration in power systems. Additionally, (Gee et al. 2013) described how SCES can extend battery life in off-grid wind systems, with similar methods successfully predicting battery lifetime in other studies (Guan et al. 2014; Li et al. 2015).

6.1 Power Electronics Integration

HESS are usually composed of two energy storage systems with different voltage, current and power ratings, and that may behave differently as voltage or current sources. Power electronics converters allows storage systems to be integrated in the same system and to control optimally the energy flow. This section provides a brief overview of the possible power electronics connection of hybrid energy storage systems, considering 4 potential connection architectures as in Fig. 6.

AC Coupling

This coupling connects two HESS separated devices in the AC side. It means, that the power- and energy-dense devices can be of different vendors, and eventually work at different voltage rates (e.g., connected with a transformer), and they are coupled only from an energy management point of view. Despite it requires several power

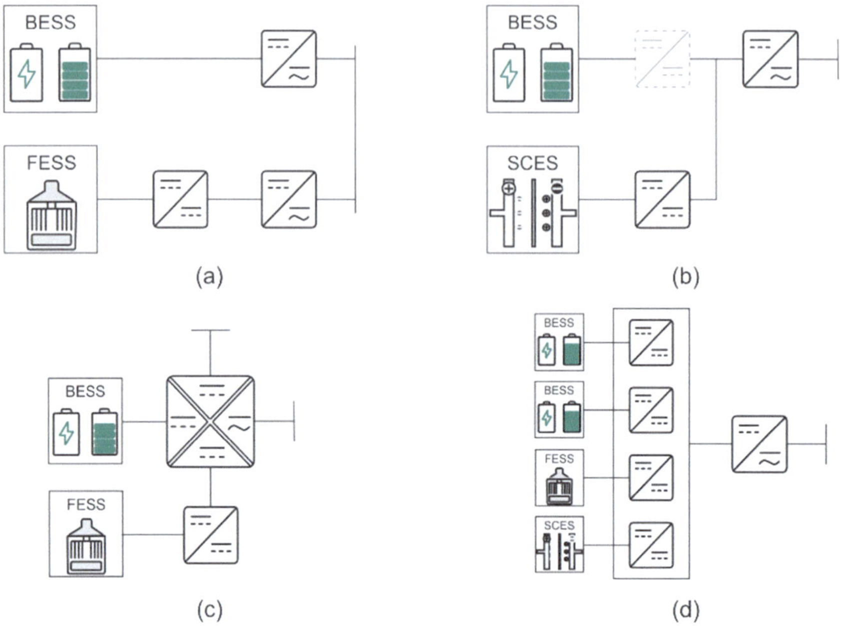

Fig. 6 HESS power electronics integration considering different energy storage technologies, including BESS, SCES and other power-intensive storage systems, such as Flywheel (FESS). Different architectures are considered: **a** AC coupling between BESS and FESS, **b** DC coupling between BESS and SCES, **c** multi-port coupling between BESS and FESS, **d** multi-level coupling with BESS, FESS, SCES

conversion stages, this coupling approach is widely spread in industrial application, due to its design flexibility (e.g., it can be easily scaled up).

DC Coupling

The DC coupling exploits the DC nature of many energy storage systems (e.g., batteries, supercapacitors, SMES, FESS) to reduce the number of dedicated conversion stages (Lo Franco et al. 2021). The HESS are interfaced in a common DC link, by means of DC/DC converters (isolated and not), that regulate the power exchange in each storage system. A common DC/AC converter interfaces the DC link to the AC system. The main advantage of using DC coupling lies in the efficiency and in the material saving: there are fewer conversion stages, leading to lower losses and a reduced need for semiconductor and passive components. This translates automatically in potentially lower bill of materials, and thus costs.

Multi-port DC Coupling

An innovative concept pursued recently is to enable the DC coupling by means of the magnetic stage of dual active bridge (DAB) DC/DC converters (Kheraluwala et al. 1992). This approach, employing usually Multiple-Active-Bridge (MAB) converters, transfers energy in AC form in medium- or high-frequency transformers (up to hundreds of kHz), and reconvert it in DC at multiple terminals. The main advantage in using this approach lies in the compactness of the solution. With respect to direct DC coupling, fewer conversion stages are required to connect several different HESS technologies. This aspect is particularly important for on-board systems, where safety is guaranteed by the galvanic insulation (Pereira et al. 2021), and a reduced weight and space enables smaller footprints.

Multi-level Coupling

Multi-level converters offer the possibility to connect directly to high voltage grids by means of low voltage components. Topologies, such as Cascaded H-Bridge and Modular Multi-level Converter, are composed of a series-connection of low voltage cells, that include usually capacitive elements. Recently, several works (Hillers et al. 2015; Jiang et al. 2021), envisioned the idea to connect the energy storage elements directly at cell level, decreasing the need for further DC/DC stages. This idea, pursued for HESS in several works, has its own challenges, particularly in the control and protection (i.e., no power flow control in case of fault).

6.2 Services and Applications

As mentioned above, HESS can manage optimally the energy delivery in the short- and long-term, allowing to extend their capability in offering services. In the following, several examples are provided, where HESS can play an important role in the energy management. A review of existing projects on the development of SCES and SMES is presented in the next section, followed by the discussion of a selected application case.

- **Renewable Energy Integration**:

Renewable energy generation, unlike conventional fossil fuel plants, is influenced by weather conditions and time of day, leading to times when power output either falls short of or exceeds demand (Deguenon et al. 2023). Grid operators must continuously manage these fluctuations in to prevent frequency imbalances, voltage instabilities, and power outages (Khalid 2024). HESSs are crucial in mitigating these issues by storing surplus energy during peak production and supplying it when production is low (Adeyinka et al. 2024; Kebede et al. 2022).

- **Electric Vehicles (EVs)**:

Due to uncertainties surrounding future energy resources and environmental, economic, and political challenges, energy efficiency policies and clean energy adoption have gained significant focus. In the private transport sector, options such as fuel cell (FC) vehicles, hybrid electric vehicles (HEVs), plug-in HEVs (PHEVs), and battery electric vehicles (EVs) are being proposed to replace or enhance internal combustion engines (ICEs), with all these alternatives incorporating electric propulsion systems to reduce dependence on ICEs. The power output of the primary sources in these vehicles, typically batteries or fuel cells, are limited by their charge/discharge cycles and may struggle to meet high power demands during short periods, such as acceleration. FESS, ultra-capacitors, and SMES are commonly recommended for power-based energy storage in EV applications (Bazzi et al. 2018; Rezaei et al. 2022).

- **Electric Trains**:

In comparison to other modes of transportation, electrified rail offers benefits such as high capacity, punctuality, and environmental sustainability. However, the growing global focus on enhancing railway use and energy efficiency has led to rising planning, investment, and construction costs of the powering infrastructure. Recently, there has been increasing interest in optimizing the use of ESSs to recover more regenerative energy and boost traction efficiency (González-Gil et al. 2013; Ise et al. 2005). Among different ESSs, batteries are still a common choice for absorbing regenerative energy in certain electrified sections. Still, they have significant limitations, including reduced energy efficiency and a limited cycle life (Vazquez et al. 2010). Given the current state of ESSs, no single type can fully meet the demands of modern electrified railway systems, such as long cycle life, cost-effectiveness, and balanced power and energy densities. As a result, HESSs offer a superior solution (Babu et al. 2020).

- **Marine Transportation**:

Traditional ships typically rely on diesel-powered internal combustion engines for propulsion. Still, these engines emit pollutants that harm the environment and human health. With growing environmental regulations, electric ships have gained prominence in the future of shipbuilding and maritime operations. Strategies such as optimized vessel routing, increased electrification supported by ESSs, renewable energy integration, and alternative fuels have been identified as effective ways to significantly reduce emissions from electric ships. Currently, most electric ships use lithium-iron phosphate batteries for energy storage (Banaei et al. 2020). However, power fluctuations during voyages can result in high current impacts that accelerate battery aging. Furthermore, the reliance on a single energy source can compromise the ship's power grid reliability. However, the potential of HESS to reduce operating costs is a reassuring factor. By combining different energy storage technologies, HESS can leverage their complementary strengths to create more efficient systems. SCES, with their fast charging, long cycle life, and high-power density, are ideal for handling high current charging and discharging. Effective energy management strategies can further extend battery life and lower operating costs, making the economic viability of all-electric ships a promising prospect (He et al. 2021).

7 Existing Projects with Power-Intensive HESS

As the energy landscape continues to evolve, numerous projects are exploring the potential of power-intensive HESS to meet the growing demand for robust and efficient energy solutions (Babu et al. 2020). These projects, from early planning stages to ongoing developments, address critical challenges such as grid stability, renewable energy integration, and peak load management. This section provides an overview of some key projects either underway or in the planning phase, offering insights into their objectives and anticipated outcomes to harness the potential of HESS in power-intensive applications.

- **Duke Energy HESS at Rankin Substation in Gaston County, N.C. (2016)**: Featuring a 100-kW/300-kWh battery and ultracapacitors. This system integrates Aquion Energy's Aqueous Hybrid Ion batteries with Maxwell Technologies' high-speed ultracapacitors for rapid energy response. The HESS aims to enhance solar power management by smoothing fluctuations, extending operational life, and enabling load shifting (Duke Energy 2016).
- **Skeleton Technologies' SuperBattery HESS (2023)**: Energy storage solution combining the benefits of supercapacitors and batteries. This hybrid technology leverages supercapacitors' high-power density and rapid charge/discharge capabilities with the high energy density and long-term storage capabilities of batteries. The SuperBattery is designed for applications requiring high-power energy storage, such as automotive and renewable energy in industrial sectors. Key features include excellent cycle life, fast charging and discharging, and robust performance in varying temperatures (Skeleton Technologies n.d.).
- **Beyonder and ABB Lithium-ion Capacitors (LiC) HESS (2024)**: Beyonder and ABB partnering to enhance product development, production, and market strategies for the large-scale production of Beyonder's LiC, utilizing key components from both companies. They aim to discover more cost-effective and sustainable methods for delivering energy storage solutions to the power grid while supporting emissions-free solutions at airports, ports, heavy transport, and fast charging stations (Beyonder 2024).
- **V-ACCESS—Vessel Advanced Clustered and Coordinated Energy Storage Systems**: The EU funded V-ACCESS project aims to integrate, in addition to battery, a 100 kW SCES system and 100 kW SMES system to match vessel's operative conditions. The project, involving 14 EU companies and research institutions with SKELETON and ASG companies as SCES and SMES technology providers respectively, investigates how these short-term energy storage technologies can operate together with batteries and how it can be ensured that they can effectively mitigate battery degradation. Suitable control of the power flows and DC integration strategies, as well as safety assessment and systems' marinization, are fully developed in the project to reach the objectives.

8 Case Study: Power Management in Hybrid Energy Storage Systems

This section offers a case study for hybrid energy storage systems, where a dynamic energy storage system, i.e., a SCES or SMES studied in this chapter, is associated with a battery energy storage system. The complementary power and energy characteristics of SCES or SMES and lithium-ion batteries enable advanced grid services with enhanced power response, energy capacity, and cycling capability while extending the system's lifetime. SCES or SMES can handle continuous, high-power, short-duration demands efficiently, protecting the battery from excessive stress while the battery provides the energy density required for sustained, longer-duration energy supply. Careful monitoring and management of State of Charge (SoC) and ramp rate are important because they directly affect system performance and longevity. High ramp rates can cause stress on the components, while extreme SoC levels (either too low or too high) can reduce efficiency or risk depletion, respectively. Simplifying the SoC and ramp rate categories to critical thresholds ensures effective real-time decisions without unnecessary computational complexity. Therefore, effective real-time power management and carefully considering storage components' SoC and ramp rates are essential to optimize performance.

As a reference case for investigating the cooperation between SCES/SMES and batteries, we consider the work Maroufi et al. (2025) which explores a novel Moving Average (MA) and Fuzzy Logic-based power management approach for a hybrid energy storage system designed to provide grid services. In Maroufi et al. (2025), a flywheel system is considered a power-intensive storage system. The developed methodology is extended to supercapacitors and SMES here. This method optimally distributes power between the two technologies by addressing key limitations: the supercapacitor's and SMES SoC and the battery's ramp rate. By employing Moving Averages for dynamic energy demand adjustments and Fuzzy Logic controllers for precise power redistribution, the system minimizes imbalances in SoC and ramp rates.

As shown in Fig. 7 the controller integrates fuzzy logic and moving average techniques to optimize power set-points, dynamically adjusting to the rate of change in the battery's power output and the SoC of the flywheel (can be SCES or SMES). Fuzzy logic is particularly suited for this application due to its flexibility and intuitive handling of nonlinear variables, enabling the determination of the moving average window length based on real-time conditions. This adaptability ensures that the system can respond effectively to rapid fluctuations in power demand while maintaining SoC and ramp rate levels within optimal ranges. Furthermore, by carefully managing the trade-off between prioritizing power response and SoC control, the HESS remains available for additional grid services. Unlike conventional methods, this integrated approach directly manages the HESS's output power, ensuring optimal overall performance under dynamic grid conditions.

In order to obtain the HESS power profile, real-world data from southern Germany, spanning three months and summarized into a representative day profile through an

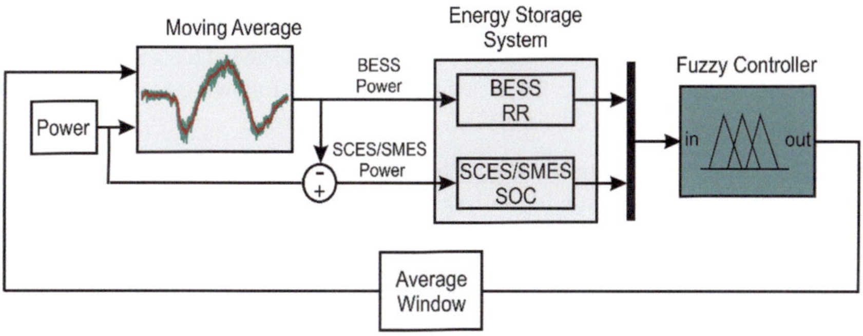

Fig. 7 Control scheme for HESS including the moving average and fuzzy logic stages

enhanced motif discovery algorithm, were employed. This power profile, derived from reference (Karrari et al. 2022), identifies the most recurrent daily consumption patterns within the target time series with a 1-s time step. The input data, collected during the summer of 2018 from four distinct 10/0.4 kV substations, captured high levels of PV generation critical for the sizing analysis. These substations were selected based on their relative voltage sensitivity to changes in active power as proposed in (Karrari et al. 2020). Figure 8 demonstrates the division of this input power using a 10-min moving average. For this particular input dataset, the moving average output varies within a range of ± 160 kW, while the difference between the input signal and the moving average filter output falls within a range of ± 30 kW.

In the following performance metrics, when the high-energy density storage (Battery Energy Storage) is fixed at 500 kWh, but the high-power density storage (e.g., SCES or SMES) varies in capacity, are compared. The three cases were

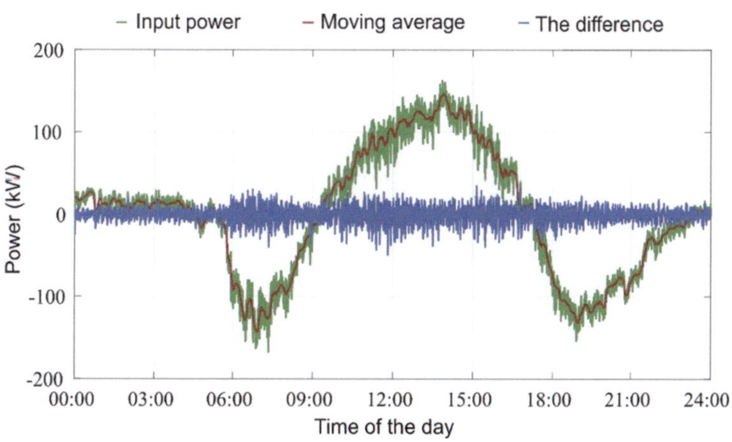

Fig. 8 The profile of the input power, the moving average filter, and the difference between them adapted from (Karrari et al. 2022)

conducted when the high-power density storage had capacities of 3.5, 7.0, and 10.5 kWh. These ratings can be met based on the present state of the art for the SCES and SMES (mostly in the 3.5 kWh case) systems. The following key observations highlight the impact of the SCES/SMES energy capacity:

1. **Effect on Average Window**: Fig. 9 shows how the size of the high-power density storage influences the Moving Average window length used in the power management strategy during a time interval of one day Systems with lower energy capacity in the high-power density storage required shorter average windows, as they have limited ability to buffer energy fluctuations. For instance, SCES or SMES necessitate more frequent power adjustments to accommodate rapid changes in demand. It is also noteworthy that SCES and SMES, with their higher power density compared to flywheels, are more effective in handling sharp power spikes. However, their lower energy capacities mean that the Moving Average window must adapt dynamically to prevent overloading.
2. **Effect on State of Charge (SoC)**: Fig. 10 shows how the SoC dynamics of the high-power density storage are impacted by its energy capacity during a time interval of one day. Smaller capacity devices reach lower SoC levels during the day because they deplete faster under equivalent power demands. While supercapacitors and SMES discharge more rapidly due to their higher power density, their ability to recharge quickly ensures they remain viable for high-frequency power adjustments. This characteristic differentiates them from flywheels, which may have slower response times in comparison.
3. **Effect on Ramp Rate of High-Energy Density Storage (Battery)**: Fig. 11 shows how a decrease in the energy capacity of the high-power density storage impacts the power ramp rate of the battery. With smaller high-power density storage, the system relies more heavily on the battery to respond to fast power fluctuations, leading to higher ramp rates. The higher power density of SCES and SMES helps to mitigate this effect to some extent by absorbing rapid fluctuations more effectively than flywheels. However, their limited energy capacity means that their ability to sustain prolonged support is reduced.

This case study highlights the intricate balance required in designing hybrid energy storage systems. Supercapacitors and SMES, with their superior power density compared to flywheels, excel in handling rapid power fluctuations but are limited by their lower energy capacities. By leveraging power management strategies like Moving Average and Fuzzy Logic, it is possible to mitigate these limitations and ensure system efficiency.

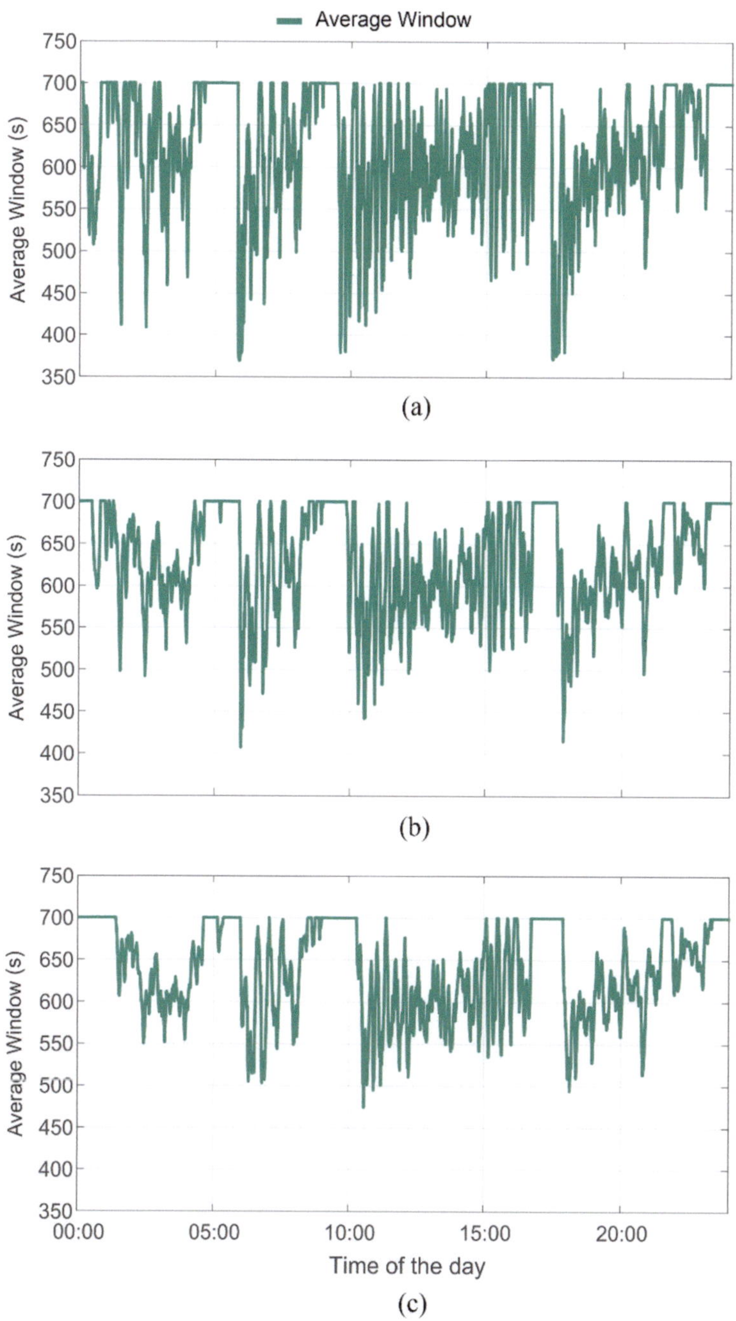

Fig. 9 Average window length of the moving average filter for different energy capacities of the high-power density storage **a** 3.5 kWh **b** 7.0 kWh **c** 10.5 kWh

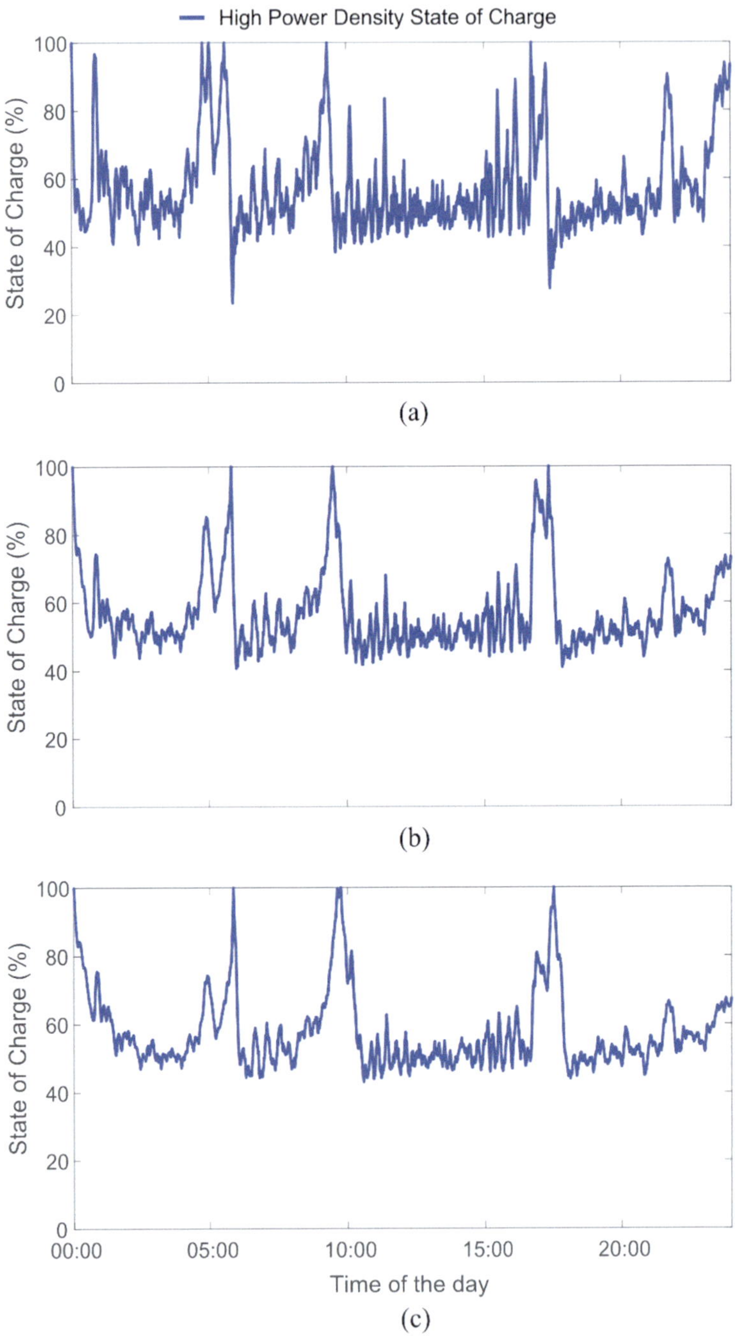

Fig. 10 State of charge of the high-power density storage for different energy capacities of the high-power density storage **a** 3.5 kWh **b** 7.0 kWh **c** 10.5 kWh

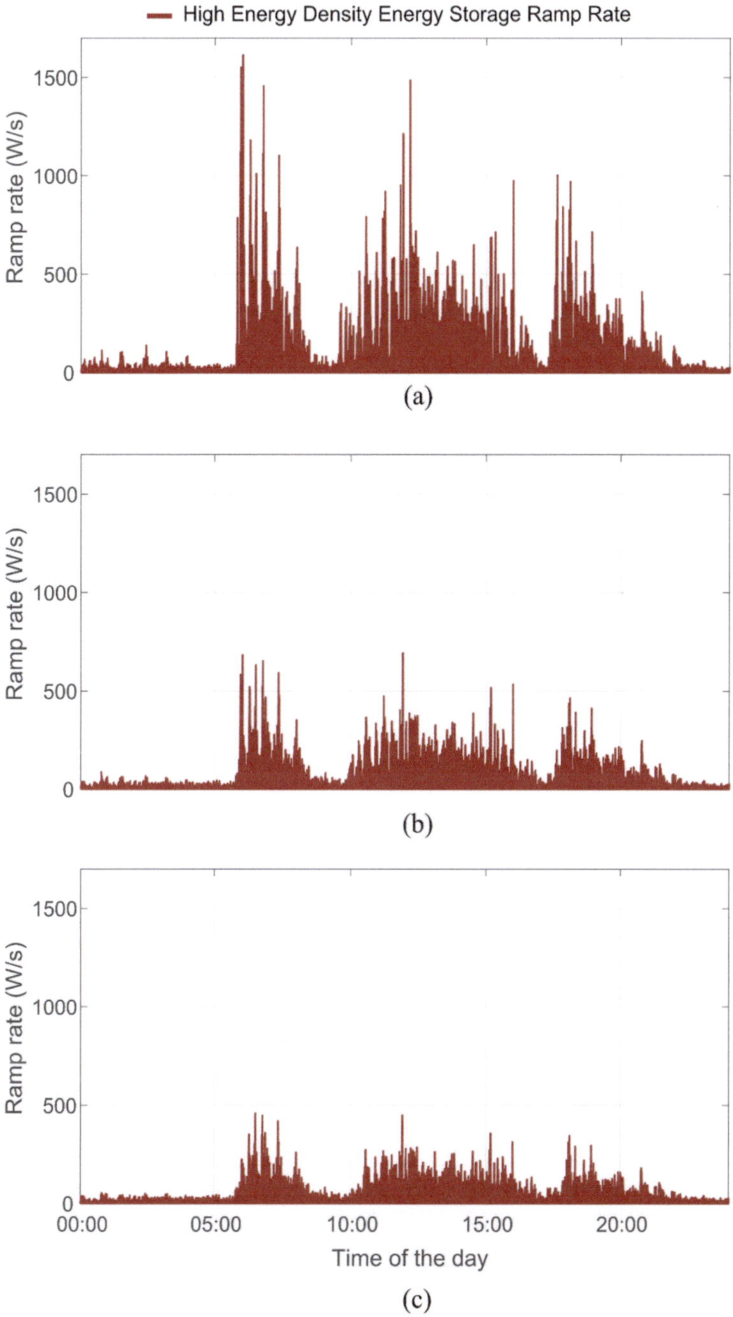

Fig. 11 Ramp rate of battery for different capacities for high-power density storage **a** 3.5 kWh **b** 7.0 kWh **c** 10.5 kWh

Acknowledgements The work of Giovanni De Carne and Seyede Masoome Maroufi was supported by the Helmholtz Association under the program "Energy System Design" and the Helmholtz Young Investigator Group "Hybrid Networks" (VH-NG-1613).

References

Adetokun BB, Oghorada O, Abubakar SJ (2022) Superconducting magnetic energy storage systems: prospects and challenges for renewable energy applications. J. Energy Storage 55:105663. https://doi.org/10.1016/j.est.2022.105663

Adeyinka AM, Esan OC, Ijaola AO, Farayibi PK (2024) Advancements in hybrid energy storage systems for enhancing renewable energy-to-grid integration. Sustain Energy Res 11:26. https://doi.org/10.1186/s40807-024-00120-4

Babu TS, Vasudevan KR, Ramachandaramurthy VK, Sani SB, Chemud S, Lajim RM (2020) A comprehensive review of hybrid energy storage systems: converter topologies, control strategies and future prospects. IEEE Access 8:148702–148721. https://doi.org/10.1109/ACCESS.2020.3015919

Banaei M, Rafiei M, Boudjadar J, Khooban M-H (2020) A comparative analysis of optimal operation scenarios in hybrid emission-free ferry ships. IEEE Trans Transp Electrific 6:318–333

Barelli L, Bidini G, Cherubini P, Micangeli A, Pelosi D, Tacconelli C (2019) How hybridization of energy storage technologies can provide additional flexibility and competitiveness to microgrids in the context of developing countries. Energies 12. https://doi.org/10.3390/en12163138

Barelli L, Pelosi D, Longo M, Zaninelli D (2022) Energy storage integration into fast charging stations installed on e-highways. In: 2022 IEEE power & energy society general meeting (PESGM). IEEE, pp 01–05

Barton JP, Infield DG (2004) Energy storage and its use with intermittent renewable energy. IEEE Trans Energy Convers 19:441–448

Bazzi AM, Liu Y, Fay DS (2018) Electric machines and energy storage: over a century of technologies in electric and hybrid electric vehicles. IEEE Electrific Mag 6:49–53. https://doi.org/10.1109/MELE.2018.2849900

Bensmaine F, Bachelier O, Tnani S, Champenois G, Mouni E (2015) LMI approach of state-feedback controller design for a STATCOM-supercapacitors energy storage system associated with a wind generation. Energy Convers Manag 96:463–472

Beyonder (2024) Beyonder and ABB to collaborate on battery technology of the future

Boenig MJ, Hauer JF (1985) Commissioning tests of the bonneville power administration 30 MJ superconducting magnetic energy storage unit. IEEE Trans Power Appar Syst U. S. PAS-104:2. https://doi.org/10.1109/TPAS.1985.319044

Boicea VA (2014) Energy storage technologies: the past and the present. Proc IEEE 102:1777–1794. https://doi.org/10.1109/JPROC.2014.2359545

Bus CE (2024) 12m ultracapacitor chariot E-bus

Carne GD, Maroufi SM, Beiranvand H, Angelis VD, D'Arco S, Gevorgian V, Waczowicz S, Mather B, Liserre M, Hagenmeyer V (2024) The role of energy storage systems for a secure energy supply: a comprehensive review of system needs and technology solutions. Electr Power Syst Res 236:110963. https://doi.org/10.1016/j.epsr.2024.110963

Chatzivasileiadi A (2012) Electrical energy storage technologies and the built environment

Cicéron J, Badel A, Tixador P, Forest F (2017) Design considerations for high-energy density SMES. IEEE Trans Appl Supercond 27:1–5. https://doi.org/10.1109/TASC.2017.2655627

Cicéron J, Badel A, Tixador P, Pasquet R, Forest F (2018a) Test in strong background field of a modular element of a REBCO 1 MJ high energy density SMES. IEEE Trans Appl Supercond 28:1–5. https://doi.org/10.1109/TASC.2018.2820906

Cicéron J, Badel A, Vialle A-J, Pasquet R, Forest F, Chaud X, Tixador P (2018b) Status of the BOSSE project : a 12 T rare earth-BaCuO solenoid used as compact 1 MJ pulse-power SMES. In: Applied superconductivity conference 2018 (ASC'18). Seattle, United States

De Carne G, Morandi A, Karrari S (2022) Supercapacitor modeling for real-time simulation applications. IEEE J Emerg Sel Top Ind Electron 3:509–518. https://doi.org/10.1109/JESTIE.2022.3165985

Deguenon L, Yamegueu D, Gomna A et al (2023) Overcoming the challenges of integrating variable renewable energy to the grid: a comprehensive review of electrochemical battery storage systems. J Power Sources 580:233343

Delille G, Francois B, Malarange G (2012) Dynamic frequency control support by energy storage to reduce the impact of wind and solar generation on isolated power system's Inertia. IEEE Trans Sustain Energy 3:931–939. https://doi.org/10.1109/TSTE.2012.2205025

Duke Energy (2016) Duke energy to put new battery and ultracapacitor system to the test in N.C

Dwyer S, Teske S (2018) Renewables 2018 global status report. In: Reprenewables 2018 global status report

Enterprises ANG (2024) Buses and coaches

Etxeberria A, Vechiu I, Camblong H, Vinassa JM (2010) Hybrid energy storage systems for renewable energy sources integration in microgrids: a review. In: 2010 conference proceedings IPEC, pp 532–537. https://doi.org/10.1109/IPECON.2010.5697053

Farhadi M, Mohammed O (2016) Energy storage technologies for high-power applications. IEEE Trans Ind Appl 52:1953–1961. https://doi.org/10.1109/TIA.2015.2511096

Farivar GG, Manalastas W, Tafti HD, Ceballos S, Sanchez-Ruiz A, Lovell EC, Konstantinou G, Townsend CD, Srinivasan M, Pou J (2022) Grid-connected energy storage systems: state-of-the-art and emerging technologies. Proc IEEE 111:397–420

Filippidis S, Bouhouras A, Poulakis N, Christoforidis G (2021) A review of the cryocooler-based cooling systems for SMES. IEEE Trans Appl Supercond 31:1–13. https://doi.org/10.1109/TASC.2021.3109799

Gee AM, Robinson FV, Dunn RW (2013) Analysis of battery lifetime extension in a small-scale wind-energy system using supercapacitors. IEEE Trans Energy Convers 28:24–33

González-Gil A, Palacin R, Batty P (2013) Sustainable urban rail systems: strategies and technologies for optimal management of regenerative braking energy. Energy Convers Manag 75:374–388

Guan T, Zuo P, Sun S, Du C, Zhang L, Cui Y, Yang L, Gao Y, Yin G, Wang F (2014) Degradation mechanism of $LiCoO_2$/mesocarbon microbeads battery based on accelerated aging tests. J Power Sources 268:816–823

Gyuk I et al (2013) Grid energy storage. Technical report. U.S. Department of Energy

Hatziargyriou N, Milanovic J, Rahmann C, Ajjarapu V, Canizares C, Erlich I, Hill D, Hiskens I, Kamwa I, Pal B et al (2020) Definition and classification of power system stability–revisited & extended. IEEE Trans Power Syst 36:3271–3281

Hawley CJ, Gower SA (2005) Design and preliminary results of a prototype HTS SMES device. IEEE Trans Appl Supercond 15:1899–1902. https://doi.org/10.1109/TASC.2005.849328

He X, Luo B, Deng X, Zhu G, Zhang G, Wang Q (2021) Research on energy management strategy of hybrid energy storage system for electric ship. In: 2021 IEEE 5th conference on energy internet and energy system integration (EI2), pp 4272–4277. https://doi.org/10.1109/EI252483.2021.9713547

Hillers A, Stojadinovic M, Biela J (2015) Systematic comparison of modular multilevel converter topologies for battery energy storage systems based on split batteries. In: 2015 17th European conference on power electronics and applications (EPE'15 ECCE-Europe), pp 1–9. https://doi.org/10.1109/EPE.2015.7309385

Horn M, MacLeod J, Liu M, Webb J, Motta N (2019) Supercapacitors: a new source of power for electric cars? Econ Anal Policy 61:93–103

Hu X, Zou C, Zhang C, Li Y (2017) Technological developments in batteries: a survey of principal roles, types, and management needs. IEEE Power Energy Mag 15:20–31

Ise T, Kita M, Taguchi A (2005) A hybrid energy storage with a SMES and secondary battery. IEEE Trans Appl Supercond 15:1915–1918. https://doi.org/10.1109/TASC.2005.849333
Jia C, Cui J, Qiao W, Qu L (2023) Real-time model predictive control for battery-supercapacitor hybrid energy storage systems using linear parameter-varying models. IEEE J Emerg Sel Top Power Electron 11:251–263. https://doi.org/10.1109/JESTPE.2021.3130795
Jiang W, Ren K, Xue S, Yang C, Xu Z (2021) Research on the asymmetrical multilevel hybrid energy storage system based on hybrid carrier modulation. IEEE Trans Ind Electron 68:1241–1251. https://doi.org/10.1109/TIE.2020.2967723
Karrari S, Vollmer M, Carne GD, Noe M, Böhm K, Geisbüsch J (2020) A data-driven approach for estimating relative voltage sensitivity. In: 2020 IEEE power & energy society general meeting (PESGM), pp 1–5. https://doi.org/10.1109/PESGM41954.2020.9281859
Karrari S, Ludwig N, De Carne G, Noe M (2022) Sizing of hybrid energy storage systems using recurring daily patterns. IEEE Trans. Smart Grid 13:3290–3300. https://doi.org/10.1109/TSG.2022.3156860
Kebede AA, Kalogiannis T, Van Mierlo J, Berecibar M (2022) A comprehensive review of stationary energy storage devices for large scale renewable energy sources grid integration. Renew Sustain Energy Rev 159:112213
Khalid M (2024) Smart grids and renewable energy systems: perspectives and grid integration challenges. Energy Strategy Rev 51:101299
Kheraluwala MN, Gascoigne RW, Divan DM, Baumann ED (1992) Performance characterization of a high-power dual active bridge DC-to-DC converter. IEEE Trans Ind Appl 28:1294–1301. https://doi.org/10.1109/28.175280
Kim J, Gevorgian V, Luo Y, Mohanpurkar M, Koritarov V, Hovsapian R, Muljadi E (2019) Supercapacitor to provide ancillary services with control coordination. IEEE Trans Ind Appl 55:5119–5127. https://doi.org/10.1109/TIA.2019.2924859
Kirby B, Hirst E (1997) Ancillary service details: voltage control (No. ONRL/CON-453). Oak Ridge National Laboratory, Oak Ridge, TN, USA
Krpan M, Kuzle I, Radovanović A, Milanović JV (2021) Modelling of supercapacitor banks for power system dynamics studies. IEEE Trans Power Syst 36:3987–3996. https://doi.org/10.1109/TPWRS.2021.3059954
Lemian D, Bode F (2022) Battery-supercapacitor energy storage systems for electrical vehicles: a review. Energies 15. https://doi.org/10.3390/en15155683
Li J, Gee AM, Zhang M, Yuan W (2015) Analysis of battery lifetime extension in a SMES-battery hybrid energy storage system using a novel battery lifetime model. Energy 86:175–185. https://doi.org/10.1016/j.energy.2015.03.132
Li X, Ma R, Yan N, Wang S, Hui D (2021) Research on optimal scheduling method of hybrid energy storage system considering health state of echelon-use lithium-ion battery. IEEE Trans Appl Supercond 31:1–4. https://doi.org/10.1109/TASC.2021.3117752
Liu Y, Han X, Xing Z, Li P, Liu H, Jiang Z (2024) Research on control strategy of hybrid superconducting energy storage based on reinforcement learning algorithm. IEEE Trans Appl Supercond 34:1–4. https://doi.org/10.1109/TASC.2024.3420314
Lo Franco F, Morandi A, Raboni P, Grandi G (2021) Efficiency comparison of DC and AC coupling solutions for large-scale PV+BESS power plants. Energies 14. https://doi.org/10.3390/en14164823
Luo X, Wang J, Dooner M, Clarke J (2015) Overview of current development in electrical energy storage technologies and the application potential in power system operation. Appl Energy 137:511–536. https://doi.org/10.1016/j.apenergy.2014.09.081
Luongo CA (1996) Superconducting storage systems: an overview. IEEE Trans Magn 32:2214–2223. https://doi.org/10.1109/20.508607
Mao Q, Gong X, Liu S, Shi D, Wang Z, Liu D (2020) Research on hybrid energy storage system with high power density and high energy density. In: 2020 IEEE sustainable power and energy conference (iSPEC), pp 2022–2027. https://doi.org/10.1109/iSPEC50848.2020.9351054

Maroufi SM, Karrari S, Rajashekaraiah K, De Carne G (2025) Power management of hybrid flywheel-battery energy storage systems sonsidering the state of charge and power ramp rate. In: IEEE Transactions on Power Electronics, vol 40, no 7, pp 9944–9956. https://doi.org/10.1109/TPEL.2025.3546013

Milano F, Manjavacas ÁO (2019) Converter-interfaced energy storage systems: need for energy storage; 2. Technical and economic aspects; 3. Energy storage technologies; Part II. Modelling: 4. Power system model; 5. Voltage-sourced converter model; 6. Energy storage system models; Part III. Dynamic analysis: 7. Comparison of dynamic models; 8. Control techniques; 9. Stability analysis; Part IV. Appendices. Cambridge University Press

Mir L, Etxeberria-Otadui I, de Arenaza IP, Sarasola I, Nieva T (2009) A supercapacitor based light rail vehicle: system design and operations modes. In: 2009 IEEE energy conversion congress and exposition. IEEE, pp 1632–1639

Mobility S (2024) Siemens mobility

Mongird K, Viswanathan V, Balducci P, Alam J, Fotedar V, Koritarov V, Hadjerioua B (2020) An evaluation of energy storage cost and performance characteristics. Energies 13. https://doi.org/10.3390/en13133307

Morandi A, Breschi M, Fabbri M, Negrini F, Penco R, Perrella M, Ribani P, Tassisto M, Trevisani L (2008) Design, manufacturing and preliminary tests of a conduction cooled 200 kJ Nb–Ti µSMES. IEEE Trans Appl Supercond 18:697–700. https://doi.org/10.1109/TASC.2008.921285

Morandi A, Fabbri M, Gholizad B, Grilli F, Sirois F, Zermeño VMR (2016) Design and comparison of a 1-MW/5-s HTS SMES with toroidal and solenoidal geometry. IEEE Trans Appl Supercond 26:1–6. https://doi.org/10.1109/TASC.2016.2535271

Morandi A, Anemona A, Angeli G, Breschi M, Della Corte A, Ferdeghini C, Gandolfi C, Grandi G, Grasso G, Martini L, Melaccio U, Nardelli D, Ribani PL, Siri S, Tropeano M, Turtù S, Vignolo M (2018) The DRYSMES4GRID project: development of a 500 kJ/200 kW cryogen-free cooled SMES demonstrator based on MgB2. IEEE Trans Appl Supercond 28:1–5. https://doi.org/10.1109/TASC.2018.2793661

Mukherjee P, Rao VV (2019) Design and development of high temperature superconducting magnetic energy storage for power applications—a review. Phys C Supercond Appl 563:67–73. https://doi.org/10.1016/j.physc.2019.05.001

Nagaya S, Hirano N, Katagiri T, Tamada T, Shikimachi K, Iwatani Y, Saito F, Ishii Y (2012) The state of the art of the development of SMES for bridging instantaneous voltage dips in Japan. Cryogenics 52:708–712. https://doi.org/10.1016/j.cryogenics.2012.04.014

Nguyen HT, Yang G, Nielsen AH, Jensen PH (2018) Combination of synchronous condenser and synthetic inertia for frequency stability enhancement in low-inertia systems. IEEE Trans Sustain Energy 10:997–1005

Nidec Conversion (2024) Supercapacitor energy storage system for an all-electric ferry—case study

Ottonello L, Canepa G, Albertelli P, Picco E, Florio A, Masciarelli G, Rossi S, Martini L, Pincella C, Mariscotti A et al (2006) The largest italian SMES. IEEE Trans Appl Supercond 16:602–607

Pelosi D, Longo M, Bidini G, Zaninelli D, Barelli L (2023) A new concept of highways infrastructure integrating energy storage devices for e-mobility transition. J Energy Storage 65:107364. https://doi.org/10.1016/j.est.2023.107364

Pelosi D, Gallorini F, Alessandri G, Barelli L (2024) A hybrid energy storage system integrated with a wave energy converter: data-driven stochastic power management for output power smoothing. Energies 17. https://doi.org/10.3390/en17051167

Pereira T, Hoffmann F, Zhu R, Liserre M (2021) A comprehensive assessment of multiwinding transformer-based DC–DC converters. IEEE Trans Power Electron 36:10020–10036. https://doi.org/10.1109/TPEL.2021.3064302

Ramirez JM, Murillo-Perez JL (2006) Steady-state voltage stability with StatCom. IEEE Trans Power Syst 21:1453–1454

Ren L, Xu Y, Liu H, Liu Y, Han P, Deng J, Li J, Chen J, Shi J, Tang Y (2015) The experimental research and analysis of a HTS SMES hybrid magnet. IEEE Trans Appl Supercond 25:1–5. https://doi.org/10.1109/TASC.2014.2368271

Rezaei H, Abdollahi SE, Abdollahi S, Filizadeh S (2022) Energy management strategies of battery-ultracapacitor hybrid storage systems for electric vehicles: review, challenges, and future trends. J Energy Storage 53:105045. https://doi.org/10.1016/j.est.2022.105045

Rouholamini M, Wang C, Nehrir H, Hu X, Hu Z, Aki H, Zhao B, Miao Z, Strunz K (2022) A review of modeling, management, and applications of grid-connected li-ion battery storage systems. IEEE Trans Smart Grid 13:4505–4524. https://doi.org/10.1109/TSG.2022.3188598

Saikia BK, Benoy SM, Bora M, Tamuly J, Pandey M, Bhattacharya D (2020) A brief review on supercapacitor energy storage devices and utilization of natural carbon resources as their electrode materials. Fuel 282:118796. https://doi.org/10.1016/j.fuel.2020.118796

Sarojini RK, Kaliannan P (2021) Inertia emulation through supercapacitor for a weak grid. IEEE Access 9:30793–30802. https://doi.org/10.1109/ACCESS.2021.3058951

Sarrias-Mena R, Fernández-Ramírez LM, García-Vázquez CA, Jurado F (2014) Fuzzy logic based power management strategy of a multi-MW doubly-fed induction generator wind turbine with battery and ultracapacitor. Energy 70:561–576

Schoenung SM, Meier WR, Hassenzahl WV (1991) A comparison of large-scale toroidal and solenoidal SMES systems. IEEE Trans Magn 27:2324–2328. https://doi.org/10.1109/20.133683

Shanghai Sunwin Bus Corporation (2024) Shanghai Sunwin Bus Corporation

Skeleton Technologies (n.d.) Superbattery

Song Z, Li J, Han X, Xu L, Lu L, Ouyang M, Hofmann H (2014) Multi-objective optimization of a semi-active battery/supercapacitor energy storage system for electric vehicles. Appl Energy 135:212–224

s.r.l E-C (2024) E-CO electric & hybrid

Sun L, Feng K, Chapman C, Zhang N (2017) An adaptive power-split strategy for battery-supercapacitor powertrain—design, simulation, and experiment. IEEE Trans Power Electron 32:9364–9375. https://doi.org/10.1109/TPEL.2017.2653842

Sutikno T, Arsadiando W, Wangsupphaphol A, Yudhana A, Facta M (2022) A review of recent advances on hybrid energy storage system for solar photovoltaics power generation. IEEE Access 10:42346–42364

Tan Y, Muttaqi KM, Ciufo P, Meegahapola L (2017) Enhanced frequency response strategy for a PMSG-based wind energy conversion system using ultracapacitor in remote area power supply systems. IEEE Trans Ind Appl 53:549–558. https://doi.org/10.1109/TIA.2016.2613074

Technologies S (2024) Supercapacitors in the Rail Industry

Torre WV, Eckroad S (2001) Improving power delivery through the application of superconducting magnetic energy storage (SMES). In: 2001 IEEE power engineering society winter meeting. conference proceedings (Cat. No.01CH37194), vol 1, pp 81–87. https://doi.org/10.1109/PESW.2001.916869

TransnetBW (2024) TransnetBW beauftragt Hitachi energy mit innovativen STATCOM-GFM-Anlagen [WWW Document]. https://www.transnetbw.de/de/newsroom/pressemitteilungen/transnetbw-beauftragt-hitachi-energy-mit-innovativen-statcom-gfm-anlagen. Accessed 22 Oct 2024

Vazquez S, Lukic SM, Galvan E, Franquelo LG, Carrasco JM (2010) Energy storage systems for transport and grid applications. IEEE Trans Ind Electron 57:3881–3895

Vulusala GVS, Madichetty S (2018) Application of superconducting magnetic energy storage in electrical power and energy systems: a review. Int J Energy Res 42:358–368. https://doi.org/10.1002/er.3773

Wang H, Zhou G, Xue R, Lu Y, McCann JA (2020) A driving-behavior-based SoC prediction method for light urban vehicles powered by supercapacitors. IEEE Trans Intell Transp Syst 21:2090–2099. https://doi.org/10.1109/TITS.2019.2912501

Wang B, Zhu S, Cai G, Yang D, Chen Z, Ma J, Sun Z (2024) Sparse measurement-based modelling low-order dynamics for primary frequency regulation. IEEE Trans Power Syst 39:681–692. https://doi.org/10.1109/TPWRS.2023.3256642

Zakeri B, Syri S (2015) Electrical energy storage systems: a comparative life cycle cost analysis. Renew Sustain Energy Rev 42:569–596. https://doi.org/10.1016/j.rser.2014.10.011

Zhang X, Wang W, He H, Hua L, Heng J (2017) Optimization of the air-cooled supercapacitor module compartment for an electric bus. Appl Therm Eng 112:1297–1304
Zhou H, Bhattacharya T, Tran D, Siew TST, Khambadkone AM (2010) Composite energy storage system involving battery and ultracapacitor with dynamic energy management in microgrid applications. IEEE Trans Power Electron 26:923–930
Zhu J, Qiu M, Wei B, Zhang H, Lai X, Yuan W (2012) Design, dynamic simulation and construction of a hybrid HTS SMES for Chinese power grid. Energy. Available Online 10

Hybrid Energy Storage Systems Coupled with Renewable Power Plants for Power Smoothing Applications

Dario Pelosi, Giacomo Alessandri, Federico Gallorini, and Linda Barelli

Abstract The growing renewable generation implies a transition from centralized to de-centralized grid infrastructure, with several low voltage local prosumers taking part as active users within the grid. In this framework, design, planning, modelling, control and optimization of multiple energy systems and grid infrastructures should be developed to manage renewable variability. Therefore, Energy Storage Systems (ESSs) could play a crucial role to provide additional grid flexibility. Grid services, such as power smoothing, load levelling and voltage and frequency sup-port, will play a crucial role to guarantee the safety and stability of future power grids. Nevertheless, single ESS technologies cannot operate over multiple time-scales. To overcome such aspects, implementing different complementary ESSs into Hybrid Energy Storage Systems (HESSs) allows to extend the operating rang-es and the lifespan of the single devices, especially when coupled with very un-predictable renewable generators. Challenging aspects, such as the HESS optimal sizing and power management strategy to maximize the hybridization benefits in power smoothing applications, are analysed in this Chapter. HESS benefits in terms of electric performance at the point of common coupling are also investi-gated. Finally, a techno-economic assessment concerning the benefits of HESS integration for performing wind power smoothing is presented.

Keywords Hybrid energy storage system · Power smoothing · Wind energy · Wave energy · Techno-economic assessment

D. Pelosi · L. Barelli (✉)
Department of Engineering, University of Perugia, Perugia, Italy
e-mail: linda.barelli@unipg.it

G. Alessandri · F. Gallorini
VGA Srl, Deruta, Italy

R. Scipioni et al. (eds.), *Hybrid Energy Storage*, Lecture Notes in Energy 47,
https://doi.org/10.1007/978-3-031-97755-8_6

1 State-of-the-Art

Future electric national and trans-national grids will face a great transition due to the massive introduction of power generation from renewable energy sources. Centralized power generation from fossil fuels will be replaced by decentralized renewable generation, with several low voltage local prosumers taking part as active users within the grid. This contributes to the expected green transition by 2050, although critical aspects on grid stability and safety of provision should be addressed.

In this framework, design, planning, modelling, control and optimization of energy systems should be continuously developed (Ciupageanu et al. 2020). Decentralized grid infrastructures should be connected to manage geographical and temporal renewable variability. It brings to an interconnected microgrids model in which Energy Storage Systems (ESSs) could play a crucial role to provide additional flexibility.

Future energy systems will be characterized by both fluctuating and intermittent renewable generation and highly dynamic loads. Renewable generators, as photovoltaic, wave and wind generators, only provide small or even no contribution to frequency stability (Barelli et al. 2021c, a, b). Such systems contribute to the degradation of the grid voltage stability due to the surplus or shortage of power.

This leads to frequent (and unpredictable) energy imbalances across the grid infrastructure, requiring flexible solutions to be implemented.

Therefore, (Jabir et al. 2017) claims the need of effective intermittent approaches for power smoothing output to increase the power quality at the grid interface.

Among the power smoothing approaches for wind generation, the pitch control of the rotor blades or power curtailment are widely used in wind generation since it is easy to implement and cheap (Lin et al. 2017).

In the view of the expected massive renewable penetration, the loss of grid inertia requires fast response power and energy reserves to maximize the production while maintaining adequate voltage and frequency levels to guarantee grid safety and stability of provision.

In this framework, ESSs represent a noteworthy option to enhance system reliability, being fully controllable and therefore flattering the adaptability of the system to sudden changes (Barelli et al. 2020).

To face with the wide operating ranges required to guarantee high grid quality standards due to the boost in non-programmable renewable generation, each energy storage technology presents intrinsic limits that should be overcome. Hybridisation of complementary ESSs operating over different time scales could be a very promising solution to bypass single storage device limitations (Pelosi et al. 2023), with enhanced overall performance and durability.

Hybrid Energy Storage Systems (HESSs) are therefore widely studied in coupling with renewables for several services, such as power smoothing, voltage and frequency support, as well as load levelling and shedding, peak shaving, etc.

Challenging aspects in the integration of HESS in renewable power plants, include the optimal sizing of the storage devices and their real time power management strategy to maximize the hybridization benefits.

2 Drivers for Installing the Hybrid System

The current electric grids, based on centralized fossil fuel power generation and high voltage transmission with passive users, needs to be renewed in accordance with the rapid boost in renewable energy penetration. This to achieve the stringent targets for climate change by 2050 and 2100. Decentralized grid infrastructures must be properly managed to avoid safety and stability concerns due to the mismatch between the non-programmable generation and load demand.

Moreover, the loss of grid inertia, previously guaranteed by large synchronous generators, due to the great use of electronic power converters, could contribute to weaken the infrastructure. To mitigate these concerns, fast response power reserves will be required.

ESS grid support services represent a promising solution to enhance system reliability, due to their high efficiency, scalability, adaptability to sudden changes and fast response (Pelosi et al. 2024). Moreover, increasing the share of renewable sources integrated in the energy mix at national and transnational level, also the energy function of grid integrated ESS is required. Therefore, the energy storage issue has to be addressed by ESS at multiple temporal scales, without penalizations in cycle efficiency and energy storage device durability. To these regards, hybridization of different and complementary (in terms of key performance indexes) technologies can play a crucial role.

3 Technology Neutral System Requirements for the Specific Use Case

ESSs are generally classified in power intensive and energy intensive technologies. Technologies suitable for short-term energy storage as flywheels and supercapacitors are considered as power intensive technologies, since they are optimized for power fluctuations management. On the other hand, technologies for mid-term and, in particular, for long-term energy storage can be considered as energy intensive. Therefore, also batteries are considered in the following within energy intensive technologies, with reference to their intra-day operation and discharge time in the order of hours. To face with the wide operating ranges required to guarantee high quality standards of provision due to the boost in non-programmable renewable generation, each energy storage technology presents intrinsic limits that should be overcome. Hybridisation of complementary ESSs, each operating over a different time scale, could provide multiple benefits, as the improvement of overall energy performance and cycling tolerance (for device affected by significant degradation depending on cycles number and related operating conditions) (Pelosi et al. 2023). For instance, in (Barelli et al. 2019a) the extension of $LiFePO_4$ battery lifespan is assessed in case of hybridization with a flywheel for a renewable-based microgrid. For the investigated application, battery lifespan is estimated to be extended by more than three times, with respect to the case of not hybrid installation of a battery of equivalent capacity.

Consequently, costs for operation and devices' replacement generally can be reduced, with respect to the case of implementation of single technologies. Moreover, in case the fluctuating component of the power profile managed by the energy storage system is relevant, technologies hybridization allows to reduce installed capacity, with respect to the power. Such advantage is of particular interest in the case power intensive technologies are coupled to storage devices which don't allow the independent sizing of capacity and power, as in the case of closed batteries. On the other hand, installation costs generally increase for HESS considering the need of two energy storage devices at least and more complex and smart converters. Also, power management becomes crucial for the real time optimization of the HESS to maximize potential benefits. Therefore, the profitability of HESS sizing and operation coupled to renewable plants should be assessed in consideration of both CAPEX and O&M costs including devices' replacement in a certain timeframe. It is usually done by applying metrics as the levelized cost of electricity (LCOE), if the economic analysis is performed for the case of RES plant coupled to a HESS, and the levelized cost of storage (LCOS) if only the storage section is considered.

4 How Can Hybrid Energy Storage Meet the Identified Needs?

4.1 Case Studies of HESS Integration in Renewable Plants for Output Power Smoothing

Since the power smoothing problem involves a high time resolution, fast responsive systems are required. Generally, power smoothing studies are approached and simulated in a daily time horizon with a time step in the order of seconds maximum, although the great variability of production from renewable energy should be considered also at the seasonal scale. Therefore, multiple timescales have to be taken into account, covering short-term functionalities typical of power smoothing applications, but also considering the different features of the daily evolution of generated power along the year.

Aiming to the HESS sizing and operation optimization, a very useful approach is based on statistical analysis of the power generation carried out on annual/decade basis. Statistical analysis allows to identify the most recurrent days or weeks over the year/decade, in reference to defined criteria useful to characterize the features of renewable power evolution profiles. Selection criteria are based on parameters, assessed for instance on daily basis, such as average produced power, bandwidth (defined as the difference between maximum and minimum instantaneous power values), bandwidth to average power ratio to highlight a greater or lower relevance of the oscillation component over energy production, evolution of the instantaneous power ramp (defined as the trend of the difference between two consecutive power values with a certain time step, e.g. one second). Relevant criteria include the selection of the days (or weeks) characterized by:

- maximum average power,
- maximum bandwidth,
- maximum mean power ramp,
- maximum bandwidth to average power ratio,
- minimum bandwidth to average power ratio.

Otherwise, for each parameter, the relative frequency can be determined, and days selection can be performed to satisfy a certain threshold in terms of number of occurrences.

Once relevant days/weeks are selected as representative of power generated evolution over a longer timeframe (year/decade), dynamic daily/weekly simulations can be performed with time steps suitable for the power smoothing application (e.g., s or ms). This procedure allows to cover a wide range of values and incidence occurrences, maintaining at the same time a reduced time step.

Another key aspect to maximize systems performance is related to the effectiveness and robustness of power management strategies. Power management and control algorithms should handle RES generation variability, aiming to determine optimal grid operating conditions, support systems resilience and stability and improve the reliability of grid supply. On the other hand, power management has to operate the HESS to maximize its average efficiency and to reduce devices' degradation rate, positively affecting the economic profitability. To this aim it is of main relevance the adoption of new regulations for the remuneration of power and capacity additional flexibility provided by energy storage systems.

In this context, the stochastic control approaches well perform in uncertain environments, overcoming deterministic strategies in achieving objective performances. However, some algorithms imply the use of complex models, which are difficult to be implemented in real-time applications. Several power management algorithms address single-objective optimization problems, although smart energy systems and hybrid architectures require a multi-objective formulation. Answering to these needs, gradient-based optimization algorithms, such as Lyapunov technique and simultaneous perturbation stochastic approximation (SPSA) approach, show the highest potential for real-time power management strategies implementation, also in solving multi-objective formulations. Lyapunov and SPSA can work with no knowledge of the future; hence, not necessarily requiring mathematical uncertainty models and considering forecast errors. Among the gradient based power management strategies, Lyapunov optimization has been widely investigated for the management of several applications; on the other hand, only recently the SPSA algorithm has been proposed for real-time power management. As matter of fact, SPSA presents several advantages, as global optimum determination, easier performance function selection and lower computational burden, make it very suitable for real-time power management applications, such as HESS integration to RES plants in a wide range of installed power. Few case studies of HESS integration to wind and wave generation including

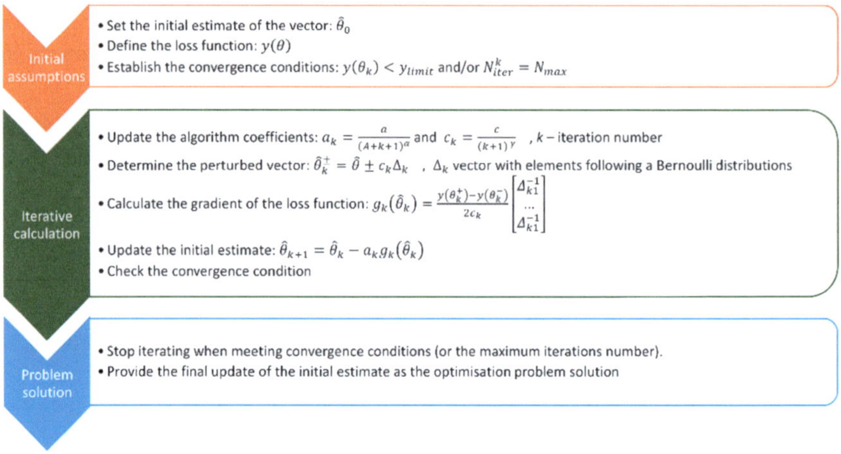

Fig. 1 Schematic flow chart of the SPSA algorithm implementation

SPSA power management strategy for real-time management in the view of power smoothing are illustrated in the following. The implementation of SPSA is illustrated in Fig. 1.

4.2 *Li-Ion Battery-Flywheel HESS for Wind Turbine Power Smoothing*

To enhance wind penetration, integrating HESS into wind turbine or wind farm can be a good solution to mitigate intermittency of production. On the other hand, a trade-off between the installation and maintenance costs of such integrated systems and the maximization of the technical benefits should be investigated to propose a feasible and widely implementable concept. Therefore, a proper sizing in HESS capacity and power and an effective and robust control of coupled storage facilities represent a milestone in obtaining maximum economic and technical benefits (Zhao et al. 2015; Amusat et al. 2018). In the study presented in (Barelli et al. 2020), a HESS consisting of a Li-ion battery coupled to a mechanical flywheel is integrated to a 2 MW wind turbine. The total yearly energy generated in the considered site for the 2 MW wind turbine reaches 4.21 GWh, with a specific energy production of 2105 kWh/kW. According to the developed methodology and statistical analysis of the annual generation data, the flywheel is sized with 21 kWh as nominal capacity and 275 kW as maximum dis-/charge power, while the selected Li-ion battery has 200 kWh capacity and 1C/3C charging/discharging C-rates, respectively. In the HESS, the flywheel accounts for high fluctuating power variations, while the battery performs a smoother charging/discharging profile.

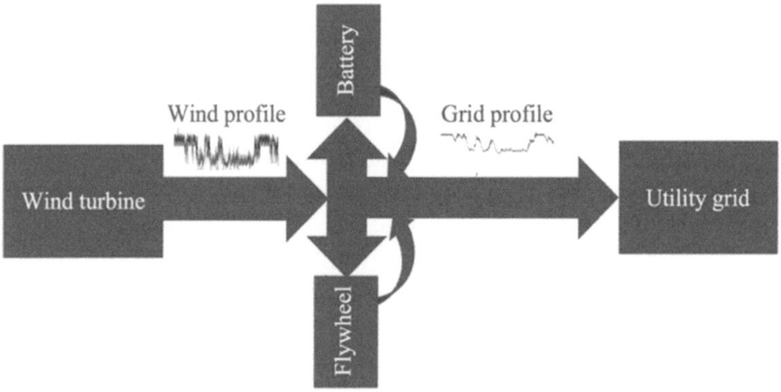

Fig. 2 Layout of the integrated system

The layout of the considered system is depicted in Fig. 2. In detail, this study focuses on the implementation of the SPSA multi-objective power management strategy, aiming at minimizing the power oscillations injected into the grid and, at the same time, provideng a peak-shaving function towards the battery by means of flywheel to increase battery lifespan.

SPSA has been implemented as follows. The initial vector θ is composed of three split factors, $\theta = \left[q_{batt}, q_{grid}, q_{fw}\right]$, representing the shares of Li-ion battery, grid and flywheel, respectively. Then, it is identified the loss function to be minimized according to the selected objectives by means of an iterative process, as illustrated in Fig. 1. In the HESS considered, flywheel provides fast power variations smoothing the profiles managed by the battery. This aims to extend battery lifetime (Barelli et al. 2018, 2019a). The two objectives of the optimization problem are:

1. To smooth the power profile at the interface with the grid (i.e., point of common coupling PCC), modeling the smoothness via the ratio between the power sent to the grid at the current timestep t and the power delivered at the previous instant t − 1 according to Eq. (1).

$$y_1^k(\theta) = \left(\frac{q_{grid} \cdot \Delta P}{P_{grid}^{\mathrm{t}-1}}\right)^2 \tag{1}$$

2. To smooth the power profile managed by the battery, modeling the smoothness as detailed by Eq. (2).

$$y_2^k(\theta) = \left(\frac{q_{batt} \cdot \Delta P}{P_{batt}^{\mathrm{t}-1}}\right)^2 \tag{2}$$

The multi-objective problem is solved as the weighted sum of these two objectives, as Eq. (3), with weights w_1 and w_2.

$$y^k(\theta) = w_1 \cdot y_1^k(\theta) + w_2 \cdot y_2^k(\theta) \quad (3)$$

Details on the statistical methodology for the case study selection and the modeling of the battery and flywheel can be found in (Barelli et al. 2018, 2019a).

Simulations have been performed on a daily basis, for all the representative days selected according to the adopted statistical procedure. The selected daily wind power profiles, covering a wide range of annual occurrences, are depicted in Fig. 3.

Figure 4 illustrates a part of the smoothed power profile at the PCC, proving the effectiveness of the SPSA power management strategy in output power smoothing.

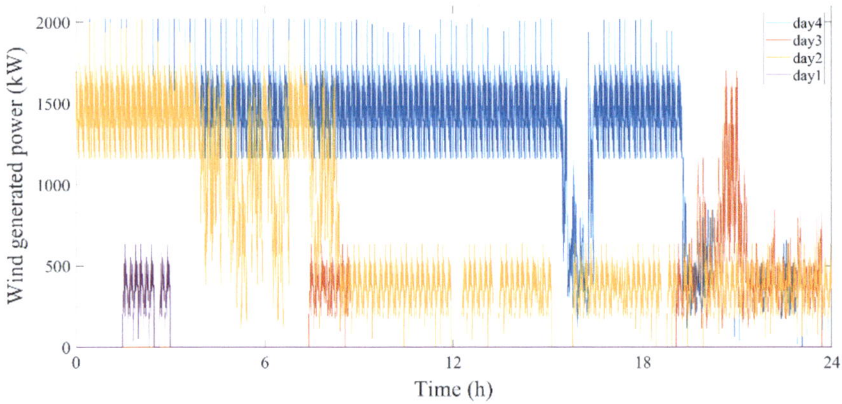

Fig. 3 Wind power profiles selected for the daily simulations

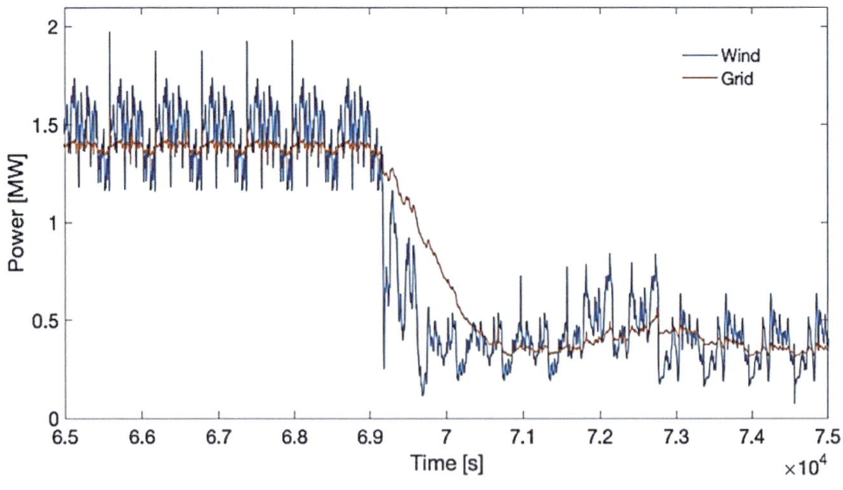

Fig. 4 Comparison of a part of the smoothed power profile injected into the grid (red line) with respect to the generated wind profile (blue line), from (Barelli et al. 2020)

Moreover, to assess the smoothing effect achieved at the PCC, the power ramp is evaluated both at the PCC and at the wind generator terminals and the corresponding cumulative density functions (CDF) are determined. Thus, considering a threshold value of 90% CDF of the power ramp of such profiles, the power ramp reductions at the PCC achieve -83%, assessed as the average value over the investigated days, with respect to the wind power profile thus achieving the first power management objective.

Moreover, by comparing the power profiles at the battery and flywheel terminals in terms of the related power ramp profiles, a reduction of -65% is assessed in terms of the 90% CDF power ramp value managed by the battery with respect to the one processed by the flywheel. The second power management target is achieved too i.e., fluctuating components are managed by the flywheel, while the battery provides energy support. This allows an extension of the battery lifespan, reducing replacement costs during the overall estimated life of the system. Such outcomes highlight the benefits of HESS integration to wind generation and the SPSA algorithm in the control and share of the instantaneous powers among the energy storage devices and the grid.

To verify the robustness of the real-time power management strategy in managing the instantaneous power shares, the developed SPSA algorithm has been also tested on a real 800 kW wind turbine of a different site, placed in Costa Rica (Barelli et al. 2021a). The energy generated on a yearly basis reaches 2.30 GWh, having a specific energy production of 2880 kWh/kW. As concerns the HESS, a 132 kW/44 kWh Li-ion battery has been coupled to a 110 kW/21 kWh mechanical flywheel for output power smoothing. By means of the HESS and the SPSA power management, a power ramp mitigation of at least − 74% is achieved, with an average value of − 80% assessed over the investigated days. Moreover, concerning the flywheel/battery comparison, an average reduction of − 48% is obtained for the power ramp managed by the battery with respect to the flywheel, thanks to the peak-shaving function acting by the flywheel.

Further considerations can be done taking into account the energy penalty (δ) produced by the HESS, defined according to Eq. (4).

$$\delta = \left(1 - \frac{E^{grid}}{E^{wind}}\right) \cdot 100 \tag{4}$$

where $E^{grid}[kWh]$ is the annual energy injected into the grid assessed at the PCC (as the sum of wind energy directly injected to the grid and energy delivered by the HESS are added) and $E^{wind}[kWh]$ is the annual generated energy by the wind plant. Therefore, analysing the two case studies, coupling the proposed HESS to the wind plants, a very small energy penalty of 5.5% is introduced integrating the HESS in the wind generators.

4.3 HESS Integration to Wave Energy Converters for Output Power Smoothing

Since the 1970s, the oil crisis pushed research efforts on studying power production from sea and ocean (Qiao et al. 2020). Physically, waves are produced by the wind blowing across the surface of seas and oceans. Energy from the waves has the highest theoretical potential among the RES (around 29,500 TWh/year) and can totally or almost cover the annual global energy consumption ([CSL STYLE ERROR: reference with no printed form.]; IRENA 2014). Energy from the waves can be exploited by wave energy converters (WECs) to produce electricity (Pecher and Kofoed 2017). Nevertheless, WECs are still in the development phase, lacking mainstream technology predominant on the others. Commercialization issues are mainly related to the difficulty in integrating power from large WECs into the electricity grid due to the intrinsically high variability of the waves (Aderinto and Li 2018). An insight on WEC technologies developed over the years is shown in (Falcão 2010).

To contribute to enhancing the power quality at the PCC for boosting power generation from the waves, two interesting studies are presented in the following. Specifically, (Barelli et al. 2022a) investigated the integration of a $LiFePO_4$ battery/flywheel HESS to a 500 kW WEC operating in grid-connected mode. As previously described, a statistical analysis performed over real wave power profiles relating to three different European sites was carried out. This to identify the most relevant days taking into account wave occurrences and noteworthy operating conditions over the year. Flywheel power and energy was sized at 33 kW and 2.1 kWh, while Li-ion battery capacity ranged from 20 to 24 kWh, with a charge and discharge rate of 1C and 3C, respectively. The SPSA multi-objective real-time power management strategy was implemented in the developed dynamic model to optimize the power shares among the WEC, the HESS and the grid. The same objectives of power oscillations minimization at both the PCC and battery terminals were applied to perform output power smoothing at the PCC, while extending battery lifespan to reduce replacement costs over the expected lifetime of the plant.

The schematic view of the developed dynamic model is depicted in Fig. 5.

Simulation results are illustrated in Fig. 6 for the day characterized by the maximum power bandwidth throughout the year.

In such study, the obtained results show how coupling the HESS to the WEC and performing the output power smoothing by means of the SPSA real-time power management, a reduction of more than − 80% is registered in terms of the power oscillations at the PCC. Moreover, battery solicitation is reduced by more than − 64% with respect to the flywheel one. Power oscillation reduction is expressed in terms of the power ramp values at the 80% CDF. This contributes to extending battery lifetime, also proving the robustness of the developed management strategy in different production sites. Furthermore, the energy penalty produced by HESS integration is about 5%. This could provide great flexibility in increasing power production from the waves, contributing in enhancing grid safety and stability by mitigating almost all the variability of wave energy generation.

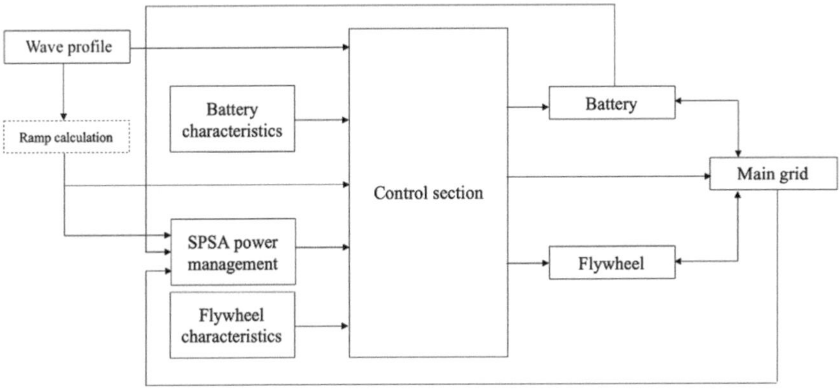

Fig. 5 Schematic view of the dynamic model for the wave generation power smoothing

The application of a different type of HESS i.e., a $LiFePO_4$ battery/supercapacitor hybrid system, was investigated in (Pelosi et al. 2024) for wave energy power smoothing. The WEC is composed by two units each of 250 kWp. The HESS was sized in order to have among all the considered days for the specific site: (i) an average instantaneous percentage reduction of the power rate at the PCC of at least − 70%, (ii) an average instantaneous percentage reduction of the power ramp managed by the battery (in the HESS case) of at least − 25% with respect to the supercapacitor absence. Consequently, as resulting from the sizing procedure, a 65 kW/0.026 kWh supercapacitor is hybridized with a 24 kWh Li-ion battery (maximum dis-/charge powers of 3C/1C respectively), achieving both targeted objectives.

The implemented multi-objectives data-driven power management strategy, based on SPSA algorithm, was implemented to minimize power fluctuations at the PCC and at the Li-ion battery terminals, thanks to the supercapacitor peak-shaving function. The implementing iterative process and the objective functions are the same of the previous studies. This proves the reliability and robustness of the SPSA management strategy in several different applications, also considering different HESS devices.

4.4 Output Power Smoothing: Electric Performance of HESS Coupling to Renewable-Based Generators at the Point of Common Coupling

From an electric point of view, what is the impact of HESS integration into renewable-based plants for output power smoothing at the point of interface with the grid? Generally, renewable power generators interface with the grid through electronics power converters. This to maximize the extraction of the generated power, while being independent from the grid because of the different operating frequencies of the rotor side phase currents with respect to the stator. As example, doubly fed induction

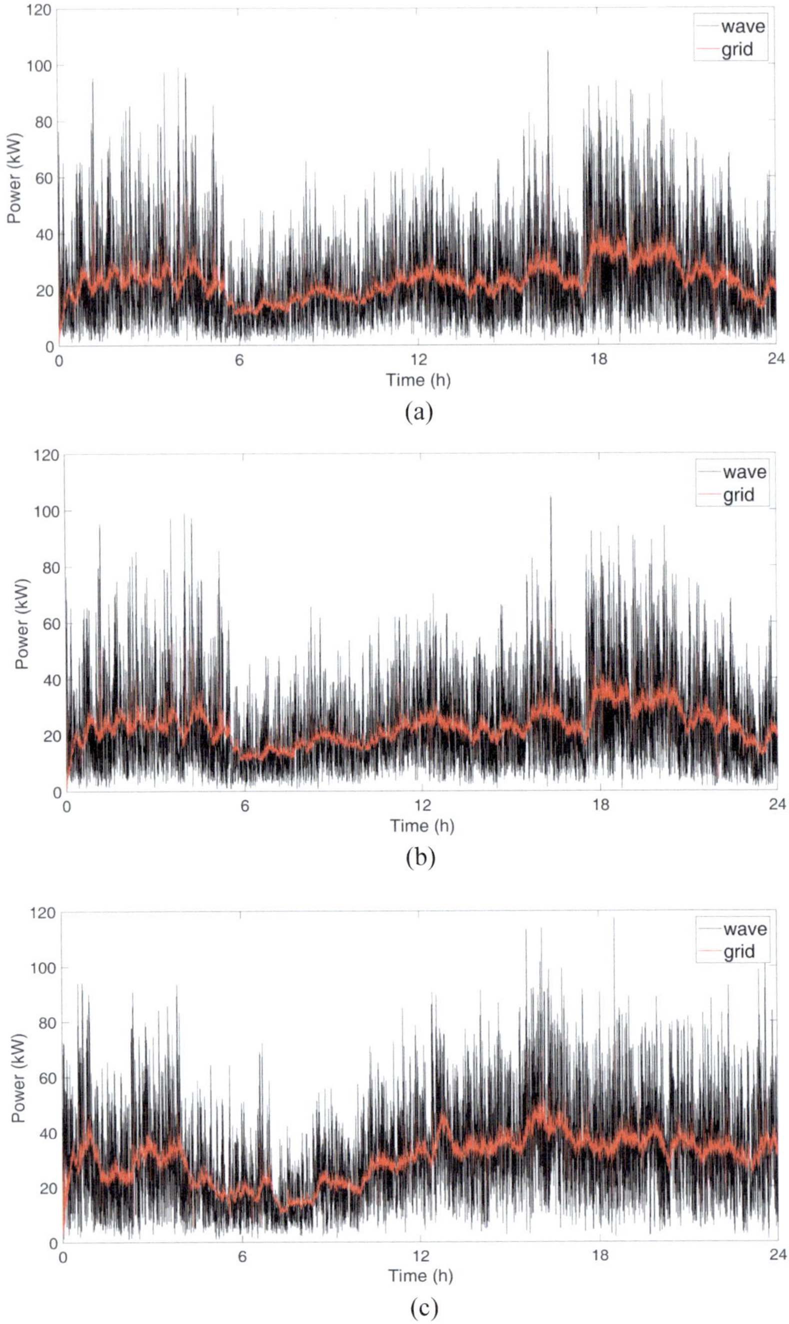

Fig. 6 Original wave profile and smoothed grid profile for all three sites in one of the selected days: **a** site 1, **b** site 2 and **c** site 3

generators (DFIGs) are connected to the grid with both the stator and the rotor, since an asynchronous generator usually provides electric power in correspondence to a narrow speed range slightly higher than the synchronism (up to 2% of the synchronous speed) (Mehta et al. 2014; Edrah et al. 2016). This implies a constant wind speed for the generation. The power converter in such configuration provides the power to the rotor to achieve the synchronism when the wind speed is not sufficient, while absorbs power when the latter pushes the rotor to speeds higher than the allowed one. Such power converter must therefore be bidirectional. Generally, it consists of a grid-side converter interfaced to a generator-side converter by means of a DC bus. Such topology is called back-to-back converter (B2B) and can be applied both to synchronous and asynchronous generators. Maintaining a constant operating value of the DC bus allows to maximize the operation of the grid-side converter, giving a higher quality of the power sent to the grid. Considering this aspect, few studies investigated the introduction of ESSs into the DC bus for voltage regulation and output power smoothing (Kenny and Kascak 2002; Dong et al. 2011; Daud et al. 2014; Brando et al. 2016).

As concerns the electric performance of HESS coupled to renewables, (Barelli et al. 2021c) analyzed the case study of a $LiFePO_4$ battery/flywheel HESS integrated to the B2B DC bus of an offshore permanent magnet synchronous generator (PMSM) wind turbine, with the purpose to prove HESS reliability to reduce active power fluctuation and voltage waveform frequency at the PCC.

Specifically, a MATLAB®/SimPowerSystems model composed by an offshore wind turbine interfaced with the grid through a full-scale B2B converter and a battery-flywheel based HESS connected to the converter DC-link was developed and compared with the case of storage absence. Figure 7 illustrates the model layout of the system considered.

This work provides insights also to validate the effectiveness of the SPSA power management strategy for smoothing the wind power generation, while comparing HESS performance with respect to the case of its absence, in terms of transient time and amplitude of the voltage waveform frequency at the PCC, under severe operating conditions of the wind turbine, determined in terms of the produced power ramp.

The main outcomes were focused on the electrical performance enhancement by means of the HESS during stressful conditions, relating to sudden changes in wind generation at the PCC, in terms of (i) active power injected into the grid and (ii) frequency of the three-phase voltage waveform. These are key parameters in evaluating the increase in grid safety and stability.

Figure 8 depicts the active power at the PCC in case of HESS presence and absence for the step-up (Fig. 8a) and step-down (Fig. 8b) scenarios. It is highlighted that HESS strongly reduces the power ramp when the wind power variation happens for both the cases.

Analyzing the voltage waveform frequency at the PCC depicted in Fig. 9, similar considerations can be made. With HESS integration, the frequency of the three-phase voltage waves is greatly reduced in terms of transient time, although the registered peak values are higher in both cases. Nevertheless, frequencies are very close to the nominal value (i.e., set at 50 Hz) and within the limits imposed by standards of

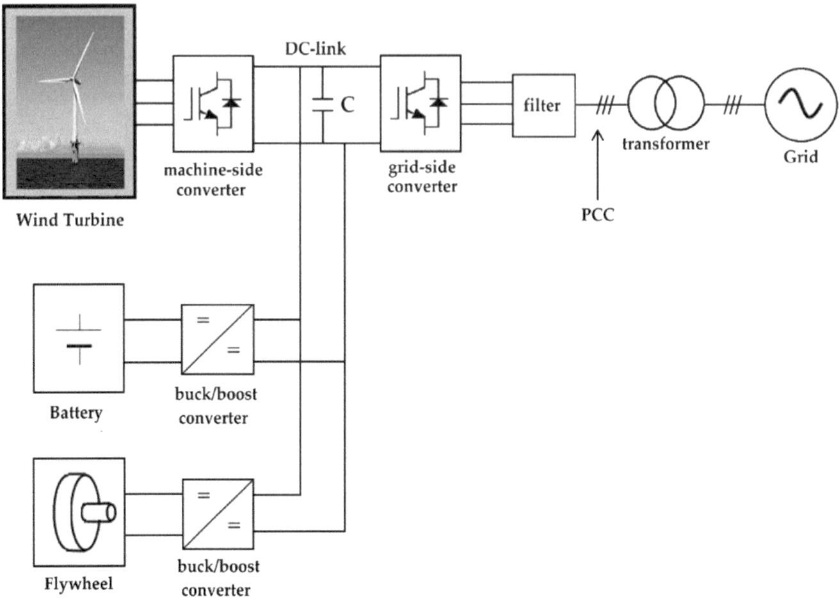

Fig. 7 Layout of the modelled system including flywheel/battery HESS integration

the Institute of Electrical and Electronics Engineers for wind turbines continuous operation (i.e., 47–52 Hz) (Rahman et al. 2021).

It was demonstrated that HESS integration can reduce the active power variation, when the wind power step is applied, to 3% and 4.8% respectively for the two simulated scenarios, in relation to more than 30% and 42% obtained for the no-storage case. Furthermore, HESS contributes to reducing the transient time of the three-phase voltage waveform frequency at the PCC of more than 89% for the analyzed scenarios.

Another noteworthy application study is presented in (Barelli et al., 2022a, b). This study deals with the development of an electrical architecture model, based on a common DC bus topology, which integrates a 24 kWh $LiFePO_4$ battery and a 33 kW/2.1 kWh mechanical flywheel HESS in coupling with a 250 kW WEC. The same purposes of the previous study involving an offshore wind turbine were investigated, in order to assess the voltage and current waveforms frequency and transient behaviour at the PCC introduced by HESS considering specific stressful generation conditions. In detail, the defined simulation scenarios demonstrated a reduction by 64–80% of the voltage wave frequency peak value at the PCC. Moreover, a faster frequency stabilization was obtained thanks to the HESS if related to storage absence, reaching the set value (50 Hz) in a shorter time (by – 10 to – 42%). These case studies demonstrate that HESS integration mitigates stability issues towards the grid, allowing to enhance the expected RES penetration in the next future.

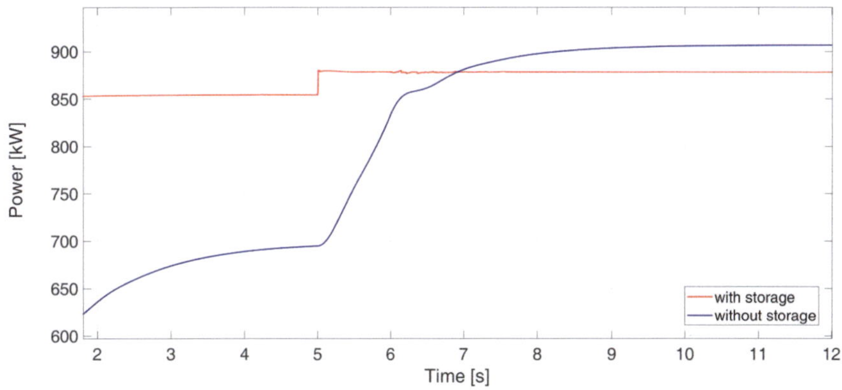

(a)

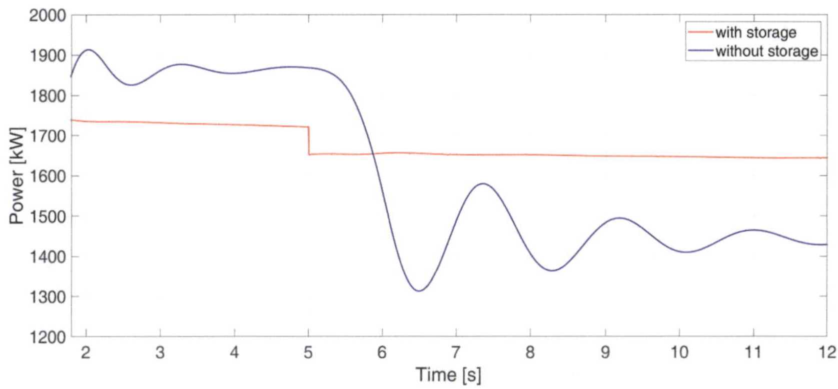

(b)

Fig. 8 Active power exchanged with the grid at the PCC for the: **a** step-up scenario and **b** step-down scenario, in case of HESS presence (red line) and absence (blue line)

5 Technical–economic and Sustainability Analysis for Identification of the Costs of the System and the Benefits Accrued

Although the technical benefits of HESS integration to RES are several, economic feasibility represents a key factor for incentivizing stakeholders to couple HESSs to renewables-based plants. As mentioned before, Li-ion batteries are the most effective storage solution due to their good capacity to power ratio, low installation costs, high efficiency, scalability and fast response. However, Li-ion batteries, like other electrochemical devices, are subjected to degradation, especially when operating at high powers. This yields to a great lifespan reduction, implying high replacements costs in long-term time horizons, as the expected 25–30 years of RES power

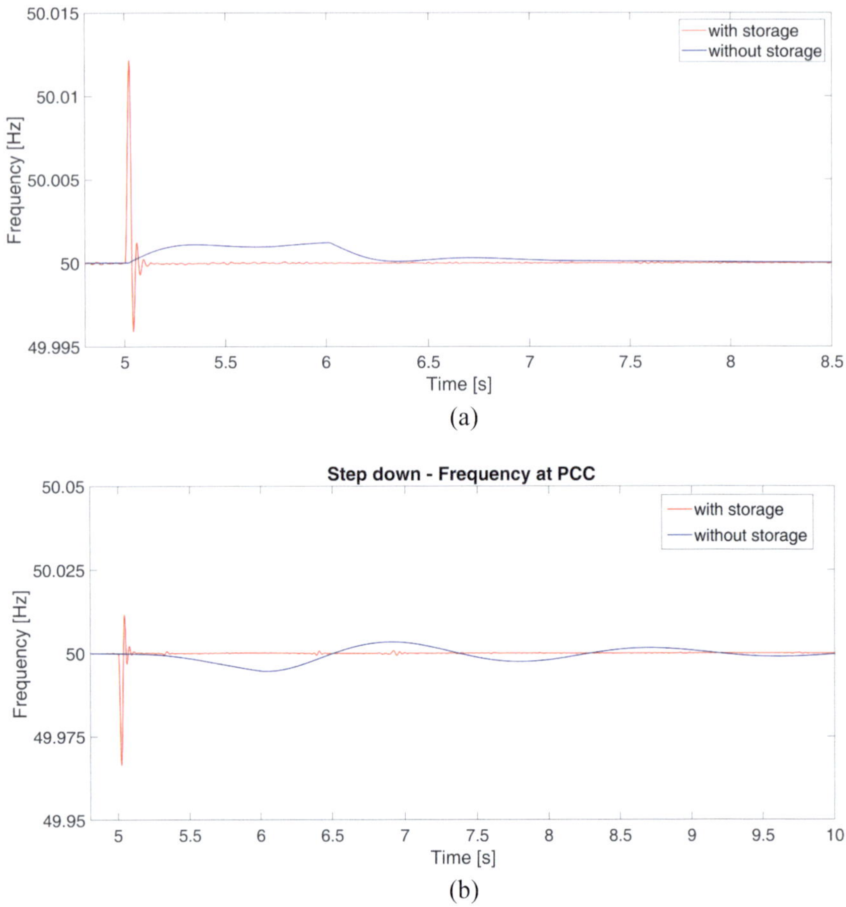

Fig. 9 Frequency of the voltage waveform at the PCC relative to HESS integration (red line) and absence of storage (blue line) in case of **a** step-up scenario and **b** step-down scenario

plants. Coupling complementary power intensive ESSs, such as supercapacitors or flywheels, allows to extend battery lifespan, as widely demonstrated in literature. In this framework, in (Barelli et al. 2021b) the economic analysis of the flywheel/Li-ion battery HESS coupling to a 2 MW wind generator is addressed. The analysis includes also the impact of any revenues in case of making fast response power available to grid support. Economic assessment is evaluated in terms of the levelized cost of electricity (LCOE) of the plant.

LCOE index can be defined as a measure of costs that allows to compare different systems of electricity generation (Lai and McCulloch 2017). Specifically, LCOE corresponds to the minimum cost at which electricity must be sold to achieve a break-even point over the time horizon of the project. International Energy Agency provides an analytical definition of LCOE (Barelli et al. 2019b), as detailed in Eq. (5):

$$LCOE = \frac{\sum_{i=1}^{n} (I_i + M_i + F_i)/(1+r)^i}{\sum_{i=1}^{n} E_i/(1+r)^i} \tag{5}$$

where

- I_i Investment costs in the year i;
- M_i Operation and maintenance costs in the year i;
- F_i Fuel costs in the year i;
- E_i Energy generation during year i;
- r Discount rate;
- n Expected lifetime of the project.

In this study, fuel costs (F_i) were not included since fossil fuel generators were absent.

The case study involves two different HESS configurations, distinguished by different chemistry of the Li-ion battery, i.e. high energy density Nickel-Manganese-Cobalt (NMC) and high-power density $LiFePO_4$ (LFP), each hybridized with a flywheel.

Firstly, dynamic model presented in (Barelli et al. 2020) was updated according to the specifications of the two considered chemistries and simulations were performed to extract battery state of charge. Then, the state of charge evolution of the two batteries is used to assess the battery lifespan by means of the rainflow cycle counting algorithm (RFC), taking into consideration the cycle-to-failure curves of the two chemistries. Estimated lifetimes were introduced into the LCOE routine, considering the replacement costs over the 25 years considered as wind plant time horizon. The two configurations were composed of: (i) a NMC battery of 200 kWh nominal capacity (1C charge/discharge powers) and a 481 kW/21 kWh mechanical low speed flywheel, (ii) a LFP battery of 200 kWh nominal capacity (1C charge/3C discharge powers) and a 315 kW/21 kWh mechanical low speed flywheel. Both are coupled with the 2 MW wind generator, able to produce 4.21 GWh annual energy. Moreover, LCOE analysis was compared for all the investigated configurations, also including the case of economic remuneration by the Italian transmission system operator Terna S.p.A. according the "Fast Reserve" call (2020).

The outcomes highlighted that employing a hybrid storage system in RES plants results economically competitive with respect to the case of wind turbine without storage unit. In particular, the assessed LCOE values were equal to 9.22 c€/kWh and 9.98 c€/kWh for NMC/flywheel and LFP/flywheel HESSs respectively, with respect to 8.73 c€/kWh of the wind turbine without storage. Moreover, if the revenue for grid ancillary services according to the "Fast Reserve" project is included, LCOE results in 8.27 c€/kWh and 8.7 c€/kWh for NMC/flywheel and LFP/flywheel, respectively. In the first configuration (i.e., NMC/flywheel HESS), a LCOE reduction of over 5% was registered, if referred to the case of energy storage absence. Therefore, the resulting analysis provides significant economic feedback aiming to accelerate and increase wind energy production in Italy.

References

Aderinto T, Li H (2018) Ocean wave energy converters: status and challenges. Energies (Basel) 11(5):1–26. https://doi.org/10.3390/en11051250

Amusat OO, Shearing PR, Fraga ES (2018) Optimal design of hybrid energy systems incorporating stochastic renewable resources fluctuations. J Energy Storage 15:379–399. https://doi.org/10.1016/j.est.2017.12.003

Barelli L, Bidini G, Bonucci F, Castellini L, Castellini S, Ottaviano A, Pelosi D, Zuccari A (2018) Dynamic analysis of a hybrid energy storage system (H-ESS) coupled to a photovoltaic (PV) plant. Energies (Basel) 11(2). https://doi.org/10.3390/en11020396

Barelli L, Bidini G, Bonucci F, Castellini L, Fratini A, Gallorini F, Zuccari A (2019a) Flywheel hybridization to improve battery life in energy storage systems coupled to RES plants. Energy 173:937–950. https://doi.org/10.1016/j.energy.2019.02.143

Barelli L, Bidini G, Cherubini P, Micangeli A, Pelosi D, Tacconelli C (2019b) How hybridization of energy storage technologies can provide additional flexibility and competitiveness to microgrids in the context of developing countries. Energies (Basel) 12(16). https://doi.org/10.3390/en12163138

Barelli L, Bidini G, Ciupageanu DA, Micangeli A, Ottaviano PA, Pelosi D (2021a) Real time power management strategy for hybrid energy storage systems coupled with variable energy sources in power smoothing applications. Energy Rep 7:2872–2882

Barelli L, Bidini G, Ciupageanu DA, Ottaviano A, Pelosi D, Gallorini F, Alessandri G, Atcheson Cruz M (2022a) An effective solution to boost generation from waves: benefits of a hybrid energy storage system integration to wave energy converter in grid-connected systems. Open Res Euro 2:40. https://doi.org/10.12688/OPENRESEUROPE.14062.2

Barelli L, Bidini G, Ciupageanu DA, Pelosi D (2021b) Integrating hybrid energy storage system on a wind generator to enhance grid safety and stability: a levelized cost of electricity analysis. J Energy Storage 34. https://doi.org/10.1016/j.est.2020.102050

Barelli L, Cardelli E, Pelosi D, Ciupageanu DA, Ottaviano PA, Longo M, Zaninelli D (2022b) Energy from the waves: Integration of a HESS to a wave energy converter in a DC bus electrical architecture to enhance grid power quality. Energies (Basel) 15(1). https://doi.org/10.3390/en15010010

Barelli L, Ciupageanu D-ADA, Ottaviano A, Pelosi D, Lazaroiu G, Ciupageanu D-ADA, Ottaviano PA, Lazaroiu G (2020) Stochastic power management strategy for hybrid energy storage systems to enhance large scale wind energy integration. J Energy Storage 31. https://doi.org/10.1016/j.est.2020.101650

Barelli L, Pelosi D, Ciupageanu DA, Ottaviano PA, Longo M, Zaninelli D (2021c) HESS in a wind turbine generator: assessment of electric performances at point of common coupling with the grid. J Mar Sci Eng 9(12):1413 (2021). https://doi.org/10.3390/JMSE9121413

Brando G, Dannier A, Del Pizzo A, Di Noia LP, Pisani C (2016) Grid connection of wave energy converter in heaving mode operation by supercapacitor storage technology. IET Renew Power Gener 10(1):88–97. https://doi.org/10.1049/iet-rpg.2015.0093

Ciupageanu D-A, Barelli L, Lazaroiu G (2020) Real-time stochastic power management strategies in hybrid renewable energy systems: a review of key applications and perspectives. Electr Power Syst Res 187:106497. https://doi.org/10.1016/j.epsr.2020.106497

Daud MZ, Mohamed A, Hannan MA (2014) An optimal control strategy for dc bus voltage regulation in photovoltaic system with battery energy storage. Sci World J 08:315–322. https://doi.org/10.1155/2014/271087

Dong B, Li Y, Zheng Z, Xu L (2011) Control strategies of microgrid with Hybrid DC and AC Buses. In: 14th European conference on power electronics and applications, no 10, pp 1–8. https://doi.org/10.1177/13675494030064005

Edrah M, Lo K, Elansari A, Mrehel OG (2016) Impact of DFIG wind turbines on transient stability of power systems. In: Proceedings of engineering & technology (PET). Hammamet, Tunisia, pp 851–866 de Falcão FAO (2010) Wave energy utilization: a review of the technologies. Renew Sustain Energy Rev 14(3):899–918. https://doi.org/10.1016/j.rser.2009.11.003

IEA Data & Statistics. https://www.iea.org/data-and-statistics?country=WORLD&fuel=Energy consumption&indicator=Electricity consumption. Accessed 21 Sep 2020

IRENA (2014) Wave energy technology brief

Jabir M, Illias HA, Raza S, Mokhlis H (2017) Intermittent smoothing approaches for wind power output: a review. Energies (Basel) 10(10). https://doi.org/10.3390/en10101572

Kenny BH, Kascak PE (2002) DC Bus regulation with a flywheel energy storage system. SAE Technical Papers. https://doi.org/10.4271/2002-01-3229

Lai CS, McCulloch MD (2017) Levelized cost of electricity for solar photovoltaic and electrical energy storage. Appl Energy 190:191–203. https://doi.org/10.1016/j.apenergy.2016.12.153

Lin FJ, Chiang HC, Chang JK, Chang YR (2017) Intelligent wind power smoothing control with BESS. IET Renew Power Gener 11(2):398–407. https://doi.org/10.1049/iet-rpg.2015.0427

Mehta B, Bhatt P, Pandya V (2014) Small signal stability analysis of power systems with DFIG based wind power penetration. Int J Electr Power Energy Syst 58:64–74. https://doi.org/10.1016/j.ijepes.2014.01.005

Pecher A, Kofoed JP (2017) Handbook of ocean wave energy. SpringerOpen

Pelosi D, Baldinelli A, Cinti G, Ciupageanu DA, Ottaviano A, Santori F, Carere F, Barelli L (2023) Battery-hydrogen vs. flywheel-battery hybrid storage systems for renewable energy integration in mini-grid: A techno-economic comparison. J Energy Storage 63(March):106968. https://doi.org/10.1016/j.est.2023.106968

Pelosi D, Gallorini F, Alessandri G, Barelli L (2024) A hybrid energy storage system integrated with a wave energy converter: data-driven stochastic power management for output power smoothing. Energies (Basel) 17(5). https://doi.org/10.3390/en17051167

Qiao D, Haider R, Yan J, Ning D, Li B (2020) Review of wave energy converter and design of mooring system. Sustainability (Switzerland) 12(19):1–31. https://doi.org/10.3390/su12198251

Rahman S, Khan I, Alkhammash HI, Nadeem MF (2021) A comparison review transmission mode for onshore integration of offshore wind farms: HVDC or HVAC—a comparison review. Electronics (Switzerland) 10(12):1–15. https://doi.org/10.3390/electronics10121489

TERNA (2020) Regolamento recante i requisiti e le modalità per la fornitura del servizio di regolazione ultra-rapida di frequenza. Regolam Fast Reserv 1–20

Zhao H, Wu Q, Hu S, Xu H, Nygaard C (2015) Review of energy storage system for wind power integration support. Appl Energy 137:545–553. https://doi.org/10.1016/j.apenergy.2014.04.103

Resilient Off-Grid Solutions

Preface

The global shift toward decentralized and renewable energy systems has highlighted the importance of reliable, efficient, and adaptable energy solutions for off-grid communities. Whether in extreme cold climates or milder southern regions, off-grid energy systems must address local challenges such as fluctuating renewable supply, heating or cooling demands, and limited access to grid infrastructure. Third Part of this book focuses on the role of Hybrid Energy Storage Systems (HESS) in overcoming these challenges across diverse environments. By combining complementary storage technologies, HESS enhance system flexibility, resilience, and sustainability. Chapter "Building Resilient Off-Grid Energy Systems: Hybrid Storage Solutions for Cold Climates" explores hybrid energy solutions in cold climates, where extreme temperatures, high heating demands, and minimal solar availability present unique obstacles. The chapter highlights a detailed case study of a dairy farm in Norway, demonstrating how integrated heat, power, and storage systems improve energy reliability and reduce costs and emissions. A comparative analysis assesses fully electric systems against hybrid electric-thermal storage solutions in terms of efficiency, carbon footprint, and energy security. Chapter "Southern Climates: Hybrid Energy Storage for Cooling and Supplying Electrical Load" shifts focus to southern climates, specifically assessing a hybrid battery-hydrogen storage system implemented in a microgrid in Gran Canaria, Spain. This case study evaluates the Power-to-Power concept for integrating photovoltaics in off-grid applications. Finally, Chapter "Benchmarking of Hybrid Thermal and Electrical Storage for Renewable Energy Communities" presents a reference benchmark model for multi-domain hybrid storage systems. It outlines a benchmark-oriented simulation approach, enabling thorough impact assessments and supporting the design of efficient electro-thermal hybrid systems. Together, these chapters provide technical, economic, and environmental insights into how hybrid storage systems can meet the evolving needs of off-grid energy systems in any climate.

Building Resilient Off-Grid Energy Systems: Hybrid Storage Solutions for Cold Climates

Roberto Scipioni, Paolo Marocco, Mari Juel, Hanne Kauko, Kyrre Sundseth, and Massimo Santarelli

Abstract Off-grid energy solutions in cold climates face unique challenges due to extreme temperatures, high heating demands, and limited solar generation during winter months. This chapter explores the role of hybrid energy storage systems in improving energy reliability, efficiency, and sustainability for off-grid applications in such environments. A particular focus is given to a comprehensive case study of a dairy farm, where hybrid heat, power, and energy storage solutions are integrated to meet diverse energy needs. The case study presents a technical, economic, and environmental assessment comparing a fully electric-driven system with a hybrid electric-thermal storage approach. The analysis evaluates system performance in terms of energy security, cost reductions, CO_2 emissions, and overall efficiency.

Keywords Hybrid storage · Off-grid resilience · Remote applications · Electric-thermal integration

1 Introduction to Off-Grid Energy Applications in Northern Europe and the Arctic Region

The move toward sustainable, decentralized energy systems is redefining the energy landscape in Northern Europe and the Arctic region. However, renewable energy systems in these areas face significant challenges due to both intraday and seasonal intermittency. In winter, solar radiation is minimal, particularly above the Arctic Circle, making solar power insufficient during the months when energy demand peaks for heating and electricity. Wind energy, while abundant and complementary

R. Scipioni (✉) · H. Kauko
Department of Thermal Energy, SINTEF Energy Research, Trondheim, Norway
e-mail: roberto.scipioni@sintef.no

P. Marocco · M. Santarelli
Department of Energy, Politecnico Di Torino, Torino, Italy

M. Juel · K. Sundseth
Sustainable Energy Technology, SINTEF Industry, Trondheim, Norway

R. Scipioni et al. (eds.), *Hybrid Energy Storage*, Lecture Notes in Energy 47,
https://doi.org/10.1007/978-3-031-97755-8_7

to solar, also fluctuates, necessitating a hybrid approach to ensure reliability and continuous supply.

Additionally, many areas in these regions are sparsely populated, and existing power grids serving remote communities often lack the capacity to meet the growing needs of a modern, decarbonized society. In the arctic region, many settlements lack access to power grid: Poelzer et al. (2016) identified nearly 1500 off-grid settlements without grid, pipelines and in many cases road connections to other communities. Traditionally, most of these communities have relied on imported fossil fuels but are now transitioning to an increased share of renewable energy. Expanding grid capacity in these areas is frequently costly, time-intensive due to challenging geographies, and can significantly impact the natural environment. As an alternative, integrating local renewable energy production with diverse energy storage technologies, such as batteries, hydrogen storage, and thermal energy storage (TES), can result in a cost-efficient and sustainable solution. These hybrid energy storage systems not only address renewable intermittency but also provide resilience, enabling the decarbonization of remote communities while reducing reliance on costly large-scale grid infrastructure.

1.1 Challenges and Technology-Neutral Requirements for Off-Grid Systems in Cold and Remote Climates

Northern Europe and the Arctic face distinct energy challenges. In Northern Europe, grid access may be limited in remote or rural areas, where long transmission distances and expensive infrastructure expansion make decentralized, renewable-based systems more viable. The Arctic region's challenges are even more pronounced due to extreme cold, long winter nights, and the logistical hurdles associated with fuel transport, especially diesel. The environmental impact of diesel generators in these remote areas is considerable, contributing to greenhouse gas emissions and pollution. Additionally, the dependency on fuel shipments, which are costly and vulnerable to supply disruptions, poses a significant risk to energy security. As a result, transitioning to off-grid renewable solutions, including wind, solar, and biomass, not only reduces carbon emissions but also mitigates the dependency on fossil fuels.

The adoption of off-grid solutions in these regions is driven by the need for energy security, environmental sustainability, and a reduction in fossil fuel reliance and costs. Off-grid systems offer energy autonomy, allowing remote communities and industries to operate independently of centralized grids or fuel imports, which can be unreliable and costly due to long distances and harsh Arctic weather conditions. For example, projects like SmartSenja ("Smart Senja" 2019) and Isfjord Radio ("SINTEF, ZEESA" 2023) in Norway showcase how renewable integration and advanced energy storage technologies can ensure consistent energy supply in remote locations. Additionally, SAFT has recently constructed one of the largest battery parks in Scandinavia, located in Longyearbyen, Svalbard ("Svalbard Energi AS" 2024; "Energy Storage News" 2022).

Off-grid systems in cold climates must meet a set of technology-neutral requirements to ensure reliability (for both power and heating), cost-effectiveness, and sustainability. These requirements define the performance criteria that any off-grid system must achieve, regardless of the specific technologies used. To assess whether these requirements are met, Table 1 presents a set of system- and application-level KPIs that provide measurable benchmarks for evaluating system performance.

One of the most critical KPIs for off-grid energy systems is Loss of Power Supply Probability (LPSP), which quantifies the annual probability of failing to meet electrical demand. Similarly, Loss of Heat Supply Probability (LHSP) measures the annual probability of failing to meet heating demand.

Table 1 List of system- and application-level KPIs for off-grid HESS in cold climates

System-level KPI	Description	Current value (and 2030 target)
Loss of power supply probability (LPSP)	Annual probability of failing to meet electrical demand	< 1%
Loss of heat supply probability (LHSP)	Annual probability of failing to meet heating demand	< 1%
Levelized cost of electricity (LCOE)	Cost of generating electricity per kWh over the system's lifetime, including capital, maintenance, and operational costs	0.15–0.40 €/kWh (0.05–0.1 €/kWh)
Levelized cost of storage (LCOS)	Cost of storing and delivering energy per kWh over the system's lifetime, accounting for capital, maintenance, operational costs, efficiency losses, and degradation	0.25–0.50 €/kWh (< 0.15 €/kWh)
Initial capital cost (CAPEX)	Upfront installation cost per kilowatt of installed capacity, including both energy storage and power generation	1500–2000 €/kW (< 1200 €/kW)
Life cycle greenhouse gas emissions (GHG)	Emissions per unit of energy generated, measured in grams of CO_2 equivalent per kilowatt-hour	150–300 gCO_2eq/kWh (< 50 gCO_2eq/kWh)
Diesel generator usage reduction	Percentage reduction in diesel reliance due to renewable integration	Target: 50–80% reduction

Additionally, economic considerations play a crucial role in the deployment of off-grid systems in remote regions. Careful optimization of both CAPEX and operational costs (OPEX) can help achieve lower LCOE and Levelized Cost of Storage (LCOS) values. The LCOE is a critical metric for evaluating the cost-effectiveness of electricity generation, in this case in hybrid off-grid systems. It includes the total costs of installation, operation, and maintenance of the energy generation system over its lifetime, divided by the total electricity generated. LCOS serves as a key metric for evaluating the cost-effectiveness of energy storage solutions in hybrid remote or off-grid energy storage applications. It covers the total costs of installation, operation, and maintenance of the energy storage systems over their lifetime, divided by the total energy stored and dispatched.

Environmental KPIs are extremely important for the assessment of the installed hybrid system. Government policies across Northern Europe and the EU, such as the EU Energy Efficiency Directive and Renewable Energy Directive, strongly support the transition to sustainable off-grid energy solutions (REGULATION (EU) 2016/1628 OF THE EUROPEAN PARLIAMENT AND OF THE COUNCIL 2016). These policies emphasize the reduction of greenhouse gas (GHG) emissions, the integration of renewable technologies, and improved system efficiency, particularly in remote regions. Nordic nations have further implemented specific measures to minimize diesel generator usage, prioritizing renewable energy as a cleaner and more reliable alternative ("DNV, Maritime Impact" 2024). Coupled with advancements in energy storage and management technologies, these initiatives are making it increasingly feasible to store and utilize energy effectively, even in the most isolated and challenging environments.

1.2 System- and Application-Level KPI Benchmarking

Table 1 groups the most important technical, economic, and environmental KPIs that an off-grid Hybrid Energy Storage System (HESS) should meet for applications in Nordic climates.

Maintaining LPSP and LHSP below 1% is essential to ensure continuous energy supply, even in extreme weather conditions. Advanced energy management systems (EMS) and storage technologies play a crucial role in optimizing energy flow and minimizing the likelihood of unmet demand. For instance, hybrid systems incorporating solar photovoltaic (PV), wind turbines, and thermal energy storage (TES) can address seasonal energy supply challenges effectively.

CAPEX for hybrid off-grid power generation systems typically ranges from $2000 to $5000 per kilowatt (kW), significantly higher than utility-scale, grid-connected systems due to transportation, installation, and operational challenges in remote environments (Stehly et al. 2023). Balance of System (BOS) costs, including site preparation and infrastructure, represent a substantial portion of total CAPEX. For instance, BOS costs in offshore wind projects can account for 22% of total CAPEX, highlighting the logistical burden even for large-scale systems (Stehly et al. 2023).

In remote installations, transportation costs alone can double or triple as a fraction of CAPEX, sometimes representing 20–30% of total costs, influenced by factors like supply hubs distance, infrastructure availability, and specialized equipment needs ("McKinsey and Company" 2023; Nunemaker et al. 2020). Smaller off-grid projects face even higher per-kilowatt costs due to a lack of economies of scale, emphasizing the need for optimized system design and logistics. Energy storage costs are also evolving. In 2022, the average CAPEX for a 4-h lithium-ion battery system was $482/kWh ($1928/kW), with projected declines to $245/kWh by 2030 and $159/kWh by 2050 (Cole and Karmakar 2030). Similarly, PV-plus-battery system CAPEX is expected to decrease 18–49% between 2022 and 2035, reaching $1354 to $842 per kW ("NREL" 2024). While these figures focus on utility-scale systems, off-grid applications face additional complexities such as enhanced storage needs and diverse renewable integration, impacting CAPEX. To ensure the long-term viability of off-grid hybrid energy systems, achieving a CAPEX target below €1200 per kW is crucial. Lowering initial capital costs, particularly for energy storage integration, will play a key role also in reducing LCOE and enhancing system affordability.

Global trends in decentralized off-grid renewable energy systems reveal significant variability in LCOE values. Some systems have achieved costs as low as $0.03 per kWh, while others exceed $1.00 per kWh, reflecting differences in system design, location, and resource availability (Weinand et al. 2023). The average LCOE for 100% renewable energy systems decreased from $0.54 per kWh in 2016 to $0.29 per kWh in 2021, driven by technological advancements and cost reductions (Weinand et al. 2023). Looking ahead, the *Solar Mini Grids* project anticipates that the cost of electricity from mini grids could decrease from an unsubsidized LCOE of $0.38 per kWh to approximately $0.20 per kWh by 2030, thanks to economies of scale and declining costs of major components ("World Bank Group" 2022). For hybrid off-grid systems, achieving an LCOE within the $0.15 to $0.40 per kWh range (Kost 2024) is feasible with well-optimized designs that maximize the share of renewables and minimize dependency on diesel generators. In addition, hybrid systems that integrate solar PV, wind turbines, and thermal energy storage offer cost-effective solutions for addressing both power and heating demands, further enhancing overall system efficiency. LCOS values for hybrid off-grid systems vary widely, depending on storage technologies, system configurations, and usage patterns. For remote or off-grid applications, typical LCOS ranges from €0.15 to €0.35 per kWh depending on the technology and scale (World Bank 2023; Lazard 2021). The Pacific Northwest National Laboratory (PNNL) emphasizes that LCOS in hybrid systems can benefit significantly from optimized energy management strategies. For off-grid systems, hybrid configurations that combine short-term storage (e.g., batteries) with long-term storage (e.g., hydrogen) have shown potential to achieve LCOS values between €0.25 and €0.50 per kWh, depending on storage efficiency and system integration ("PNNL" 2024). Hybrid energy storage systems that integrate batteries for short-term dispatch and hydrogen for seasonal storage offer a cost-effective approach to managing energy variability in remote applications. These solutions, when combined with renewables, minimize fossil fuel reliance and significantly reduce OPEX and overall system costs.

Finally, reducing greenhouse gas (GHG) emissions is a primary goal for off-grid energy systems in Northern Europe and the Arctic. Diesel generators, which have traditionally formed the backbone of off-grid power systems, emit between 800 and 1000 gCO_2eq/kWh ("NREL" 2021; "United Nations—Framework Convention on Climate Change" 2022), contributing significantly to environmental degradation. The reliance on these generators also incurs high fuel costs and poses risks to energy security, especially in remote regions where fuel transport is logistically complex and vulnerable to disruptions. Transitioning to hybrid systems that integrate renewable energy sources and advanced storage technologies offers a pathway to significantly reduce diesel generator usage and associated emissions. Hybrid systems that combine solar PV, wind turbines, and energy storage technologies such as batteries, hydrogen, and thermal energy storage can achieve a 50–80% reduction in diesel generator reliance. In specific cases, such as a renewable power system for an off-grid telescope, deploying a combination of solar PV, batteries, and green hydrogen reduced diesel consumption by an impressive 95% compared to a diesel-only baseline (Viole et al. 2023). These systems not only minimize emissions but also improve energy efficiency and resilience by leveraging locally available renewable resources to meet both power and heating demands. While transitioning to renewable energy systems, the use of alternative fuels like biodiesel and ultra-low sulfur diesel (ULSD) can serve as interim solutions to reduce emissions (Paliza 2012). Biodiesel, which is compatible with existing diesel engines, significantly lowers CO_2 and sulfur emissions. ULSD, with its sulfur content reduced to 10 ppm, further decreases particulate matter and enhances aftertreatment performance, reducing the overall environmental impact of diesel-based systems.

The European Union (EU) and Nordic countries have established stringent policies to accelerate the reduction of GHG emissions and phase out diesel generators in favor of renewable energy solutions. The EU Energy Efficiency Directive (2023/1791) ("DIRECTIVE (EU) 2023/1791 OF THE EUROPEAN PARLIAMENT AND OF THE COUNCIL" 2023) emphasizes the "energy efficiency first" principle, encouraging the adoption of efficient and renewable energy sources. In parallel, the EU has implemented strict emission standards for diesel engines, targeting a significant reduction in their environmental footprint ("COMMISSION IMPLEMENTING DECISION (EU) 2020/728" 2020). Nordic countries, such as Sweden, have taken proactive measures to phase out diesel generators by adopting national targets for energy efficiency and renewable energy ("International Energy Agency" 2019). These policies reflect a strong commitment to reducing fossil fuel reliance and aligning with international climate goals. Achieving GHG emissions below 50 gCO_2eq/kWh is an aspirational target for off-grid systems ("NREL" 2021), aligning with international climate commitments. While this goal is ambitious, especially in remote areas with high energy demands and limited infrastructure, it underscores the importance of transitioning to renewable energy sources and storage solutions. Even if emissions below 50 gCO_2eq/kWh are not immediately feasible in all cases, significant reductions, well below the 800–1000 gCO_2eq/kWh typically associated with diesel-only setups—are attainable with optimized hybrid systems.

2 State-of-the-Art and Future Trends in Hybrid Energy Storage Systems

The variety of off-grid systems in Northern Europe and the Arctic reflects the differing regional needs. Hybrid systems, which combine multiple renewable sources with different storage technologies, are particularly effective in providing reliable power. PV systems remain central to renewable energy strategies, despite the seasonal variability of sunlight. During summer months, PV systems serve as a critical energy source, particularly in Arctic regions, where long daylight hours maximize solar generation. However, during winter, the diminished solar resource limits their standalone effectiveness. Wind energy complements PV systems by providing power during these darker months, particularly in coastal areas and islands where wind resources are abundant. Together, solar and wind energy form the backbone of many hybrid renewable energy systems, providing a balanced and reliable energy supply year-round ("European Commission-Press Release" 2021).

Energy storage technologies are indispensable for hybrid off-grid systems, as they ensure a consistent energy supply despite the variability of renewable sources. To select the right storage technology for a hybrid system, it is crucial to assess key performance indicators (KPIs) such as response time, crucial for power regulation in isolated setups; discharge duration to ensure consistent supply during prolonged periods without generation; safety; sustainability; and economic factors like capital and operational expenditures, as well as levelized cost of storage (LCOS). These KPIs help determine which technologies are best suited to meet the specific energy demands of remote off-grid applications.

The primary categories of storage technologies—electrochemical, chemical, mechanical, and thermal—are mostly required, each offering unique strengths and limitations depending on the KPIs and system needs. While thermal energy storage is critical for meeting heating and cooling needs, the primary solutions for storing electricity are electrochemical, chemical, and mechanical storage technologies, each suited to specific applications based on their unique characteristics and performance metrics.

Electrochemical storage, primarily in the form of lithium-ion batteries, is a well-established solution for short-term energy storage offering round-trip efficiencies exceeding 90%, compact size, and high energy density. Lithium-ion batteries have remarkably improved over the past decade, and projections indicate that by 2030, their performance and cost will improve even further. The cost of lithium-ion battery packs, which was approximately 133 €/kWh in 2023 ("BloombergNEF" 2023), is expected to decrease down to 75 €/kWh by 2030 (60 €/kWh for the cell + 15 €/kWh for the "cell to pack" costs) ("Batteries Europe" 2025). Notably, some Chinese manufacturers have already achieved battery costs at the cell level, in the range of $50–60/kWh for advanced lithium iron phosphate (LFP) batteries ("IDTechEx" 2024), illustrating that global targets for 2030 may be realized sooner than anticipated in certain markets. These cost reductions are driven by improvements in manufacturing processes, economies of scale, and innovations in battery chemistry. In addition to cost, the longevity of lithium-ion batteries is set to improve. Current batteries typically achieve 2000–4000 equivalent full cycles (at 60–80% Depth-of-Discharge, DoD), but advancements in materials and cell design, including the development of solid-state batteries, could extend cycle life to more than 6000 cycles by 2030 ("Batteries Europe" 2025). Concern about the sustainability of raw materials like cobalt and lithium also limit their scalability for larger systems. Enhanced recycling technologies are also expected to play a significant role in reducing the environmental impact of battery production. By 2030, recycling rates for critical materials such as cobalt and nickel are projected to exceed 90%, addressing supply chain risks and reducing dependency on virgin resources ("Batteries Europe" 2025). Applications requiring long-duration storage may benefit from vanadium redox flow batteries (VRFBs), which offer exceptional durability, modularity, and lifespans exceeding 20 years. With efficiency rates of 75%–85%, they are well-suited for large-scale, stationary applications due to their ability to decouple energy storage from power output. By 2030, VRFBs are projected to achieve up to 20,000 full equivalent cycles at 60–80% depth of discharge (DoD) ("Batteries Europe" 2025). However, concerns over vanadium toxicity, high upfront costs, and substantial space requirements remain key challenges. In response, emerging flow battery technologies using more abundant, cost-effective materials with higher energy density and less critical raw materials (CRMs) are currently being developed (Shoaib et al. 2024).

Chemical storage, particularly in the form of hydrogen, is increasingly recognized as a promising solution for long-term energy storage and a versatile energy carrier. Produced via electrolysis using excess renewable electricity, it can be stored in tanks and later converted back into electricity using fuel cells or utilized directly for transport, heating, and industrial applications. While its capacity for large-scale storage and transportability is often seen as a major drawback—given the high costs, complex infrastructure, and safety-related challenges associated with integrating hydrogen into the existing grid (Alexopoulos et al. 2021)—these same characteristics can be an advantage in remote areas where electricity transmission and distribution are either unfeasible or more expensive. In such locations, hydrogen storage and transportability offer a viable alternative for ensuring a reliable energy supply. However, hydrogen storage still requires large volumes, high pressures, or cryogenic temperatures (liquid hydrogen), adding costs and technical complexity to its implementation. Beyond direct use, hydrogen can be further converted into synthetic natural gas (via power-to-gas) by reacting with CO_2 (Tommasi et al. 2024), or into liquid fuels (via power-to-liquid) such as alcohols and dimethyl ether, supporting renewable fuel production (Palys and Daoutidis 2022). Additionally, solar fuels are gaining momentum, with research focusing on natural photosynthesis, artificial photosynthesis, and thermochemical production as alternative methods for sustainable fuel generation (Mitali, Dhinakaran, and Mohamad 2022). While stationary hydrogen storage systems using fuel cells and electrolysers are less mature than batteries, rapid advancements are expected by 2030. The European Hydrogen Strategic Research and Innovation Agenda (SRIA) projects a 50% reduction in CAPEX for electrolysers

(500 €/kW) and fuel cells (900 €/kW) by 2030 ("Clean Hydrogen JU SRIA 2021–2027" 2022). Efficiency improvements are also anticipated, with proton exchange membrane (PEM) fuel cells expected to reach 60–65% efficiency (up from ~ 50%). These advancements will enhance the viability of hydrogen for long-duration energy storage, making it a key enabler for seasonal energy balancing and energy security in renewable systems.

Mechanical energy storage solutions, such as pumped hydro storage, and flywheels, play a crucial role in hybrid systems designed for off-grid applications. Pumped hydro storage, while mature and highly efficient, is typically unsuitable for flat or remote regions due to its dependence on specific geographic conditions. Flywheel energy storage systems (FESS) store energy through rotational kinetic energy and are particularly useful for their durability and rapid response capabilities, essential for meeting immediate energy demands and managing the intermittency of renewable energy sources. Their drawbacks include relatively low energy density and the need for advanced components, such as magnetic bearings and vacuum enclosures, that can complicate maintenance, especially in remote or inaccessible locations (Olabi et al. 2021).

Although thermal energy storage is not a primary solution for storing electricity, it remains essential in hybrid systems to meet heating demands. Hot water storage tanks are a widely spread, low-cost and easily applicable TES solution. Phase-change materials (PCMs) are an emerging technology with higher storage density, with high potential for applications where space is limited, as well as for cold TES. TES systems are highly efficient and can easily be integrated into district heating networks or individual heating systems. However, their suitability depends on the availability of heat sources and the infrastructure for heat distribution, which can limit their application in some areas. Hot water tanks and PCMs are suitable for short-term diurnal storage. For seasonal storage of excess renewable or surplus heat, borehole TES and other underground TES solutions have been developed for off-grid applications, with examples given in Sect. 2.2. Borehole TES stores heat at lower temperatures and requires typically a heat pump for upgrading the heat. In terms of hybridization of TES with other storage technologies, the prospected advancements are less on the technological side, but more on the side of innovative combination, control and monitoring of the hybrid storage solutions. The technologies for the most applicable short-term and seasonal TES, such as hot water tanks and borehole TES, respectively, are well developed. However, tools for optimal design and operation of TES systems when integrated with renewable energy production and other storage technologies are required to avoid over-dimensioning and thus to minimize total system costs, which is critical to promote wider implementation.

2.1 Key Sectors Benefiting from Hybrid Energy Storage

Off-grid hybrid systems in Northern Europe and the Arctic address the energy demands of various sectors, including residential, industrial, agricultural, and aquacultural applications. In small communities and remote Arctic settlements, hybrid systems integrating solar PV, wind energy, and TES ensure autonomy from centralized grids. By providing both power and heating, these systems reduce reliance on fossil fuels, significantly lowering emissions. For industrial operations, such as mining and forestry, hybrid systems deliver the consistent energy required for machinery and heating, while reducing logistical challenges associated with fuel transport (Kalantari and Ghoreishi-Madiseh 2022). Similarly, tourism operations in remote Arctic areas, such as eco-lodges and expedition camps, rely on clean, renewable energy to align with sustainability goals and attract environmentally conscious travelers. By integrating renewable energy with storage solutions, these systems not only reduce operational costs but also minimize emissions, enhancing their ecological compatibility.

Agriculture in Nordic regions benefits from hybrid systems that integrate renewable energy with storage technologies. Greenhouses, essential for year-round cultivation, rely on stable heating and electricity to maintain optimal growing conditions. TES systems ensure surplus renewable energy is stored as heat, enabling efficient greenhouse operation even during the coldest months. Additionally, agricultural waste can be converted into biogas, creating a sustainable energy loop that provides power and heat while reducing waste. Fish farming, a vital industry in Nordic countries, also benefits from hybrid off-grid systems. Fish farms require consistent power for oxygenation, feeding systems, and temperature regulation, all of which are critical for maintaining healthy aquatic environments. Hybrid systems with battery storage and TES ensure uninterrupted operations even during periods of low renewable energy production. There is also potential for collecting and utilizing the oxygen produced as a biproduct in water electrolysis for hydrogen production for aquaculture applications (Mohammadpour et al. 2021). By reducing dependency on diesel generators, these systems significantly lower operational costs and environmental impact, aligning with the sustainability goals of the aquaculture industry ("UN Global Compact" 2022).

2.2 Common Hybrid Configurations for Off-Grid Applications

Battery and Hydrogen. Off-grid systems based on both hydrogen and batteries can result in the cheapest system configuration since they can rely on the high efficiency of batteries for short duration storage and the relatively lower cost per unit of stored energy from the hydrogen tanks for long-duration and seasonal storage (Marocco et al. 2023). Figure 1 shows how the Levelized Cost of Hydrogen (LCOH) changes

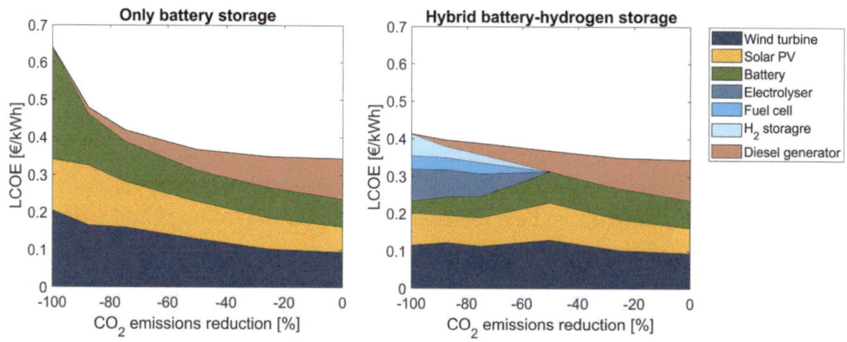

Fig. 1 Hydrogen for off-grid sites (Froan island in Northern Europe): Levelised Cost of Electricity (LCOE) as a function of CO_2 emissions reduction (compared to current scenario with only diesel). Modified from (Marocco et al. 2022)

when carbon dioxide (CO_2) emissions are reduced compared to the actual diesel-based scenario at the Froan island (Marocco et al. 2022). In particular, hydrogen becomes essential in transitioning to configurations with high RES penetration, while maintaining low electricity production costs. Indeed, a system relying solely on batteries (Fig. 1, left) would incur higher costs than the hybrid hydrogen-battery case (Fig. 1, right) due to the significant oversizing required for both renewable generators and batteries.

Trapani et al. (Trapani et al. 2024) investigated electrification solutions across 138 remote sites in Norway, concluding that hydrogen is crucial for preventing the oversizing of renewable generators and batteries while ensuring long-term storage capacity. Richards et al. (Richards and Conibeer 2007) also demonstrated that the economic viability of hydrogen-based technologies improves in hybrid energy storage systems located at more extreme latitudes, where the seasonal variation in solar radiation is relevant. Additionally, the potential of wind-hydrogen plants was investigated in the off-grid Arctic communities of Grimsey (Iceland) (Chade, Miklis, and Dvorak 2015) and Mykines (Faroe Islands) (Enevoldsen and Sovacool 2016). Both studies found that the long-term hydrogen storage capacity is essential for improving wind power utilization and addressing the variability and seasonality of the electrical load and renewable production.

Batteries and thermal energy storage. Space heating and hot water production take up a major share of the total energy use in European households: 64% and 15%, respectively (Eurostat 2024). TES is therefore crucial to reduce peak energy demands and increase energy system flexibility. Hot water demand has a regular, diurnal variation throughout the year, and hot water storage tanks are readily available in most houses. Active use of these storages is hence a cost-efficient way to store renewable energy. Space heating demand, on the other hand, has a seasonal variation, which raises the need for larger, seasonal TES (STES) solutions.

The feasibility of different TES options for off-grid energy systems depends on (1) the share of the thermal energy demands of the total energy demand as well as

(2) the variability in these demands on a diurnal and a seasonal level. Hot water tanks are a low-cost option for short-term storage, and can significantly reduce the required battery capacity, as was shown by e.g. (Häring et al. 2019). STES systems can improve the energy-independency of off-grid energy systems, but are associated with high investment cost and thermal losses, and require an abundant, low-cost heat source, such as solar energy or industrial excess heat. A prerequisite for utilizing a STES system is a waterborne heat distribution system, in buildings and within the area.

At Svalbard, Store Norske has reduced the CO_2 emissions from their off-grid energy system at Isfjord Radio by 70% by adding PV, battery and TES to the originally diesel driven energy system ("Store Norske" 2023).

A new neighborhood in Trondheim, Norway, will have an integrated energy system consisting of a local heating network, STES utilizing waste-heat and a seawater-based heat pump, as well as hot water TES tanks and PV at the buildings. Kauko et al. (2024) carried out a simulation study showing that a low (45 °C) distribution temperature in the local heating network is essential for efficient utilization of local energy resources and STES. The local TES units at the buildings were shown to be efficient in absorbing daily variations in PV production and electricity prices, reducing the demand for batteries.

Batteries, hydrogen, and thermal energy storage. Hybrid energy storage systems integrating batteries, hydrogen and TES are a less studied combination that could efficiently improve the energy-independency of remote communities in cold climates. Kalantari et al. (2022) proposed a stand-alone net zero energy system for open-pit mining operations combining local wind power production with different configurations of energy storage in hydrogen, batteries and rock-pile TES. TES combined with an electric heater was included in all configurations to store surplus wind energy and supply heating demands. Hybrid battery / fuel cell storage and a hydrogen-powered fleet was shown to be the most profitable alternative for the mine.

Electrolyzers and fuel cells generate heat during operation, which can be recovered to meet part of the thermal demand of the end-user (van der Roest et al. 2023). Puranen et al. (2021) assessed the feasibility of a PV-based hydrogen-battery system with a heat pump-based heating system under northern climate conditions, suggesting that recycling fuel cell waste heat could reduce electricity consumption during winter. This, in turn, could allow for a reduction in the size of the hydrogen storage system.

Sun et al. (2022) simulated an off-grid zero energy building with a hydrogen, battery and TES system for two cities in Iran with different climates. The electricity supply was based on rooftop-integrated PV panels or wind turbines. Heating and cooling were supplied with a heat pump/chiller, and domestic hot water demands were covered with a solar collector, with excess heat from the fuel cell and electrolyser as an auxiliary heat supply. In both cases, the local energy system was able to satisfy all the energy demands over a year, with the required configuration and capacity of PV panels, solar collectors and wind turbines depending on the site.

To summarize, while batteries together with diurnal and seasonal TES can enable year-round off-grid operation at least in terms of thermal energy demands, the combination with hydrogen storage is often required to be able to cover the electricity specific demands as well, or in industrial applications for fueling heavy machinery. Moreover, the combination of hydrogen storage with TES and batteries is crucial for reducing the capacity required for hydrogen production and storage, as will be shown in Sect. 3.

2.3 Projected Technological Advancements

The future of off-grid HESS will be shaped by advancements in energy management systems (EMS), real-time control strategies, and high-efficiency energy conversion and storage. As renewable energy penetration increases, ensuring integration of diverse storage and generation technologies requires intelligent control architectures capable of dynamic optimization. The complexity of hybrid controls grows exponentially as multiple energy storage and conversion technologies—electrochemical, chemical, thermal, and mechanical—are hybridized to improve resilience and efficiency. Developing robust, adaptive control strategies is critical to enabling fully autonomous off-grid energy systems.

Advancements in EMS will be at the core of future HES, enabling real-time decision-making through predictive analytics, resource forecasting, and adaptive dispatch. Leveraging artificial intelligence and machine learning will improve both the development and real-time operation of hybrid controls, enhancing energy flow optimization and system resilience (Murphy and Mills 2021). AI-driven EMS can dynamically balance supply and demand across multiple time scales, incorporating weather predictions, load variability, and market conditions to refine dispatch strategies. This capability is especially critical for long-duration storage applications, where optimizing storage cycling and minimizing degradation directly impact performance and cost-effectiveness. Integrating advanced computational methods will further enhance cyberphysical control strategies, ensuring stable system operation across varying load conditions (Murphy and Mills 2021). Smart grid-compatible EMS will allow off-grid HES to operate efficiently, whether in islanded mode or within larger decentralized microgrids. Grid-forming and grid-following capabilities will be essential for coordinating the interactions between inverters, storage, and generation assets, maintaining frequency stability, and preventing system imbalances. Additionally, microgrid controllers must evolve to support hybrid configurations, allowing smooth transitions between different power sources and enabling modular scalability. Resource forecasting will play a crucial role in next-generation HES controls (Murphy and Mills 2021; Yuan, Kocaman, and Modi 2017). By integrating real-time data analytics with EMS, hybrid systems can proactively adjust energy storage and dispatch strategies to accommodate fluctuations in renewable generation. Improved forecasting of solar and wind availability will reduce reliance on backup generation, enhancing overall system efficiency. For long-duration storage solutions, predictive algorithms will optimize the timing of energy conversions, such as power-to-gas or thermal storage utilization, to align with expected energy demands and available resources.

Field demonstrations and pilot projects will be essential for validating these control advancements (Murphy and Mills 2021; Blackburn et al. 2020; US Dept of Commerce 2022). Real-world testing in remote, islanded, and disaster-prone regions will provide critical insights into system behavior, allowing for iterative improvements in EMS, forecasting algorithms, and cyberphysical security measures. By 2030, these innovations will make off-grid HESS more intelligent, cost-effective, and resilient, ensuring reliable and sustainable energy access in areas where traditional grid infrastructure remains impractical (Murphy and Mills 2021).

3 Toward Sustainable Off-Grid Farming: A Case Study on Hybrid Heat, Power, and Storage Solutions for Comprehensive Dairy Farm Energy Needs

The energy demands of modern dairy farming extend beyond basic electricity, including significant heating needs for water, milk processing, and maintaining comfortable conditions for livestock. Addressing these demands sustainably is critical for reducing greenhouse gas emissions and operational costs, while enhancing resilience in rural agricultural operations. This case study explores the implementation of a hybrid energy system tailored to meet the comprehensive heat and power needs of a dairy farm, integrating renewable sources such as solar and wind with thermal energy storage (TES) and advanced battery and hydrogen systems.

The significance of this approach goes beyond agriculture, as it also serves as a model for rural housing and community energy solutions. By demonstrating how integrated hybrid systems can simultaneously power homes and farming operations, this case highlights the potential for scalable, sustainable energy solutions in agricultural settings.

3.1 Case Study Description

Reference case study. The case study under analysis is a remotely located milk farm in Rye, Norway. Within the framework of the REMOTE project ("REMOTE Project Official Website" 2018), a PV system of 86.4 kW and a wind turbine of 225 kW have been installed to power the farm. Moreover, in order to enhance the exploitation of the local solar and wind energy, a hybrid hydrogen-battery storage system was implemented and operated. As shown in Table 2, a Li-ion battery of 550 kWh capacity (5 racks of 110 kWh) was selected to provide short-term energy storage. Hydrogen-based technology (including electrolyzer, hydrogen storage and

Table 2 Sizes of the components of the renewable hydrogen-battery system installed in Rye in the framework of the REMOTE project

Component	Size
PV system	86.4 k W
Wind system	225 kW (1 turbine)
PEM electrolyzer	50 kW
PEM fuel Cell	100 kW
Hydrogen storage	3333 kWh or 100 kg (max. 30 bar)
Li-ion battery storage	550 kWh
Biofuel generator	48 kW

fuel cell) was instead adopted to achieve cost-effective longer-term storage capability. Specifically, a 50-kW proton exchange membrane (PEM) electrolyzer, with operating pressure of 30 bar, was chosen for the production of hydrogen based on excess renewable energy. The PEM typology was considered given its maturity, wide modulation range and good dynamic behavior, making it very suitable for integration with fluctuating renewable energy production. A PEM fuel cell was also considered for the reconversion of hydrogen into electricity when the wind/solar generator and the battery are not sufficient to cover the electrical demand. It should be noted that the maximum pressure of the hydrogen storage tank (100 kg capacity) was set at 30 bar, thus avoiding the need for a compressor to store the hydrogen coming from the electrolyzer (where hydrogen is already generated at 30 bar pressure). Additionally, a biofuel generator was integrated as a final back-up unit to make the site entirely off-grid. Pictures of the renewable hydrogen-battery storage system in Rye are shown in Fig. 2, while Fig. 3 presents a map of the installation location.

Extended Case Study. Observations from the case study indicate that most of the site's energy demand arises from heating requirements rather than electricity consumption. To build upon the reference case study, the extended system design incorporates thermal energy storage to address the significant share of energy demand associated with heating requirements, such as hot water and space heating. The extended hybrid energy storage system is a conceptual design, developed solely to explore the potential for future installations.

The incorporation of TES directly targets this thermal demand, offering a more energy-efficient, cost-effective, and sustainable alternative. By storing thermal energy in a hot water tank, the need for energy-intensive electricity storage and conversion processes is reduced. Additionally, systems equipped with surplus heat recovery from hydrogen-based technologies, further minimize energy waste and operational costs.

This chapter lays the foundation for a detailed comparison between fully electric and hybrid electric-thermal system configurations, analyzing their energy efficiency, environmental impact, and economic performance to identify the optimal solution for meeting the site's combined thermal and electrical demands.

Fig. 2 Installed hybrid energy storage system in Rye, Norway. Photographs courtesy of Bernhard Kvaal

3.2 System Configurations and Control Strategy

This case study explores three distinct configurations of a hybrid energy system, each designed to balance electrical and thermal energy demands under varying conditions. The configurations range from a simple, fully electric setup to advanced systems integrating heat recovery for enhanced efficiency. These configurations are analyzed to evaluate their performance, energy efficiency, and environmental impact.

Configuration 1 *Electric-Driven Hybrid System*: this system exclusively relies on electricity for meeting both power and thermal energy needs. In this setup, renewable energy sources, such as PV and wind turbines, directly supply electricity for electric specific uses and space heating, as well as to power an electric boiler that converts electricity into hot water. Excess electricity generated during periods of high renewable availability is stored either in batteries for direct electrical use or

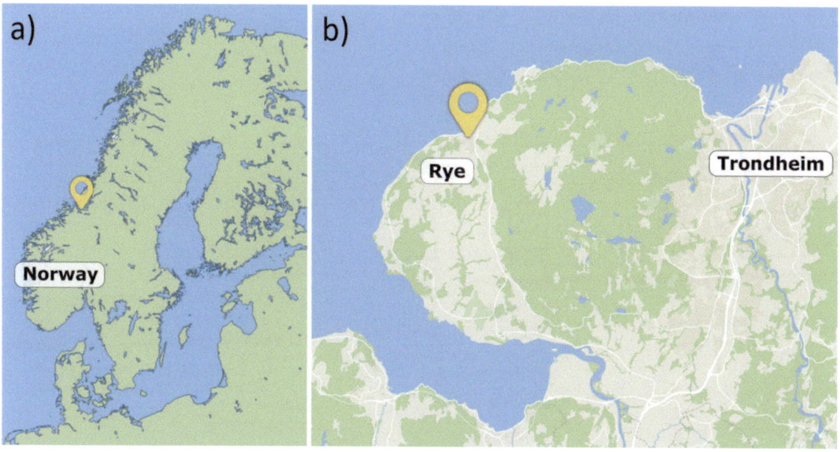

Fig. 3 Location of the installed HES system in **a** Norway (Made with Natural Earth. Free vector and raster map data @ naturalearthdata.com); and **b** zoomed view of Rye, Trondheim (© OpenStreetMap contributors. Data licensed under the Open Database License, via MapTiler map tiles)

converted into hydrogen via electrolysis, which acts as a fuel for long-term storage. This configuration offers a simple, fully electric solution for integrating thermal and electrical energy (Fig. 4).

Configuration 2 *Heat Pump-Driven Hybrid System*: this configuration also relies on renewable electricity but adds a heat pump for more efficient thermal energy production. Excess thermal energy is stored in a hot water tank, providing heat when renewable energy is scarce. Renewable electricity powers the heat pump and other electrical loads, with surplus stored in batteries or converted to hydrogen for long-term use. This configuration efficiently combines electrical and thermal energy storage and production (Fig. 5).

Configuration 3 *Advanced Heat Recovery Hybrid System*: building on the heat pump-driven setup, this configuration incorporates a heat exchanger to recover

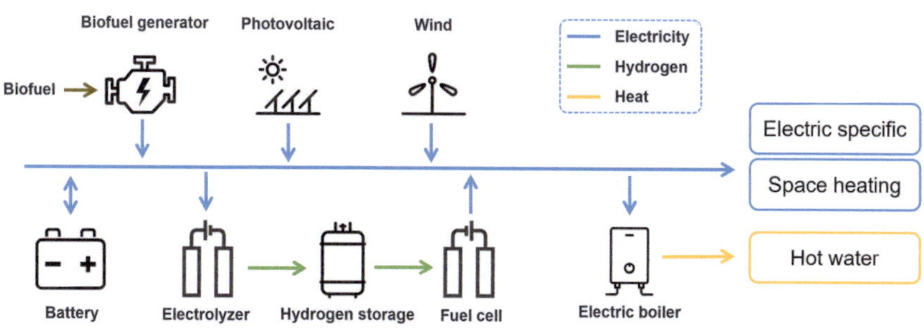

Fig. 4 General layout of the electric-driven hybrid system. Similar to the configuration of the hybrid renewable-based P2P system installed in Rye in the framework of the REMOTE project

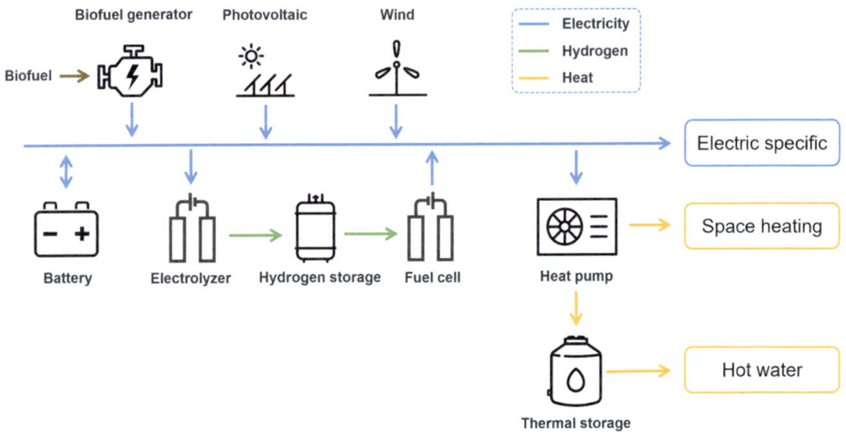

Fig. 5 General layout of the heat pump-driven hybrid system

surplus heat from the fuel cell and electrolyzer, trying to maximize system efficiency by utilizing the extra heat for space heating and hot water. Heat recovery from the biofuel generator was not included due to its more intermittent operation compared to the fuel cell and electrolyzer, which provide a more stable and predictable heat source; additionally, the generator is intended for minimal use only as a backup in case of power supply shortages and is planned for future phase-out (Fig. 6).

Control Strategy. The control strategy of the hybrid energy storage system is designed to efficiently balance the renewable energy supply with electrical and thermal energy demands, while minimizing reliance on external energy sources and

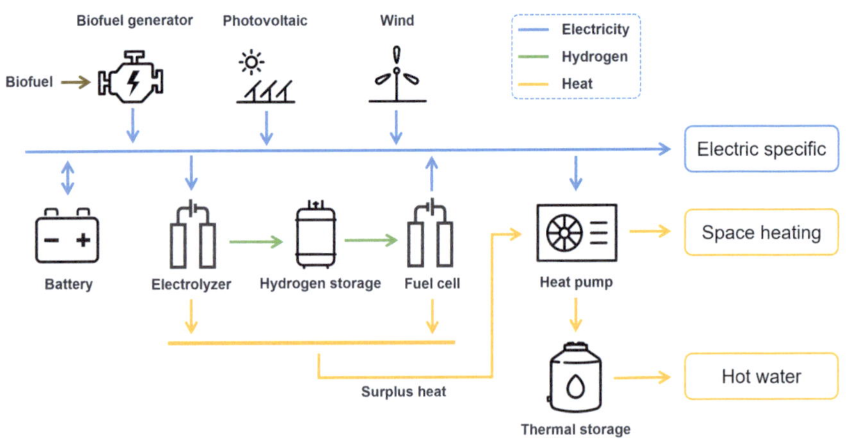

Fig. 6 General layout of the advanced heat recovery hybrid system

optimizing energy storage utilization. The system dynamically manages the distribution of energy between various components (renewable energy sources, energy storage units, and loads) on a timestep basis.

Dynamic Load Allocation. The first step of the control strategy consists in evaluating, at a given time, the net power balance, calculated as the difference between the renewable energy output and the total energy demand. This balance determines whether the system experiences a surplus (renewable energy exceeds demand) or a deficit (renewable energy is insufficient to meet demand). The energy demand includes both electrical loads and thermal loads, such as space heating and hot water requirements. The control strategy dynamically allocates renewable energy between electrical and thermal loads, optimizing energy distribution based on the availability of resources and system priorities. Proportional redistribution of surplus or energy deficit ensures a balanced approach to meet all energy demands.

In deficit scenarios, where renewable energy is insufficient to meet the load, the control strategy prioritizes available energy sources based on a combination of availability, thermal and electrical efficiency to minimize disruptions:

1. **Allocation of Renewable Energy**. Renewable energy is initially allocated to meet both electrical and thermal demands. If heat recovery from previous operations, such as surplus heat from the fuel cell or the electrolyzer, is available, it is utilized to reduce the thermal load on renewables and improve efficiency.
2. **Deficit Distribution**. When a deficit persists, the system calculates the shortfalls in electrical and thermal demands separately. The available renewable energy is distributed proportionally to the energy demands.
3. **Battery Discharge**. The battery is the first energy storage system utilized to cover the remaining deficit. The control logic ensures that electrical and thermal deficits are addressed proportionally to their size, optimizing the battery's usage and ensuring balanced load coverage.
4. **Hydrogen Storage and Fuel Cell Operation**. If the battery cannot fully cover the deficit, the hydrogen storage system is activated. Through the fuel cell, stored hydrogen is converted into additional power. Surplus heat generated during fuel cell operation is recovered, when possible, and directed toward reducing thermal load demands.
5. **Biofuel Generator**. In extreme deficit scenarios, when neither the battery nor hydrogen storage can meet the demand, the biofuel generator is used as a last resort to ensure uninterrupted energy supply.

When renewable energy supply exceeds the load, the system manages the surplus energy to maximize its utilization:

1. **Battery Charging**. The surplus energy is first directed to charge the battery until it reaches its maximum capacity.
2. **Hydrogen Production via Electrolyzer**. Any remaining surplus energy is used to produce hydrogen via the electrolyzer.
3. **Heat Pump or Electric Boiler Operation**. Once the battery is fully charged and the hydrogen tank is refilled, any further surplus energy is directed toward

heating water to its maximum temperature using either a heat pump or an electric boiler.

4. **Curtailment**. If surplus energy cannot be stored or utilized, it is curtailed as a final measure.

Thermal Load Management and Heat Recovery. The system incorporates thermal load management strategies and, when available, heat recovery mechanisms to reduce the thermal load demand:

1. **Thermal Storage Utilization**. A hot water tank stores surplus thermal energy for later use. The control strategy ensures that the tank is recharged during periods of surplus and discharged during deficit scenarios to meet thermal demands.
2. **Optional Heat Recovery**. When available, surplus heat recovery systems capture heat from the fuel cell or electrolyzer and use it to preheat water or assist space heating via a heat exchanger. This optional feature reduces the thermal load demand on primary energy systems, enhancing overall efficiency.

3.3 Components in Energy Demand

The site consists of three residential buildings, as well as a barn for milk cows. The main energy demand at the site consists of equipment at the barn (feed blower, milk robot), as well as space heating, and hot water production for the residential houses. Figure 7 presents weekly electricity demand profiles for the farm for a cold winter and a warm summer week. In both profiles, four distinct peaks are observed daily, appearing as two closely spaced pairs occurring at regular intervals. These peaks are related to the operation of the barn, and particularly the milk robot, which is washed regularly. The high peaks during the winter are related to the use of the feed blower.

As only the total electrical demand of the site is available, estimating the space heating and hot water demand required an analysis based on the type of equipment in use at the barn and the size of both the barn and the residential houses. In order to evaluate the contribution of the thermal energy supply and TES in contributing to

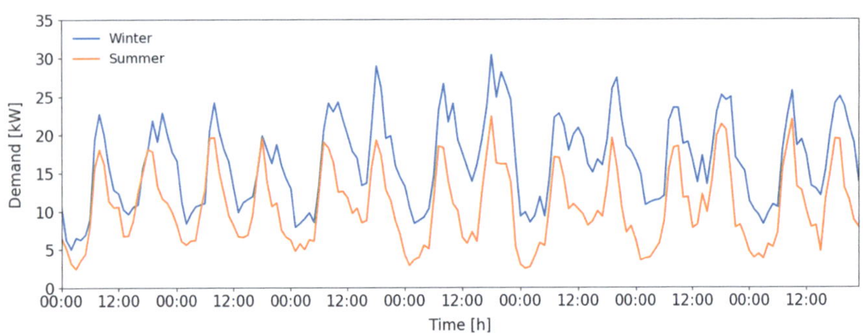

Fig. 7 Weekly load profiles for a winter week (2nd–8th Jan) and a summer week (3rd–9th July)

flexibility and reduce the total costs of an off-grid energy system, it was necessary to separate the demands for space heating and hot water production from the total demand data.

Space heating demand is the component of the load with highest dependency on ambient temperature. This component was therefore obtained by applying 2nd order polynomial regression on the total load as a function of ambient temperature for temperatures ranging from − 15 to 15 °C (Fig. 8), following an approach similar to (Heimar Andersen et al. 2021). The space heating demand is then simply calculated using the resulting 2nd order polynomial coefficients and the ambient temperature data for each hour of the year. As the load covers the total load for the farm, including the feed blowers used sporadically over the winter, the ambient temperature dependency is less clear than when using load data for a residential building, for instance, and the summer heat load was unrealistically high. The load profile for spring, summer and autumn was therefore scaled down. Figure 9 shows the resulting space heating load together with the ambient temperature.

The hot water load was assumed to be regular and temperature independent, similar to the approach in (Heimar Andersen et al. 2021), thus set equal to the summer load minus the minimum load for the week to extract any other loads at the farm. Figure 10 shows the resulting loads for a cold winter and a warm summer week, divided into three components: hot water production (37% of the total annual load), space heating (34%) and electricity specific (29%). The electricity specific load calculated from the initial total load minus the loads for space heating and hot water production.

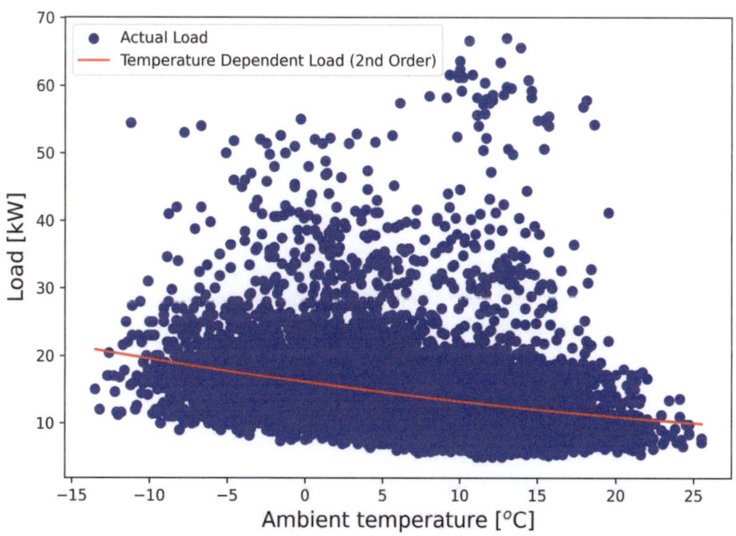

Fig. 8 Total load as a function of ambient temperature and the applied polynomial fit

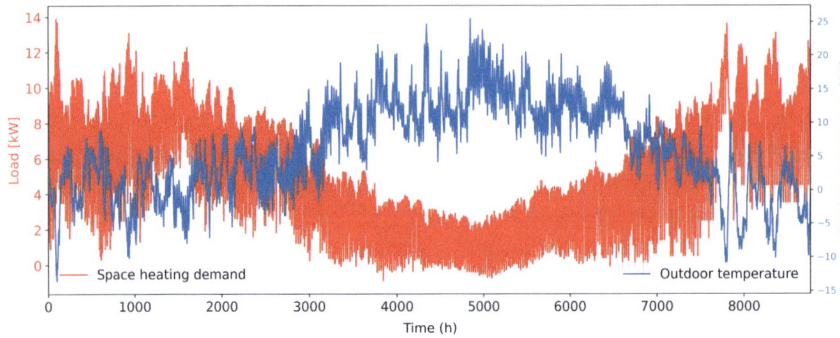

Fig. 9 Ambient temperature together with the space heating load extracted from the total load

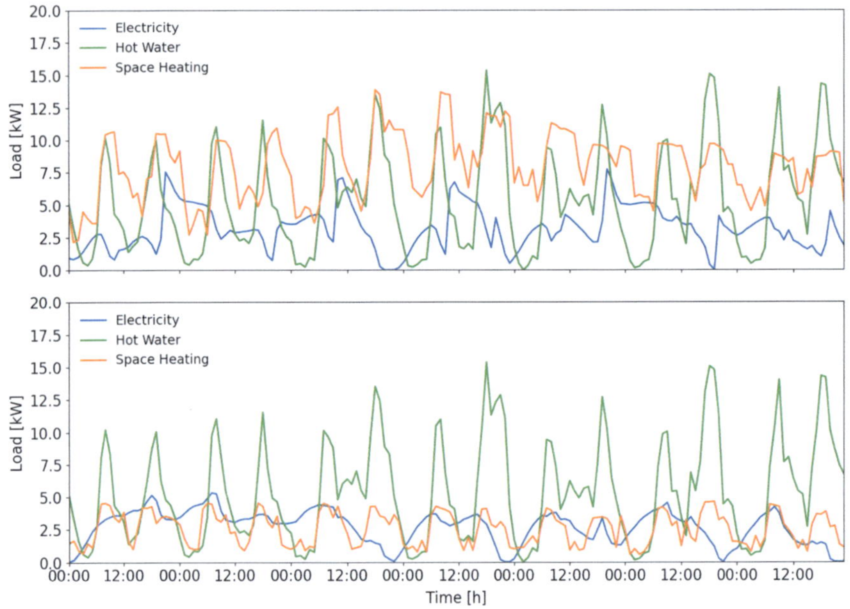

Fig. 10 Weekly load for a winter week (2nd– 8th Jan) and a summer week (3rd– 9th July) split into three components: hot water production, space heating and electricity specific

3.4 *Optimisation Strategy*

The optimization approach used in this case study focuses on minimizing a weighted objective function that evaluates both techno-economic and environmental metrics for the system. An in-house Python code was employed, which includes the `scipy.optimize` library and specifically the **Powell method** (Virtanen et al. 2020), a derivative-free optimization method that does not rely on gradients or the Jacobian matrix.

The optimization problem for hybrid energy systems involves both convex and non-convex components. While some constraints, such as energy balance and component limits, are linear and thus convex, the objectives (e.g., NPC, LCOE, LCOS, and GHG emissions) are generally non-convex due to the system's complex dynamics and interactions. Reliability metrics (e.g., LPSP and LHSP) derived from simulation results also exhibit non-convex behavior.

The Powell method is particularly well-suited to such problems because it is a derivative-free optimization algorithm that does not rely on convexity or smoothness. Instead, Powell systematically explores the parameter space, iteratively refining the solution using line search techniques. While it does not guarantee a global minimum for non-convex problems, its robustness makes it effective for identifying high-quality solutions in complex, noisy objective functions, such as those derived from yearly demand simulations.

By complementing it with a **weighted-sum approach** or the **Tchebycheff method**, the Powell method effectively handles multiple objectives, optimizing the balance between technical, economic, and environmental constraints. This method is well-suited for the complex, non-linear response surfaces characteristic of the analyzed hybrid energy systems. The multi-objective nature of the analyzed problem requires that a scalarization technique is employed to combine multiple objectives into a single scalar objective, making the Powell method applicable in this case.

The **weighted sum approach**, employed to find the optimal size of the investigated hybrid energy system, combines individual objectives $f_1(x), f_2(x), \ldots, f_{1k}(x)$ as:

$$F(x) = w_1 \cdot f_1(x) + w_2 \cdot f_2(x) + \cdots + w_k \cdot f_k(x)$$

where w_i are weights representing the relative importance of each objective. While this approach simplifies the optimization process, it provides a single solution rather than a Pareto front (Feng et al. 2014), and it is here applied to find the optimal size (capacity, kWh, or power, kW, of the hybrid energy system components) required to match the power and heat demand of the case study.

The **Tchebycheff method**, used to explore the trade-offs between objectives, minimizes the maximum weighted deviation from an ideal solution, expressed as:

$$L_\infty(x, \lambda) = \max_i \left[\lambda_i \cdot \left| f_i(x) - z_i^* \right| \right]$$

where x represents the decision variables, $f_i(x)$ is the ith objective function, z_i^* is the ideal value for the ith objective, and λ_i is the weight for the i-th objective (Feng et al. 2014).

Problem Formulation The process begins with defining the decision variables, which include parameters such as RES power, battery energy, fuel cell and electrolyzer power, boiler power, tank capacities, and the inclusion of a heat exchanger. These parameters are fed into the simulation model along with system metadata, which specifies the configuration (electric- or heat pump-driven) and governs the activation of specific components. The simulation then calculates techno-economic results, such as Net Present Cost (NPC), Levelized Cost of Electricity (LCOE), Levelized Cost of Storage (LCOS), Loss of Power Supply Probability (LPSP) and Loss of Heat Supply Probability (LHSP), as well as environmental metrics like greenhouse gas (GHG) emissions.

Net Present Cost (NPC) is calculated as:

$$C_{NPC,tot} = \sum_i C_{inv,i,0} + \sum_{j=1}^{n} \frac{\sum_k C_{rep,k,j} + \sum_i C_{OM,i,j}}{(1+d)^j} - \sum_k \frac{C_{sal,k,n}}{(1+d)^n}$$

where

- n is the lifetime of the project (set to 20 years in this study) and d is the real interest rate (which is a function of the nominal interest rate and the annual inflation rate)
- $C_{inv,i,0}$. is the investment cost of component i $C_{rep,k,j}$ is the replacement cost of component k in year j.
- $C_{OM,i,j}$ is the operation and maintenance cost of component i in year j.
- $C_{sal,k,n}$ is the salvage value of component k at the project's end.

LCOE and LCOS are calculated as:
Where:

- $C_{NPC,gen}$ is the net present cost of electricity generation components (€).
- $C_{NPC,stored}$ is the net present cost of the storage system (€).
- $E_{tot,j}$ is the total energy demand covered in year j (kWh).
- $E_{stored,j}$ is the total energy stored and later used in year j (kWh).

LPSP and LHSP are calculated as:

$$LPSP = \frac{\sum_{j=1}^{n} E_{u,j}}{\sum_{j=1}^{n} E_{d,j}} \quad LHSP = \frac{\sum_{j=1}^{n} H_{u,j}}{\sum_{j=1}^{n} H_{d,j}}$$

where

- $E_{u,j}$ and $H_{u,j}$ are the unmet electricity and heat demands in year j.
- $E_{d,j}$ and $H_{d,j}$ are the total electricity and heat demands in year j.

Specific GHG emissions are calculated as:

$$GHG = \frac{\sum_i \left(G_{con,i} + G_{rep,i} - G_{sal,i}\right) + G_{fuel,prod} + G_{fuel,cons}}{\sum E_{demand}}$$

where

- The first summation represents the total GHG emissions from the power generation and storage i components, including construction, replacement, and salvage value adjustment.

- $G_{fuel,prod}$ and $G_{fuel,cons}$ represent GHG emissions from biofuel production and consumption.
- E_{demand} is the total energy demand covered in the project (kWh).

The **weighted objective function** aggregates these outputs, rescaling them for uniformity and incorporating penalty terms to ensure feasibility. For instance, penalties are applied if the LPSP or LHSP exceed acceptable thresholds, or if fuel-based GHG emissions are non-zero. Mathematically, the problem is formulated as:

$$\min_x F(x) = w_1 \cdot \text{NPC}(x) + w_2 \cdot \text{LCOE}(x) + w_3 \cdot \text{LCOS}(x) + w_4 \cdot \text{LPSP}(x) \\ + w_5 \cdot \text{LHSP}(x) + w_6 \cdot \text{GHG}(x) + \text{P}(x)$$

where

- x is the vector of decision variables, i.e.:

$$[PV_{power}, Wind_{power}, FC_{power}, EL_{power}, Hydrogen_{capacity}, \\ Battery_{capacity}, Generator_{power}, ElectricBoiler_{power}, \\ HeatPump_{power}, WaterTank_{capacity}, HeatExchanger_{power}]$$

- $P(x)$ is the penalty term (e.g., for unmet demand, infeasibility).
- $w_1, w_2, w_3, w_4, w_5, w_6$ are the weights for each objective, reflecting relative importance.

The **Powell method** systematically explores the parameter space and iteratively adjusts decision variables to minimize the scalarized objective function. This ensures robustness in identifying the global minimum, even for non-smooth or noisy objective functions. The method balances competing objectives by minimizing the worst-case deviation from ideal values. For example, mismatches in power and heat supply are penalized, leading to optimal sizing of the components. The optimization reshapes supply curves to align with demand, ensuring that the energy balance constraints are respected. Here is the list of types of constraints and bounds on variables that are implemented to maintain realistic and feasible solutions:

1. **Energy Balance Constraints**:

$$\int_0^T S_p(t)dt \geq \int_0^T D_p(t)dt, \int_0^T S_h(t)dt \geq \int_0^T D_h(t)dt$$

   where $S_p(t)$ and $S_h(t)$ are the power and heat supplied by the system at time t, while $D_p(t)$ and $D_h(t)$ are the respective demands.
2. **Component Constraints**:

$$x_{min} \leq x \leq x_{max}$$

Ensuring component sizes in vector x remain within operational limits.

3. **System Reliability Constraints**:

$$LPSP \leq LPSP_{max}, LHSP \leq LHSP_{max}$$

Combination of Powell and Tchebycheff methods: Specifically for this case study, the Tchebycheff method was used to evaluate combinations of LPSP and LHSP with respect to a desired ideal solution (z^*):

$$L_\infty(x, \lambda) = \max\left[\lambda_1 \cdot \left|LPSP(x) - LPSP^*\right|, \lambda_2 \cdot \left|LHSP(x) - LHSP^*\right|\right]$$

where $LPSP^*$ and $LHSP^*$ are the ideal values (e.g. 0.1% or 1%), while λ_1 and λ_2 are weights for prioritizing LPSP and LHSP.

The optimization code computes NPC, LCOE, LCOS, and GHG for each point in the LPSP, LHSP space, generating a Pareto front by systematically varying weights in the scalarized objective function, enabling exploration of trade-offs between NPC, LCOE, LCOS, and GHG emissions. At convergence, the set of **optimized parameters** and corresponding **objective metrics** are calculated and output. The optimized system design reflects the best possible trade-off between technical, economic, and environmental performance, while ensuring that power and heat demands are met efficiently.

3.5 Technical, Economic, and Environmental Input Parameters

The parameters for the different electric and thermal components presented in Table 3 were used as inputs to parameterize the model for optimizing the three hybrid energy storage systems (HESS) configurations. These values, including performance metrics such as efficiencies, lifetime, flow rates, or defined operational ranges (e.g., state-of-charge limits), ensure accurate representation of each component behavior within the system. This detailed parameterization enabled the model described in Sect. 3.4 to evaluate technical feasibility, system reliability, and the optimal configuration for achieving performance targets.

Table 4 consists of the main economic parameters, including investment costs, replacement costs, and operational and maintenance expenses, to be used as an input into the optimization model to assess the economic feasibility of HESS configurations.

Finally, Table 5 includes the greenhouse gas (GHG) emissions data for each component as critical inputs for the model to evaluate the environmental sustainability of the HESS configurations (Bionaz et al. 2022). These lifecycle emissions metrics enabled the optimization process to prioritize low-emission technologies and assess the overall environmental footprint of each configuration.

Table 3 Technical input parameters

Component	Parameter	Value
Li-ion battery	Minimum SOC	0.2
	Maximum SOC	0.9
	Charge efficiency	0.95
	Discharge efficiency	0.95
	Lifetime	15 y[1]
PEM electrolyzer	Avg. efficiency (LHV basis)	0.63
	Modulation range (% of rated power)	10–100%
	Lifetime	67 000 h
PEM fuel cell	Avg. efficiency (LHV basis)	0.5
	Modulation range (% of rated power)	6–100%
	Lifetime	40,000 h
Hydrogen tank	Minimum SOC	0.2
	Maximum SOC	0.9
	Lifetime	25 y
Biofuel generator	Fuel consumption	220 g/kWh
	Lifetime	20,000 h
Converter	Efficiency	0.955
Rectifier	Efficiency	0.93
Water tank	Minimum temperature	40 °C
	Maximum temperature	80 °C
Electric boiler	Efficiency	0.95
	Heat loss coefficient	0.1 kWh/h
Heat pump	COP	3.5
Heat exchanger	Efficiency	0.85
Pump	Efficiency	0.85
	Flow rate	0.01 m^3/s
	Head	10 m

3.5.1 Results and Discussion

The performance of hybrid renewable energy systems was evaluated across three configurations: Electric-driven (ED), Heat Pump-driven (HPD), and Advanced Heat Recovery (AHR) systems. This evaluation, based on the yearly load profile of the dairy farm, incorporates economic, environmental, and operational considerations.

[1] Unlike fuel cells, electrolyzers, and generators (which are rated in operating hours due to continuous operation), the battery lifetime is input in years in the model, as it degrades based on both time (calendar aging) and usage (cycle aging).

Table 4 Economic input parameters

Component	Investment	Replacement	OM
PV system	1545 €/kW	–	24 €/kW/y
Wind system	1175 €/kW	–	3%/y of Inv. cost
PEM electrolyzer	4600 €/kW	35% of Inv. cost (67,000 h)	3%/y of Inv. cost
PEM fuel cell	1978 €/kW	35% of Inv. cost (40,000 h)	3%/y of Inv. cost
H_2 storage	470 €/kg	–	2%/y of Inv. cost
Li-ion battery	275 €/kWh	50% of Inv. cost (15 y)	10 €/kWh/y
Biofuel generator	420 €/kW	420 €/kW (20,000 h)	2 €/l
Electric boiler	100 €/kW	–	3%/y of Inv. cost
Heat pump	1500 €/kW	–	2%/y of Inv. cost
Water tank	500 €/unit[a]	–	2%/y of Inv. cost
Heat exchanger	200 €/kW	–	1%/y of Inv. cost

[a] Unit = 1000 L

Table 5 Environmental input parameters

Component	GHG
PV system	304.0 $kgCO_2eq/m^2$
Wind system	226,741.7 $kgCO_2eq$/turbine
PEM electrolyzer	190.5 $kgCO_2eq$/kW
PEM fuel cell	405.5 $kgCO_2eq$/kW
H_2 storage	67,820.6 $kgCO_2eq$ (capacity of 100 kg of hydrogen)
Li-ion battery	130.7 $kgCO_2eq$/kWh
Biofuel generator	17.4 $kgCO_2eq$/GJ (diesel production) 2.9 $kgCO_2eq$/l (diesel combustion) 107.7 $kgCO_2eq$/kVA (gener. manufacturing)
Electric boiler	50 $kgCO_2eq$/kW
Heat pump	100 $kgCO_2eq$/kW
Water tank	2.5 $kgCO_2eq$/L
Heat exchanger	150 $kgCO_2eq$/kW
Pump	75 $kgCO_2eq$/kW

Insights were drawn from the results presented in Table 6 and additional analysis from Figs. 11 and 12.

Yearly load profile. The analysis was conducted by assessing the yearly load profile of the dairy farm, taking into account economic, environmental, and operational factors. Figure 11 illustrates the state of charge (SOC) of the battery, hydrogen, and thermal storage throughout the year for the advanced heat recovery (AHR) system. Figure 12 presents the load demand, split into power and heat load demand, for a typical winter and summer week.

Table 6 Technical, economic, and environmental results for the three HES configurations: Electric-driven (ED), Heat Pump-driven (HPD), and Advanced Heat Recovery (AHR)

	unit	ED	HPD	AHR
Power generation				
PV	kW	84.1	45.4	39.6
Wind	kW	219.3	77.3	71.5
Biofuel generator	kW	19.8	4.9	4.7
Electrical storage				
Battery	kWh	392.5	190.3	189.5
Fuel Cell	kW	72.3	52.3	51.7
Electrolyzer	kW	41.3	39.2	38.8
Hydrogen tank	kWh	3251	2950	2953
Thermal storage				
Electric boiler	kW	10.6	–	–
Heat pump	kW	–	17.8	5.7
Water tank	l	1000	1000	1000
Heat exchanger	kW	–	–	4.9
Results				
NPC	EUR	1 033 860	632 278	604 181
LCOE	EUR/kWh	0.51	0.46	0.45
LCOS	EUR/kWh	0.50	0.58	0.55
LPSP	%	0.12	0.12	0.16
LHSP	%	0.09	0.03	0.00
GHG specific	gCO_2eq/kWh	241.19	293.46	290.42
GHG-system	$kgCO_2eq/yr$	32 187.99	17 130.51	16 951.79
GHG-fuel	$kgCO_2eq/yr$	2 870.21	794.46	793.88
Heat demand covered by surplus heat	%	–	–	9.12%
RES curtailment	%	38.59%	19.04%	18.29%

As observed, the high utilization of storage components during autumn, winter, and early spring highlights that renewable energy system (RES) production is significantly lower during these months due to reduced solar radiation in Nordic regions, resulting in lower PV output. This creates a net power deficit, as shown by the negative power balance, requiring continuous discharge of the battery, hydrogen system, and water tank to meet power requirements. This is particularly evident in Fig. 12, where the typical winter profile shows a high reliance on hydrogen and battery storage to cover power demands.

In contrast, during the summer, the system frequently experiences surplus energy conditions, with the exception of the last weeks of June and August where the intense discharge peaks observed in both the battery and the hydrogen tank SOC are related to the use of the feed blower. During the surplus periods, the battery, hydrogen,

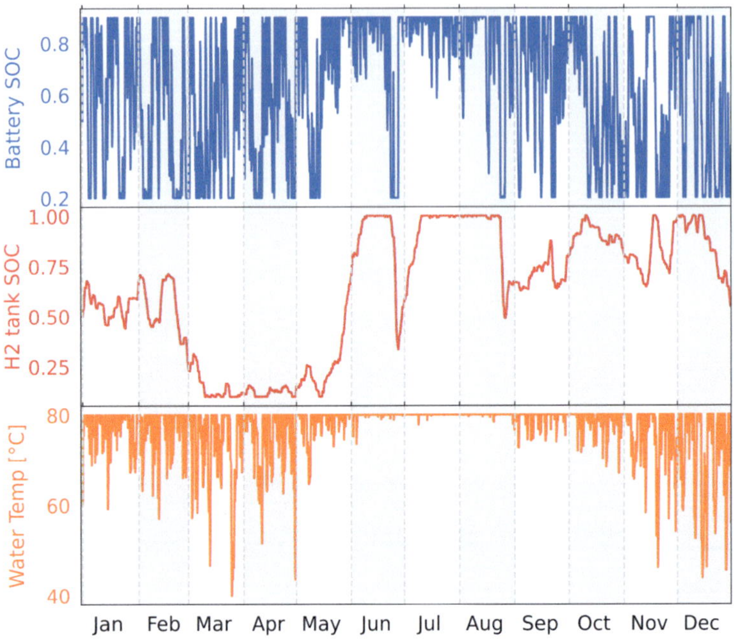

Fig. 11 Advanced heat recovery system: yearly state-of-charge (SOC) profile for **a** battery, **b** hydrogen tank. **c** Yearly temperature profile for the water tank

and water tank remain at maximum state-of-charge (or maximum temperature), and renewable generation almost entirely meets the power demand. Heat load demand, which includes both space heating and hot water, is primarily fulfilled by the heat pump. However, a notable fraction, approximately 9% as shown in Table 6, is covered by surplus heat recovered from the hydrogen system.

This heat recovery remains consistent year-round, as illustrated in Fig. 12. The surplus heat is recovered either during deficit conditions, when the fuel cell is in operation, or during surplus conditions, when the electrolyzer is active.

Technical, economic, and environmental analyses were conducted, revealing key interactions between energy generation, storage, and thermal recovery systems. These findings offer a detailed understanding of their impacts on cost, emissions, and system reliability. The outcomes of the three system optimizations are summarized in Table 6.

The Electric-driven system demonstrates a reliance on larger generation capacities and storage components to meet energy demands. For example, it requires 219 kW of wind energy, supplemented by 84 kW of PV power, to support a battery capacity of 392 kWh and a hydrogen tank size of over 3200 kWh. In addition to the high resource investment, this configuration produces substantial greenhouse gas (GHG) emissions of 32,188 kg CO_2-eq/year, of which 2870 kg CO_2-eq/year only attributed to the biofuel consumption from the generator. This system also incurs the highest Net

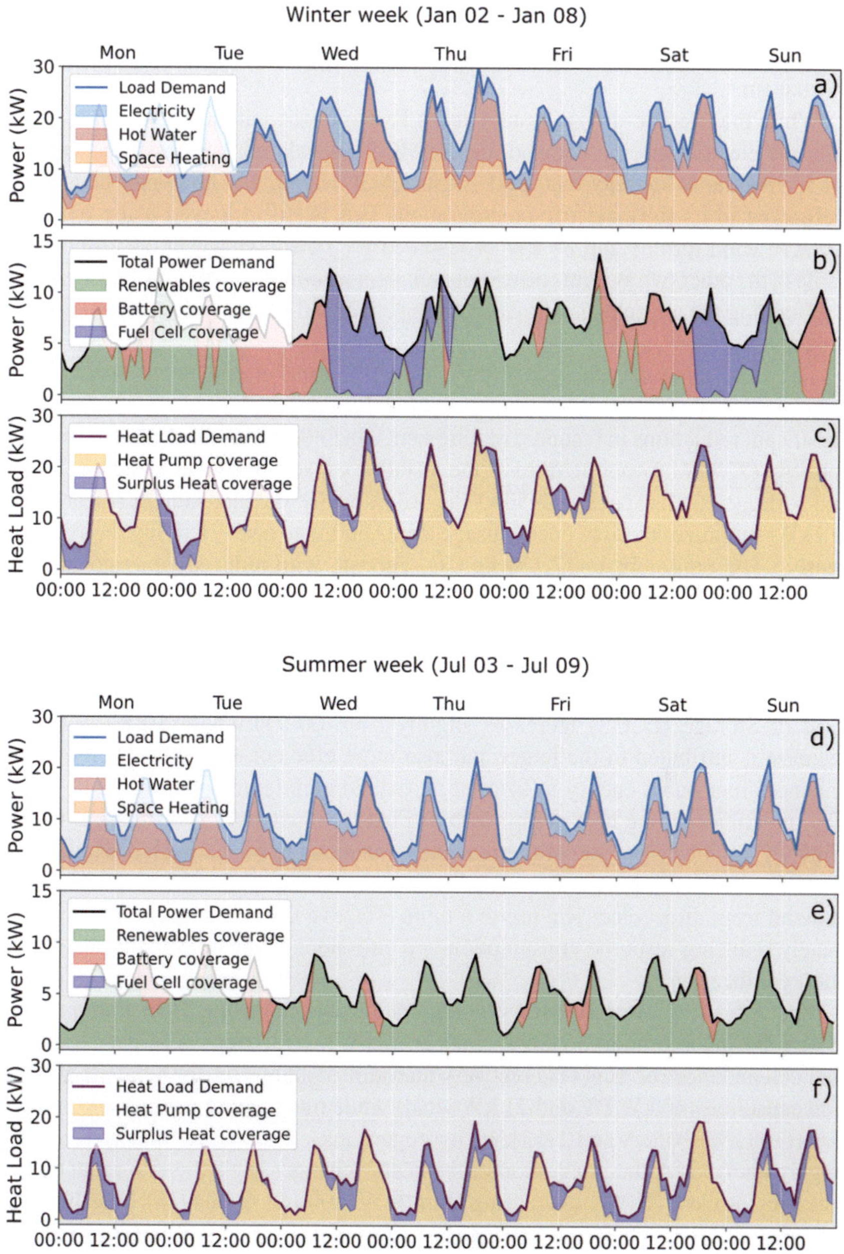

Fig. 12 Typical **a** winter and **d** summer weekly load demands. Total power demand and coverage (**b**, **e**). Heat load demand and coverage (**c**, **f**)

Present Cost (NPC) of € 1,033,860, underscoring the financial burden of achieving almost zero Loss of Power Supply Probability (LPSP) through purely electric-driven mechanisms.

While the Electric-driven system at the Rye site was initially designed to cover only the electric-specific demand, the REMOTE installation successfully provides for a potential case study that implements space heating and hot water demand as well, even in its fully electric configuration. This is still achieved using only one 225 kW wind turbine and 86 kW of installed PV power. However, as observed in Table 6, the other two systems offer more efficient solutions for meeting heat demand, highlighting the advantages of incorporating advanced thermal energy management systems.

The Heat Pump-driven system introduces a thermal management strategy that reduces dependency on electricity for meeting heat demands. This system achieves significant reductions in resource requirements, including a smaller wind generation capacity (77 kW, about one third of the installed capacity), battery size (190 kWh), and hydrogen tank capacity (2950 kWh). Additionally, the inclusion of a heat pump (18 kW) optimizes thermal energy usage, contributing to nearly halving the system's yearly GHG emissions to 17,130 kg CO_2-eq/year, with reduced generator utilization and fuel consumption yearly emissions falling to just 794 kg CO_2-eq/year. These improvements are achieved with a substantially lower NPC of € 632,278. The Levelized Cost of Electricity (LCOE) decreases from $0.51/kWh in the *Electric-driven* system to $0.46/kWh in the *Heat Pump-driven* system. While the Levelized Cost of Storage (LCOS) increases slightly from $0.50/kWh to $0.58/kWh, this increase is attributed to the larger, but also more efficient, storage capacity which enhances renewable energy utilization by reducing curtailment (from 38 to 19%). The integration of the heat pump, with its high Coefficient of Performance (COP), allows a significant portion of thermal energy demand to be met with minimal electrical input. This improves the overall efficiency of the system by shifting energy demand from direct electrical use to a more effective thermal energy pathway, ultimately reducing resource consumption and emissions while optimizing renewable energy utilization.

The Advanced Heat Recovery configuration builds on the Heat Pump-driven system by incorporating a heat exchanger (5 kW) to recover waste heat from the fuel cell and electrolyzer. This recovery mechanism allows for the smallest generation capacities (40 kW PV and 71 kW wind) while maintaining comparable storage sizes (190 kWh battery and 2953 kWh hydrogen tank). While GHG emissions remain largely unchanged, the integration of the heat exchanger leads to a 5% reduction in Net Present Cost (NPC), decreasing from € 632,000 to € 604,000. Additionally, this configuration achieves a lower LCOS, dropping from € 0.58/kWh to € 0.55/kWh. The ability to recover surplus heat, which meets 9% of the system's heat demand, alleviates the need for additional heating energy input. With a total NPC of $604,000 and an LCOE of $0.45/kWh, this configuration delivers both economic and environmental advantages, achieving optimal performance.

Heatmap Profiles from Tchebycheff Optimization. The optimization process conducted using the Tchebycheff method, computes LCOE, LCOS, and yearly GHG emissions (tonCO_2eq/yr) for each point in the LPSP, LHSP space, generating a Pareto front by systematically varying weights in the scalarized objective function. At convergence, the final set of **optimized parameters** and corresponding **objective metrics** (reported in Table 6) are calculated and output.

When modeling the system using the Tchebycheff method, the impact of achieving low LPSP and LHSP on LCOE, LCOS, and specific GHG emissions (gCO_2eq/kWh) has been analyzed. The results, presented as 2D heatmaps in logarithmic scale (Fig. 13), align with the Pareto front obtained from the Tchebycheff method and provide insights into the trade-offs between system reliability, costs, and environmental impacts. The heatmaps compare the LCOE, LCOS, and GHG emissions for two different system configurations: the electric-driven (ED) system (left column) and the advanced heat recovery (AHR) system (right column).

When aiming for extremely low LPSP and LHSP, the higher costs of LCOE are largely driven by the need to oversize renewable energy sources. This oversizing is essential to ensure that both electrical and thermal energy demands are met consistently, even during periods of low resource availability. Achieving such reliability from the system significantly impacts the LCOE. More generation capacity and infrastructure are required, which increases the overall cost. The effect is more pronounced in the ED system (Fig. 13a), where the higher absolute values of LCOE, compared to the AHR system (Fig. 13b), reflect the increased dependency on oversized electrical generation and storage due to the absence of an efficient thermal recovery system. While the contour slopes are similar, the higher baseline costs in the ED system indicate a stronger economic impact when achieving low LHSP.

For LCOS, the dynamic is slightly different. In the AHR system (Fig. 13d), while the cost is still influenced by achieving low LPSP and LHSP, it is primarily affected by the size of the electrical energy storage systems, such as batteries and hydrogen storage. As suggested by the almost vertical contour lines, thermal energy storage (and LHSP) contributes relatively little to the overall cost, as it tends to be more cost-effective compared to its electrical counterparts. This distinction highlights that while LCOE and LCOS are interconnected, the key cost drivers differ in the AHR system. LCOS values in the ED system (Fig. 13c) follow a similar trend to LCOE, meaning both power and heat reliability contribute to cost. However, the impact of LHSP on LCOS is still lower than LPSP, suggesting electrical storage remains the dominant driver.

When it comes to the yearly GHG emissions from the hybrid systems, they are heavily influenced by the sizing of electrical storage components in both the ED and AHR systems, particularly at very high LPSP. The large-scale deployment of batteries and hydrogen systems, which may involve emissions from production and infrastructure, adds to the carbon footprint.

However, when attempting to achieve both low LPSP and LHSP, the oversizing of electrical storage remains the primary driver of GHG emissions, particularly in the ED system. In the AHR system (Fig. 13f), the thermal energy storage does not significantly contribute to emissions; rather, the additional renewable capacity needed

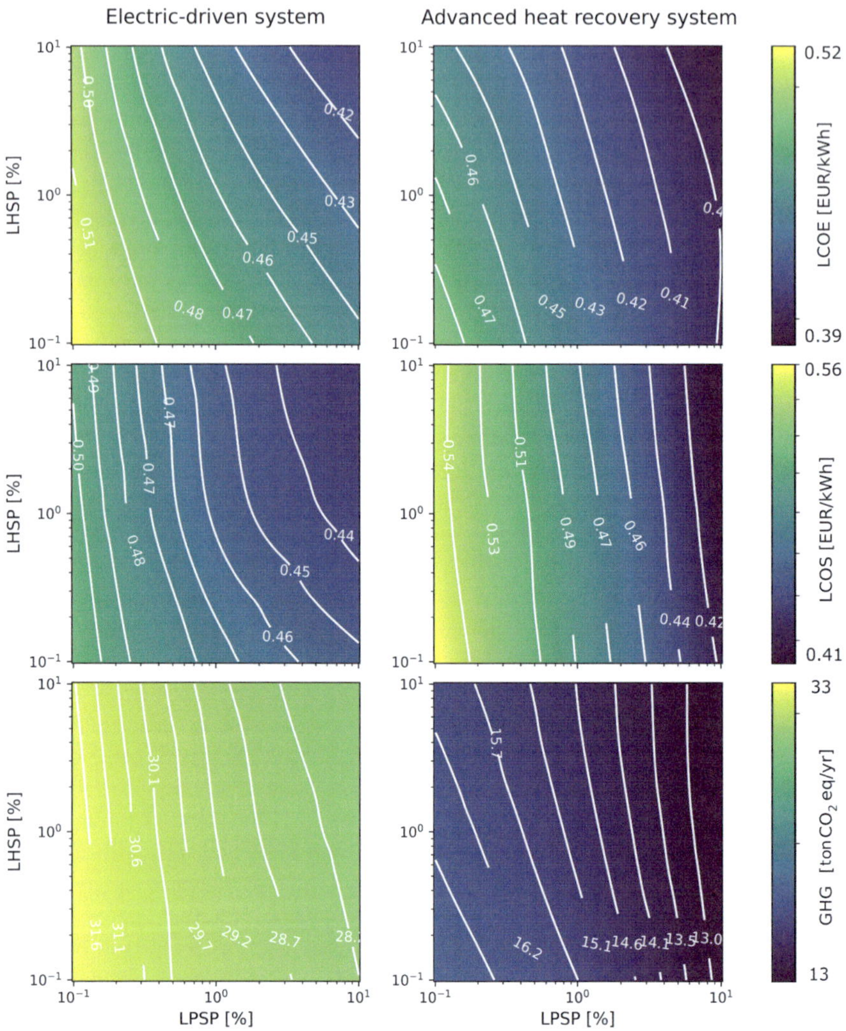

Fig. 13 Heatmap Profiles from Tchebycheff Optimization: LCOE, LCOS, and yearly GHG emissions for the electric-driven system (**a**, **c**, **e**) and for the advanced heat recovery system (**b**, **d**, **f**)

to ensure ultra-low LHSP may introduce embodied emissions related to material use and infrastructure expansion.

4 Conclusions

This study builds on the original installation at a remote milk farm in Rye, Norway, as part of the REMOTE Project. The original system, designed to meet the site's electrical demands, consisted of a hybrid renewable energy system integrating an 86 kW photovoltaic (PV) system and a 225 kW wind turbine, supported by a hydrogen-battery storage solution. The initial design focused on optimizing the use of local renewable resources, employing a 550 kWh lithium-ion battery for short-term energy storage and a 50 kW proton exchange membrane (PEM) electrolyzer, integrated with a 100-kg hydrogen storage and a 100-kW PEM fuel cell for long-term energy storage and conversion. Hydrogen was stored at 30 bars, eliminating the need for compression, while a biofuel generator served as a final backup to ensure complete off-grid operation.

Observations from the original case study highlighted the critical role of heating requirements, such as hot water and space heating, as a significant component of the site's overall energy demand. While the initial system was primarily designed to address electrical energy demands, it was recognized that optimizing the system to account for thermal energy needs could significantly enhance its overall efficiency and sustainability. Building on these insights, this study analyzed an extended system configuration as a potential future case study.

The extended system design incorporates thermal energy storage (TES) and heat pump technologies to address the heating demand more effectively. Although these components were not physically installed, their integration was modeled to explore how they could improve energy efficiency, economic performance, and environmental impact. Specifically, TES was evaluated as a cost-effective solution for meeting thermal energy demands, while heat pumps with a high Coefficient of Performance (COP) were analyzed for their ability to shift thermal demand from direct electrical energy use to more efficient pathways. Additionally, advanced surplus heat recovery mechanisms, such as waste heat utilization from hydrogen-based technologies, were incorporated to assess their potential to minimize energy waste and reduce operational costs.

This analysis serves as a foundation for identifying strategies to enhance the original system's performance and provides a roadmap for potential future upgrades to address the combined thermal and electrical energy needs of similar off-grid sites.

The results of the analysis highlight the critical advantages of hybrid systems that integrate thermal and electrical energy solutions. The *Electric-driven* configuration demonstrated the ability to meet both electrical and thermal demands using existing PV and wind installations, supplemented by large-scale energy storage. However, this system requires substantial resource investments, including a 392-kWh battery and a 3251-kWh hydrogen tank, leading to the highest Net Present Cost (NPC) of €1,033,860 and greenhouse gas (GHG) emissions of 32,188 kg CO_2-eq/year. While the system could provide reliable energy, its cost and environmental performance underscored the inefficiency of relying solely on electrical-driven solutions for combined energy demands.

The *Heat Pump-driven* and *Advanced Heat Recovery* configurations were designed to address the site's significant heating requirements more efficiently. The *Heat Pump-driven* system incorporates an 18-kW heat pump to reduce reliance on electrical energy for thermal demands, achieving significant reductions in NPC (€632,278) and GHG emissions (17,130 kg CO_2-eq/year). The system's LCOE of €0.46/kWh, lower than the Electric-driven system, reflects the shift from electrical energy to more efficient thermal pathways enabled by the heat pump's high COP. Although the Levelized Cost of Storage (LCOS) increased slightly from €0.50/kWh to €0.58/kWh, this increase was offset by improved renewable energy utilization, reducing curtailment from 38 to 19%.

Building on this, the *Advanced Heat Recovery* configuration introduces a heat exchanger to recover waste heat from hydrogen-based technologies, covering 9% of the heat demand with surplus heat. This system achieved the lowest NPC (€604,000), GHG emissions (16,952 kg CO_2-eq/year), and a balanced LCOE of €0.45/kWh, demonstrating its superior economic and environmental performance. By recovering heat and reducing energy waste, the *Advanced Heat Recovery* system achieved the optimal balance between cost and energy efficiency.

Seasonal performance analysis further emphasized the benefits of integrating thermal solutions. During winter, when solar resources were limited, systems incorporating TES and heat recovery effectively utilized wind energy and maintained consistent performance, while the Electric-driven system relied heavily on oversized storage to meet seasonal demands. The exponential relationship between Loss of Power Supply Probability (LPSP) and Loss of Heat Supply Probability (LHSP) with economic and environmental parameters, illustrated in the heatmap analysis, shows that achieving extremely low LPSP and LHSP significantly increases costs and GHG emissions due to the need for oversizing renewable energy sources and electrical storage systems. While LCOE is primarily impacted by RES oversizing, LCOS is driven by the scale of electrical energy storage, with thermal storage playing a minor role.

In conclusion, this study demonstrates that extending the original Rye case study to include hybrid configurations with TES, heat pumps, and waste heat recovery significantly improves energy efficiency, economic performance, and environmental impact. The *Advanced Heat Recovery* system, in particular, emerges as the optimal solution for meeting the site's combined thermal and electrical demands. These findings offer valuable insights for designing future renewable energy systems, highlighting the importance of integrating advanced thermal management technologies to achieve sustainable, cost-effective, and reliable off-grid energy solutions.

Acknowledgements The authors acknowledge financial support from the EU-Projects StoRIES (Storage Research Infrastructure Eco-System) (Grant Agreement No. 101036910) and REMOTE (Remote area Energy supply with Multiple Options for integrated hydrogen-based TEchnologies) (Grant Agreement No. 779541), as well as Trønder Energi/ANEO for providing the data and Lars Hoem for granting access and sharing insights on the buildings and energy usage. Additionally, the authors thank Bernhard Kvaal for providing pictures of the hybrid energy storage system installed in Rye.

References

Alexopoulos DK, Anastasiadis AG, Vokas GA, Kaminaris SD, Psomopoulos CS (2021) A review of flexibility options for high RES penetration in power systems—focusing the Greek case. Energy Rep 7(November):33–50. https://doi.org/10.1016/j.egyr.2021.09.050

Batteries Europe (2025). https://batterieseurope.eu/results/kpis-benchmarking-2/kpis-benchmarking-iii-february-2025/

Bionaz D, Marocco P, Ferrero D, Sundseth K, Santarelli M (2022) Life cycle environmental analysis of a hydrogen-based energy storage system for remote applications. Energy Rep 8(November):5080–5092. https://doi.org/10.1016/j.egyr.2022.03.181

Blackburn CJ, Flowers ME, Matisoff DC, Moreno-Cruz J (2020) Do pilot and demonstration projects work? Evidence from a green building program. J Policy Anal Manage 39(4):1100–1132. https://doi.org/10.1002/pam.22218

BloombergNEF (2023). https://about.bnef.com/blog/lithium-ion-battery-pack-prices-hit-record-low-of-139-kwh/.

Chade D, Miklis T, Dvorak D (2015) Feasibility study of wind-to-hydrogen system for arctic remote locations—Grimsey island case study. Renewable Energy 76:204–211. https://doi.org/10.1016/j.renene.2014.11.023

Clean Hydrogen JU SRIA 2021–2027 (2022)

Cole W, Karmakar A (2021) Cost projections for utility-scale battery storage: 2023 update. www.nrel.gov/publications

COMMISSION IMPLEMENTING DECISION (EU) 2020/728 (2020)

DIRECTIVE (EU) 2023/1791 OF THE EUROPEAN PARLIAMENT AND OF THE COUNCIL (2023)

DNV, Maritime Impact 2024 (2024). https://www.dnv.com/expert-story/maritime-impact/strategies-for-meeting-upcoming-decarbonization-targets/

Energy Storage News (2022). https://www.energy-storage.news/saft-wins-project-for-largest-bess-in-the-arctic/

Enevoldsen P, Sovacool BK (2016) Integrating power systems for remote island energy supply: lessons from Mykines, Faroe Islands. Renew Energy 85:642–648. https://doi.org/10.1016/j.renene.2015.06.065

European Commission-Press Release (2021) A stronger EU engagement for a greener, peaceful and prosperous Arctic

Eurostat (2024) Energy consumption in households. https://ec.europa.eu/eurostat/statistics-explained/index.php?title=Energy_consumption_in_households#Energy_consumption_in_households_by_type_of_end-use

Feng Z, Zhang Q, Tang Q, Yang T, Ge J (2014) Control-structure integrated multiobjective design for flexible spacecraft using MOEA/D. Struct Multidiscip Optim 50(2):347–362. https://doi.org/10.1007/s00158-014-1053-7

Häring T, Rosin A, Biechl H (2019) Using common household thermal storages to support the PV- and battery system in nearly zero energy buildings in off-grid mode. Sustain Energy Technol Assess 35(October):12–24. https://doi.org/10.1016/j.seta.2019.05.014

Heimar Andersen K, Krekling Lien S, Byskov Lindberg K, Taxt Walnum H, Sartori I (2021) Further development and validation of the 'PROFet' energy demand load profiles estimator. In: Proceedings of building simulation 2021: 17th conference of IBPSA. https://doi.org/10.26868/25222708.2021.30159

IDTechEx (2024). https://www.idtechex.com/en/research-article/how-catls-us-57-kwh-battery-would-transform-electric-cam-machines/31204

International Energy Agency (2019). www.iea.org/t&c/

Kalantari H, Ghoreishi-Madiseh SA (2022) Hybrid renewable hydrogen energy solution for remote cold-climate open-pit mines. Hydrogen 3(3):312–332. https://doi.org/10.3390/hydrogen3030019

Kauko H, Brækken A, Askeland M (2024) Flexibility through power-to-heat in local integrated energy systems with renewable electricity generation and seasonal thermal energy storage. Energy no. Under review

Kost C (2024) Study: levelized cost of electricity-renewable energy technologies. www.ise.fraunhofer.de

Lazard (2021) Lazard's levelized cost of storage analysis—version 7.0

Marocco P, Ferrero D, Lanzini A, Santarelli M (2022) The role of hydrogen in the optimal design of off-grid hybrid renewable energy systems. J Energy Storage 46:103893. https://doi.org/10.1016/j.est.2021.103893

Marocco P, Novo R, Lanzini A, Mattiazzo G, Santarelli M (2023) Towards 100% renewable energy systems: the role of hydrogen and batteries. J Energy Storage 57:106306. https://doi.org/10.1016/j.est.2022.106306

McKinsey & Company (2023). https://www.mckinsey.com/industries/infrastructure/our-insights/the-plant-as-a-product-hyperscaling-green-capex/.

Mitali J, Dhinakaran S, Mohamad AA (2022) Energy storage systems: a review. In: Energy storage and saving. Elsevier B.V. https://doi.org/10.1016/j.enss.2022.07.002

Mohammadpour H, Cord-Ruwisch R, Pivrikas A, Ho G (2021) Utilisation of oxygen from water electrolysis—assessment for wastewater treatment and aquaculture. Chem Eng Sci 246 (December). https://doi.org/10.1016/j.ces.2021.117008

Murphy C, Mills A (2021) Hybrid energy systems: opportunities for coordinated research. https://www.nrel.gov/docs/fy21osti/77503.pdf

NREL (2021). https://data.nrel.gov/submissions/171

NREL (2024). https://atb.nrel.gov/electricity/2024/utility-scale_pv-plus-battery

Nunemaker J, Shields M, Hammond R, Duffy P (2020) ORBIT: offshore renewables balance-of-system and installation tool. www.nrel.gov/publications

Olabi AG, Wilberforce T, Abdelkareem MA, Ramadan M (2021) Critical review of flywheel energy storage system. Energies. MDPI AG. https://doi.org/10.3390/en14082159

Paliza GF (2012) NRMM emissions regulations in Europe: what they mean for diesel powered generating systems white paper. www.cumminspower.com

Palys MJ, Daoutidis P (2022) Power-to-X: a review and perspective. Comput Chem Eng 165 (September). https://doi.org/10.1016/j.compchemeng.2022.107948

PNNL (2024). https://www.pnnl.gov/projects/esgc-cost-performance/lcos-estimates

Poelzer G et al (2016) Developing renewable energy in arctic and sub-arctic regions and communities. Fulbright Arctic Initiative

Puranen P, Kosonen A, Ahola J (2021) Technical feasibility evaluation of a solar PV based off-grid domestic energy system with battery and hydrogen energy storage in Northern climates. Sol Energy 213:246–259. https://doi.org/10.1016/j.solener.2020.10.089

REGULATION (EU) 2016/1628 OF THE EUROPEAN PARLIAMENT AND OF THE COUNCIL (2016)

REMOTE project official website (2018). https://www.remote-euproject.eu/

Richards BS, Conibeer GJ (2007) A comparison of hydrogen storage technologies for solar-powered stand-alone power supplies: a photovoltaic system sizing approach. Int J Hydrogen Energy 32(14):2712–2718. https://doi.org/10.1016/j.ijhydene.2006.09.013

van der Roest E, Bol R, Fens T, van Wijk A (2023) Utilisation of waste heat from PEM electrolysers—unlocking local optimisation. Int J Hydrogen Energy 48(72):27872–27891. https://doi.org/10.1016/j.ijhydene.2023.03.374

Shoaib M, Vallayil P, Jaiswal N, Iyapazham Vaigunda Suba P, Sankararaman S, Ramanujam K, Thangadurai V (2024) Advances in redox flow batteries—a comprehensive review on inorganic and organic electrolytes and engineering perspectives. Adv Energy Mater Wiley. https://doi.org/10.1002/aenm.202400721

SINTEF, ZEESA (2023). https://www.sintef.no/en/publications/publication/2297622/

SMART SENJA (2019). https://smartsenja.no/

Stehly T, Duffy P, Mulas D. Hernando, (2023) 2022 Cost of wind energy review

Store Norske (2023). https://www.snsk.no/nyheter/9198/verdens-nordligste-solcellepark-er-i-drift-pa-isfjord-radio

Sun K, Chen X, Dastjerdi SM, Yang Q (2022) Dynamic simulation of hydrogen-based off-grid zero energy buildings with hydrogen storage considering Fanger model thermal comfort. Int J Hydrogen Energy 47(62):26435–26457. https://doi.org/10.1016/j.ijhydene.2022.03.248

Svalbard Energi AS (2024). https://www.svalbard-energi.no/milepael-i-energiomstillingen.6653251-586879.html

Tommasi M, Degerli SN, Ramis G, Rossetti I (2024) Advancements in CO_2 methanation: a comprehensive review of catalysis, reactor design and process optimization. Chem Eng Res des 201(January):457–482. https://doi.org/10.1016/j.cherd.2023.11.060

D. Trapani, P. Marocco, D. Ferrero, K.B. Lindberg, K. Sundseth, M. Santarelli (2024) The potential of hydrogen-battery storage systems for a sustainable renewable-based electrification of remote islands in Norway. J Energy Storage 75:109482. https://doi.org/10.1016/j.est.2023.109482

UN Global Compact (2022) Setting science-based targets in the seafood sector: best practices to date

United Nations—Framework Convention on Climate Change (2022) Methodological tool—default values for common parameters

US Dept of Commerce (2022). Understanding energy storage

Viole I, Valenzuela-Venegas G, Zeyringer M, Sartori S (2023) A renewable power system for an off-grid sustainable telescope fueled by solar power, batteries and green hydrogen. Energy 282 (November). https://doi.org/10.1016/j.energy.2023.128570

Virtanen P, Gommers R, Oliphant TE, Haberland M, Reddy T, Cournapeau D, Burovski E, Peterson P, Weckesser W, Bright J, Van Der Walt SJ (2020) SciPy 1.0: fundamental algorithms for scientific computing in python. Nat Methods 17(3):261–72. https://doi.org/10.1038/s41592-019-0686-2

Weinand JM, Hoffmann M, Göpfert J, Terlouw T, Schönau J, Kuckertz P, McKenna R, Kotzur L, Linßen J, Stolten D (2023) Global LCOEs of decentralized off-grid renewable energy systems. In: Renewable and sustainable energy reviews. Elsevier Ltd. https://doi.org/10.1016/j.rser.2023.113478

World Bank (2023) Energy storage for mini grids—status and projections of battery deployment. http://documents.worldbank.org/curated/en/099121323112040367

World Bank Group (2022). https://www.worldbank.org/en/news/press-release/2022/09/27/solar-mini-grids-could-power-half-a-billion-people-by-2030-if-action-is-taken-now

Yuan S, Kocaman AS, Modi V (2017) Benefits of forecasting and energy storage in isolated grids with large wind penetration—the case of Sao Vicente. Renew Energy 105:167–174. https://doi.org/10.1016/j.renene.2016.12.061

Southern Climates: Hybrid Energy Storage for Cooling and Supplying Electrical Load

Domenico Ferrero and Massimo Santarelli

Abstract In this chapter it is addressed the application of hybrid energy storage (HES) combining batteries and hydrogen energy storage in the off-grid/microgrid scenario within southern climates. Hybrid battery-hydrogen Power-to-Power electricity storage concept has been thoroughly assessed under the EU-funded REMOTE project that demonstrated the feasibility of the hybrid storage concept through on-site demonstrations and techno-economic evaluations. The HES demonstration plant installed in southern Europe in Spain (Gran Canaria) within REMOTE project has been selected as case study. This demonstrator is an energy storage solution for photovoltaic (PV) integration in the microgrid of a cattle farm. The chapter provides the detailed description of demonstration plant, including performance indicators calculated from its initial months of operation, the estimation of its environmental performance compared to base-case internal combustion engine generators, and the economic estimation of the Levelized Cost of Electricity (LCOE) and Storage (LCOS).

Keywords Power-to-power · Hybrid storage · PV integration · Energy storage demonstration · LCOE · LCOS

1 State-of-the-Art

In this chapter, we address the application of hybrid energy storage (HES) in the off-grid/microgrid scenario within southern climates, focusing particularly on insular cases. By "southern climate," we refer to Southern Europe, which is predominantly characterized by a Mediterranean climate.

The Mediterranean climate features mild winters and cool summers, especially in island regions. Winter temperatures generally stay above 5 °C and rarely fall below 0 °C, while summer daily peak temperatures typically reach around 30 °C, moderated by sea breezes. These mild seasonal variations lead to moderate but predictable

D. Ferrero (✉) · M. Santarelli
Department of Energy, Politecnico di Torino, Corso Duca degli Abruzzi, 24, 10129 Torino, Italy
e-mail: domenico.ferrero@polito.it

R. Scipioni et al. (eds.), *Hybrid Energy Storage*, Lecture Notes in Energy 47,
https://doi.org/10.1007/978-3-031-97755-8_8

heating and cooling demands in buildings, confined to winter and summer, respectively. In spring and autumn, outdoor conditions often align with indoor thermal comfort levels, minimizing the need for active climate control (Katsaprakakis et al. 2020). In the European islands located in the Atlantic Ocean at the same latitudes of the Mediterranean sea, the climate is even milder showing little seasonal temperatures variation over the course of the year; as a result, there is only a minor requirement for summer-time cooling or space-heating in winter, and the majority of thermal energy demand in buildings is for domestic hot-water heating.

The climate-driven heating and cooling requirements affect not only residential buildings but also agricultural, commercial, industrial, and institutional buildings. Despite the favorable climate, energy consumption in Mediterranean buildings remains disproportionately high. This is primarily due to inadequate insulation in building envelopes, affecting both opaque and transparent surfaces. A comparative analysis of annual heating energy consumption per square meter of conditioned space revealed that uninsulated buildings in Greece consume more energy than insulated buildings in colder countries like Denmark, Finland, and Switzerland (Balaras et al. 2005).

On the generation side, the Mediterranean region has considerable potential for developing renewable energy sources (RES), most of it remaining untapped. The region's current installed capacity includes 90 GW of solar PV and 82 GW of wind power, but its estimated solar and wind potential exceeds 3 TW (ECCO 2025). As the share of RES in the region is projected to increase, there is a growing need for reliable and efficient energy storage systems (ESS) to manage the intermittent nature of RES. Islands, in particular, offer prime opportunities for investment in renewable energy and storage systems due to their abundant wind and solar resources and the high cost of electricity generation from thermal power plants. To maintain grid stability, however, island network operators impose strict limitations on renewable energy expansion. ESS provides an effective solution to overcome these restrictions, enabling greater integration of RES.

Hybrid energy storage systems (HESS), which combine two or more storage technologies, emerge as cost-effective solutions by leveraging the complementary strengths of different storage methods. As introduced in Chap 7, HESS was thoroughly assessed under the EU-funded REMOTE project (2023). This project demonstrated the feasibility of hybrid hydrogen-battery storage systems in both northern and southern Europe through on-site demonstrations and techno-economic evaluations. Hydrogen proved cost-effective for long-term storage, while batteries were more efficient for short-term cycling. For further details on the advantages and costs of hybrid battery-hydrogen storage systems, readers are referred to Chap. 7.

In the context of insular Mediterranean applications, Karapidakis et al. (2023) demonstrated the effectiveness of hybrid hydrogen-battery storage to significantly reduce electricity production costs in Crete. The hybrid hydrogen-battery storage can be improved by incorporating also thermal energy storage (TES), enabling the use of heat pumps for the conversion energy from electricity to thermal energy to be stored in TES. Reversible heat pumps are a highly efficient option for renewable heating and cooling in buildings. For instance, Martinopoulos et al. (2016) compared various heating systems in Greece, finding heat pumps to be the most cost-effective solution for insular buildings based on total lifecycle costs. During summer, reversible heat pumps play a critical role in microgrid systems in Mediterranean climates by providing efficient cooling. Urbanucci et al. (2019) highlighted the role of heat pumps in trigeneration systems for renewable district heating and cooling microgrids. Similarly, Vourdoubas (2019) investigate the possible use of systems driven by solar electricity for cooling in the mediterranean region, with heat pumps identified as the most efficient solution.

Within the REMOTE project, two demonstration plants were installed in southern Europe: one in Agkistro, Greece, and the other in La Aldea de San Nicolás, Gran Canaria, Spain. The Greek demonstrator served as a backup electricity system for a small agri-food processing facility, specifically supporting the drying process of aromatic herbs. The Spanish demonstrator, on the other hand, functioned as an energy storage system for photovoltaic (PV) integration in the microgrid of a cattle farm.

The Spanish demonstrator notably showcased the application of hybrid energy storage for RES integration in southern climates. Although heat pumps were not included in the site's microgrid, the cattle farm's electrical load included significant cooling demands (e.g., from a chiller). The hybridization of the site could be further enhanced by incorporating heat pumps and TES.

In the following section, we provide a detailed case study of the demonstration plant, including performance indicators calculated from its initial months of operation.

2 Case Study: Integrated Battery and Hydrogen Storage for Power-to-Power Application in Southern Climates. The Demonstration Plant of the REMOTE Project in Gran Canaria (ES)

2.1 Plant Description

Within the REMOTE project, a microgrid has been installed in Agropecuaria Furel, located in La Aldea de San Nicolas, Gran Canaria (Spain). A photovoltaic plant provides renewable electricity to the microgrid. The demonstration plant, provided by the project partner Inycom, consists of a power-to-power (P2P) system based on hybrid battery and hydrogen energy storage. The P2P plant is composed by: a Li-Ion battery, an alkaline electrolyzer, a PEM fuel cell stack, compressed hydrogen storage, and a backup diesel genset. The micro-grid configuration for the energy supply to the farm is depicted in Fig. 1.

The PV plant has a nominal power of 100 kW (peak 110 kW), and the hours of equivalent sunshine are 1684 h/year at the selected site. The microgrid electricity storage system consists of a 100 kW inverter (Norvento GridMaster, Spain) developed for isolated systems or microgrids and a 100 kWh battery bank of Lithium-LFP

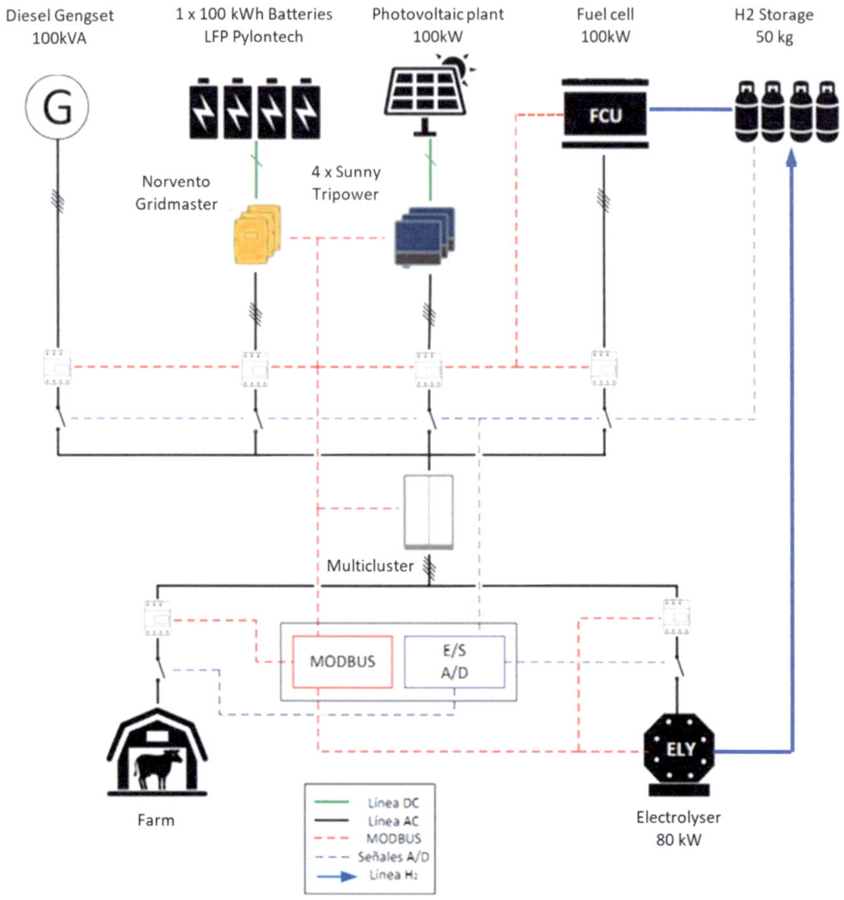

Fig. 1 Schematic architecture of the microgrid

technology (Pylontech, China). The hydrogen plant is composed of an 80 kW electrolyser (Idroenergy, Italy), one hydrogen storage of 50 kg (Calvera Hydrogen, Spain) and 100 kW of fuel cell (Ballard, Denmark). A 100 kVA backup diesel generator is also installed onsite.

Electrolyzer data are summarized as follows:

- Alkaline technology
- Nominal size: 80 kW
- Hydrogen compressor: 5 kW
- Operation Temperature: 40–60 °C
- Operating pressure: 4.5–6 bars
- Max H_2 Production: 13.4 Sm^3
- Hydrogen Purity: 99.5% (after purifier: 99.999%).
- Pressure outlet: 200 bar.

Fuel cell data:

- PEM fuel cell technology
- Nominal size: 100 kW
- H_2 consumption nominal: 2.6 g/s
- Operating pressure: 7 bar
- Operating temperature: 50–70 °C.

Hydrogen storage data:

- Compressed hydrogen, cylinders
- Nominal storage size: 50 kg
- Storage elements: 68 cylinders (50 L/cylinder), 3400 L total volume
- Working pressure: 200 bar.

SCADA systems have been installed for the control and monitoring of the plant.

Before installing the plant, a demand measurement campaign was conducted to assess the energy requirements of the facility and establish baseline consumption levels. This analysis revealed that the consumption profile is cyclical, aligning with the livestock farm's milking cycles. Each day consists of three milking cycles occurring every 8 h, resulting in nearly identical demand profiles throughout the day, as illustrated in Fig. 2 (upper panel). These cycles generate electricity demand peaks of up to 65 kW. The profiles remained consistent throughout the week and exhibited no seasonal variation.

Subsequently, a new cooling system was installed, significantly increasing the energy demand. Unlike the milking cycles, this demand proved to be much less stable and more challenging to manage. The new cooling facilities rely on instantaneous cooling, leading to dynamic and less predictable energy requirements, as shown in Fig. 2 (lower panel). With this updated configuration, the peaks increased up to 120 kW.

The operation strategy implemented is based on prioritizing the charge of the battery during excess PV production, while during the ESS discharge phase the fuel cell is used to cover the base load and the battery the electricity peaks.

The operation strategy is summarized as follows:

1. In case the generation of the PV is lower than the demand, all the PV is used to cover the loads. Then, batteries and genset are used to supply the remaining part of the load.
2. In case there is an excess of RES, the system charges the batteries.
3. In case there is an excess of RES, and the SOC of the batteries is above 90%, the electrolyser is operated in order to produce hydrogen. The electrolyser operation stops when batteries reach SOC of 70% and the excess renewable energy is shifted towards the battery recharge again.
4. The fuel cell and the batteries work always in parallel. The fuel cell has a constant setpoint and the batteries are used to generate whenever there is a peak of demand, and to absorb the fuel cell energy in case the demand suddenly decreases.

Fig. 2 Schematic of the electric load profile of the farm, without cooling (upper panel), with cooling (lower panel)

5. Whenever the energy that could give the batteries plus the fuel cell is not able to cover the load, a diesel genset would be used.
6. The fuel cell and the batteries supply the energy until the system runs out of hydrogen or until the SOC of the battery goes under 15%. In case the battery runs out of energy, the diesel generator starts and supplies the energy for the loads, and it also recharges the battery. In case the SOC of the battery is over 30% and there is still hydrogen to be used, the fuel cell and the battery cover the loads until there is no hydrogen left (Fig. 3).

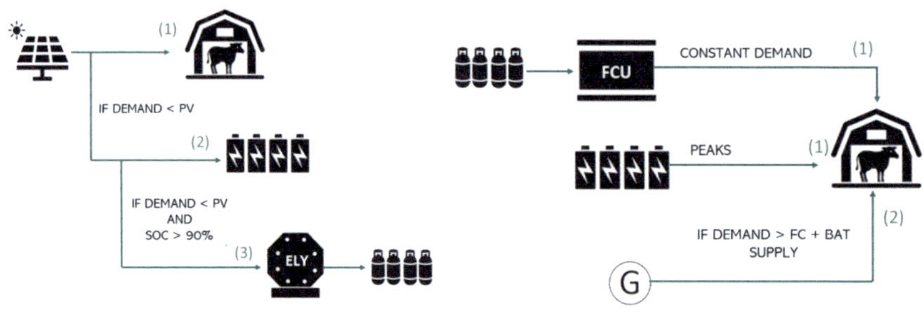

Fig. 3 Operation strategy: charging phase (left), discharging phase (right)

2.2 Key Performance Indicators

The selected KPIs are the ones that express how the renewable generation is provided to the load by direct coupling or through the P2P system, and how efficiently the storage chain through batteries and hydrogen has performed.

The KPIs have been divided in three categories:

- Load-related KPIs
- RES-related KPIs
- P2P-related KPIs.

Other indicators have been estimated to analyze the system performance:

- Environmental indicators
- Economic indicators.

2.2.1 Load-Related KPIs

The electricity demand of the final user can be covered directly by the RES installed on-site when the demand and production are matching, or by the P2P system when the electricity demand exceeds the renewable production. In general, the load coverage over the time horizon of the analysis is expressed as:

$$LD_{tot} = LD_{RES} + LD_{BT} + LD_{FC} + LD_{G} \tag{1}$$

where LD_{tot} is the total load, LD_{RES} is the load directly covered by renewable electricity (e.g. the load directly covered by photovoltaic generation LD_{PV}), LD_{BT} is the load covered by battery, LD_{FC} is the load covered by the fuel cell and LD_{G} is the load covered by the genset.

The different contributions to the total load of Eq. (1) are calculated as KPIs, as follows:

- Percentage of the total load covered by the photovoltaic (PV) generation over the time horizon of the analysis:

$$\%LD_{PV} = \frac{LD_{PV}}{LD_{tot}} \cdot 100 \tag{2}$$

- Percentage of the total load covered by the battery (BT) over the time horizon of the analysis:

$$\%LD_{BT} = \frac{LD_{BT}}{LD_{tot}} \cdot 100 \tag{3}$$

- Percentage of the total load covered by the fuel cell (FC) over the time horizon of the analysis:

$$\%LD_{FC} = \frac{LD_{FC}}{LD_{tot}} \cdot 100 \tag{4}$$

- Percentage of the total load covered by the genset over the time horizon of the analysis:

$$\%LD_{G} = \frac{LD_{G}}{LD_{tot}} \cdot 100 \tag{5}$$

2.2.2 RES-Related KPIs

The RES usage over the time horizon of the analysis has been estimated as:

$$RES_{tot} = RES_{LD} + RES_{BT} + RES_{EL} + RES_{AUX} \tag{6}$$

where RES_{LD} is the renewable production directly provided to the load, RES_{BT} is the RES fraction used to charge the batteries, RES_{EL} is the RES fraction converted to hydrogen in the electrolyzer and RES_{AUX} is the fraction of RES generation consumed by the auxiliary systems of the P2P plant. The RES-related KPIs calculated are the fractions of the different terms of Eq. (6) to the total RES generation.

- Percentage of RES sent to load over the time horizon of the analysis:

$$\%RES_{LD} = \frac{RES_{LD}}{RES_{tot}} \cdot 100 \tag{7}$$

- Percentage of RES sent to battery (charging) over the time horizon of the analysis:

$$\%RES_{BT} = \frac{RES_{BT}}{RES_{tot}} \cdot 100 \tag{8}$$

- Percentage of RES sent to electrolyzer over the time horizon of the analysis:

$$\%RES_{EL} = \frac{RES_{EL}}{RES_{tot}} \cdot 100 \tag{9}$$

- Percentage of RES sent to auxiliaries over the time horizon of the analysis:

$$\%RES_{AUX} = \frac{RES_{AUX}}{RES_{tot}} \cdot 100 \tag{10}$$

2.2.3 P2P-Related KPIs

The efficiency of the P2P system has been estimated for the different sections of the system (i.e., electrolyzer, fuel cell, battery, hydrogen storage) and for the entire plant. The efficiency of electrolyzers and fuel cells can be estimated both in terms of power (instantaneous efficiency) or energy (integrated efficiency over time), as these technologies are continuous machines that perform the electricity-hydrogen conversion. In contrast, the efficiency of batteries, hydrogen storage, and the overall system can only be evaluated over a specified time horizon, as these involve energy storage processes (i.e., within the battery or hydrogen storage system). For this reason, the efficiency of the battery, hydrogen storage, and the entire system is expressed as round-trip efficiency—calculated as the ratio of energy released to energy input during charging—over the analysis time horizon.

Electrolyzer efficiency over the time horizon of the analysis has been calculated as:

$$\eta_{P2G} = \frac{H2_{prod}}{EL} \tag{11}$$

where $H2_{prod}$ is the energy/power content (LHV basis) of the hydrogen produced, and EL is the electric energy/power input to the electrolyzer.

Fuel cell efficiency over the time horizon of the analysis has been calculated as:

$$\eta_{G2P} = \frac{FC}{H2_{cons}} \tag{12}$$

where $H2_{cons}$ is the energy/power content (LHV basis) of the hydrogen consumed by the fuel cell, and FC is the electric energy/power output of the fuel cell.

The round-trip efficiency of the battery storage system over the time horizon of the analysis has been calculated as:

$$\eta_{RT,BT} = \frac{BT_{dc}}{BT_{ch} - \frac{(SOC_{BT,end} - SOC_{BT,start}) \cdot Cap_{BT}}{\eta_{BT,ch}}} \tag{13}$$

where BT_{dc} is the energy released by the battery during discharge, BT_{ch} is the energy in input to the battery during charging, Cap_{BT} is the battery capacity (energy) and $\eta_{BT,ch}$ is the charging efficiency of the battery. The terms $SOC_{BT,end}$ and $SOC_{BT,start}$ are the State of Charge (SOC) of the battery at the end and at the beginning of the time horizon considered for the analysis. The denominator includes a correction term to take into account the SOC variation during the time horizon. Indeed, if the SOC decreases in the time period of the analysis, the amount of energy that has been retrieved from the storage must be accounted in the balance, also including the charging efficiency to take into account that the input energy for storing the amount of energy corresponding to the SOC variation has been higher due to battery charging losses. In the case of a positive SOC variation, the correction term excludes from the balance the energy that has been used to increase the SOC, including also the charging efficiency.

The round-trip efficiency of the overall hydrogen storage system (i.e., from electrolyzer to fuel cell) over the time horizon of the analysis has been calculated as:

$$\eta_{RT,H2} = \frac{FC}{EL - \frac{(SOC_{H2,end} - SOC_{H2,start}) \cdot Cap_{H2}}{\eta_{EL}}} \tag{14}$$

In this case, the correction term at the denominator has the same meaning of the battery SOC correction described above, but it is calculated using the SOC and capacity (in terms of energy) of the hydrogen storage and the electrolyzer efficiency.

The round-trip efficiency of the hybrid storage system over the time horizon of the analysis is calculated as:

$$\eta_{RT,tot} = \frac{BT_{dc} + FC}{BT_{ch} - \frac{\Delta SOC_{BT}}{\eta_{BT,ch}} + EL - \frac{\Delta SOC_{H2} \cdot Cap_{H2}}{\eta_{EL}}} \tag{15}$$

This KPI is the combination of the KPIs of battery and hydrogen storage.

2.2.4 Environmental Indicators

The evaluation of the environmental performance of the system has been performed by calculating the CO_2 emissions reduction compared to electricity generation by Diesel-fueled internal combustion engines (ICEs), which are the state-of-the-art for the off-grid electricity generation in remote locations.

An efficiency of 30% has been assumed for the Diesel-fueled ICEs, with a lower heating value of 10 kWh/liter for the fuel. Both the emissions related to Diesel production and combustion have been considered, to calculate the equivalent CO_2 emission per liter of fuel.

The EU average value of 17.4 $kgCO_{2eq}$ per GJ of produced diesel was assumed for the fuel production phase (European Commission, Directorate-General for Energy 2015). This value refers to a well-to-tank approach, from extraction up to the final consumer (with the exclusion of the combustion phase). A value of 2.9 $kgCO_{2eq}$ per liter of diesel burnt was hypothesized for the combustion process (Fleck and Huot 2009), which lies in the range of 2.4–3.5 $kgCO_{2eq}$/liter reported by Jakhrani et al. (2012).

2.2.5 Economic Indicators

The economic performance has been evaluated by calculating: (1) the Levelized Cost of Electricity (LCOE), to estimate the cost of the electricity provided by the complete system (i.e. including PV and genset) to the user, and (2) the Levelized Cost of Storage (LCOS) to show the cost of the electricity provided to the user by the electricity storage system.

The LCOE (in €/kWh) was evaluated as follows:

$$LCOE = \frac{C_{NPC,tot}}{\sum_{j=1}^{L_{PR}} \frac{E_{tot,j}}{(1+d)^j}} \quad (16)$$

where $C_{NPC,tot}$ is the net present cost (in €) of the complete system (i.e. PV, storage and genset), $E_{tot,j}$ is the total electrical demand (in kWh) during the j-th year and L_{PR} is the project lifetime, assumed equal to 20 years. A discount rate d of 3% has been used in the calculation.

The net present cost includes the capital investment $C_{inv,i}$ (in €), the operation and maintenance costs $C_{O\&M,i,j}$ (in €/year), the energy cost for the genset fuel $C_{fuel,j}$ (in €/year), and the replacement expenditures $C_{rep,i,j}$ (in €/year) incurred throughout the lifetime of the system:

$$C_{NPC,tot} = \sum_i C_{inv,i} + \sum_{j=1}^{L_{PR}} \left(\frac{\sum_i C_{O\&M,i,j}}{(1+d)^j} + \frac{\sum_i C_{fuel,j}}{(1+d)^j} + \frac{\sum_i C_{rep,i,j}}{(1+d)^j} \right) \quad (17)$$

The LCOS (in €/kWh) was evaluated as follows:

$$LCOS = \frac{C_{NPC,storage}}{\sum_{j=1}^{L_{PR}} \frac{E_{storage,j}}{(1+d)^j}} \quad (18)$$

where $C_{NPC,storage}$ is the net present cost (in €) of the hybrid storage system (i.e. battery, electrolyzer, hydrogen storage and fuel cell), and $E_{storage,j}$. (in kWh) is the electricity provided to the user by the battery and the fuel cell.

The economic data required to evaluate the LCOE and LCOS are listed in Table 1. The economic data were assumed according to previous techno-economic analysis performed by the authors (Marocco et al. 2021). Among the investment costs, those of the electrolyzer anfuel cell have been scaled compared to reference values using the cost exponents reported in the table.

2.3 Results of the Operation

The results presented are based on the data collected during the first 6 months of operation of the demonstration plant. During this period, the microgrid provided 164 MWh to the farm and the electrolyzer. The recorded PV generation was 81 MWh and the diesel genset produced 87.5 MWh (values of the microgrid sections, net of losses). During the period of the data analysis, the electrolyzer was always operative from the beginning of the monitoring period, while the fuel cell was operative from the second month with a limitation to 50% of the nominal power, due to the ongoing commissioning and optimization of the plant.

The photovoltaic plant together with the batteries have been in operation before the installation of the hydrogen section of the ESS plant, and the photovoltaic plant

Table 1 Economic parameters

Parameter	Value
PV plant	
Investment cost	1547 €/kW
Fixed O&M	24 €/(kW•year)
Lifetime	Project lifetime
Li-ion battery	
Investment cost	550 €/kWh
Replacement cost	50% of investment
Fixed O&M	10 €/(kWh•year)
Lifetime	10 years
H2 tank	
Investment cost	500 €/kg
Fixed O&M (% of inv. cost)	2%/yr
Lifetime	Project lifetime
Alkaline electrolyser	
Ref. specific investment cost, $c_{inv,ref}$	2000 €/kW
Ref. rated size, $P_{rated,ref}$	300 kW
Cost exponent, n	0.65
Replacement cost of the stack	25% of investment
Fixed O&M (% of inv. cost)	4%/yr
Lifetime (stack)	10 years
PEM fuel cell	
Ref. specific investment cost, $c_{inv,ref}$	3947 €/kW
Ref. rated size, $P_{rated,ref}$	10 kW
Cost exponent, n	0.7
Replacement cost of the stack	25% of investment
Fixed O&M	4%/yr
Lifetime (stack)	10 years

was almost never reaching the maximum nominal power (100 kW), as the installation was not able to provide power into the electricity grid. Typical operating profiles of a day are shown in Fig. 4.

After the installation of the H_2 plant and the activation of the new cooling facilities, the demand increased significantly. In Fig. 5 it is shown the corresponding operating profiles for a day.

In order to analyze the operation strategy implemented, two specific cases are shown.

The first case is representative of the operation when excess RES are available (Fig. 6). In Fig. 6 (upper panel) it is possible to see how, as soon as the SOC of the batteries (in red) is over the 90%, the PV power (orange) drops as there are no

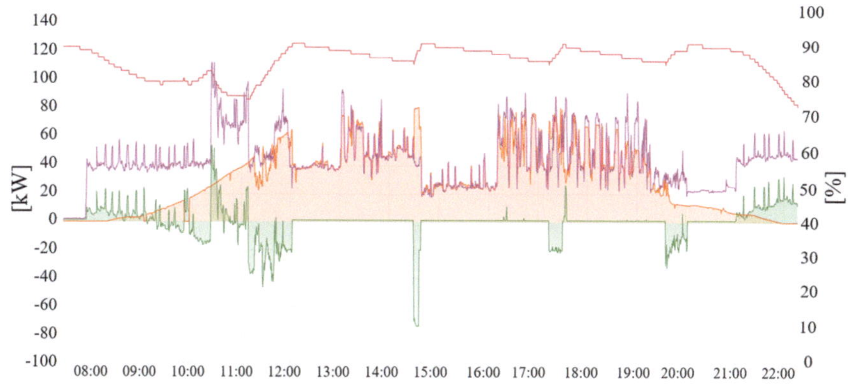

Fig. 4 Power and SOC profiles during operation (day) before the H_2 plant installation (cooling facilities not active). Batteries (kW) (green), demand (kW) (purple), PV (kW) (orange), SOC [%—right vertical axis] (red)

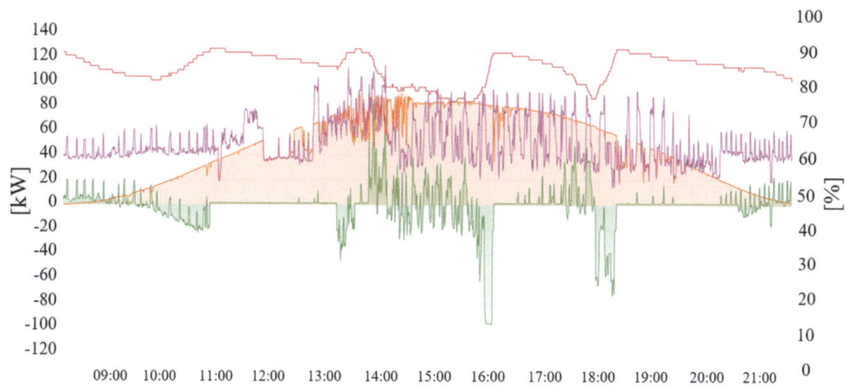

Fig. 5 Power and SOC profiles during operation (day) after H_2 plant installation (cooling facilities active). Batteries (kW) (green), demand (kW) (purple), PV (kW) (orange), SOC [%—right vertical axis] (red)

loads to supply, and then the electrolyser (blue) is powered on to store the excess of RES by producing hydrogen. Batteries are used to maintain the electrolyzer power at the set-point when there are fluctuations of the PV power production. As soon as the SOC of the batteries goes under 70%, as shown in Fig. 6 (lower panel), the electrolysis is stopped, letting the batteries (green) charge again.

The operation profiles during night operation are shown in Fig. 7. It is possible to see how the fuel cell (purple) covers the constant demand and the batteries (green) only the peaks.

A summary of the energy balances of the electrolyzer over the testing period is shown in Table 2.

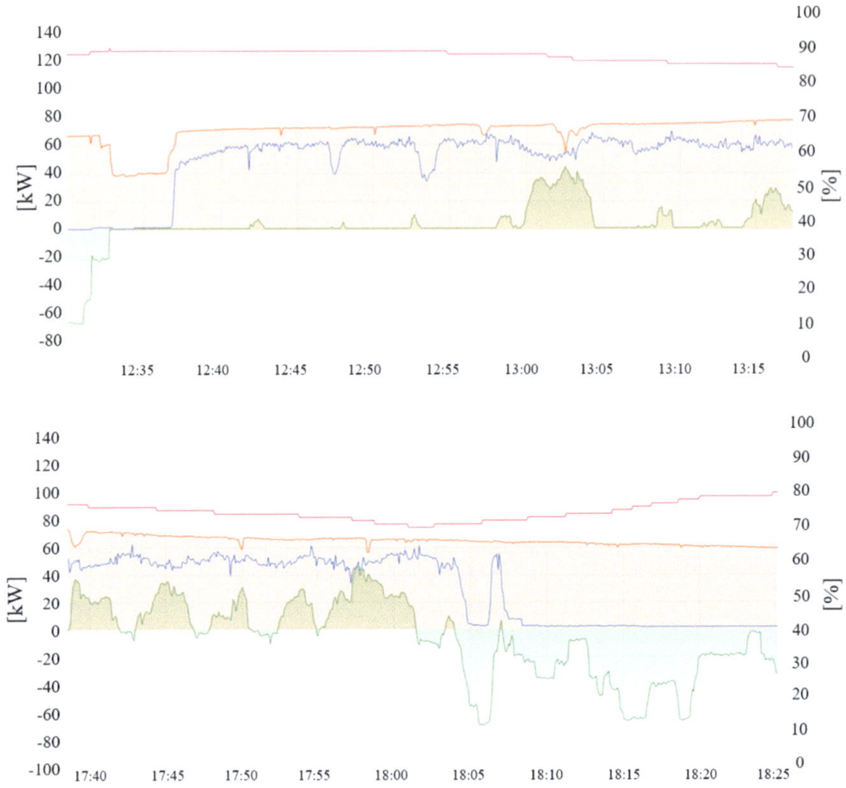

Fig. 6 Case of excess RES. Power and SOC profiles. Batteries (kW) (green), electrolyser (kW) (blu), PV (kW) (orange), SOC [%—right vertical axis] (red)

Fig. 7 Case of operation during the night (one hour of operation). Batteries (kW) (green), Fuel Cell (kW) (purple), SOC [%—right vertical axis] (red)

Table 2 Electrolyzer measured operative data

Total electricity consumption of the electrolyzer system	(kWh)	6541
Total electricity consumption of the electrolyzer system (without stand-by consumption)	(kWh)	3011
Hydrogen production	(kg)	55.3
Average efficiency (LHV) of the electrolyzer system	(%)	28.2%
Average efficiency (LHV) of the electrolyzer system (without stand-by consumption)	(%)	61.2%

The efficiency of the electrolyzer was always in the range from 60 to 70%, in line with state-of-the-art data for commercial alkaline electrolyzers. Of the total electrical energy consumed by the electrolyzer system, approximately 50% were due to electricity consumption in stand-by conditions, which are responsible for a power consumption corresponding to the 3.5% of the rated power of the electrolyzer. The total amount of hydrogen generated was about 55 kg and the average electrolyzer efficiency was 61.2% (without taking into account stand-by electricity consumed).

A summary of the energy balances of the fuel cell over the testing period is shown in Table 3. Overall, the total electrical energy produced by the stack system amounted to 235 kWh, of which approximately 150 kWh were supplied to the micro-grid. The efficiency of the Fuel Cell from field operation was always in the range from 50 to 65%, with an average value of 60.8%.

The KPIs calculated are summarized in Table 4.

In the Table, the η_{P2G} is the efficiency of the electrolyzer system (excluding the standby consumption), η_{G2P} is the Fuel Cell system efficiency, and the $\eta_{RT,BT}$ is the round-trip efficiency of the battery calculated using Eq. (13). During the period of the analysis, the RES provided directly 39% of the load of the farm (155 MWh), the batteries the 13% and the Genset the 48%. As previously stated, the fuel cell operation was limited during the plant testing period. For this reason there is an higher utilization of the Genset.

Overall, the off-grid hybrid microgrid demonstrated to be effective in reliably integrating solar PV production on the site. The battery energy storage showed an high round-trip efficiency of 87%. The RES generation accounted for 81 MWh, 76% of which was used to provide electricity to the farm and the remaining part to charge the batteries (13%) and to produce hydrogen (11%).

In order to estimate the available margin for increasing the effectiveness of the storage system to integrate the PV production in the microgrid of the site, the PV curtailment has been estimated. The theoretical production of the PV plant (110 kW

Table 3 Fuel cell measured operative data

Total electricity produced by the Fuel Cell stack	($kWh_{DC,stack}$)	235
Total electricity delivered by the Fuel Cell system to the micro-grid	($kWh_{AC,grid}$)	150
Average AC efficiency (LHV) of the Fuel Cell system	(%)	60.8%

Table 4 Results: KPIs

Load (MWh)	155.0
Electrolyzer (MWh)	8.8
Estimated PV curtailment (MWh)	28
LD_{RES} (%)	39%
LD_{BT} (%)	13%
LD_{G} (%)	48%
RES (MWh)	81
RES_{BT} (%)	13%
RES_{LD} (%)	76%
RES_{EL} (%)	11%
η_{P2G} (%)	61.2%
η_{G2P} (%)	60.8%
$\eta_{RT,BT}$ (%)	87%

peak) on the site has been calculated using the software PVGIS (EU Science Hub, Joint Research Center 2024), which amounts to 109 MWh in the 6 months of the plant operation, and the difference between the theoretical PV production and the effective measured production corresponds to the curtailment, that is 28 MWh. This value, even if calculated from a theoretical estimation, indicates that there is a not negligible fraction of PV production that could not be directly delivered to the load or stored. Hence, the storage system could potentially store and integrate more renewable electricity on the site with an optimized management. It is worth noting that the first 6 month of operation had been affected by the setting-up of the system and by limitations on the hydrogen part of the storage system, and it is expected that after the starting phase the storage system will increase its utilization factor and consequently its effectiveness to integrate the PV source.

In general, the results also show that there is a limited PV excess to be used by the P2P system, this fact causing a low utilization factor for the storage technologies installed. In order to overcome this limitation, it is foreseen to increase the PV plant size to 150 kW onsite. It is worth mentioning that batteries were partially recharged using genset energy during the night, and that with an increased PV availability for hydrogen production and battery storage, the genset use for battery recharge would be also reduced.

2.4 *Environmental and Economic Performance*

The measured data cover a limited period of 6 month. In order to evaluate the performance of the system from an environmental and economic perspective, the analysis has been extended on a yearly scenario and considering a total lifetime of the plant of 20 years.

Different scenarios have been analyzed:

- Base-case: genset only
- Case 1: current operation
- Case 2: improved PV integration (current PV size: 100 kW)
- Case 3: increased PV size (150 kW)

Base Case

The base case is developed as reference for the environmental performance. In this case, all the load is provided by the genset. On annual basis, the fuel consumption is estimated to be around 100,000 L of diesel, corresponding to 729 tons of CO_{2eq} per year.

Case 1 Current Operation

In this scenario, the plant operation is assumed to continue with the current performance for all the lifetime (20 years). This scenario reflects a non-optimized performance of the storage system. In this case, the yearly emissions are 350 tons of CO_{2eq} and the economic performance indicators estimated are 0.57 €/kWh for the LCOE and 1.50 €/kWh for the LCOS. As expected, the LCOS value is very high, reflecting the low utilization of the storage system. The LCOE value is lower due to the high contribution of the direct PV utilization ($RES_{LD} = 76\%$). The value of the LCOE is coherent with the values of 0.5–0.8 €/kWh estimated by the authors within the REMOTE project for off-grid remote PV-fed Power-to-Power systems (Marocco et al. 2021).

Case 2 Improved PV Integration

This case investigates a theoretical scenario in which the microgrid management is optimized and all the estimated PV curtailment—28 MWh—can be used. It is assumed that 50% of the additional PV production can be directly coupled with the load, while the remaining 50% is stored (35% to the battery and 15% to the electrolyzer) and provided to the load by the battery and by the fuel cell. Under this assumption, the load covered by the genset in the first six months reduces from 74.4 MWh to 50.8 MWh, decreasing the emissions (on a yearly basis) to 294 tons of CO_{2eq}/year. The economic performance indicators estimated are 0.47 €/kWh for the LCOE and 1.02 €/kWh for the LCOS. Compared to the current case, the relevant LCOS reduction is due to the better utilization of storage, with 57 MWh provided from the battery—41% increase from the current case—and 2.52 MWh from fuel cell compared to 0.3 MWh (estimated yearly) of the current case.

Case 3 Increased PV Size

In the last case investigated, the PV size is increased to 150 kW. The PV production has been estimated by PVGIS (EU Science Hub, Joint Research Center 2024)—around 300 MWh yearly—and it has been assumed that 50% of the PV production is directly covering the load, while the remaining part is stored in the battery (35%) and in hydrogen (15%) and later used to cover the load. In this case, the emissions are reduced to 135 tons of CO_{2eq}/year, and the electricity from genset is reduced from

Table 5 Environmental and economic performance of the system

	Emissions (ton CO_{2eq})	*LCOE* (€/kWh)	*LCOS* (€/kWh)
Base-case	729	–	–
Case 1: current operation	350	0.57	1.50
Case 2: improved PV integration	294	0.47	1.02
Case 3: increased PV size	135	0.39	0.59

74 MWh of the current case to 29 MWh on 6 months of operation. The economic performance indicators estimated are 0.39 €/kWh for the LCOE and 0.59 €/kWh for the LCOS, clearly showing the improved effectiveness in the utilization of the investment done for the storage system.

The environmental and economic indicators are summarized in Table 5.

3 Conclusions

This chapter has shown the analysis of the energetic, economic and environmental performance of hybrid energy storage (HES) applied in a off-grid/microgrid scenario within southern climates. The case study analyzed is a hybrid battery-hydrogen Power-to-Power demonstration plant installed in Gran Canaria within the REMOTE project. The real operation data showed that hybrid storage demonstrated to be effective in reliably integrating solar PV production on the site, halving the emissions compared to a reference case based on fossil-fueled genset. The scenarios investigated showed that an effective management of the storage system installed in the micro-grid could be able to increase the PV integration and further reduce the reliance on the genset, and consequently emissions. With an improved storage use, the LCOE and LCOS value of the current installation could be reduced from the estimated 0.57 €/kWh and 1.5 €/kWh to 0.47 and 1.02, respectively. The increase of the renewable PV generation on the site—without changing the storage capacity—could be an effective strategy to further reduce emissions to less than 20% of the reference case and to reduce LCOE and LCOS to more competitive values of 0.39 €/kWh and 0.59 €/kWh, respectively.

References

Balaras CA, Droutsa K, Dascalaki E, Kontoyiannidis S (2005) Heating energy consumption and resulting environmental impact of European apartment buildings. Energy Build 37(5):429–442. https://doi.org/10.1016/j.enbuild.2004.08.003

ECCO (2025) Setting the scene for an interconnected, renewable Mediterranean energy system. https://eccoclimate.org/setting-the-scene-for-an-interconnected-renewable-mediterranean-energy-system/.

European Commission, Directorate-General for Energy (2015) Study on actual GHG data for diesel, petrol, kerosene and natural gas. Directorate-General for Energy, Brussels
EU Science Hub, Joint Research Center (2024) Photovoltaic geographical information system. https://ec.europa.eu/jrc/en/pvgis
Fleck B, Huot M (2009) Comparative life-cycle assessment of a small wind turbine for residential off-grid use. Renew Energy 34(12):2688–2696. https://doi.org/10.1016/j.renene.2009.06.016
Jakhrani AQ, Rigit AR, Othman AK, Samo SR, Kamboh SA (2012) Estim carbon FootprS Diesel Gener Emiss. In: 2012 International conference on green and ubiquitous technology. IEEE, pp 78–81. https://doi.org/10.1109/GUT.2012.6344193
Karapidakis E, Kalogerakis C, Pompodakis E (2023) Sustainable power generation expansion in island systems with extensive RES and energy storage. Inventions 8(5):127. https://doi.org/10.3390/inventions8050127
Katsaprakakis DA, Zidianakis G, Yiannakoudakis Y, Manioudakis E, Dakanali I, Kanouras S (2020) Working on buildings' energy performance upgrade in Mediterranean climate. Energies 13(9):2159. https://doi.org/10.3390/en13092159
Marocco P, Ferrero D, Lanzini A, Santarelli M (2021) Optimal design of stand-alone solutions based on RES + hydrogen storage feeding off-grid communities. Energy Convers Manage 238(June):114147. https://doi.org/10.1016/j.enconman.2021.114147
Martinopoulos G, Papakostas KT, Papadopoulos AM (2016) Comparative analysis of various heating systems for residential buildings in Mediterranean climate. Energy Build 124(July):79–87. https://doi.org/10.1016/j.enbuild.2016.04.044
REMOTE (2023) REMOTE project. https://www.remote-euproject.eu/remote-project/
Urbanucci L, Testi D, Bruno JC (2019) Integration of reversible heat pumps in trigeneration systems for low-temperature renewable district heating and cooling microgrids. Appl Sci 9(15):3194. https://doi.org/10.3390/app9153194
Vourdoubas J (2019) Possibilities of using solar energy in district cooling systems in the Mediterranean region. Open J Energy Effic 08(02):21–34. https://doi.org/10.4236/ojee.2019.82002

Benchmarking of Hybrid Thermal and Electrical Storage for Renewable Energy Communities

Marcos Blanco, Seyede Zahra Tajalli, Sridevi Krishnamurthi, Gabriella Ferruzzi, Raffaele Liberatore, Jorge Nájera, and Kai Heussen

Abstract Renewable energy communities (RECs) facilitate the local synergy between different energy forms and offer an opportunity to couple electrical and thermal energy demands with local renewable energy sources and locally sited thermal and electrical storage. This application presents a complex integration setting for the assessment of hybrid energy storage, motivating the development of a new assessment and benchmarking framework. This chapter will revisit relevant electrical and thermal storage technologies suitable for REC integration and hybrid energy and storage systems. A benchmarking framework for application-level assessment of hybrid energy storage systems is proposed. Requirements and relevant KPIs for the REC application of a multi-domain hybrid storage system are identified. To demonstrate the benchmarking framework on the REC application, a complete reference implementation is presented, including configurable system inpucts, optimisation and simulation modules as well as a calculation module for performance indicators. A showcase of the reference benchmark is computed and analysed.

Keywords Electro-thermal hybrid · Application benchmark · Benchmarking pipeline · Open-source model · Renewable energy community (REC)

M. Blanco · J. Nájera
Unidad de Accionamientos Eléctricos, CIEMAT, Madrid, Spain

S. Z. Tajalli · K. Heussen (✉)
Department of Wind and Energy Systems, Technical University of Denmark, DTU, Lyngby, Denmark
e-mail: kheu@dtu.dk

S. Krishnamurthi
Battery Technology, SINTEF Industry, Trondheim, Norway

G. Ferruzzi
Portici Research Centre, ENEA—Italian National Agency for New Technologies, Energy and Sustainable Economic Development, Portici, Naples, Italy

R. Liberatore
Department of Energy Technologies and Renewable Sources, ENEA—Italian National Agency for New Technologies, Energy and Sustainable Economic Development, Rome, Italy

R. Scipioni et al. (eds.), *Hybrid Energy Storage*, Lecture Notes in Energy 47,
https://doi.org/10.1007/978-3-031-97755-8_9

1 Introduction and Motivation

Energy Storage Systems (ESS) have a key role in enhancing renewable energy sources' ability to contribute to grid stability, flexibility and resilience. In a fully renewable energy system, energy storage is required on several time scales to buffer and balance between intermittent and variable energy supply and the non-flexible energy demand. The integration of energy storage systems with renewable energy power plants in bulk energy systems has been extensively studied, with their benefits well-documented. Especially remote and island systems as well as microgrids, also called "electrical islands", depend on energy storage to complement the intermittent and variable nature of renewable energy sources (RES) for their operation in isolation from the bulk energy system. Since the fluctuations of RES availability appear both on short and longer time scales, the balancing energy from ESS must also be available on several scales. Hybrid Energy Storage Systems (HESS) offer the vision to combine different types of ESS to optimally match the balancing needs.

As part of a HESS, thermal energy storage systems often offer more cost-effective long-term energy storage than the often-dominant electrochemical ESS. The efficient combination of electrical with thermal storage further motivates investigation of potential thermal energy consumers in addition to purely electrical consumption, increasing the effective energy efficiency. The integration of production and consumption of both thermal and electrical energy domains is economically possible, e.g. under the regulatory paradigm of Renewable Energy Communities (RECs).[1]

As with any storage solution, the eventual benefits of hybrid energy storage must be assessed in such an application context. For any given storage technology, there are clear indicators of technological progress: efficiency, lifetime (cycle life), capacity cost (€/MWh), response time, which apply to a broad range of applications. For individual storage applications, a simple demand profile typically is sufficient to assess how key performance indicators quantified in the procurement of the appropriate technology. For *hybrid* storage solutions, however, the quantification is much less apparent: the indicators of several technologies would have to be combined, the decision on how the individual storage subsystems are used will depend on a control system with dispatch optimization considering application objectives, and the optimal integration also depends on application considerations, where complementarity and flexibility of demand, energy conversion technologies and energy transport infrastructure need to be taken into account.

The main question to be addressed in this book chapter is: *How can benefits and technology trade-offs of within hybrid energy storage solutions be benchmarked for applications where electrical and thermal loads are to be served as well as renewable energy production is available for local consumption?*

Hybrid energy storage can be beneficial when there is complementarity between several storage technologies, so that the most advantageous combinations can be expected to beat a single technology. However, since the technology parameters of

---

[1] RECs were first legally defined in the European Union Directive 2018/2001 on the promotion of the use of energy from renewable sources (Dec. 2018).

hybrid storage are not restricted by any specific storage technology, and the technologies used will interact in a given application context, it is not meaningful to assess competitiveness on a set of standard indicators that are quantified independent of the application. Instead, the idea would be to assess the technologies in an application context using benchmark systems that can accommodate a variety of storage technologies. Such benchmark systems should represent relevant applications and allow for tuning of all relevant parameters, enabling the investigation of combinations of various storage technologies.

This chapter presents the concept of this application-oriented assessment approach and illustrates it on a reference application of renewable energy communities. The renewable energy community application is a complex application model, involving thermal and electrical energy production, consumption and storage. Such applications are also expected to bring hybridization benefits in terms of the complementarity between thermal and electrical energy processes, e.g. by recovering and storing renewable or waste heat sources in addition to supplying renewable heat from electricity-based infrastructure.

As a benchmarking approach, the objective of the given framework is to enable a FAIR and open approach to benchmarking. The presented model and calculation pipeline is therefore aimed at an open-source approach, and the development of both models and calculation codes is well- documented.

1.1 A Benchmarking Framework for Hybrid Energy Storage Systems with Complex Application Integration

The proposed benchmarking and assessment framework addresses the suggested requirements, following a simple structure as illustrated in Fig. 1. The main purpose of the assessment framework is to integrate assessment of key performance indicators with a configurable benchmark system. The notion of "benchmark" comprises a reference application, including an application model, with a preselected set of key performance indicators (KPIs).

The framework supports the calculation and reporting of both static (fixed by technology choice) and dynamic (computed by a simulation model) performance indicators. Static KPI can be derived directly from the system structure definition,

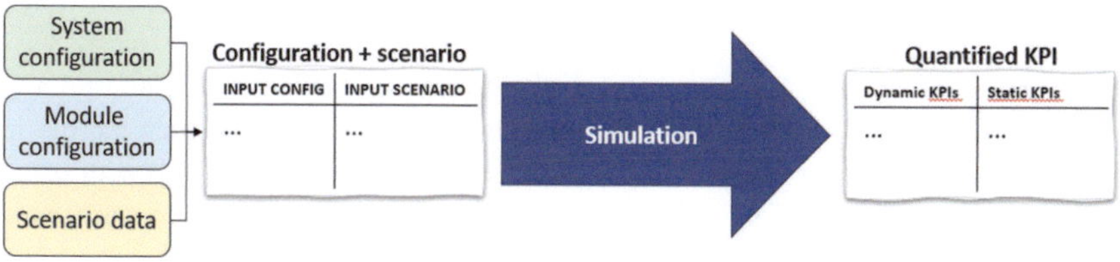

Fig. 1 Conceptual outline of assessment framework

such as the devices and materials used; static KPIs need to be updated when a user selects a different set of system parameters or replaces modules. Dynamic performance indicators require a dynamic simulation of the system's operation and can then be calculated based on system information and the time traces of the simulation.

The main idea of this approach is to harmonize the calculation principles and to deliver an integrated system view of all relevant KPIs for a hybrid energy storage application. For example, operating costs, lifetime and efficiency of a hybrid storage solution require a dynamic simulation, since the mix and contributions of each technology will vary on the operating demands and control strategies employed; the critical raw materials used to compose the respective storage solution, in contrast, are based on static look-up in respective databases. The focus of this chapter is to outline the benchmarking concept and to illustrate its configurable application to a renewable energy community case.

To perform reasonably in an application context, the hybrid energy storage or energy system also requires some form of dispatch optimization as well as logic for the real-time coordination of the respective storage systems. Such an application-oriented assessment method enables a technology-neutral approach to performance comparison. From a technology assessment point-of-view, we would require a larger set of such benchmark models, (at least) one for each of the applications considered. As renewable energy production as well as consumption profiles can vary widely depending on geographical conditions, the formulation of a benchmarking approach should eventually also consider this geographical dependency.

1.2 A Renewable Energy Community as Reference Case for Electro-Thermal Hybrid Energy Storage

A reference case was conceived to demonstrate the assessment of a complex hybrid energy storage system, integrating multiple electrical and thermal storage systems in one application. The application is considered as grid-connected Renewable Energy Community, in part because of the additional challenges of evaluating the performance of grid-connected systems as opposed to fully islanded energy system.

To construct a realistic system, the parameters of an existing electrical system were used and amended with suitable thermal consumption, energy harvesting, energy conversion, and thermal energy storage technologies. A schematic of a sample renewable energy community with hybrid electrical and thermal energy storage is illustrated in Fig. 2.

The electrical grid and resources for the reference case of an energy community is based on an existing electrical system available at the CEDER research facility in Soria, Spain. It is a micro-grid composed of a medium voltage ring (15 kV) with seven low-voltage substations. Each of these substations has combinations of renewable energy sources, energy storage systems, and consumption (electric load) connected. The complete microgrid is monitored and can be managed in real time.

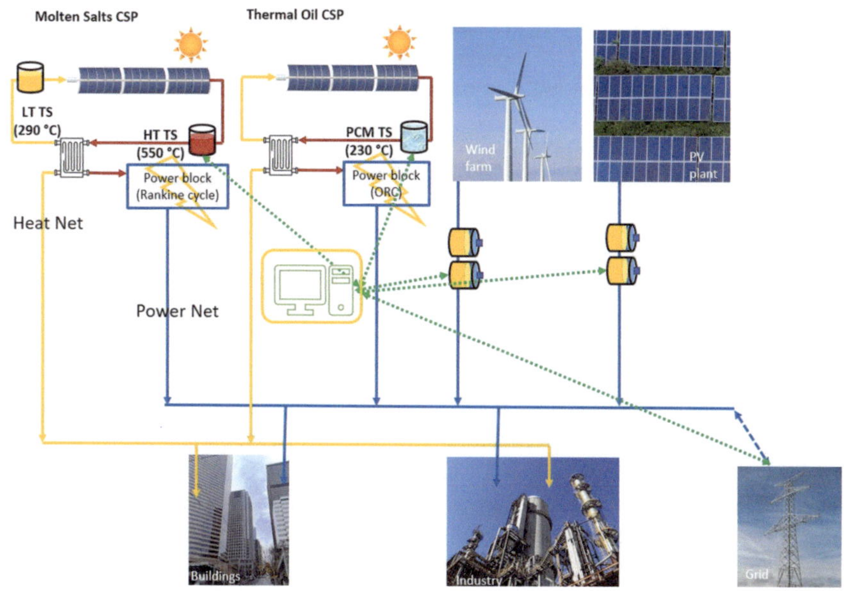

Fig. 2 Concept of a renewable energy community with hybrid electrical and thermal energy storage

To complete the CEDER-based system with a thermal network aspect, the given electrical systems is coupled with a fictional thermal network which includes two Concentrated Solar Power (CSP) plants at a different temperature level parametrised for the irradiance of the selected location, a thermochemical storage, a phase change material (PCM) storage, a boiler, a molten salt energy storage system integrated with, a Rankine cycle system for the electrical production. This system has a classic Rankine cycle using compressed steam superheated at 550 °C and an Organic Rankine cycle using an organic fluid to manage the lower temperature of the second CSP. Thermal and electrical loads include hospital, residential buildings, offices, hotels and some industrial facilities such as a pelletizer. Several of these thermal components are based on structures and research results associated with a research facility in Portici, Italy.

2 Electrical, Thermal and Hybrid Energy Storage for Renewable Energy Communities

With a growing share of discontinuous renewables on the electricity grid, such as wind and solar, needed to meet international decarbonization targets, it is becoming increasingly difficult to balance supply and demand, ensure grid stability and avoid distortions in electricity markets, so there is a need for increased research and development of energy storage and system flexibility. This section provides background on existing thermal and electrical storage technologies as well as the literature on integrating storage and hybrid storage in renewable energy systems applications, including the application of multi-domain renewable energy communities.

2.1 Thermal Energy Storage

Thermal storage, together with other forms of energy storage, can provide significant help, especially by integrating its various types of storage technologies.

Thermal Energy Storage and Coupling with Renewable Energy Production.

Thermal storage technologies are in most cases less expensive than electrical energy storage and, especially at high temperatures, they can power Rankine or Bryton cycles for the production of electrical energy and chemical processes for the production of synthetic fuels that represent a further efficient storage solution for the coupling not only of the electrical sector, but also of mobility.

The fields of application, considering different temperature levels, concern:

- Residential heating and cooling, including domestic hot water;
- Waste heat valorisation and continuity in the thermal and/or electrical supply of powered industrial processes;
- Coupling with CST/CSP (Concentrating solar thermal/power) plants, both to produce electrical energy and for the powering of industrial processes (SHIP);
- Carnot batteries (CB) include a set of multiple technologies having the common basic principle of converting electricity into thermal energy, storing it in thermal energy storage systems (TES) and reconverting the heat into electricity, when needed. In the charging phase, the electrical energy is used to heat a thermal storage medium. This task can be performed by a traditional heat pump (HP), an electric heater or any other technology, such as direct conversion by Joule effect (P2H: power to heat). Similarly, in the discharge phase, any heat engine technology can be used, using Rankine, Brayton, Stirling or other different thermodynamic cycles, such as thermoelectric, thermionic and thermo-photovoltaic systems. Pumped Thermal Energy Storage (PTES) is a type of Carnot battery.

The main types of thermal energy storage are:

- Sensible (SHTES): energy is stored or released through an increase or decrease in the temperature of the storage medium. Therefore, these systems use the heat capacity of the material to store energy, and this is always present in a single phase, usually solid or liquid. The most common example of a SHTES with a liquid medium is water, while a solid medium would be a rock-type one. Both have the advantage of being cheap storage materials. The specific heat capacity of water is about four times higher than the rock-type one. However, water needs high pressures to reach temperatures above 100 °C, while rock-type material can easily reach temperatures of 700 °C. For a SHTES system, the efficiency strongly depends on the effectiveness of the insulation provided against heat losses. Depending on this, a SHTES can achieve an efficiency included in the range between 50 and 90%. Generally, STHES systems have a low specific energy, in the range of 10–50 Wh/kg. This leads to very large storage tank sizes. The capital costs associated with a SHTES system are in the range of 3400–4500 \$/kW, while the price per unit of stored energy is in the range of 0.1–10 \$/kWh.

- Latent (LHTES): the materials used are called PCMs. The energy released or absorbed during the phase change is known as latent heat. These transitions occur at an approximately constant temperature, thus facilitating the stabilization of the temperature over which the heat transfer occurs. They have the advantage of having a high specific energy (50–150 Wh/kg) compared to SHTES systems. The capital costs required for LHTES are in the range of 6000–15,000 \$/kW, and the price per unit of stored energy is in the range of 10–85 \$/kWh, significantly higher than for SHTES systems.
- Thermochemical (TCES): these storage systems are based on the principle of feeding chemical reagents with thermal energy, resulting in their dissociation into other components that can be separated, allowing them to store the supplied thermal energy. When needed, these components are made to react with each other, to return the energy previously supplied. The capital cost required for a TCES system is the lowest of the three types of storage technologies, in the range of 1000–3000 \$/kW, and, in addition, the energy density, of 120–250 Wh/kg, is the highest. However, the price per unit of stored energy is 8–100 \$/kWh, making it the most expensive of the three storage technologies, because it has currently a low TRL, especially for the high temperature applications.

Benefits from Combination of Different Thermal Storage Technologies

An emerging method for an energy storage system combines latent and sensible thermal energy storage systems.

For instance, (Laing et al. 2012) developed a combined storage solution for direct steam generation in CSP plants with a concrete storage for superheating steam, and a LHTES for evaporating water. A 700 kWh prototype plant was built at Litoral of Endesa in Carboneras, Spain.

In Rome (Italy) a hybrid TES prototype is under construction including a module able to store 40 kWh of thermal energy by PCMs, followed by a concrete module of about 150 kWh and another 40 kWh PCM TES with a higher phase change temperature. The overall operating temperature range of the system is 290 $\div$ 450 °C (Liberatore 2024).

The benefits of these kind of systems are expected in a positive effect both to facilitate the management of the system due to stabilization of the Heat Transfer Fluid outlet temperature for many hours and to provide high-quality heat to the user.

Novotny et al. (2022) highlighted the importance of utilizing excess electricity to drive a Power-to-Heat (P2H) system, generating a temperature difference (thermal exergy). This can involve both hot and cold storage. During discharge, this stored thermal exergy is reconverted into electricity through a Heat-to-Power (H2P) system.

Dumont et al. (2020) illustrated an emerging method for a large-scale energy storage system, along with a charging cycle involving an Organic Rankine Cycle (ORC) and a discharging cycle involving a heat pump. It combines latent and sensible thermal energy storage systems. The authors demonstrate the potential of this combination through a temperature-entropy graph.

Some authors have reported the coupling of low temperature heat sources also with PTES (Pumped Thermal Electricity Storage) (Rehman et al. 2015). Other hybridization methods for medium–high temperatures used to obtain more compact cement-based TES consist of adding a shape-stabilized amount of PCM to their mix-design, so as not to compromise their thermo-mechanical performance. Such a hybrid sensitive/latent storage medium has the advantages of low concrete cost combined with a higher energy density (Miliozzi et al. 2021).

2.2 *Electrical Storage for Renewable Energy Systems*

Electrical energy storage offers the capacity to convert available surplus electricity into a storage form and to retrieve it in form electricity again. Bulk mechanical electricity storage has been available in form hydroelectric or pumped storage (gravitational) or compressed air (thermodynamic), which both depend on geographical features. Electrical and electrochemical energy storage systems, such as batteries and supercapacitors, are more flexible in terms of geographic deployment.

Energy storage mechanisms depend on the chemical and physical properties of the materials involved. In these systems, energy is typically stored in chemical form within the electrode materials (batteries) or through charge accumulation at interfaces (supercapacitors).

Batteries

Batteries store energy through chemical processes, specifically faradaic reactions or redox reactions, which take place at the anode and cathode. These reactions involve the transfer of electrons during oxidation at the anode and reduction at the cathode, enabling the conversion between chemical and electrical energy.

Battery storage technologies come in various forms, each characterized by attributes like energy efficiency, specific energy, and cycle duration. Lead-acid batteries, one of the oldest and most widely used technologies, are favoured for bulk energy storage due to their low cost and reliability (Mustafizur et al. 2020). However, they suffer from limitations such as short lifetimes, low energy density, and a restricted depth of discharge (DOD), which impacts their efficiency and usability, and they face environmental challenges due to the toxic nature of lead. In contrast, lithium-ion batteries provide significantly higher energy efficiency (90–95%), longer lifespans, and superior power density, making them ideal for electric vehicles. These advantages come at a higher initial cost and with concerns around the sourcing of critical raw materials like lithium and cobalt (Mustafizur et al. 2020; Mitali et al. 2022). Flow batteries, on the other hand, stand out for their flexibility, allowing power and energy capacity to be scaled independently. This makes them well-suited for large-scale grid applications, though their lower energy density and larger physical footprint limit their use in space-limited applications, Additionally, the most widely used vanadium flow batteries entail higher costs compared to lithium-ion batteries (Poli et al. 2024). Finally, nickel–cadmium batteries offer high energy density, durability, and low maintenance requirements, performing well under extreme conditions. However, they face environmental challenges due to the toxic nature of cadmium, leading to a decline in usage as more sustainable alternatives, such as nickel-metal hydride and lithium-ion batteries, have become available. Each of these technologies presents trade-offs in cost, efficiency, and environmental impact, influencing their application in specific use cases (Mustafizur et al. 2020; Mitali et al. 2022).

According to the review article by (Mustafizur et al. 2020), electro-chemical battery storage systems rank third in installed capacity, with a total power of 2.03 GW. The most widely used utility-scale electro-chemical batteries are lead-acid, lithium-ion, sodium-sulphur, nickel–cadmium, and flow batteries. Among battery technologies, Li-ion holds the largest market share, with a power production capacity of 1.66 GW, followed by sodium-based batteries (204.32 MW) and flow batteries (71.94 MW).

Supercapacitors

Supercapacitors are considered a complementary technology to batteries due to their unique characteristics, including high power density and exceptional cyclability, although they exhibit low energy density. These devices store energy by separating negative and positive charges. This separation allows supercapacitors to store static electricity for later use, enabling rapid energy release when needed.

There are three main types of supercapacitors: electric double-layer capacitors (EDLCs), pseudo capacitors, and hybrid supercapacitors. EDLCs store energy through charge accumulation on both sides of the dielectric layer, a process that is highly reversible and provides high power density. However, their energy density is relatively low compared to batteries. Pseudo capacitors combine this mechanism with Faradaic redox reactions, enabling improved energy storage through features like a large contact area and short paths for electron and ion transport. Despite these advantages, their performance heavily depends on precise control of electrode properties such as morphology, structure, and composition. Deviations can lead to reduced electroactive surface area and diminished performance. Hybrid supercapacitors integrate both Faradaic and non-Faradaic processes in their electrodes. Even though some electrode materials like carbon and metal oxides may exhibit poor conductivity, hybrid supercapacitors achieve high cycling stability, power density, and energy density. This is often accomplished using advanced 3D mesoporous electrode structures, which enhance their performance and affordability. These features make supercapacitors an invaluable addition to energy storage systems, especially in applications requiring quick energy discharge and high durability (Zhang et al. 2023).

In a HESS, supercapacitors are effectively utilized to mitigate short-term voltage and frequency fluctuations in the grid. Their high-power density and rapid charge–discharge capability make them well-suited for handling transient disturbances. When integrated with larger energy storage systems, such as batteries, supercapacitors act as a complementary component (Zhang et al. 2023).

Bocklisch (2015) highlights the advantages of combining storage technologies, emphasizing how hybrid systems can enhance overall efficiency. This approach is particularly valuable in mitigating the drawbacks or damages associated with reliance on a single technology, offering a more balanced and resilient energy storage solution.

2.3 Hybrid Electrical and Thermal Energy Storage in Energy Systems Applications

The European Union Renewable Energy Directive (RED II), part of the Clean Energy for All Europeans Package, defines "Renewable Energy Communities" (RECs), establishes a governance framework for them, and enables energy sharing within these communities. RECs are citizen-driven projects designed to advance the clean energy transition through various forms of consumer co-ownership. These initiatives empower local communities to engage in renewable energy production and consumption, fostering resilience, enhancing energy efficiency, and supporting decentralized renewable energy generation.

Storage systems are considered vital components of RECs. These systems play a crucial role by storing variable energy and deploying it as dispatchable generation during periods of high demand within the electrical grid (Lowitzsch et al. 2020). This raises the question of hybridizing these storage systems: can combining different forms of electrical and thermal energy storage provide enhanced benefits compared to using them individually?

Whereas electrical and thermal energy forms are relevant to quite different applications and require different infrastructure. Hybridization of electrical and thermal energy storage solutions is typically only relevant in applications where both forms of energy are eventually required by an end user. From a physical perspective, thermal storage at lower temperatures has lower quality, since it cannot be converted to electricity. Depending on the infrastructure involved and the system boundary, the services required, and the applicable legal and commercial conditions, the performance criteria and applicable solution concepts vary widely, also in terminology and technological maturity.

Storage System Integration with Renewable Energy in Applications

A report from the National Renewable Energy Laboratory (NREL) (Reilly et al. 2022) examines state-of-the-art wind-storage configurations and system controls, along with the techno-economic sizing of wind-battery hybrid systems for different grid types. Extensive research has been conducted on the design, regulation, and management of energy within storage systems, with a strong emphasis on optimization algorithms for sizing individual components. In their review article, Nkwanyana et al. (2023). Examine the optimization techniques used in the design, configuration, and deployment of various RESs and different storage systems.

Future economic feasibility studies have explored the impact of configuration-location combinations in HES hybrid systems. Schleifer et al. (2023) demonstrated that smaller batteries can achieve similar economic performance to larger ones when paired with complementary PV-wind systems. They also analysed how the energy and capacity values of PV-wind-battery hybrid systems might evolve over time across locations with differing levels of solar and wind complementarity. Aykut Fatih Guven et al. (2024) take a nuanced approach by exploring a range of scenarios involving load demand and generation fluctuations due to the variability of RESs. By utilizing real-world data on load, wind speed, and solar irradiance, alongside multiple storage options, they examine various algorithms for convergence and optimize the system to identify the most efficient configuration. As can be seen, a lot of these studies focused on specific storage technologies and their optimal integration into a RES system.

Sector Coupling in Hybrid Energy Systems

Sector coupling refers to the exploitation of flexibility options and associated benefits (e.g. increased renewable energy adoption) across energy usage sectors, such as electricity and heating or electricity and transportation.

The studies mentioned above do not include sector coupling. Sector coupling between electricity and thermal systems is recognized as a crucial strategy for maximizing the full potential of RESs and supporting the achievement of green energy transition goals6. This approach enables excess renewable energy to be stored and converted into heat using boilers or heat pumps, which can then be stored in thermal energy storage systems, such as pumped thermal energy storage (PTES), for later use (Steinmann et al. 2019).

Vehicle-to-Grid (V2G) and Vehicle-to-Home (V2H) technologies have been proposed as innovative solutions to support decarbonization efforts. The core concept involves utilizing the batteries (Lithium-ion) of electric vehicles (EVs) to store energy during periods of low demand and discharge it during peak hours to assist with peak shaving. To evaluate their feasibility and effectiveness, various techno-economic models have been developed. Feasibility studies have also been conducted on integrating HRES into renewable energy hybrid microgrids and vehicle-to-grid/home systems. García-Vázquez et al. (2022) studied the feasibility of a hybrid renewable energy system with an energy storage system for a house, where the vehicle-to-home (V2H) option is considered as a backup/support system and focused on the sizing of PV panels and storage devices. The widespread adoption of these technologies by the public remains to be seen in the future. A study conducted in Norway—one of the countries with the highest EV adoption rates—highlighted key strategies to encourage uptake. These include providing financial incentives such as tax credits, launching educational campaigns to emphasize the benefits of EVs with V2G capabilities, expanding EV charging infrastructure, and fostering the development of a robust public–private V2G ecosystem (Mehdizadeh et al. 2024).

Goyal et al. (2024) propose a design for a sector-coupled microgrid that integrates the electric, thermal, hydrogen, and transport sectors. Their design incorporates battery storage for electricity and hydrogen storage tanks for hydrogen. The study evaluates the techno-economic impacts and total emissions, indicating that a decarbonized sector-coupled microgrid is both feasible and effective in advancing sustainability goals.

Electro-Thermal Hybrid Storage Energy Systems Applications

Sihvonen et al. (2024). investigated an electrical-thermal hybrid storage system, involving an underground pumped hydro storage (PHS) system and a sand-based high-temperature thermal energy storage (HTTES), an electrical thermal energy storage, which is coupled with power-to-heat and discharged into district heating (DH) in an island energy system. Their findings highlight a clear need for storage capacity from diverse sources. However, this study did not consider electrical storage components, such as batteries and supercapacitors, revealing a gap in comprehensive research on integrating both thermal and electrical storage systems into RESs grids, particularly when coupled with power-to-heat technologies.

Geographical Variability Effects on Hybrid Energy Storage Applicability

Most studies investigate the application of hybrid energy systems assessments for a specific given geographic location or area. Zhu et al. (2024) conducted a comprehensive review of the findings and limitations associated with various REss and storage solutions implemented at different sites worldwide.

Our review identified several studies that address different types of consumption data and load profiles based on seasonality. Enrique Rosales-Asensio et al. (2024) examined two distinct use cases of buildings in different locations—New York and California—using REopt to determine the optimal sizing of storage systems for minimal costs and emissions. The study revealed a high level of complexity, with outcomes heavily influenced by factors such as solar irradiance, load profiles, and the energy tariffs applied. Kobashi et al. (2022) investigated the integration of EVs in commercial and residential districts across cities in Japan. Their study concluded that such strategies could play a significant role in advancing urban decarbonization efforts. Lysenko et al. (2023) modelled the variable nature of RES and consumption as a random process in their study of hybrid power systems with storage. This approach was used to analyse and optimize the sizing of the storage battery system, accounting for the stochastic behaviour of energy generation and demand. Mazzeo et al. (2020) investigated the worldwide techno-economic mapping and optimization of standalone and grid connected PV-wind HRES to supply the electrical demand of an office building district. These studies, however, neither account for sector coupling nor offer a comprehensive framework for analysing alternative use cases.

3 Benchmarking Framework for Hybrid Energy Storage

Hybrid energy storage technology performance is characterized by three aspects that are each influenced by the application they operate under: (a) storage technology performance, (b) interaction and integration of multiple storage technologies, and (c) the control and coordination strategies. To assess the potential impact of variations in material, device, integration and control system parameters at an application level, it is of interest to offer a benchmarking approach that accounts fully for the relevant constraints and performance criteria for the given application.

Since modelling and simulation of energy storage in energy systems applications is critical for planning and technical assessment of potential energy solutions, there exists a wealth of planning and technical design tools and models in the field, each aimed at different purposes. However, benchmarking contrasts with the requirements planning and technical design, which typically include a case-by-case adaptation of model and analysis approach: the performance estimation via a benchmark instead requires a stable but configurable and easily adaptable benchmark model.

Another dimension of interest would then be to consider adaptation to the geographic and associated environmental conditions which induce a systematic and consistent variation of input parameters that would substantially affect outcomes, given different demand and resource availability. In this section, we will outline a benchmarking framework that supports the above-listed requirements.

3.1 Requirements for Application-Level Performance Assessment for Hybrid Energy Storage

To develop a framework for benchmarking for hybrid energy storage at application-level, we shall consider the key ingredients of the benchmark model: an appropriate system application model (system boundary and dynamics), associated modelling tools (potential modularization and configurability), assessment scenarios (ensuring consistent input data), and the identification of relevant performance assessment criteria.

3.1.1 Functional Aspects of an Application-level Benchmark

An effective benchmark case should consider and represent the following aspects:

Dynamic Simulations Covering the Full Application Range Including Partial Charge Performance

Storage systems must be able to respond quickly to verify the flexibility to be provided to users and grid services needed for RES integration, including the time scale. For thermal systems, partial charge studies quantify how much performance can degrade when operating temperatures are dynamically varied as compared to the thermodynamic design.

Application-Level Cross-Domain Integration of Storage with Non-Storage Sub-Systems

Hybrid thermal/electrical integrated systems should have significant advantages in terms of reducing energy losses. Such application-level benefits cannot be assessed at the level of storage technology.

Multi-Model Capability

Hybrid energy storage applications, accrue benefits from operating over a wide range of timescales and behaviours relevant to application-level performance. A single model cannot capture the full range of these application-aspects. A benchmark requires a set of jointly configurable models.

Control Strategies

Control strategies determine the system behaviour and performance in view of a control objective, concerning, for example, power/capacity range, fluctuations in electricity prices, electricity demand, power production and any waste heat integration as well as produce a clear environmental benefit especially for lower greenhouse gas emissions. Several control strategies may be combined to cover several time scales.

Performance Indicators and Cost Estimates

A benchmark is defined by its ability to summarize performance of a specific technology in reference to specific performance indicators. Such indicators should cover the full range of relevant technological, economic and sustainability indicators, consistent across sub-modules. E.g. cost reduction should be comparable sustainability trade-offs in the use of critical raw materials.

Assessment Reference Scenarios and Geographical Adaptation

Benchmark-style assessment requires a reference set of input operating scenario data. In RES based systems, the geographical location used as reference has significant effects toward the evaluation of a given system design. In principle it is possible to relocate a given model to a distinct geographical location by altering the environmental variables and associated input time series.

3.1.2 Limitations of Existing Modeling Tools in View of Benchmarking Requirements

Modelling tools used for performance assessment are not designed to meet all of the above listed requirements for benchmarking. Distinguished by the modelling purpose, we can revisit typical capabilities of energy system planning/optimisation tools and technical design and analysis tools.

Modelling and Planning Tools for Hybrid Energy Systems and Assessment Criteria

Since both renewable energy and demand tend to be fluctuating, uncertain and uncontrollable, the performance of a storage solution needs to be assessed in context of its application, e.g. to balance energy supply and demand, or provide the minimum cost of energy, given electricity market prices and varying tariffs, and possibly also to offer flexibility back to the energy system/electricity grid. Planning tools contain models of storage technologies and consider the expected behaviour over time of renewable energy supply and energy demand, a planning tool typically offers to optimize a system (configuration, sizing) for a given performance indicator (e.g. revenues, total system cost, CO_2 emissions). A challenge for most planning tools is, however, that electricity storage operates across time and across time scales, that revenues from multiple sources can be stacked, but also that the resource allocation at different time scales interact. An additional challenge in the planning of hybrid storage solutions is that there is an interaction between storage types, depending on the optimal forward-looking allocation as well as on the real-time reactive control algorithms, operating on a timescale of microseconds to minutes. It is therefore challenging to determine an appropriate problem reduction without too many (over-)simplifying assumptions.

There is a wide range of energy planning and modelling tools available that are capable of modelling hybrid energy systems, both as open source, free software (e.g. EnergyPlan[2]) and commercial software (e.g. Plexos,[3] Homer Pro[4]). These software packages are widely used for sizing, simulating operational scenarios, and optimizing multi-domain energy systems—covering thermal and electrical components—while integrating hybrid renewable and conventional energy sources, as well as multiple storage technologies. However, given that some of these tools come with limitations. For example, Plexos and Homer Pro are not free to use and typically require paid licenses, which can be a barrier for many users. While EnergyPlan is a free software for modelling smart energy systems, its hard-coded heuristics make it difficult to alter performance objectives, assess technology variants or to integrate it in a pipeline with other tools. A common challenge with the adoption of commercial planning tools is the lock-in situation since the models are typically not transferrable or open to modifications, and the performance objectives are built into the system. Limited configurability and openness become a barrier when more complex adaptations are required, as in the case of hybrid energy storage for grid-connected renewable energy communities.

Alternative to commercial planning tools are open-source energy system models. There is a wide range of open software models (https://wiki.openmod-initiative.org/wiki/Open_Models), each with a specific modelling, planning or assessment scope. Most models are built around an economic optimization capability, enabling optimization over time scales of days to decades.

Assessment of Non-Stationary Behaviour and Dynamic Technical Performance

In contrast with the energy systems modelling tools, a separate category of technical modelling software aims to represent the detailed dynamic (transient) behaviour of the technical system, typically in time scales from hours to microseconds, depending on the domain relevant physical phenomena and technology investigated. Here the use of commercial modelling software is dominant (e.g. MATLAB/Simulink, TRNSYS, Dymola, PSS/e), but also open-source platforms exist (e.g. Octave, Modelica/OpenModelica). Here, modelling tools can be divided in all-purpose (e.g. Simulink, Modelica) and domain-specific tools (e.g. PSS/e for Power systems, TRNSYS for Thermal applications).

[2] https://energyplan.eu/

[3] https://www.energyexemplar.com/plexos

[4] https://homerenergy.com/products/pro/index.html

System Boundaries and Their Effect on Model Detail and Assessment Criteria

Depending on the size, geographical extent, or electrical grid connection type, hybrid energy systems vary widely in the applicable technologies as well as the relevant services to be accounted for. Practical planning tools typically abstract the technologies and simplify the application service to allow for a general tool implementation, foregoing technology details for the benefit of a generic and simple planning tool. Detailed technical models increase time-resolution and thus enable the investigation of non-stationary physical phenomena that can affect the integration technologies and performance of the integrated system. Such detailed models limit the system boundary and accuracy of representation to the relevant time scale and relevant transient interactions.

The system boundary also affects the objectives, incentive structures and technical operating parameters of an energy system design. For example, an isolated (island) energy system is simpler to model in a planning tool (e.g. EnergyPlan), but the tool faces limitations when applied to scenarios where the system is connected to the grid and influenced by e.g. dynamic market prices and external grid interactions.

Geographical Adaptation of Assessments

Data sources like SECURES-Met (Europe) and TRNSYS (World) offer precise meteorological information, such as radiation and wind speed, which are vital for estimating solar and wind energy production. Similarly, electricity prices and associated tariffs and taxes vary significantly across regions; as most electricity networks in Europe are fully transparent on electricity prices, so such data can be located and integrated for several world regions.[5] Finally, consumption data, which also varies geographically and seasonally, is equally essential for developing a comprehensive assessment tool. Comprehensive, geographically organized and aggregated data on electricity, heat and fuel consumption is available, e.g. via (Hidalgo and Uihlein 2023). However, for assessing storage performance, detailed and disaggregated time-series are required, based on the modelling assumptions and time resolution; this aspect is difficult to achieve based on geographical databases alone. Here representative time series may be chosen from public databases,[6] considering an approximate match of e.g. climate, culture and business operating hours, but also synthetic load profile generation should be considered as a valid approach to represent characteristic load behaviour.

[5] An overview of useful data sources for electricity prices is listed here: https://wiki.openmod-initiative.org/wiki/File:Electricity_prices

[6] https://wiki.openmod-initiative.org/wiki/Data

Challenges with Existing KPIs and Performance Assessments

Key performance indicators (KPIs) are intended to enable technology-neutral assessment to differentiate technologies by performance for the same application. It is often desired to identify optimal performance or a performance trade-off between alternative technologies. Optimization methods may use a cost function to represent the performance indicators directly—generating an optimal design for the considered "cost". Above-mentioned planning tools will optimize typically by minimizing the total system cost (investment and operational), considering other technical criteria in the form of constraints (e.g. reduce/eliminate exchange with the power grid). For planning, where a wide range of alternatives need to be considered and traded off, such simple criteria are useful.

The suitability of a design, however, can in most applications not be reduced to a single straightforward cost (function) or constraint. The need to reduce complexity for optimization purposes also implies that the model details cannot completely capture the full range of criteria, so detailed dynamic models may be required to extend the assessment. It is therefore meaningful to formulate additional performance criteria that may be derived from the available application models.

3.2 *The Proposed Application-Level Assessment Framework*

This work focuses on developing a technology-neutral, open-source framework to quantify the application-oriented KPIs of configurable HESS, including electro-thermal storage. The framework is intended to be adaptable to alternative geographic locations, enabling seamless integration of generation and consumption data to facilitate comprehensive analysis based on the application-level system performance. Figure 3 outlines the main components and interactions in the proposed application-level assessment framework. Notably, the framework imposes a fixed benchmark theme, which then determines the content of the configuration (light blue), model (red) and KPI calculation (green) modules. Following the principle of benchmarking, a user interacts with the configuration, not the models or the KPI selection.

Application Benchmark Definition and Models

For a given application, the benchmark defines a specific application with a reference system configuration, relevant performance indicators and a complete reference implementation of the required models. Baseline scenarios and valid KPI ranges are identified for the given application. Static computations and data are available for configuration-driven calculations of static KPIs. The computational modules that perform the calculations for the associated KPIs are also included to avoid interpretation issues in text-based definitions of performance indicators. The model is documented fully so that a user can interpret and interact with the system and

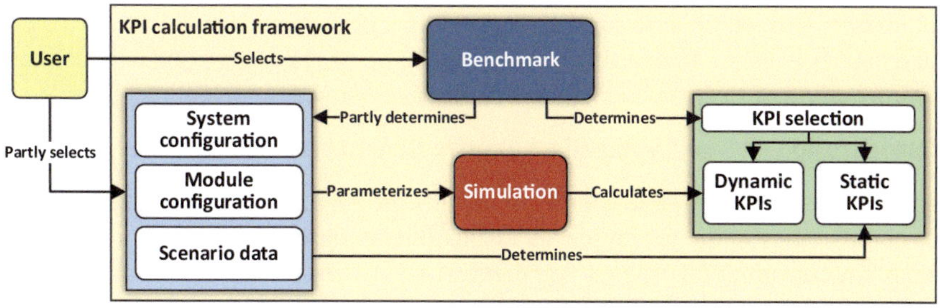

Fig. 3 Schematic of assessment framework with benchmark context

scenarios configuration. An application benchmark typically comprises several sub-models, such as for example a scheduling model and a dynamic operational model in order to represent relevant dynamics in all timescales associated with the quantified performance indicators.

System and Scenario Configuration

For the benchmark model, it is essential that the calculations are harmonized and any modifications to the benchmark parameters are traceable. It is therefore paramount to enable a file-based parametric configuration of the benchmark model, enabling to document and trace parameter changes to observed effects and to enable comparison of results across the literature. The user interface of the benchmark is therefore mainly in the specification of system configuration and scenario parameters, enabling the otherwise consistent and automated assessment of a given benchmark across several sub-models. Examples of adjustable parameters is the selection of size or numbers and types of storage modules of different types of storage included in the benchmark. Users may be able to investigate alternative efficiency among the technologies. An extreme case of configurability would be the possibility to alter the geo-location of the application, as this would result in the need to automatically update the location-associated time series and parameters, as discussed above.

Module Configuration and Model Adaptation

The application benchmark should ideally be formulated in such a way that it includes and is enabled to operate with a wide range of storage technologies, which may or may not be activated. An intended but more comprehensive adaptation of the benchmark could be performed to integrate and alternative storage technology. The benchmark application model would have to be updated with a drop-in replacement of the respective model components in all the required sub-models.

Assessment Pipeline Automation

An implementation of this proposed benchmarking framework would be greatly facilitated by means of a configuration-driven software pipeline, where the input parameters, simulation model results and resulting performance indicators are calculated based on an automated and parallelizable process. This approach will facilitate the integration of the proposed framework in automated assessment pipelines.

Consideration of Geographically Adaptive Benchmarking

Since datasets for renewable energy production are available based on geo-location, the complexities in the geographical adaptation are more present in the representation of the consumption profiles. Another feature in a benchmarking approach would be flexibility on integration options that match the local build environment, regulatory requirements and so on. These complexities cannot be addressed all at once. However, a framework that facilitates the geographical adaptation of the benchmark in a computational sense (i.e. by parameterization of the relevant variables) would greatly facilitate the future adaptation of the given benchmark case to various local conditions. The suggested approach is therefore a clustered adaptation to specific design-geolocations, where the complete dataset is available and the benchmark has a meaningful representation for the given application.

3.3 Considerations on Performance Indicators

Key performance indicators are the main device of comparison stipulated by a benchmark. A benchmark setup should in principle include many pre-defined performance indicators to ensure consistency of comparisons. On the other hand, the prioritization of a small number of "key" performance indicators facilitates the discourse.

From Performance Indicators to Application-Oriented KPIs

In view of designing a valid application-oriented benchmark system for hybrid energy storage, it is important to consider a selection of key performance indicators that are relevant to be quantified by means of the application-oriented benchmark model. There are technical considerations on the requirements for computing aggregate, dynamic KPIs for a hybrid storage application, there are criteria to select and prioritize KPIs.

On Computation Requirements for KPI Calculation and Aggregation

A recent analysis of performance indicators for hybrid energy storage systems in Heussen et al. (2025a) and Scipioni (2025) suggests a classification of by several characteristics, that are important to establish a basis for comparison of hybrid to non-hybrid energy storage solutions. The two main considerations for a given KPI are (a) the relevant method of aggregation of KPIs from single-technology KPIs to KPIs representing the HESS performance, and (b) the variability subject to dynamic behaviour of the storage in operation.

Assessing overall system performance is to *aggregate* the KPIs using different methods:

- **Weighted Average KPIs:** These are KPIs calculated based on the proportional contributions of each component in the hybrid system.
- **Cumulative KPIs:** KPIs representing the total contributions of all components, reflecting the overall system performance.
- **Dominant- or Limiting- Technology Influenced KPIs**: These are KPIs that are influenced by either the strongest or the weakest technology in the system.
- **Application-level KPI**: KPI to benchmark the performance (enhancement) of an integrated system design, for a given application (requires a benchmark system implementation)

The second distinction can be made between dynamic and static KPIs. **Static KPIs** are primarily determined by the inherent material or device properties of the system components. In contrast, **dynamic KPIs** require evaluation of time-dependent behaviour in an application; for hybrid energy storage, this behaviour is also influenced by the interactions among storage technologies. These dynamic metrics require a representation of the controlled system behaviour under representative operating conditions.

A further distinction relevant to computation of KPIs is whether the indicator can be computed independently of (**absolute KPI**) or with reference to a baseline value (**relative KPI**). In application-based benchmarks Dimensioning-based vs. Fixed-sizing-based KPI: KPIs that consider system sizing based on operational dimensions versus fixed-sizing approaches.

On the Formulation of Key Performance Indicators for Application-Level Assessment

Firstly, performance indicators should reflect a sufficiently holistic view on technology performance so that important trade-offs are directly apparent from an analysis. For example, if economic improvements come at the cost of sustainability performance this should become immediately evident. Thus, a relevant set of performance indicators should include at least one indicator from each "technical", "economic" and "sustainability" categories. Secondly, the importance of performance indicators should be weighed by the interest of relevant stakeholders in those indicators. Finally, the prioritization of KPI is driven by the considered application and its associated technology-neutral service requirements. To select effective KPI the authors therefore suggest the following criteria for the formulation and selection of KPIs: quantitative, technology-agnostic, reproducible, mathematically robust, and well-defined across the system's entire operating range. Additionally, they must effectively capture key objectives.

In summary, we suggest the following criteria for discussing and selecting KPIs:

- *Quantitative and Comparable*: KPIs that are measurable and can be compared across different systems or technologies.
- *Replicable/Reproducible*: KPIs that can be consistently replicated across various scenarios or studies.
- *Feasible to quantify with the given benchmark*: KPIs that can be effectively quantified using the established benchmark.

- *Mathematically Sound*/Well-defined: KPIs that are defined clearly and are supported by sound mathematical principles.
- *Relevant to the Application*: KPIs that directly apply to and reflect the objectives of the specific system use cases, motivated by.
- *Absolute, if possible*: Give preference to KPI defined in absolute terms instead of relative to a reference scenario since citation of reference baselines weakens the general interpretability outside the application context.

The proposed criteria assisted in the below reported KPI selection process.

4 Technology Neutral System Requirements for REC

In this section performance indicators and functional requirements for a REC are outlined.

4.1 Storage Services in a REC Application

The foundation for RECs is the "Community" setting and its identification with a renewable energy supply. Renewable energy production (bio, thermal, electrical) is volatile and does not meet consumption needs for thermal, chemical or electrical energy in time. Storage can shift this energy in time and thus may enable a fully autonomous energy supply if the community were entirely based on an islanded or isolated energy system. However, the REC concept also applies to grid-connected communities, where at least the electricity exchange with the environment is possible.

Regarding the non-functional requirements for the integration of multi-domain hybrid energy storage systems in a Renewable Energy Community, three main groups of services can be facilitated: a) cross-domain energy balancing services, b) slow response grid services, and c) fast response grid services.

The first set of services is associated with optimizing energy efficiency, energy cost and minimizing CO_2 emissions and operates over timescales of hours to days. Here the thermal storage systems and cross-domain energy conversion units come to bear. The second and third group of services are focused on the balancing of the electricity flows. In case of exchange with the electricity grid it would be concerned with electricity prices and ancillary grid services, or, in case of a fully islanded/isolated system, it would be the local frequency and power balancing.

Slow-response grid services ramp in the range of tens of seconds to minutes, for a step equal to nominal power) and last in the range of one or several hours at nominal power. Examples of objectives in this group of services are minimizing consumption from the grid, reducing the curtailment of renewables, or participation in the electricity market for maximizing profit. Fast response grid services operate in the range of hundreds of milliseconds to a few seconds and last in the range of minutes at nominal power, such as participation in ancillary services (frequency regulation) or other fast response grid services (angle stability, reactive, etc.) current or future.

4.2 Performance Indicators for a Renewable Energy Community

Performance indicators should enable the assessment of hybrid energy storage contributions in an 'open' energy system application, i.e. a REC may exchange energy with its surrounding infrastructure. This section reports on the KPI selection for RECs, considering the criteria introduced in Sect. 3.3.

Relevant Performance Indicators for RECs

To quantify the benefits of hybridizing various energy storage technologies in a technology-neutral manner, the application-level performance should measure the effectiveness of the HESS in supporting the above listed services. If we further consider the stakeholders setting expectations to the REC, additional KPIs may emerge:

- End Users: Both residential, municipal, and industrial (SME) end users, who have evolved from passive participants to active contributors in the energy chain due to liberalization.
- Transmission and Distributed System Operator (TSO/DSO): Responsible for managing the high, medium and low voltage networks, ensuring the security and reliability of the energy system.
- Regulatory Authorities and Institutions: These stakeholders establish the rules and regulatory frameworks, ensuring that all parties adhere to the appropriate guidelines. They are also responsible for transposing EU regulations into national laws.
- Other Market Operators: This category includes electricity producers, aggregators, suppliers, Energy Service Companies (ESCOs), DR aggregators, and prosumers

It is self-evident that the services listed above are dynamic in nature and only a dynamic representation of the REC can offer the information needed to quantify them. Therefore, dynamic KPIs are required to evaluate system performance.

From a literature review on KPIs in energy system management (Scipioni 2025) three categories of relevant KPIs emerged: technical, economic, and sustainability indicators. Indicators within each category often overlap in scope and not all are relevant to the REC application.

Selected KPIs for This Case Study

In Table 1, a list of selected KPIs is presented to evaluate the REC from different aspects. The selection of KPIs for assessing RECs is driven by the need to evaluate their sustainability, economic, and operational performance. System Energy Efficiency ensures optimal use of renewable resources, reflecting energy waste. Meanwhile, System Operational CO_2 Emissions quantify the sustainability aspect, highlighting decarbonization efforts. The Flexibility Factor measures the system's ability to adjust consumption, generation, or ESS based on market/system status.

Economic assessment is a crucial aspect of REC development, highlighting Capital Expenditure (CAPEX) and Operational Expenditure (OPEX) as essential

Table 1 List of KPI for this case study

Name	Description	Expected range for REC from literature	Classification
System energy efficiency	The ratio of consumed energy to the total generated energy. *This KPI is mainly relevant to regulatory authorities*	85–95% (Manso-Burgos et al. 2022; M et al. 2023)	Dynamic, absolute, application-level, technical
System operational CO_2 emission	The sustainability performance of an energy system by comparing its carbon emissions to a baseline scenario	0.08–0.2 $kgCO_2$/kW*1 (Bianco et al. 2021; Manso-Burgos et al. 2022; M et al. 2023; Angelakoglou et al. 2020)	Dynamic, relative, application-level, sustainability, SH: all
Flexibility factor	Quantifies the technical capability of the system to adjust consumption, generation, or energy storage systems to import energy at the lowest price and export energy at the highest price within the overall trade market. *Relevant to consumer & DSO*	–*2	Dynamic, absolute, application-level, technical
Capital expenditure (CAPEX)	Initial investment costs required to set up renewable energy infrastructure and related assets, defined here as Initial installation cost per kilowatt of installed capacity	1500–2000 €/kW *3	Static, absolute, application-level, economic
Operational expenditure (OPEX)	Costs associated with operation and maintenance after the initial setup	– €/KWh	Static, absolute, application-level, economic
Levelized cost of electricity (LCOE)	The total discounted costs incurred over the lifetime of a power generating system—including CAPEX and OPEX—by the total electricity generated during its lifespan	0.15–0.40 €/KWh *3	Static, absolute, application-level, economic

Table Notes: *(1) Range estimated for "fully self-supplied"—"fully grid-supplied"; *(2) New KPI definition; *(3) Ref. Chap. 6

KPIs. CAPEX represents the initial investment required for energy generation, storage, and infrastructure, while OPEX covers ongoing costs such as maintenance and management. These costs help in evaluating financial decision-making. Additionally, Levelized Cost of Energy (LCOE) provides standardized metrics for quantifying costs of electrical energy, considering instalment, operation, and management costs over the lifetime of the system. Together, these metrics provide a comprehensive assessment framework, supporting various key aspects of the system.

4.3 Functional Requirements for Hybrid Energy Storage Integration

To enable the services and integration of hybrid energy storage across electrical and thermal domains, we can establish a set of functionalities that need to be established and available for a REC to actively manage and benefit from the integration of electrical and thermal energy storage into a multi-domain hybrid energy storage system.

Power Grid

In terms of voltage levels, a wide range of voltages would be valid, from low voltage to medium voltage. Presumably it will be low voltage (400 V) for small RECs but could have an accessible medium voltage connection point. For larger RECs it is possible to have medium voltage distribution (e.g. 15 kV). Regarding power, the maximum rated power (determined by the rated power of the connection transformer) could be sufficient with the sizing already done with the renewable generation systems for cases oriented to consumption management and generation maximization, however, for an operation in the electricity markets to maximize profit, it may be necessary to have a contracted power equal to the power of the renewables plus the installed ESS.

Storage systems could be connected in a distributed manner along the grid, or at the point of connection of the REC to the grid, depending on the service to be provided. For example, a large ESS plant connected at the REC PCC could provide electricity market operation services, while a distributed system could provide voltage stability services within the RES distribution grid.

Communication for monitoring and control

The communications network interconnects each device with the overall REC energy management system. This network must be capable, in terms of hardware, of being able to interconnect all the devices in a robust way and be able to manage the frequency of the commands and the variables to be monitored. It is possible to use a dedicated grid such as an Ethernet network or to wire an RS485 type communication network, or to integrate into the electrical distribution network by means of PLC (Power Line Communication) type systems. The storage systems should be able to be integrated into the REC's communication grid.

Control and monitoring system

The overall REC energy management control system would be responsible for sending power commands to each system so that the objectives/services to be implemented are fulfilled. Through the communications grid, it would communicate via an internal communications protocol, presumably a field bus (e.g. MODBUS) with each of the generation systems, storage, and—if any—manageable loads. Depending on the size of the power grid, this system could have an additional distributed monitoring system, so that in addition to monitoring information from each of the generation and storage devices, monitoring is available with dedicated devices at different points in the internal REC network.

Electric power generation

These are the elements that inject electrical power into the grid and can be conventional generation directly connected to the grid—synchronous power generating module (e.g. hydroelectric plants, diesel generators) or connected through electronic converters—power park module (e.g. wind generators). In the following, this section describes converters as a type of equipment or system.

The generation systems should be able to be controlled by the overall energy management system, so that they share the same communications protocol, and exchange the required setpoints and monitoring signals.

Electrical energy consumption

Electrical energy consumption can be manageable or unmanageable. The unmanageable ones will be desirable to be monitored, and the manageable ones can be connected to the grid through electronic converters (e.g. an electric vehicle charger) or not (such as an industrial process that can be interrupted, like a pelletizer). Similar to generation, certain types of loads could be associated with storage systems for various purposes, such as ensuring uninterrupted power supply to such loads.

Electronic converters

Power electronics devices are responsible for interconnecting the internal AC distribution grid of the REC with, and forming part of, storage systems, renewable generation systems, and—if any—manageable loads. These power electronics systems are usually connected at low voltage (e.g. 400 V) but it would be possible to connect them at medium voltage (either using multilevel configurations or by means of transformers integrated in the systems).

The converters usually integrate the power management control of each individual device, which must integrate appropriate communication protocols to integrate into the overall power management system. The amount of data and its frequency may depend on the service in question to be provided by the REC, for example, participating in the electricity market (one day-ahead) may involve fortnightly setpoints, while participating in secondary frequency regulation may involve setpoints every 4 s. These requirements may define the necessary communications protocol, with the most common being the use of how buses (e.g. MODBUS, CANOPEN, etc.).

4.3.1 Heat Networks

Heat networks deliver heat from one or several central sources through insulated pipes to buildings for space and water heating. This heat often comes from cogeneration plants powered by fossil fuels or biomass. However, heat-only boilers, geothermal energy, heat pumps, solar thermal, industrial waste heat, and even nuclear power plants can also be used. District heating plants are generally more efficient and have better pollution control than individual boilers. If combined with heat and power they can be an effective way to reduce carbon emissions.

Control and monitoring of the thermal system

Control and monitoring of thermal systems includes the optimized management of heat production (using different energy sources efficiently), the distribution network (reducing heat losses and ensuring an adequate heat supply to all users), and demand (monitoring consumption and encouraging efficient energy use). Various technologies, such as sensors (namely thermocouples), valves, control units, management software, and communication systems, are used to achieve these goals.

Heat generators

Heat generators are essential components in various applications, from heating residential buildings to powering industrial processes. The most important ones could be summarized in the following categories: combustion-based, such as furnaces, boilers, and water heaters; electric-based, which use electricity to generate heat; solar thermal, that can capture solar energy and convert it into heat, like solar water heaters, solar air heaters or for higher temperatures: concentrated solar heat plants. Besides, geothermal generators are a useful way to adopt the Earth's natural heat to produce thermal renewable energy.

Thermal power generators

Thermal power generators are devices that convert heat energy into electricity, so they function as sector coupling devices. Typical technologies are Rankine cycle (steam) or using heated compressed air in case of Bryton cycle.

Thermal energy storage

A thermal energy storage t is a component that allows heat or cold to be stored for later use in a variety of applications, such as heating and cooling buildings, generating electricity, and powering industrial processes. Its application is essential for ensuring supply continuity for users of renewable energy, particularly solar. These components can operate for different durations: short (hours), medium (days), or long (seasons). Depending on the application, they can utilize sensible, latent, or thermochemical heat.

Heat Exchangers

They are devices designed to transfer heat between two or more fluids, used in a wide range of applications, from heating and cooling buildings to industrial processes. These devices can be realized in various configurations, including shell and tube (with plain or finned tubes), plate, spiral, and coil. Heat exchangers are essential components in many engineering systems, and their proper design and operation are crucial for ensuring efficiency and safety.

5 How Hybrid Energy Storage Meets Identified Needs and Its Expected Benefits

The requirements for the case of a renewable energy community have been outlined in the previous section. This section provides the specific application and model where the suitable hybrid energy storage configuration is defined.

5.1 the Renewable Energy Community with Hybrid Energy Storage

This case study of a renewable energy community with multi-domain hybrid energy storage solution is based on the physical characteristics of an existing microgrid facility, complemented with a virtual thermal power setup. The reference system for this complex renewable energy community is illustrated in Fig. 4. The energy storage solutions are added as modules that can be scaled to fit.

Electrical Subsystem

The given electrical system is an experimental facility of CIEMAT dedicated to develop and test renewable energy systems, located at Soria (Spain). This facility is a micro-grid composed of a medium voltage ring (15 kV) with 7 substations for low voltage transformation as shown in Fig. 4.

Electric power generation, energy storage and electricity consumption are distributed among the substations.

In particular, the given microgrid has the following power generations systems:

- 5 wind turbines with a total installed capacity of 150 kW
- 1 photovoltaic power plant and photovoltaic panels installed on the roofs of the offices with total installed capacity of 360 kW
- A diesel generator (for emergency events) with an installed capacity of 100 kW

The electrical consumers of this microgrid are the following:

- 3 different office buildings

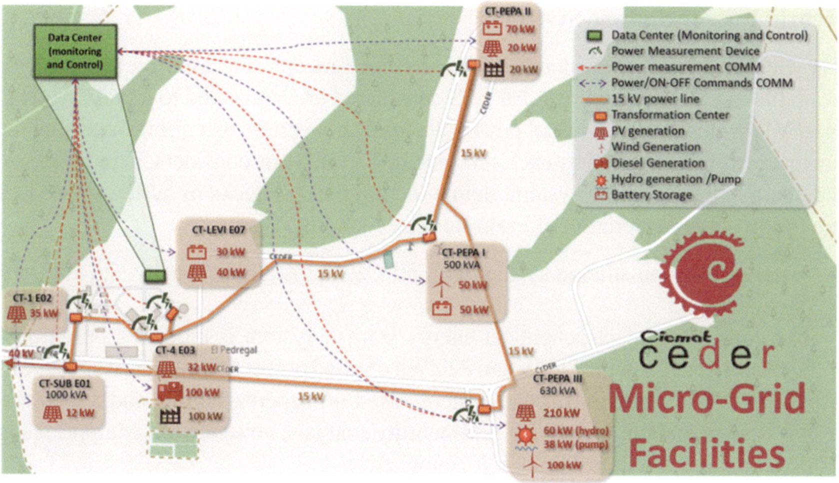

Fig. 4 Scheme of the CEDER facility microgrid

- 1 pelletizer facility (linked with a biomass power plant under commissioning), capacity of 100 kW electric.
- 8 electric vehicle chargers.

The microgrid has two relevant energy management systems, i.e. a complete DC grid to manage DC loads and battery-based storage systems, and a grid decoupling system to manage power within the power ring and emulate specific power frequency behaviours (not part of this model).

Thermal Subsystem

In addition to the physically present electrical system, the model includes a complementary fictional thermal network, based on components available in another lab facility.

The thermal power will be generated from renewable sources such as:

- A concentrated solar power plant (CSP) with an internal ESS based on molten salts. The output power of the CSP + ESS system has been evaluated for the irradiance of CEDER locations to maximize efficiency and power production.
- The CSP has the particularity that has a Rankine Cycle to allow heat conversion to electricity a part of the generated thermal power. This element permits to link both grids, electric and thermal, and it is considered essential in the energy management of the complete system.

The thermal consumption has been generated from real data and from standard consumptions of open databases. The available power consumption profiles to be considered has been hospitals, offices, residential buildings, hotels, and industrial loads. The consumptions have been scaled to the same scale as the real electric grid.

Hybrid Energy Storage System

The electrical ESSs considered in the reference model are based on real systems available in CEDER facilities. All ESS considered are connected to low AC voltage (400 V) by means of individual power electronic converters that allow managing the power delivered or stored by each subsystem. The analysis considers, in order to size the energy storage in the system, define the number of stacks or systems based on the following consideration of basic stack of each ESS type:

- A pumped hydro storage system with 50 kW of installed capacity and 2000 m^3 of deposit.
- The considered electrochemical energy storage system is modelled based on a stack of $LiFePO_4$ with 82 kW and 82 kWh of rated power and energy, respectively.
- A 150 kW fast energy storage system based on supercapacitors and a flywheel has been modelled, supporting grid stability analysis; however, this example study uses only a 1 h time resolution, it is not active.

In the thermal grid, molten salt-based systems are already integrated into the CSP and TSP and are governed as integrated part of the power plants themselves. An additional ESS based on PCM is integrated as pure thermal storage module to actively contribute to energy management.

The complete complex multidomain system will be controlled by a single energy management system to optimize performance metrics, as described further in Heussen et al. (2025a).

5.2 *Expected Benefits from Hybridization*

This section explains how energy storage contributes to benefits in the REC and how it is expected to enhance KPIs.

Expected Areas of Benefits

Benefits from the hybrid energy storage in the Renewable Energy Community application arise in two fundamental ways. Firstly, the introduced thermal energy storage systems *enable and facilitate* the integration of thermal renewable energy into the REC, thereby replacing fossil sources of energy, such as a combined heat and power plant or direct oil burners. Secondly, the cross-domain hybrid energy storage system *optimizes* the operation of the available energy storage systems with respect to efficiency and grid-oriented flexibility, which adds financial benefits.

To quantify this range of benefits, a set of four KPIs is proposed. The benefit of enabling solar energy in meeting thermal energy demands massively offsets carbon emissions, measured in terms of a relative KPI, the *Operational* CO_2 *emissions*, since the carbon offset is relative to a reference technology. The optimization benefit is associated with reducing energy losses (KPI: *System Energy Efficiency*), reducing cost of grid-procured electricity (KPI: *Operating cost*) and creating additional revenues by means of flexible operation with respect to the electricity grid (KPI: *Flexibility Factor*). With reference to the criteria outline in Sect. 3, these KPI cover technical, economic and sustainability criteria.

Quantification of Benefits Using Selected KPI

As outlined in Sect. 4.2, a set of KPIs has been identified to quantify the benefits of hybrid storage systems in the REC: system energy efficiency, system operational CO_2 emissions, and the flexibility factors. These KPIs have been selected to provide a comprehensive assessment of sustainability, technical and economic performance of hybrid storage systems. This subsection provides specific definitions and explanations of these KPIs.

In evaluating hybrid storage systems, it is essential to not only quantify performance improvements but also conduct a trade-off analysis to understand potential compromises, which further motivates the need for these KPIs. In this regard, system energy efficiency highlights how effectively the system minimizes energy losses during conversion and storage processes. System operational CO_2 emissions provide insight into sustainability performance, emphasizing the reduction in carbon emissions relative to a baseline scenario. Meanwhile, the flexibility factor and its variations help evaluate both technical and economic trade-offs, such as the system's ability to technically shift energy imports and exports during favourable price periods to reduce costs and economically maximize revenues.

Considering these KPIs, we can assess the system performance and quantify trade-offs like cost reductions and potential energy losses, that lead to an informed decision-making in the configuration of hybrid storage systems.

System Energy Efficiency

This metric measures how effectively an energy system utilizes the energy it generates. In essence, it is the ratio of the energy that is consumed to the total energy produced. The energy efficiency can be calculated for any types of energy in a hybrid power system as follows:

$$\begin{aligned} \mathit{Efficiency} = &(\mathit{Consumed\ energy} \\ &+ \mathit{Energy\ exported})/(\mathit{Total\ generated\ energy} \\ &+ \mathit{Energy\ imported}) \end{aligned}$$

For a hybrid system with thermal and electrical subsystems, thermal and electrical efficiency can be calculated as follows, respectively:

$$\begin{aligned} \mathit{Efficiency\_th} = &(\mathit{Consumed\ thermal\ energy} \\ &+ \mathit{Thermal\ energy\ converted\ to\ electricity}) \\ &/(\mathit{Total\ generated\ thermal\ energy} \\ &+ \mathit{Thermal\ energy\ converted\ from\ electricity}) \end{aligned}$$

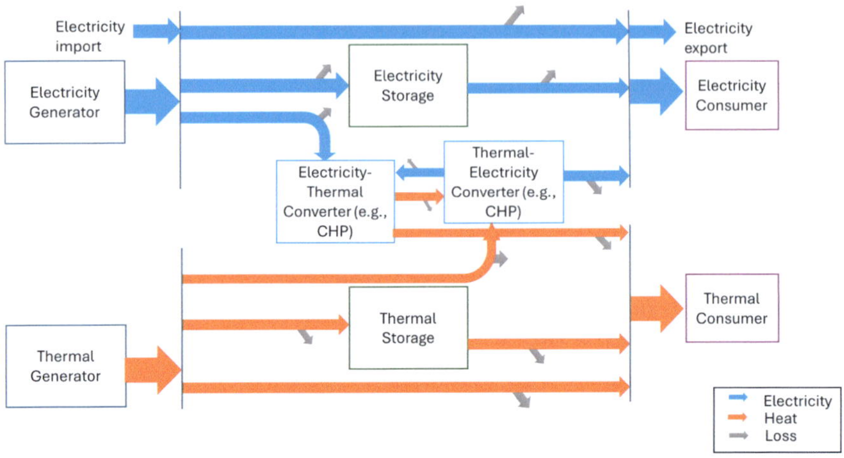

Fig. 5 Thermal and electrical energy flows in a hybrid energy system

$$\begin{aligned} Efficiency\_el = & (Consumed\ electricity + Exported\ electricity \\ & + Electricity\ converted\ to\ heat) \\ & /(Total\ generated\ electricity + Imported\ electricity \\ & + Electricity\ converted\ from\ heat) \end{aligned}$$

Figure 5 illustrates the interaction between different components of a thermal-electrical energy system, emphasizing how these interactions influence overall system energy efficiency. Energy efficiency in this context refers to how well the system minimizes energy losses and maximizes the use of both electricity and heat throughout the energy conversion, storage, and consumption processes.

System Operational CO_2 Emissions

This indicator quantifies the sustainability performance of an energy system by comparing its carbon emissions to a baseline scenario (Bianco et al. 2021). The baseline is defined as a situation in which all energy demands are met using traditional energy sources, such as fossil fuels. The metric expresses the reduction ratio in CO_2 emissions achieved by employing cleaner or more efficient technologies. The formulation is as follows:

$$ECO_2 = \frac{(Baseline\ \mathrm{CO_2}\ emission - Actual\ \mathrm{CO_2}\ emission)}{Baseline\ \mathrm{CO_2} emission}$$

A higher value reflects a more significant reduction in the system's carbon footprint.

Flexibility Factor

Inspired by established flexibility KPIs outlined in review paper (Clauß- et al. 2017), an enhanced flexibility factors has been devised in this section to comprehensively address the dynamics of energy systems. In this chapter, we just calculate the following flexibility factor and refer to the documentation for more factors (Heussen et al. 2025a).

The proposed Flexibility Factor represents the combined significance of imported energy at the lowest price time slots (LPT) and exported energy at the highest price time slots (HPT) within the overall trade market. It quantifies the system's technical ability to import energy during LPT and export energy during HPT. Thus, it is formulated as follows:

$$\begin{aligned} FF = & (Energy\ imported\ during\ LPT \\ & + Energy\ exported\ during\ HPT) \\ & /(Total\ energy\ imported + Total\ energyexported) \end{aligned}$$

In this study, price slots are categorized based on the median price: prices above the median are classified as high-price time slots, while prices below the median are classified as low-price time slots. It is worth noting that the factor depends on import/export volumes and price variability. Besides, the calculation assumes that the total amounts of energy import/export remain constant. Special cases include fixed pricing or when transactions are absent during asymmetrical pricing. For simplicity, we have not delved into further details. The detailed formula and other flexibility factors can be found in Heussen et al. (2025a).

Cost Analysis: CAPEX, OPEX, and LCOE

This section presents key economic indicators for evaluating the financial viability of an energy system. These indicators include capital expenditure (CAPEX), operational expenditure (OPEX), and levelized cost of energy (LCOE). The CAPEX, OPEX, and LCOE are calculated as shown in the following.

- **Capital Expenditure (CAPEX)** accounts for the total investment required to establish an energy system, including the costs of equipment and installation.
- **Operational Expenditure (OPEX)** represents the fixed annual costs incurred in maintaining system operations. These costs are assumed to remain constant throughout the project's lifetime.
- **Levelized Cost of Energy (LCOE)** quantifies the cost of generating energy over the project lifetime, incorporating investment, operational, and fuel costs. It is calculated using the following formula:

$$LCOE = \sum_{t=1}^{n} \left[C_t + O_t + F_t)/(1+r)^t \right] / \sum_{t=1}^{n} \left[E_t/(1+r)^t \right]$$

where C_t represents the capital expenditure in year t, O_t denotes the operational expenditure in year t, F_t accounts for fuel costs in year t, and E_t represents the energy generated in year t. r is the discount rate, and n is the project lifetime in years. Key assumptions include a plant lifetime of 25 years, which is standard for this type of facility, and an annual discount rate of 3.41% (Mongird et al. 2019). A lower LCOE indicates improved cost-efficiency in energy generation. The metric provides a comparative basis for evaluating different energy systems under varying economic and operational conditions.

6 Technical–Economic and Sustainability Analysis of REC—Proof of Concept Study

A system model of the fully integrated renewable energy community with hybrid electro-thermal energy storage and conversion has been developed in parallel by using MATLAB Simulink as modelling environment, and an optimization module in the CPLEX environment. MATLAB Simulink was chosen to enable the potential investigation of faster time scales relevant to the electrical system integration and operation and due to the availability of models suitable for investigating this timescale. The optimization module is used for sizing optimization of components in the system and performing daily economic energy dispatch among HESSs and other dispatchable components. Simulink and optimization models are presented in detail in Heussen et al. (2025a), and the source code is available for evaluation at Heussen et al. (2025b).

For basic usage of the benchmark system, the following parameters are intended for adjustment by users:

- Day—Represents the simulated day of the year (0–365).
- P nominal—Defines power profiles for production and consumption, measured in peak kWs.
- Type of ESS—Specifies the type of electrical storage unit, selectable from options 1 to 3.
- Ness—Sets the number of parallel storage stacks (0–500).
- SoCi—Determines the initial state of charge for storage systems on a scale from 0 to 1.
- Location—Identifies the system's geographical placement, such as Soria or Portici.
- Profile Case—Defines the type of electric vehicles, building consumption, and pelletizer processes (see documentation for details).

Due to the modularity of the provided benchmark and the modelling framework, additional changes are straightforward to implement for advanced users, such as:

- Extension of the benchmark to other input data profiles and other locations
- The addition of alternative consumption profiles

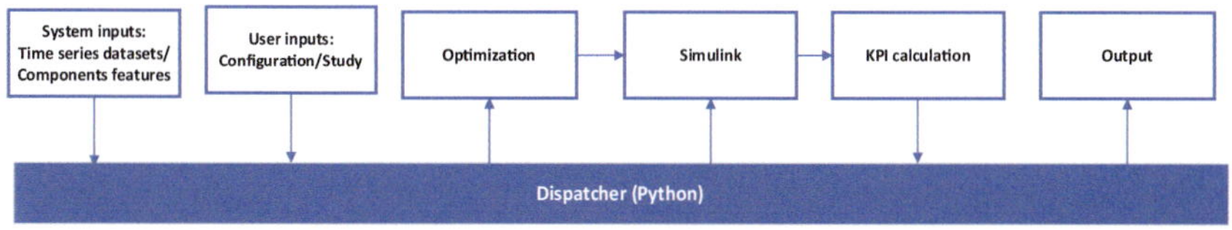

Fig. 6 Pipeline dispatcher and different modules of the system

The modular design of the system also facilitates the alteration of the system to serve more advanced study goals:

- Alternative electrical storage models (via drop-in replacement of existing storage models)
- Extension of simulation models to investigate performance criteria at faster time scales.

The calculation pipeline orchestrates the execution of a given study maintaining consistent parameters and configurations across optimization and simulation modules, ensuring traceable documentation of the process. The Dispatcher, as shown in Fig. 6 and as the core coordination module, manages the flow of data and computations across different components of the framework using Python.

The pipeline consists of the following key stages:

System Inputs: This stage includes time series datasets and component features, providing the foundational data required for simulations and analyses.

User Inputs: Users define the study configuration, specifying parameters and settings to tailor the simulation according to specific research or design objectives.

Optimization: This module processes input parameters and performs optimization computations, improving system performance based on predefined constraints and objectives.

Simulink: The framework interfaces with Simulink to execute dynamic simulations, validating and refining system behaviour through time-domain analysis.

KPI Calculator: KPIs are calculated based on simulation results, offering quantitative insights into system performance and efficiency.

Output: The results, including optimized parameters, simulation outputs, and performance metrics, are documented and stored for further analysis.

The Dispatcher integrates these components, ensuring seamless data exchange and execution order. This structured approach enables the automatic documentation of input data, parameters, and results, facilitating reproducibility and traceability. Additionally, it supports integration into larger data pipelines, such as impact assessments of design parameters or statistical design of experiments, making it a robust tool for complex system analyses.

The source code and data for the REC Benchmark are available in the public dataset (Heussen et al. 2025b), and further use cases as well as the user interface are documented in Heussen et al. (2025a).

Baseline Configuration Design

A basic system design was chosen based on heuristics that enable the REC to operate in a fully self-sufficient mode. This is an extreme case for a grid-connected REC and leads to high system costs but aligns well with the remote off-grid cases studied in previous book chapters.

To evaluate the performance of the hybrid storage system in various conditions, first, four representative days were selected through analysis, with one day chosen from each quarter of the year. The selection was performed using the k-medoids clustering method (Kaufman and Rousseeuw 1990), leading to the selected days for design: 20 February 2022, 23 May 2022, 28 July 2022 and 16 December 2022.

After selecting days, a basic dispatch is run through a simple rule-based approach used to allocate energy resources for each day, ensuring that energy storage is available for the next day. It consists of a thermal dispatch, where the Rankine Cycle system converts excess thermal energy into electricity for peak-saving, and the PCM system manages the remaining thermal energy. The electric dispatch prioritizes maximizing grid power delivery while minimizing grid power intake, using pumped hydro storage to reduce power exchange and batteries to manage residual energy. Shiftable loads such as EVs are not considered.

The dispatch process results in specific daily storage requirements, evaluated in three key metrics:

- Power-based requirement: Storage capacity needed to meet power constraints.
- Energy-based requirement: Storage capacity required to satisfy energy demands.
- Final storage requirement: The maximum of the power-based and energy-based requirements, ensuring system feasibility.

The results give the storage units' requirements for each day presented in Table 2.

Based on the analysis, the baseline storage capacities for the hybrid system are determined as average value of final storage requirement of the four main representative days as follows:

- 83 units of Battery Stacks
- 9 units of Hydro-Pump systems
- 28 units of PCM storage systems

Table 2 The storage units' requirements for each representative day

nDay	dateDay	nBAT			nHyP			nPCM		
51	20-Feb-2022	4	94	94	7	9	9	14	26	26
143	23-May-2022	3	77	77	7	11	11	13	26	26
209	28-Jul-2022	4	107	107	7	11	11	19	31	31
350	16-Dec-2022	4	53	53	7	5	7	13	27	27

Although one unit of Supercapacitor Energy Storage appears in the results, it is not utilized, so it has been removed from the study. The features of the ESSs are illustrated in Table 3 (values represent one unit/module).

Scenario Definition

To evaluate the impact of hybrid storage system variations on the REC, three different scenarios are considered, listed in Table 4, where variations between PCM and Battery storage module numbers were considered for approximately equal investment and maintenance cost. The idea of this scenario choice is to investigate trade-offs between electrical and thermal storage options in the REC application. Each scenario is investigated twice, using a self-sufficiency objective function, and an economic objective function. The economic scenario means that we minimize the total energy costs, while the self-sufficiency scenario means that we minimize the energy drawn from the grid.

Scenario Assessment Results

To assess the economic KPIs, CAPEX LCOE, and OPEX are listed in Table 5, calculated for the baseline configuration showing the initial capital cost of €757.65/kWh, which reflects the investment needed for full self-sufficiency in energy supply. Besides, the LCOE is 0.097€/kWh, and operational cost is obtained €1.54/kW, supporting the economic viability of the system.

For the performance evaluation, two days 51 and 143 were selected from the four original k-medoids-clustered design days to represent a range of seasonal and operational conditions. Table 6 compares system performance under self-sufficiency and economic objectives. The KPIs shown in the table reflect the average values calculated from these two days.

Table 3 Features of the storage systems

Features	Pumped-hydro energy storage	Battery	PCM
Maximum output power (kW)	450	6806	1400
Efficiency	75%	95%	90%
Capacity (kWh)	2575	6806	5740
Operation and maintenance cost (c£/kWh)	0.002	0.005	0.003

Table 4 Scenarios for hybrid storage system configurations and capital costs

Scenarios	Number of battery stacks	Number of PCM modules	Total capital cost of storages
Scenario 1	83	28	3,373,545
Scenario 2	88	20	3,380,925
Scenario 3	78	36	3,366,165

Table 5 Economic indicators of hybrid storage systems, including CAPEX, LCOE, and OPEX

KPIs	Value
Initial capital cost (CAPEX)	757.6476 €/kWh
Levelized cost of electricity (LCOE)	0.097€/kWh
OPEX	1.54€/kW

Across all scenarios, the flexibility factor remains relatively stable, ranging between 0.58 and 0.62, indicating that flexibility is not a major differentiator in this comparison. The more significant variations are observed in efficiency and CO_2 emission reduction.

It is observed that Scenario 1 offers strong overall performance, with high efficiency (0.93–0.94) and the greatest CO_2 emission reduction[7] under the economic objective (0.45 kg/kWh). Besides, Scenario 3 (economic) achieves the highest efficiency (0.98) while maintaining moderate emissions reduction (0.35 kg/kWh), making it suitable for performance-driven applications. In contrast, Scenario 2 (economic) shows the lowest efficiency (0.86) and minimal emissions reduction (0.02 kg/kWh), suggesting it is the least favourable configuration among the three.

Finally, the LCOE remains constant at 0.097–0.098€/kWh across all scenarios, suggesting that this difference in battery and PCM module configurations does not significantly affect the overall cost of electricity generation.

[7] The reduction in CO_2 emissions in our system—comparable to a fully renewable setup—can only be evaluated based on the energy exchanged between thermal and electrical systems. This assessment relies on the emission factors of the electrical and thermal grids. The emissive factor represents the greenhouse gas emissions (primarily CO_2) associated with the production and distribution of electrical and thermal energy. The emissive factors used in this study are 0.354 $kgCO_2/kWh$ for the electrical grid and 0.202 $kgCO_2/kWh$ for natural gas.

Table 6 Performance comparison of hybrid storage scenarios under self-sufficiency and economic objectives

KPIs	Scenario 1 (self-sufficiency)	Scenario 1 (economic)	Scenario 2 (self-sufficiency)	Scenario 2 (economic)	Scenario 3 (self-sufficiency)	Scenario 3 (economic)
Flexibility factor	0.62	0.62	0.60	0.60	0.58	0.58
Efficiency day	0.93	0.94	0.93	0.86	0.93	0.98
CO_2 Emission/kWh reduction	0.33	0.45	0.17	0.02	0.27	0.35
LCOE	0.097	0.097	0.097	0.097	0.097	0.098

References

Angelakoglou K, Kourtzanidis K, Giourka P, Apostolopoulos V, Nikolopoulos N, Kantorovitch J (2020) From a comprehensive pool to a project-specific list of key performance indicators for monitoring the positive energy transition of smart cities—an experience-based approach. Smart Cities 3(3):705–735. https://doi.org/10.3390/smartcities3030036

Bianco G, Bonvini B, Bracco S, Delfino F, Laiolo P, Piazza G (2021) Key performance indicators for an energy community based on sustainable technologies. Sustainability 13(16):8789. https://doi.org/10.3390/su13168789

Bocklisch T (2015) Hybrid energy storage systems for renewable energy applications. Energy Procedia 73(June):103–111. https://doi.org/10.1016/j.egypro.2015.07.582

Clauß J, Finck C, Vogler-Finck P, Beagon P (2017) Control strategies for building energy systems to unlock demand side flexibility—a review. https://doi.org/10.26868/25222708.2017.462

Dumont O, Frate GF, Pillai A, Lecompte S, Lemort V (2020) Carnot battery technology: a state-of-the-art review. J Energy Storage 1(32):101756. https://doi.org/10.1016/j.est.2020.101756

García-Vázquez CA, Espinoza-Ortega H, Llorens-Iborra F, Fernández-Ramírez LM (2022) Feasibility analysis of a hybrid renewable energy system with vehicle-to-home operations for a house in off-grid and grid-connected applications. Sustain Cities Soc 86:104124. https://doi.org/10.1016/j.scs.2022.104124

Goyal A, Bhattacharya K (2024) Optimal design of a decarbonized sector-coupled microgrid: electricity-heat-hydrogen-transport sectors. IEEE Access 12:38399–38409. https://doi.org/10.1109/ACCESS.2024.3375336

Guven AF, Abdelaziz AY, Samy MM, Barakat S (2024) Optimizing energy dynamics: a comprehensive analysis of hybrid energy storage systems integrating battery banks and supercapacitors. Energy Convers Manag 15(312):118560. https://doi.org/10.1016/j.enconman.2024.118560

Heussen K et al (2025a) D3.4—Framework for hybrid energy storage modelling. StoRIES Consortium. https://doi.org/10.5281/zenodo.15553044

Heussen K et al (2025b) Hybrid energy storage renewable energy community—reference case. Available: https://doi.org/10.5281/zenodo.15552600

Hidalgo GI, Uihlein A (2023) High-resolution energy atlas. Luxembourg, JRC136080. https://doi.org/10.2760/83784

Jimenez Martinez M, Igualada L, Valdés Martín R, Farriol Salas A, Heredia Julbe P, Noris F, Garcia Munoz F, Corchero C (2023) MODECO—modelling study on the role of energy communities in the energy transition. Luxembourg (Luxembourg): Publications Office of the European Union. https://doi.org/10.2760/118421

Kaufman L, Rousseeuw PJ (1990) Finding groups in data. Wiley Series in Probability and Statistics. Wiley. https://doi.org/10.1002/9780470316801

Kobashi T, Choi Y, Hirano Y, Yamagata Y, Say K (2022) Rapid rise of decarbonization potentials of photovoltaics plus electric vehicles in residential houses over commercial districts. Appl Energy 306:118142. https://doi.org/10.1016/j.apenergy.2021.118142

Laing D, Eck M, Hempel M, Steinmann WD, Meyer-Grünefeldt M, Eickhoff M (2012) Analysis of operation test results of a high temperature phase change storage for parabolic trough power plants with direct steam generation. In: ASME 2012 6th International conference on energy sustainability, Parts A and B, 273–80. American society of mechanical engineers. https://doi.org/10.1115/ES2012-91056

Liberatore R (2024) JCA Eni ENEA project. In Proceeding of Solar Paces, Rome

Lowitzsch J, Hoicka CE, van Tulder FJ (2020) Renewable energy communities under the 2019 European clean energy package—governance model for the energy clusters of the future? Renew Sustain Energy Rev 122:109489. https://doi.org/10.1016/j.rser.2019.109489

Lysenko O, Kuznietsov M, Hutsol T, Mudryk K, Herbut P, Vieira FMC, Mykhailova L, Sorokin D, Shevtsova A (2023) Modeling a hybrid power system with intermediate energy storage. Energies 16(3):1461. https://doi.org/10.3390/en16031461

Manso-Burgos Á, Ribó-Pérez D, Gómez-Navarro T, Alcázar-Ortega M (2022) Local energy communities modelling and optimisation considering storage, demand configuration and sharing strategies: a case study in Valencia (Spain). Energy Rep 8:10395–10408. https://doi.org/10.1016/j.egyr.2022.08.181

Mazzeo D, Matera N, De Luca P, Baglivo C, Congedo PM, Oliveti G (2020) Worldwide geographical mapping and optimization of stand-alone and grid-connected hybrid renewable system techno-economic performance across Köppen-Geiger climates. Appl Energy 276:115507. https://doi.org/10.1016/j.apenergy.2020.115507

Mehdizadeh M, Nayum A, Nordfjærn T, Klöckner CA (2024) Are Norwegian car users ready for a transition to vehicle-to-grid technology? Transp Policy 146:126–136. https://doi.org/10.1016/j.tranpol.2023.11.014

Miliozzi A, Dominici F, Candelori M, Veca E, Liberatore R, Nicolini D, Torre L (2021) Development and characterization of concrete/PCM/diatomite composites for thermal energy storage in CSP/CST applications. Energies 14(15):4410. https://doi.org/10.3390/en14154410

Mitali J, Dhinakaran S, Mohamad AA (2022) Energy storage systems: a review. Energy Storage Saving 1(3):166–216. https://doi.org/10.1016/j.enss.2022.07.002

Mongird K, Viswanathan V, Balducci P, Alam MJ, Fotedar V, Koritarov V, Hadjerioua B (2019) Energy storage technology and cost characterization report. Richland, WA (United States). https://doi.org/10.2172/1573487

Mustafizur RM, Oni AO, Gemechu E, Kumar A (2020) Assessment of energy storage technologies: a review. Energy Convers Manage 223:113295. https://doi.org/10.1016/j.enconman.2020.113295

Nkwanyana TB, Siti MW, Wang Z, Toudjeu I, Mbungu NT, Mulumba W (2023) An assessment of hybrid-energy storage systems in the renewable environments. J Energy Storage 72:108307. https://doi.org/10.1016/j.est.2023.108307

Novotny V, Basta V, Smola P, Spale J (2022) Review of Carnot battery technology commercial development. Energies 15(2):647. https://doi.org/10.3390/en15020647

Poli N, Bonaldo C, Moretto M, Guarnieri M (2024) Techno-economic assessment of future vanadium flow batteries based on real device/market parameters. Appl Energy 362:122954. https://doi.org/10.1016/j.apenergy.2024.122954

Rehman S, Al-Hadhrami LM, Mahbub Alam M (2015) Pumped hydro energy storage system: a technological review. Renew Sustain Energy Rev 44:586–598. https://doi.org/10.1016/j.rser.2014.12.040

Reilly J, Poudel R, Krishnan V, Anderson B, Rane J, Baring-Gould I, Clark C (2022) Hybrid distributed wind and battery energy storage systems. Golden, CO (United States). https://doi.org/10.2172/1874259

Rosales-Asensio E, de Loma-Osorio I, Palmero-Marrero AI, Pulido-Alonso A, Borge-Diez D (2024) Optimal microgrids in buildings with critical loads and hybrid energy storage. Buildings 14(4):865. https://doi.org/10.3390/buildings14040865

Schleifer AH, Harrison-Atlas D, Cole WJ, Murphy CA (2023) Hybrid renewable energy systems: the value of storage as a function of PV-wind variability. Front Energy Res 11. https://doi.org/10.3389/fenrg.2023.1036183

Scipioni R et al (2025) D3.5—Roadmap for hybridisation of energy storage. StoRIES Consortium. Available: https://zenodo.org/communities/h2020-stories/

Sihvonen V, Riikonen J, Price A, Nordlund E, Honkapuro S, Ylönen M, Kivioja V, Hedman Å, Tullberg R (2024) Combined utilization of electricity and thermal storages in a highly renewable energy system within an island society. J Energy Storage 89:111864. https://doi.org/10.1016/j.est.2024.111864

Steinmann W-D, Bauer D, Jockenhöfer H, Johnson M (2019) Pumped thermal energy storage (PTES) as smart sector-coupling technology for heat and electricity. Energy 183:185–190. https://doi.org/10.1016/j.energy.2019.06.058

Zhang J, Min G, Chen X (2023) Supercapacitors for renewable energy applications: a review. Micro Nano Eng 21:100229. https://doi.org/10.1016/j.mne.2023.100229

Zhu Y, Siqi W, Li J, Jia Q, Zhang T, Zhang X, Han D, Tan Y (2024) Towards a carbon-neutral community: integrated renewable energy systems (IRES)–Sources, storage, optimization, challenges, strategies and opportunities. J Energy Storage 83:110663. https://doi.org/10.1016/j.est.2024.110663

Future Mobility and Transport Solutions

Preface

The decarbonization of the transportation sector is crucial to mitigate global climate change and entails several fields, from road and rail to maritime and aviation. To decarbonize named sectors, technology must balance efficiency, energy density, and environmental impact. Chapter "Hybrid Energy Storage System for BEV and FCEV Charging Stations—Use Case for Aluminum as Energy Carrier" provides insights to the integration of aluminum as an energy carrier as a novel approach to hybridizing battery electric vehicle (BEV) charging and fuel cell electric vehicle (FCEV) hydrogen refueling. Aluminum helps improve the flexibility and reliability of EV infrastructure and to cover further grid related services. Maritime applications are in the focus of Chapter "Waterborne Transport. Hybrid Power Supply for Electrification of Port Infrastructures, Shore-to-Ship Power, and Ship Power and Propulsion". Here, hybrid propulsion systems and shore-to-ship power solutions demonstrate how different energy hybridization strategies can reduce greenhouse gas emissions. Chapter "Hybrid Energy Sstorage Systems in Rail Transport" aims at the electrification and hybridization of railway systems. Such systems offer a promising alternative to diesel-powered trains, supporting the transition toward a greener rail network. In Chapter "Battery Systems for Air Transport Climate Neutrality", the ongoing developments of airborne battery technologies, hybrid-electric propulsion, and hydrogen-based systems are highlighted and their potential to align air travel with European Green Deal objectives is discussed. Chapter "On-Board Integration of Hybrid Energy Storage Systems in Heavy Duty Vehicles: The Electric Buses Use Case" presents hybrid propulsion systems for urban electric buses, comparing three configurations—LiFePO4 battery (LFP) with Proton Exchange Membrane Fuel Cell (PEMFC), LFP with Vanadium Redox Flow Battery (VRFB), and Ni-NaCl battery with PEMFC—against a conventional LFP-powered electric bus. The findings highlight the advantages of hybridization in extending range and increasing battery lifespan, exploiting the potential for a more efficient and sustainable transition. In sum, this section addresses the interplay of emerging technologies across multiple modes of transportation. Each chapter provides a comprehensive perspective on the role of hybrid energy systems in the broader sustainability transition for the transport sector.

Hybrid Energy Storage System for BEV and FCEV Charging Stations—Use Case for Aluminum as Energy Carrier

Nicola Musicco, Hüseyin Ersoy, Linda Barelli, Manuel Baumann, and Stefano Passerini

Abstract The development of electric vehicle (EV) charging infrastructure and load management remains a significant challenge in the transition to sustainable mobility. This chapter explores the use of aluminum (Al) as an energy carrier to enable a hybrid management of BEV charging and fuel cell electric vehicle (FCEV) hydrogen (H_2) refueling. The use of aluminum enables on-site power and flexible H_2 generation, enhancing flexibility and versatility in EV charge management strategies. The study introduces this emerging concept, providing a theoretical foundation for its techno-economic implications and presenting a formulated use case that examines the potential of the Al wet-combustion process for large hybrid charging stations. By leveraging aluminum's high energy density, recyclability, and multi-functionality, this approach offers a promising pathway to improve charging infrastructure resilience and energy efficiency.

Keywords Aluminum · Power-to-X · EV charging · Mobility · Energy transition · Metal fuels

N. Musicco
Department of Mechanical and Industrial Engineering, University of Brescia, Brescia, Italy

H. Ersoy · M. Baumann
Institute for Technology Assessment and Systems Analysis (ITAS), Karlsruhe Institute of Technology (KIT), Karlsruhe, Germany

H. Ersoy
Center for Environmental and Sustainability Research (CENSE), NOVA School of Science and Technology, Caparica, Portugal

L. Barelli
Department of Engineering, University of Perugia, Perugia, Italy

S. Passerini (✉)
Center of Transport Technologies, Austrian Institute of Technology (AIT), Vienna, Austria
e-mail: stefano.passerini@ait.ac.at

R. Scipioni et al. (eds.), *Hybrid Energy Storage*, Lecture Notes in Energy 47,
https://doi.org/10.1007/978-3-031-97755-8_10

1 Introduction

There is a high need of energy storage systems to mitigate flexible generation and charging behavior of a growing number of electric vehicles (IEA 2024). Here, hybrid energy storage systems (HESS) can play a crucial role in providing such flexibility services. These services are needed on several levels, starting from private owners using home charging stations, up to large fast charging stations located at a highway level. Other larger applications are electric buses or trucks where charging power can reach multiple MW-scales. However, considering a high equality factor of charging can lead to local grid parameter violation or worse, a breakdown of power supply (Mahmud et al. 2023). There are several energy storage alternatives to support the charging of electric vehicles via renewable energies, in particular hybrid energy storage solutions (Yadav et al. 2023; Al Wahedi and Bicer 2020; Gonzalez-Rivera et al. 2021). Here, a broad variety of combinations of storage solutions is proposed, starting from batteries with supercapacitors, thermal storage, H_2 production and batteries and more recently the use of reactive metals. The latter represents a rather new way of storing energy for both short and long term with the possibility of offering multiple services for electric mobility and are considered as hybrid energy storage systems.

Moreover, the realization of a decarbonized economy requires the development of large energy storage solutions capable of buffering the daily, weekly and seasonal fluctuations of energy (electricity) generated from renewable energy sources (RES), as, solar and wind power (Denholm and Mai 2019). The implementation of such sustainable, low-cost, and large-scale storage systems is urgently required, but still full of challenges.

Reactive metals as aluminum, magnesium, iron have both high volumetric and gravimetric energy densities and can be converted in different ways on multiple power levels. In addition, they are highly suitable for long term storage and can be transported using existing infrastructures (trains, ships, road transport) (Barelli et al. 2020). After their use, metal oxides can be collected and then be recycled (i.e., reduced to the metal state) using processes powered by renewables (Bergthorson 2018). Aluminum is one of the most promising candidates due to its favorable properties. It is as a functional construction and energy material with vital significance for achieving the determined sustainable development goals. Mainly, Al demand is dominated by the transport (27%), construction (24%), and packaging (15%) sectors. One notable point is the anticipated demand growth in the transport and energy applications of aluminum. Until 2050, expected demand growth of the transport sector corresponds to 55% in relation to the year 2017 (Aluminium 2019). This increased demand may support the sector coupling of aluminum production for a case where aluminum is used as an energy storage material at the same time.

A kg of aluminum under theoretical considerations has an energy density of 8.6 kWh, as 4.2 kWh in the form of heat and about 4.4 kWh of energy in the form of H_2, i.e., 0.111 kg equivalent based its lower heat value (LHV = 33,3 kWh kg^{-1}) when it is oxidized with water (Petrovic and Thomas 2011). The volumetric energy density equals 23.5 kWh l^{-1} versus 2.3 kWh l^{-1} of liquefied H_2, making aluminum also a potentially viable solution as a H_2 storage carrier.

This chapter provides an overview of current strategies for using HESS for the integration and charge management of EV charging stations. The focus is set on Aluminum as energy carrier as a possible solution to do so. Different Al-conversion paths and current system designs are briefly introduced. Then a use case for an Al wet combustion system and its techno-economic performance is provided. Finally, relevant KPIs and sustainability implications are discussed.

2 Hybrid Energy Storage System for Integration Charge Management of EV Charging Stations

In the following some selected examples are provided to display how HESS can support the charge management of EV-charging stations. (See Table 1) The aim of the section is to provide an overview of current potential solutions for hybrid energy storage systems and how they can support the transition towards electric vehicles.

The work of (Yadav et al. 2023), presents a power management scheme for EV-charging on AC and DC side via the combination of a supercapacitor with a generic EV battery. Here the supercapacitor reduces the stress on the non-defined EV battery due to sudden changes in generation and normal operation and allows to reduce overall power consumption.

A broader HESS concept has been proposed by (Al Wahedi and Bicer 2020) for the off grid-based charging of up to 80 electric vehicles per day through renewables, in particular with PV, wind turbines, and biomass based Rankine cycle located in Qatar. The system consists of lithium-ion batteries, H_2, Ammonia, and a phase change based thermal storage unit. The H_2, and NH_3, produced via renewables and stored on site is combined with fuel cells. The latter are supported by lithium-ion batteries. The main advantage of the hybrid solution is a high exergy efficiency, with positive impacts on the single components during their operation.

A Model Predictive Control-Based Optimized Operation of a Hybrid Charging Station for Electric Vehicles has been investigated by (Gonzalez-Rivera et al. 2021). Here a combination of a PV-system, a battery, fuel cells and electrolyzers to support six fast charging units is modelled. The results indicate a positive impact of HESS regarding overall operation cost, NPV, and reduces grid utilization.

Roslan et al. proposed a combination of a H_2 storage system including an electrolyzer and fuel cell in combination with a AC/DC conversion multiport network and a Li-ion battery for EV charging in Malysia (Roslan et al. 2024). Here a set of economic (COE, NPC and LCOH,) and one environmental KPIs (e.g., CO_2 emission reduction) is calculated using HOMER®. The calculations indicated promising results of the proposed HESS system, helping to achieve a stable electricity supply

Table 1 Comparison of selected studies on HESS for EV-charging support

Source	Scope	Technology	KPIs	Hybrid storage benefit
Yaday et al. (2023)	Power management system for PV based charging	Supercapacitor with battery	Power consumption reduction	Avoiding battery stress, lower power consumption
Wahedi et al. (2020)	Proposition of a stand-alone fast EV charging station running on purely hybrid RES	PV, Li-Ion batteries, Phase Change Material thermal storage, wind turbine, H_2 and ammonia electrolyzers and fuel cells, biomass conversion	Energy efficiency, Exergy efficiency	Improvement of overall exergy efficiency
Gonazles-Rivera et al. (2021)	Provision of an energy management system based on a novel approach using model predictive control	PV system, battery, H_2 system based on a fuel cell, electrolyzer, and tank as an energy storage system	Grid use reduction, efficiency, utilization cost, net present cost	Lower utilization cost (−25.3%), grid use reduction (−60%) and efficiency improvement
Roslan et al. (2024)	Techno-economic analysis of hybrid energy storage system for electric vehicles charging stations using renewables	H_2 electrolyzer, tank and fuel cells. Li-Ion batteries, Wind and PV-system, and an AC/DC conversion multiport network	COE, NPC and LCOH, Emissions: CO_2, CO, SO_2, NOx, particulate matter, and unburned H_2	Provision of environmental and economic benefits
Güven et al. (2025)	Identify hybrid systems to ensure the electricity supply to EV charging stations	PV, wind turbine, biomass, electrolyzer, H_2 tank, fuel cell, batteries, inverter	LCOE, NPC, Payback Period, OPEX, CAPEX, Emissions: CO_2, CO, SO_2, NOX, particulate matter	Provides favorable economic performance due to high renewable potential, emissions are reduced

with reduced CO_2 emissions. Also, several recommendations for EV-charging infrastructure are provided in terms of modelling, optimization or potential of exploring further technologies for electricity conversion.

Güven et al., investigate hybrid systems to ensure electricity supply to EVs in the Çukurova region of Adana, Turkey (Güven et al. 2025). A combination of a biomass gasifier, a H_2 electrolyzer, tank and fuel cell, photovoltaics and wind turbines to support local EV charging stations has been analyzed via six different design scenarios. In sum, the system is regarded as beneficial in terms of net present cost, and levelized cost of electricity in areas with a high solar irradiation. Again, as in Roslan et al., HOMER® has been used for modelling.

There are several studies on HESS in the specific application field of supporting EV-charging infrastructure. Interestingly, all systems include batteries, and mostly H_2 to exploit the benefits of short and long terms storage. The used KPIs are mainly techno-economic in the selected studies, with the inclusion of different emission factors.

3 State-of-the-Art: Use of Aluminum as an Energy Carrier

3.1 Electrochemical Conversion

The combination of Al production via inert-anode smelting and Al conversion to electricity via Al-air batteries is a potential option to achieve cost-effective and zero-carbon-emission seasonal/annual energy storage is highly required for the Zero Emission Scenario (ZES) by 2050. (See Fig. 1) Although Al-air batteries may play a significant role, two main issues of this battery technology need to be addressed for the realization of Al production/conversion systems (APCSs) with high round-trip energy efficiency (RTE) (Xu et al. 2024). The first one is the limited energy conversion efficiency of Al metal into $Al(OH)_3$ (later transformed into Al_2O_3 for reuse in Al production), which is determined by the effective Al utilization and the cell discharge voltage. The spontaneous chemical reaction of Al metal in alkaline electrolytes leads, in fact, to H_2 evolution and thus low coulombic efficiencies (although the evolving H_2 could be collected and utilized). Additionally, the polarization occurring at both the Al metal anode and the air cathode leads to low cell discharge voltage, i.e., low voltage efficiency. Both hurdles contribute to the low specific energy and RTE of Al-air batteries. The second issue is the difficulty in collecting the discharge product (Ersoy et al. 2022), e.g., $MAl(OH)_4$ ($M = Na, K$) and/or $Al(OH)_3$. In alkaline electrolytes, the typical discharge product $MAl(OH)_4$ is highly soluble, converting to the $Al(OH)_3$ precipitate only at very high concentrations, i.e., when its solubility limit is reached. However, reaching the solubility limit inside the cell results in the precipitation of the solid product in the cell itself, causing the formation of an inert coating on the Al electrode as well as the clogging of the positive air electrode, which reduces the RTE even further. Therefore, these two aspects crucially affect the RTE of an APCS.

To solve these obstacles, Al-air batteries have been extensively studied in the past decades, but mainly from the materials aspects, including cathode catalysts for oxygen reduction reactions, doping of the Al metal anode to suppress self-corrosion, and electrolytes additives for more protective electrolyte/electrodes interphase (Liu et al. 2022). However, little attention has been paid on factors beyond materials, that actually affect the energy density delivered by Al-air batteries, particularly the RTE

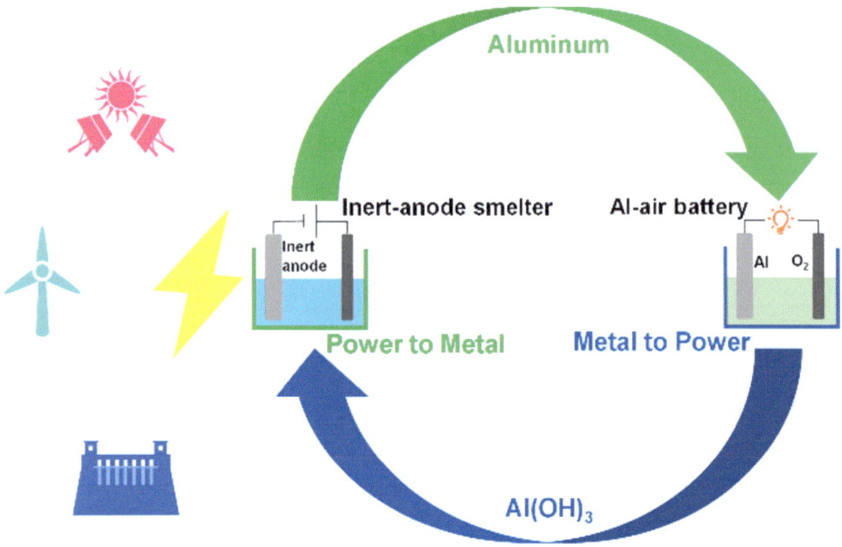

Fig. 1 Al production/conversion (P2Al2P) system. Schematic of the combination of Al-air batteries and inert-anode based Al electrolysis from Ref. (Xu et al. 2024). (Copyright of the authors)

of an APCS. In a recent publication (Xu et al. 2023), the accumulation of aluminate in the electrolyte has been identified as one of the causes of the poor efficiency. To address this problem, a seeded precipitation process has been demonstrated, allowing the aluminate removal and electrolyte regeneration. A wider operation temperature range is required to make it more efficient. A new cell design is also proposed, integrating more functions. Since self-corrosion and energy inefficiency during the electrochemical operation led to simultaneous heat release, cell design to maintain the operating temperature at the optimal state and even utilize the released heat is vital for overall energy utilization.

Figure 2 illustrates the operation of the cells employing 10 mL electrolytes upon long-term discharge at 100 mA cm^{-2} and 50 °C. The Al foil electrode was nearly consumed after 8 h and therefore changed with a new Al foil, while the same cathode was used for the whole measurement. The initial electrolyte, 4 M KOH aqueous solution with 5 g L^{-1} $Na_2SnO_3 \cdot 3H_2O$, was used for the initial 24 h discharge. Afterward, the electrolyte was regenerated via seeded precipitation at 20 °C. Operating under these conditions, the cell delivered the highest specific energy (4.29 kWh kg^{-1}) at 50 mA cm^{-2} resulting from the best combination of conversion efficiency and average voltage.

Besides the electrochemical conversion of aluminum into alumina, generating electricity, there are two main approaches to exploit the high energy content of aluminum through thermodynamic conversion as schematized in Fig. 3.

These approaches are (Bergthorson 2018):

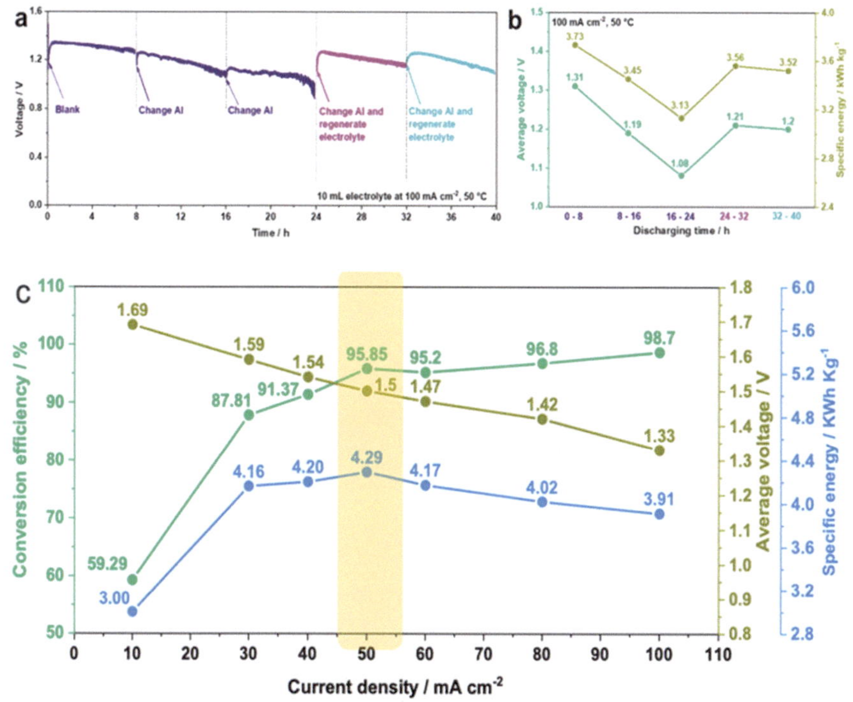

Fig. 2 **a** Voltage evolution and **b** average voltage and specific energy of a mechanically recharged Al-air cell including electrolyte regeneration by seeded precipitation. **c** Conversion efficiency, Average voltage and Specific energy of the same cell subjected to discharge at different current densities. Figure redrawn from Ref. (Xu et al. 2023) (Copyright of the authors)

Dry cycle: the energy of metallic powders is harnessed through direct combustion with air and utilized by external combustion thermal engines.

Wet cycle: the reaction between aluminum powders and H_2O, and consequently the conversion of H_2O into H_2, can occur with either activated (activation methods) or non-activated powders (conversion methods).

This distinction can be categorized as it follows.

3.2 *Dry Cycle*

3.2.1 Direct Metal-Air Combustion

Under the right conditions, aluminum reacts vigorously with oxygen to form aluminum oxide (alumina). The fundamental reaction is: $4\ Al + 3\ O_2 \rightarrow 2\ Al_2O_3$ (Bergthorson et al. 2015). This oxidation is highly exothermic, releasing about 31 MJ per kilogram of aluminum. However, the activation of Al by removal of the thin

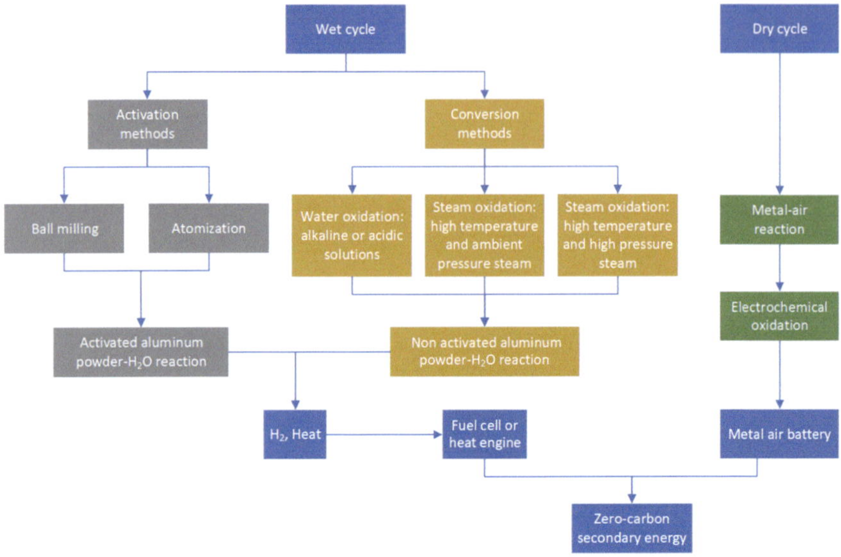

Fig. 3 Diagram of the different methods for water-aluminum and metal-air reactions to produce zero-carbon energy

alumina layer requires extreme temperatures (> 2,300 K) or Al needs to be fed into the combustion chamber as fine powder for efficient ignition at lower temperatures (around 930 K) (Puri and Yang 2010). Benefitting from the substantial heat release, Al is used in solid rocket fuels, offering high thrust and compact energy storage. Its use in combustors for thermal energy and power supply purposes appears to be only viable in external combustion engines where the extracted heat via reactor cooling is utilized (Bergthorson 2018). Its utilization in external combustion engines is expected to enable fine Al power combustion without sticking and wearing issues, similar to the use of pulverized coal in thermal power plants. Nevertheless, the combustion stabilization issue remains as the main hurdle in a combustor to transfer the heat while capturing the alumina combustion products (Bergthorson et al. 2015).

3.3 *Wet Cycle*

Techniques aimed at overcoming the oxide barrier on aluminum, thereby enabling reactions with water or steam, fall into two principal classifications: activation, which necessitates a preparatory treatment that endows aluminum powders with the capacity to react in the presence of water, i.e., conversion with water, and conversion with steam, which is achieved through a single-phase steam oxidation process.

3.3.1 Activation Methods

Activation methods prepare the material, rendering it reactive upon contact with water. It is, therefore, upon contact with water that the effective production of H_2 is enabled. To achieve this, activation methods involve a mechanical treatment with the addition of additives (salts) in the case of ball milling, and a thermal and pressurization treatment in the case of atomization. In both instances, there is a significant energy expenditure that impacts the energy balance.

Ball Milling

Mechanical ball milling emerges as an efficient method for the activation of aluminum particles, optimizing their reactivity, particularly pertinent in the reaction with water. This process conducted under inert atmosphere, combines mechanical alloying and milling, resulting in the dimensional reduction of particles and the augmentation of specific surface area. Such structural modifications expedite induction times and increment H_2 production (Zhang et al. 2014; Alinejad and Mahmoodi 2009).

The reduction of particles to nano-powders, via high-speed milling, introduces lattice defects and dislocations, fracturing the Al_2O_3 passivation layer. The introduction of specific additives, such as salts, metals, or oxides, during milling, facilitates the formation of reactive composites and creates pathways for water penetration, enhancing contact between aluminum and reactant (Du Preez and Bessarabov 2021).

The resulting composites exhibit irregular morphologies, with fissures and rough surfaces, which amplify the reaction area. This increment, coupled with structural defects, boosts hydrolysis kinetics, promoting a more efficient and sustained H_2 generation. In summary, ball milling, through mechanical and chemical modifications, transforms aluminum into a highly reactive material, optimizing H_2 production (Dossi 2024; Irankhah et al. 2018).

Atomization

Gas atomization represents a technique for the production of metallic powders activated for the reaction with water to produce H_2. The process is based on the melting of the metal in an inert atmosphere, followed by its fragmentation into micro-droplets via high-pressure gas jets. The rapid solidification of these droplets generates spherical particles, collected in a controlled environment to prevent contamination (Yang et al. 2019; Wang et al. 2015).

This method is distinguished by its capacity to obtain powders with homogeneous composition, essential for consistent hydrolysis performance and high H_2 yields. The reduction of grain size and the use of inert gases minimize oxidation, increasing reactivity. Atomization also promotes the disruption of the oxide layer, ensuring complete reaction of the material. In summary, this technique offers precise control over the microstructure and reactivity of the powders, optimizing H_2 production (Chen et al. 2021; Deng et al. 2024).

3.3.2 Conversion Methods

Alkaline or Acidic Water Solutions

The utilization of acidic or alkaline solutions represents an alternative methodology for the conversion of water into H_2 through the reaction with aluminum powders. Specifically, alkaline solutions demonstrate a greater efficacy compared to acidic ones in the dissolution of the oxide layer, which impedes the reaction (Alviani et al. 2019). The efficiency of the conversion process is influenced by a multiplicity of parameters, including the molarity and pH of the solution, the porosity of the aluminum powders, the dimensions of the particles, and the temperature of the water (Yang et al. 2019).

Sodium hydroxide (NaOH) is extensively utilized in alkaline solutions due to its low cost, simplicity of using, and its nature as a strong base (Bolt et al. 2020; Testa et al. 2024). This chemical compound proves particularly effective in disrupting the superficial oxide layer, accelerating the reaction kinetics and reducing the induction period. NaOH acts as a catalyst, promoting the exposure of fresh metallic surfaces to contact with water and facilitating the production of gaseous H_2, with the formation of sodium aluminate as a byproduct.

Steam Oxidation

H_2 and heat production through reaction between aluminum powders and steam is influenced by several factors, including steam temperature and pressure, purity, specific surface area (the presence of pores facilitates the reaction), and particle size of aluminum (Gao et al. 2023; Setiani et al. 2018).

The presence of high temperatures and pressures provides the advantage that no additives or catalysts are required for the combustion of the powders to occur. However, the energy used to superheat the steam can be recovered, for example, using a steam turbine, resulting in a positive energy balance for the overall system (Farmani and Eskandari Manjili 2024). The reaction products are $Al(OH)_3$ and AlO(OH) at temperatures ranging between 120 and 200 °C (Gao et al. 2023), and only AlO(OH) up to 370 °C (Setiani et al. 2018; Kirton et al. 2024; Gao et al. 2024).

Typically, the system comprises several components: a steam generator, an inert gas flushing system, a high-temperature furnace or reactor (ceramic or quartz) (Etminanbakhsh and Reza Allahkaram 2023), a condensation system, a condensate collector, a dryer, and a H_2 analyzer (Gao et al. 2023; Li et al. 2017). For high-pressure systems, the reactor is usually cylindrical and made of stainless steel (Setiani et al. 2018; Trowell et al. 2022).

The reaction can be divided into three phases. The first phase, known as the induction phase, involves the weakening of the oxide layer due to temperature and pressure, but the reaction has not yet started. The second phase begins with the penetration of steam through the pores and the initiation of the reaction. In this phase, the highest H_2 production occurs alongside the concurrent production of $Al(OH)_3$ or AlO(OH). In the third and final phase, H_2 production slows due to the deposition and clogging of pores by $Al(OH)_3$ or AlO(OH) (Gao et al. 2023; Trowell et al. 2020).

High temperature and ambient pressure steam oxidation

For micrometric aluminum powders (25 μm), experiments have shown that at ambient pressure and temperatures between 130 °C and 200 °C, a maximum H_2 production yield of 14–31% can be achieved (Gao et al. 2023).

At higher temperatures, ranging from 450 °C to 650 °C, the effects of additives such as $NaBH_4$ (Li et al. 2017) and NaF (Zhu et al. 2019) were studied to reduce ignition temperature and time. However, in the case of $NaBH_4$, a very low hydrogen yield was observed, varying from 0.8% to 1.6%, with a slight increase linked to higher temperatures or the presence of the additive. Conversely, adding 10 wt.% NaF can lower the ignition temperature from 960 °C to 743 °C and reduce the ignition time from 60 to 27 s.

An alternative method to facilitate the reaction at high temperatures (600 °C) within seconds, without requiring high pressures, involves electrically heating a graphite rod (to temperatures above the melting point of alumina) inside the reactor. This approach breaks the alumina layer (Etminanbakhsh and Allahkaram 2023).

A highly promising method is the aluminum steam oxidation proposed in (Barelli et al. 2022a, b). Such a process allows to achieve in a fixed-bed reactor, at 900 °C and ambient pressure, very high Al conversion rate to H_2, using non-activated Al micrometric powder (particle size < 44 μm). No additives or catalysts are used. Produced powder consists mostly of spherical micro particles γ-Al_2O_3 (83.4%w). The remaining portion is Al (16.6%w). In a further phase of the study, the addition in the reactor of alumina itself as inert material is proposed to avoid agglomeration and reactor clogging. The tendency of alumina clumping is proved to be hindered (Barelli et al. 2024), thus enabling a continuous process. Moreover, since only Al and Al_2O_3 are added in the reactor, also the direct use of produced oxides in the smelting process for fully recyclability is enabled as implemented in the power-to-X framework described in Sect. 4.

High temperature and high-pressure steam oxidation

Setiani et al. studied the effect of temperature in the range from 230 °C to 340 °C (Setiani et al. 2018), showing that higher temperatures corresponded to higher H_2 yields. This study also examined the effect of pressure, which significantly increased H_2 yield from 51% (270 °C, 55 bar) to 94% (280 °C, 62 bar). Trowell et al. demonstrated that at higher pressures (130–250 bar) and temperatures between 280 °C and 330 °C, yields of 100% can be achieved (Trowell et al. 2022, 2020). Gao et al. observed that increasing the yield from 60 to 100% required raising the temperature from 264 °C at 50 bar to 343 °C at 137 bar (Gao et al. 2024).

3.3.3 Activation and Conversion Methods Comparison

Considering these methods of water-to-vapor conversion or activation of aluminum powders for H_2 production, it is essential to acknowledge the limitations inherent to each technique. In particular, according to a recent review focused on the subject (Musicco et al. 2025), the following have to be considered:

- Atomization emerges as a technique for the rapid production of H_2 from ultrafine aluminum particles; however, it necessitates specific alloys containing bismuth, tin, and iron, which are subsequently present in the byproducts, and entails a high energy consumption.
- The activation of aluminum powders via ball milling with salts is distinguished by its simplicity and the cost-effectiveness of additives, offering environmental advantages due to the absence of metals. However, the energy consumption of the milling process limits its overall energy balance.
- The utilization of acidic or alkaline solutions, while simple and low-cost, presents environmental and safety challenges related to byproduct management and plant corrosion.
- The reaction with water steam, despite its high energy demand and severe operational conditions, offers a favorable environmental profile due to the production of readily recyclable alumina, i.e., γ-Al_2O_3 can be used in the smelting process without additional steps, and the potential recovery of thermal energy utilized for steam generation.

In summary, the selection of the optimal methodology depends on the application context. Milling, atomization, and acidic/alkaline solutions exhibit limitations related to byproducts and energy costs, whereas steam oxidation, although requiring critical conditions, proves promising for sustainability and circular economy, owing to the potential for energy recovery. Therefore, in the context of wet combustion, steam oxidation emerges as the most promising technique for the utilization of aluminum as an energy carrier, due to its potential energy sustainability and minimization of environmental impact.

4 Use Case: Al Wet Combustion System

Herein we propose a flexible operating energy storage system to cope with the energy demand using Al as a renewable electro-fuel (Barelli et al. 2020; Baumann et al. 2020). In particular, the wet combustion of Al yields heat, to generate electricity, as well as H_2, which can be used either for generating additional electricity or its direct use for refueling FCEVs (Shkolnikov et al. 2011; Vlaskin et al. 2011). According to the steam oxidation process of non-activated Al powder at 900 °C and ambient pressure (Barelli et al. 2022a, b), the solid product (Al_2O_3) can be directly used in the existing commodity-scale production of aluminum, allowing for the full recycling

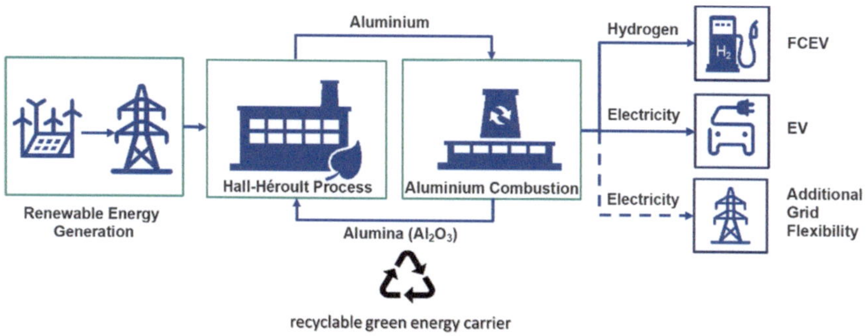

Fig. 4 Circular approach enabled by Al as a green energy carrier in power-to-X applications (Copyright of the authors)

of the energy carrier without additional investments and leading to the full decarbonization using renewable energy surplus for the Al production (World Aluminium 2021).

For what above, the use case here investigated is the one represented by a large scale multiservice station for FCEV/BEV in the framework of hybrid multi-functional refueling/recharging stations for EVs. These stations, with an installed electric power in the MW range, are usually installed along or close highways infrastructure. The overall concept complies strongly with the circular economy considering the whole life cycle of the material avoiding any need of intermediate transformation (Gislev et al. 2018) as depicted in Fig. 4.

Regarding the reduction phase from the oxide to the metal, improvements in the conventional Hall-Héroult aluminum production process towards the use of inert electrodes are considered in the following. This to enable the CO_2- and other GHG-free production (Reverdy and Potocnik 2020; ELYSIS 2019), considering their potential implementation at the industrial scale in the decarbonized European scenario.

Moreover, the high volumetric energy density and the consequent high locally storable capacity, enable the proposed concept of a multiservice station for EVs as a source of flexibility for the grid, rather than a further load. Thus, the proposed concept can provide an infrastructure development aligned with the electric grid's capabilities, aiming to potentially serve as a flexible component by integrating energy storage systems to exploit surplus renewable energy generation, as well as for energy self-sufficiency (Pelosi et al. 2023; Barelli et al. 2022a, b).

Regarding the potential implementation impact of the proposed concept, it is highlighted as fast-charging infrastructure is not yet implemented in an extended way in the continental Europe, except for Germany, Netherlands and France (European Automobile Manufacturers' Association (ACEA) 2024). As regards H_2 refueling stations, only a few points are present in the middle of Europe, mostly in Germany. Therefore, an extended fast charging and H_2 refueling infrastructure has to be developed to support the transition to EVs. To this regard, the Alternative Fuel

Infrastructure Regulation (AFIR) targets 1,100 H_2 refueling stations across Europe by 2030, even if such a target is considered significantly inadequate. Moreover, to speed up infrastructure deployment and the transition to a sustainable mobility, a multi-technology approach is recommended, creating hybrid stations to supply both FCEVs and BEVs (Hydrogen Council 2021; IEA 2021; Bernard 2023).

4.1 System Configuration

According to these requirements, this use case considers a refueling/recharging station for EVs consisting of 26 points of fast charging (150 kW each). This design corresponds to the mean number of charging points per fast charging station assessed in (Jochem et al. 2019) for the highway networks in France and Germany.

Therefore, an overall nominal electrical power of about 3.9 MW is considered to be installed as illustrated in Fig. 5. Such a power is produced by exploiting both H_2 and heat released by aluminum steam oxidation process.

Steam for the reactor feeding is produced by internal heat recovery, while aluminum is fed in the powder form. The reactor temperature is maintained through a further cooling by a secondary pressurized circuit feeding a steam turbine for electricity generation.

Produced H_2 is used to fed a solid oxide fuel cell (SOFC). Downstream the SOFC, operated at 750 °C and with a fuel utilization factor of 0.8, unburned H_2 is oxidized in a suitable afterburner. The exhausts from afterburning are expanded in a gas turbine, producing additional electric power, which activates the compressor to

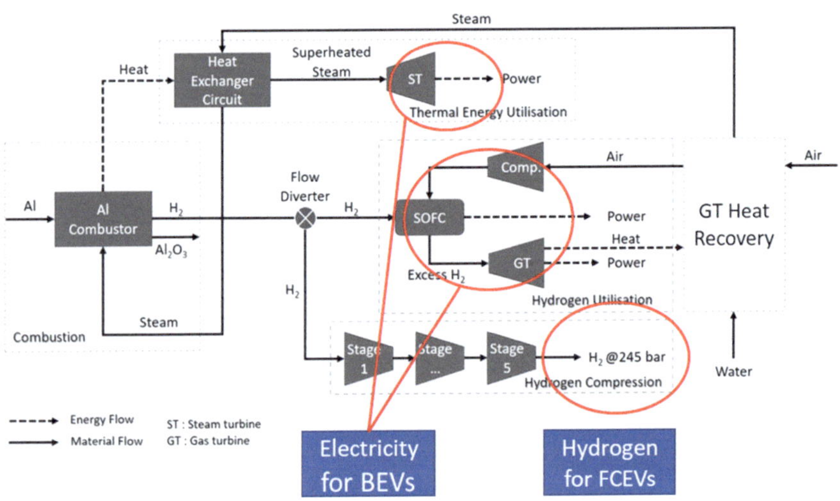

Fig. 5 Simplified layout of the multi-functional refuelling/recharging stations for EVs. (Copyrights of the authors)

supply pressurized air to the SOFC. The cathode feeding is subsequently preheated by the turbine exhaust and then heated up to 800 °C through the cathodic thermal regeneration.

This is the operation of the simplified layout of Fig. 5 when only electricity production occurs. For the complete system layout and details on operating conditions at both plant and components' levels refer to ref. It is highlighted as the SOFC polarization curve implemented in the model has been determined by tests at the laboratory scale under the specific operating feeding conditions resulting from the coupling with the aluminum steam oxidation process.

Moreover, to allow H_2 production for FCEVs, SOFC part load operation at 80% and 65% of the nominal power (2 MW) is implemented (see Table 2), according to the possible SOFC regulation at constant operating temperature within the 65–100% range investigated and presented in (Barelli et al. 2017). Therefore, in relation to the electricity and H_2 demands, part of the produced H_2 stream is delivered to a 245-bar storage tank via a 5-stage compression section.

Performance is assessed through simulation at the different required electric loads. Obtained results, summarized in Table 2, are processed to determine the Al-to-Power ($\eta_{M\text{-}P}$) and Metal-to-X ($\eta_{M\text{-}X}$) efficiencies. Increasing the part load degree, $\eta_{M\text{-}X}$ increases while $\eta_{M\text{-}P}$ decreases. The $\eta_{M\text{-}X}$ increase is reflected also in the cycle efficiency ($\eta_{P\text{-}X}$), determined in reference to the Power-to-X overall cycle.

To this aim, 11 kWh $kg^{-1}{}_{Al}$ specific energy consumption is considered. It corresponds to a 15% reduction of the specific consumption typical of current best practice Hall–Héroult electrolysis cells (ca. 13 kWh $kg^{-1}{}_{Al}$) due to the implementation of the wettable drained cathode technology, as remarked by Moya et al. (2015). Moreover, further improvements are expected to result from the implementation of inert and dimensionally stable non-carbon anodes. Finally, it is observed that the assumed specific energy consumption for Al production (11 kWh $kg^{-1}{}_{Al}$) is only slightly lower than the one already achieved in Norway (Segatz et al. 2016).

For these reasons, the assumption made on energy intensity of CO_2-free Al smelting process is considered comparable with reference to the decarbonized scenario by 2050.

4.2 Techno-Economic Evaluation

Defossilisation of the mobility sector is a complex problem, the offered solutions aim to respond to this demand with various environmentally-friendly solutions within techno-economic limits. Consideration of techno-economics in this context is vital for supporting decisions for the wide deployment and development of the infrastructure to mitigate the environmental and economic burdens associated with the energy demand of the mobility sector while ensuring the grid stability. In this sense, use of Al as an energy carrier in this context is potentially a suitable alternative. Hence, to be able to have a better understanding if the proposed conversion path is applicable in techno-economic terms, a 3.9 MW power capacity Al wet combustion system

Table 2 Plant simulated performances at different electric load conditions

% P_{SOFC} (%)	P_{SOFC} [kW]	P_{GT} [kW]	P_{ST} [kW]	P_C [kW]	P_{TOT} [MW]	H_2 [kg h^{-1}]	$\eta_{M\text{-}P}$ (%)	$\eta_{M\text{-}X}$ (%)	$\eta_{P\text{-}X}$ (%)
100	2.000	906	1.064	0.0	3.9	–	81	81	35.6
80	1.600	683	916	53.2	3.1	28	65	88	38.8
65	1.300	520	884	89.0	2.6	46.8	54	93	40.7

is considered for simultaneous supply of H_2 and electricity based on (Ersoy et al. 2022). The most important findings are briefly summarized here. The formulated business case is assessed from a techno-economic perspective utilizing deterministic and probabilistic estimation approaches.

First, the capital expenditures (CAPEX) are estimated considering equipment cost (i.e., steam turbine, gas turbine, solid-oxide fuel cell (SOFC), heat exchangers, pumps and other process equipment) utilizing learning curves. Furthermore, the installation factors are introduced to estimate installed system costs. Additionally, other CAPEX costs (i.e., engineering and procurement and construction (EPC) costs and working capital) are included. Under these considerations the analysis estimates a CAPEX of 4,200–6,200 € kW^{-1}. The depreciable capital is then used for estimation of fixed asset costs and salvage value of the system using Modified Accelerated Cost Recovery System (MACRS) rates.

Accordingly, operational expenditures (OPEX) are included considering fixed operation and maintenance (O&M) cost and insurance costs. The variable OPEX consists of variable O&M cost of system, fuel costs and transportation costs (aluminum or oxides) assuming a 400 km distance based on the annual operation scenarios. Considering historical prices, probabilistic approaches are implemented to incorporate the market price of Al ranging from 1.2 to 1.95 € kg^{-1} (from 2014 to 2019). Considering the presented average price breakdown of Al in Table 3, the share of different cost components is also incorporated to the economic evaluation model.

Since the Al is not consumed but converted to its oxide, in this circular context it makes more sense to consider a net price of Al accounting the reduced price due to return of alumina to the smelter. The net cost of Al is estimated by deducting the cost of alumina from the price of the Al, and it is assumed the other cost shares except energy costs remain constant. Since the study exploratively evaluates the techno-economics, energy costs scenarios are implemented considering the low price of electricity during hours where renewable generation exceeds the demand, resulting in an electricity price of 0–50 € MWh^{-1}. Considering that the provided price breakdown refers to Norway where electricity is provided at a cost-effective price (*ca.* 35 € MWh^{-1}) (Statistics Norway 2021), it is assumed that this scenario approximates the case as summarized in the scenario overview in Table 4.

Table 3 Aluminum price breakdown from Norsk Hydro (Norsk Hydro 2020)

Price cost components	Share of cost components (%)
Aluminum oxide	43
Energy costs	24
Carbon anodes	17
Fixed costs	11
Other process related costs	5

Table 4 Scenario-specific parameters, Al price, electricity price and energy intensities (Norsk Hydro 2020)

Scenario	Al price [€ kg_{Al}]	Al_2O_3 price [€ kg_{Al}^{-1} eq.]	Energy intensity [kWh $kgAl^{-1}$]	Electricity price [€ MWh^{-1}]
Scenario-I	1.44–1.86 (μ = 1.65)	0.52–0.83	14.25	50
Scenario-II	1.06–1.48 (μ = 1.26)	($\mu = 0.67$)	11	30
Scenario-III	0.73–1.15 (μ = 0.93)		11	0

Partial operation loads introduced in the system configuration part, and 4,000 full load hours annual operation duration are considered for the economic evaluation similar to what is expected from other Power-to-X technologies. Hence, the system will supply 10.4–16 GWh of electricity and 112–187 tons of H_2 annually depending on the chosen partial operation load. Under given scenarios, discounting the expenditures and amount of energy supplied during system's lifetime a levelized cost analysis has been conducted for electricity (LCOE) and H_2 (LCOH). The results indicate an LCOE ranging from 152 to 334 € MWh^{-1}, while the LCOH of Al-based H_2 corresponds to 5.4–11.8 € kg^{-1} based on various operation modes in the reference scenario. The breakdown of costs highlights that the cost is mainly driven by the net cost of Al followed by CAPEX. Also, to analyze the sensitivity of the full load hours, a sensitivity analysis is conducted as shown in Fig. 6.

Both LCOE and LCOH indicate a declining trend with the increasing full load hours. The evaluation results suggest that the system remains viable under various electricity pricing conditions, especially in low-cost or surplus renewable energy scenarios. Findings indicate that aluminum-based energy storage can be a cost-competitive alternative while improving energy security. In the context of EV

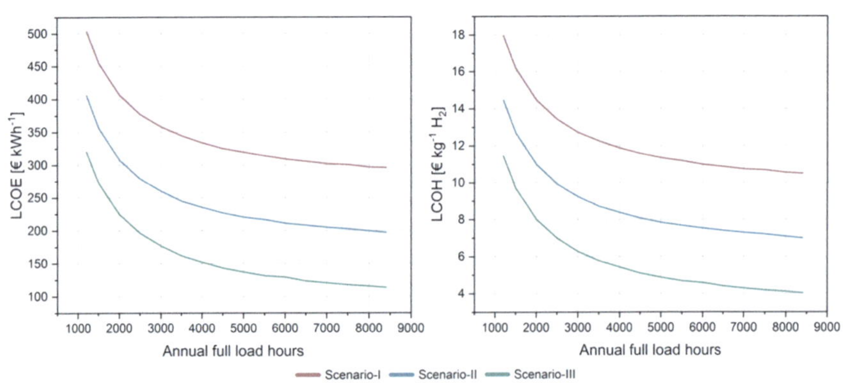

Fig. 6 LCOE and LCOH estimations reference values (median of probability distributions) for the considered scenarios

charging infrastructure, this system supports both BEV charging and FCEV refuelling by stabilizing grid demand through on-site energy generation, reducing dependence on large-scale H_2 transport and storage. In particular, on the system level, hybrid aluminum-powered EV charging stations can reduce peak load stress.

In conclusion, aluminum-based hybrid storage presents a flexible, scalable, and prospectively sustainable solution for H_2 production and energy storage. While challenges remain in production efficiency and recycling, this technology has the potential to complement conventional batteries and H_2 storage, with strong potential for EV charging and broader energy applications.

4.3 Sustainability Aspects

Reactive metals such as Al offer a potential solution for improving the efficiency and sustainability of EV charging stations, which are crucial for the transition to clean energy and the reduction of global warming. However, several challenges need to be addressed before these technologies can be fully utilized. These technologies, including the solutions presented here, are still in the early stages of development, making it difficult to assess their long-term viability and sustainability. The sustainability considerations of using Al as an energy carrier are mainly associated with its supply chain and circular use. Considering the entire supply chain, from raw materials to final Al products, the sustainability hotspots can be summarized as follows:

1. **Mining**: Conventionally, Al is extracted from bauxite ore. Bauxite is primarily mined in Australia, Guinea, China, Brazil, and India. The regional environmental and social impacts in these mining locations need to be assessed within their specific context.
2. **Bauxite refining:** Bauxite refining is one of the most significant sustainability hotspots in the entire supply chain, mainly due to bauxite residue (red mud) and its typical waste treatment approach. The residue is stored in large bauxite lakes, where the alkaline, iron-rich red powder is mixed with water to prevent airborne dispersion. However, this leads to soil alkalinity, water and air pollution, and long-term human health risks (Healy 2022). Although alternative treatment methods are under development, they have not yet provided a techno-economically feasible solution due to the low economic value of the extracted minerals (Ujaczki et al. 2018).
3. **Transportation:** The Bayer process (bauxite refining) is usually conducted at the mining site, after which alumina is transported over long distances to countries where high-value-added Al products are manufactured.
4. **Hall-Héroult Process:** The conventional aluminum smelting process is highly energy-intensive. Additionally, the use of carbon anodes in electrolysis leads to CO_2 and perfluorocarbon (PFC or PFAS) emissions. Several mitigation strategies are being developed, including:

- Electrolyzer optimization and alternative designs to improve energy efficiency and reduce greenhouse gas emissions.
- Integration of inert anodes and drain cathodes, which eliminate direct carbon and PFC emissions. Significant progress has been made in these areas, and wider adoption of these solutions is expected in the near future (Light Metal Age 2025).

The sustainability hotspots outlined above are critical for ensuring the sustainable implementation of the proposed concept. Since aluminum is used in a circular system within the considered conversion pathway, most of these impacts, except those from the smelting process are minimized, as aluminum can theoretically and practically be recycled indefinitely. However, the Hall-Héroult process requires special attention, as it serves as the charging phase in the energy storage cycle, whereas discharging occurs via wet combustion. The energy intensity of this process significantly influences the round-trip efficiency of the overall system. However, when benchmarked against Power-to-X conversion efficiencies, the findings indicate high potential for Al as an energy carrier, particularly due to its high volumetric energy density. If ongoing decarbonization efforts in the Hall-Héroult process achieve the desired advancements, a breakthrough adoption of aluminum-based energy storage could occur.

Another critical consideration is ensuring that the oxides produced during the combustion process are of smelter-grade purity, as impurities could hinder aluminum's circular reuse. The feasibility of this approach will depend heavily on experimental demonstrations to validate the purity and recyclability of post-combustion aluminum oxides.

5 Relevant KPIs

For the further development of the charging infrastructure and consideration of Al-based hybrid energy storage, key technical, economic, environmental, and social KPIs are identified.

From an energetic perspective, energy density and round-trip efficiency remain as crucial parameters. In particular, the high volumetric energy density of Al makes it an attractive energy carrier, allowing large amounts of energy to be stored in a stable, inert solid. Theoretical assessments suggest that the round-trip efficiency of the proposed concept has the potential to outperform or compete with other Power-to-X technologies. Additionally, the system's versatility in providing both electricity and H_2 is a notable advantage. Another key KPI is the reaction time, ensuring the system delivers energy promptly. While experimental validation is required, a maximum reaction time of minutes must be ensured for practical implementation.

From an economic perspective, critical KPIs include the levelized cost of electricity/ H_2, cost of Al, operation and maintenance costs, and recycling cost and value recovery. These factors are further influenced by parameters such as charging station energy capacity, refueling and regeneration cycle time, H_2 storage and handling capacity, and scalability.

Given the sustainability objectives of the proposed concept, evaluating environmental KPIs such as global warming potential, water consumption, acidification, and resource depletion is essential. As circularity is a major advantage of this approach, waste by-products and material circularity index are also key considerations.

Reliability and safety are fundamental, particularly in the context of charging stations. Key KPIs include system stability and performance lifetime, fire and explosion risks, resilience to grid disruptions, and compliance with public and industrial safety standards. From a social perspective, social acceptance plays major role regarding visual impact, health and safety considerations and overall sustainability. Addressing these factors will be crucial for the successful deployment and public adoption of the proposed Al-based HESS.

Acknowledgements H.E., M.B. and S.P. acknowledge the basic support of the Helmholtz Association and the Karlsruhe Institute of Technology for the project ALU-STORE (Aluminium Metal as Energy Carrier for Seasonal Energy Storage) within the KIT Future Fields framework.

References

Al Wahedi A, Bicer Y (2020) Development of an off-grid electrical vehicle charging station hybridized with renewables including battery cooling system and multiple energy storage units. Energy Rep 6:2006–2021. https://doi.org/10.1016/j.egyr.2020.07.022

Alinejad B, Mahmoodi K (2009) A novel method for generating hydrogen by hydrolysis of highly activated aluminum nanoparticles in pure water. Int J Hydrogen Energy 34(19):7934–7938. https://doi.org/10.1016/j.ijhydene.2009.07.028

European Aluminium (2019) Vision 2050—European aluminium's contribution to the EU's mid-century low carbon roadmap. European Aluminium. https://www.european-aluminium.eu/media/2545/sample_vision-2050-low-carbon-strategy_20190401.pdf

Alviani VN, Setiani P, Uno M, Oba M, Hirano N, Watanabe N, Tsuchiya N, Saishu H (2019) Mechanisms and possible applications of the Al–H_2O reaction under extreme pH and low hydrothermal temperatures. Int J Hydrogen Energy 44(57):29903–29921. https://doi.org/10.1016/j.ijhydene.2019.09.152

Barelli L, Bidini G, Ottaviano A (2017) Integration of SOFC/GT hybrid systems in micro-grids. Energy 118:716–728. https://doi.org/10.1016/j.energy.2016.10.100

Barelli L, Baumann M, Bidini G, Ottaviano PA, Schneider RV, Passerini S, Trombetti L (2020) Reactive metals as energy storage and carrier media: use of aluminum for power generation in fuel cell-based power plants. Energ Technol 8(9):2000233. https://doi.org/10.1002/ente.202000233

Barelli L, Trombetti L, Di Michele A, Gammaitoni L, Asenbauer J, Passerini S (2022b) Aluminum steam oxidation in the framework of long-term energy storage: experimental analysis of the reaction parameters effect on metal conversion rate. Energ Technol 10(9):2200441. https://doi.org/10.1002/ente.202200441

Barelli L, Pelosi D, Longo M, Zaninelli D (2022) Energy storage integration into fast charging stations installed on e-highways. In: 2022 IEEE Power & energy society general meeting (PESGM), 01–05. Denver, CO, USA: IEEE. https://doi.org/10.1109/PESGM48719.2022.9916974

Barelli L, Trombetti L, Zhang S, Ersoy H, Baumann M, Passerini S (2024) Potential of aluminum as a metal fuel for supporting EU long-term energy storage needs. Adv Mater Technol 2302206. https://doi.org/10.1002/admt.202302206

Baumann M, Barelli L, Passerini S (2020) The potential role of reactive metals for a clean energy transition. Adv Energy Mater 2001002

Bergthorson J (2018) Recyclable metal fuels for clean and compact zero-carbon power. Prog Energy Combust Sci 68:169–196. https://doi.org/10.1016/j.pecs.2018.05.001

Bergthorson J, Goroshin S, Soo MJ, Julien P, Palecka J, Frost DL, Jarvis DJ (2015) Direct combustion of recyclable metal fuels for zero-carbon heat and power. Appl Energy 160:368–382. https://doi.org/10.1016/j.apenergy.2015.09.037

Bernard MR (2023) European union alternative fuel infrastructure regulation (AFIR). POLICY. Int Counc Clean Transp

Bolt A, Dincer I, Agelin-Chaab M (2020) Experimental study of hydrogen production process with aluminum and water. Int J Hydrogen Energy 45(28):14232–14244. https://doi.org/10.1016/j.ijhydene.2020.03.160

Chen X, Wang C, Liu Y, Shen Y, Zheng Q, Yang S, Huanming L et al (2021) Popcorn-like aluminum-based powders for instant low-temperature water Vapor hydrogen generation. Mater Today Energy 19:100602. https://doi.org/10.1016/j.mtener.2020.100602

Deng R, Wang M, Zhang H, Yao R, Zhen K, Liu Y, Liu X, Wang C (2024) Development of cost-effective Sn-free Al-Bi-Fe alloys for efficient on board hydrogen production through Al–water reaction. Materials 17(20):4973. https://doi.org/10.3390/ma17204973

Denholm P, Mai T (2019) Timescales of energy storage needed for reducing renewable energy curtailment. Renew Energy 130:388–399. https://doi.org/10.1016/j.renene.2018.06.079

Dossi S (2024) Using Al-based activated powders for on-demand hydrogen production: possibilities and perspectives of the 'white' hydrogen. In: Recent advances in environmental science from the euro-mediterranean and surrounding regions (3rd ed), edited by Mohamed Ksibi, Abdelazim Negm, Olfa Hentati, Achraf Ghorbal, Arturo Sousa, Jesus Rodrigo-Comino, Sandeep Panda, José Lopes Velho, Ahmed M. El-Kenawy, and Nicola Perilli, 415–18. Advances in science, technology & innovation. Springer Nature, Cham. https://doi.org/10.1007/978-3-031-43922-3_94

Du Preez SP, Bessarabov DG (2021) On-demand hydrogen generation by the hydrolysis of ball-milled aluminum composites: a process overview. Int J Hydrogen Energy 46(72):35790–35813. https://doi.org/10.1016/j.ijhydene.2021.03.240

ELYSIS (2019) Rio Tinto and Alcoa announce world's first carbon-free aluminium smelting process. ELYSIS. April 1, 2019. https://www.elysis.com/en/rio-tinto-and-alcoa-announce-worlds-first-carbon-free-aluminium-smelting-process

Ersoy H, Baumann M, Barelli L, Ottaviano A, Trombetti L, Weil M, Passerini S (2022) Hybrid energy storage and hydrogen supply based on aluminum—a multiservice case for electric mobility and energy storage services. Adv Mater Technol. https://doi.org/10.1002/admt.202101400

Etminanbakhsh M, Allahkaram SR (2023) Reaction of aluminum particles with superheated steam to generate hydrogen gas as a readily usable clean fuel. Fuel 332:126011. https://doi.org/10.1016/j.fuel.2022.126011

European Automobile Manufacturers' Association (ACEA) (2024) Charging ahead: accelerating the roll-out of EU electric vehicle charging infrastructure. ACEA. https://www.acea.auto

Farmani A, Manjili FE (2024) Modelling and assessment of hydrogen combined cycle power plant using aluminum-water reaction as renewable fuel. Int J Hydrogen Energy 50:276–291. https://doi.org/10.1016/j.ijhydene.2023.08.239

Gao X, Wang CA, Bai W, Hou Y, Che D (2023) Experimental investigation of the catalyst-free reaction characteristics of micron aluminum powder with water at low and medium temperatures. J Energy Storage 59:106543. https://doi.org/10.1016/j.est.2022.106543

Gao X, Wang CA, Hou Y, Zhao L, Bai W, Che D (2024) Experimental study on catalyst-free high-temperature and high-pressure aluminum-water reaction characteristics using metal as energy carrier. Int J Hydrogen Energy 50:1338–1357. https://doi.org/10.1016/j.ijhydene.2023.09.216

Gislev M, Grohol M, Mathieux F, Ardente F, Bobba S, Nuss P, Blengini GA et al (2018) Report on critical raw materials and the circular economy. http://publications.europa.eu/publication/manifestation_identifier/PUB_ET0418836ENN

Gonzalez-Rivera E, Garcia-Trivino P, Sarrias-Mena R, Torreglosa JP, Jurado F, Fernandez-Ramirez LM (2021) Model predictive control-based optimized operation of a hybrid charging station for electric vehicles. IEEE Access 9:115766–115776. https://doi.org/10.1109/ACCESS.2021.3106145

Güven AF, Ateş N, Alotaibi S, Alzahrani T, Amsal AM, Elsayed SK (2025) Sustainable hybrid systems for electric vehicle charging infrastructures in regional applications. Sci Rep 15(1):4199. https://doi.org/10.1038/s41598-025-87985-7

Healy S (2022) Sustainable bauxite residue management guidance. Int Alum Inst

Hydrogen Council (2021) Roadmap towards zero emissions–the complementary role of BEVs and FCEVs. Summary Document

IEA (2021) Net zero by 2050: a roadmap for the global energy sector. OECD. https://doi.org/10.1787/c8328405-en

IEA (2024) Outlook for electric mobility—global EV outlook 2024—analysis. IEA. 2024. https://www.iea.org/reports/global-ev-outlook-2024/outlook-for-electric-mobility

Irankhah A, Fattahi SMS, Salem M (2018) Hydrogen generation using activated aluminum/water reaction. Int J Hydrogen Energy 43(33):15739–15748. https://doi.org/10.1016/j.ijhydene.2018.07.014

Jochem P, Szimba E, Reuter-Oppermann M (2019) How many fast-charging stations do we need along European highways? Transp Res Part D Transp Environ 73:120–129. https://doi.org/10.1016/j.trd.2019.06.005

Kirton T, Saceleanu F, Mobarakeh MS, Reza Kholghy M (2024) Cogeneration of hydrogen, alumina, and heat from aluminum-water reactions. Int J Hydrogen Energy 68:115–127. https://doi.org/10.1016/j.ijhydene.2024.04.038

Li F, Zhu B, Sun Y, Tao W (2017) Hydrogen generation by means of the combustion of aluminum powder/sodium borohydride in steam. Int J Hydrogen Energy 42(6):3804–3812. https://doi.org/10.1016/j.ijhydene.2016.07.015

Light Metal Age (2025) Carbon free aluminum production with inert electrodes for clean energy storage and production. Light Metal age magazine (blog). February 10, 2025. https://www.lightmetalage.com/news/industry-news/smelting/carbon-free-aluminum-production-with-inert-electrodes-for-clean-energy-storage-and-production/

Liu X, Jiao H, Wang M, Song W-l, Xue J, Jiao S (2022) Current progresses and future prospects on aluminium-air batteries. Int Mater Rev 67(7):734–764

Mahmud I, Medha MB, Hasanuzzaman M (2023) Global challenges of electric vehicle charging systems and its future prospects: a review. Res Transp Bus Manag 49:101011. https://doi.org/10.1016/j.rtbm.2023.101011

Moya RJ, Boulamanti A, Slingerland S, van der Veen R, Kuenen JJP, Visschedijk AJH (2015) Energy efficiency and GHG emissions: prospective scenarios for the aluminium industry. https://doi.org/10.2790/263787. JRC Scientific and Policy Reports. Brussels: European Commission. https://ec.europa.eu/jrc/en/publication/eur-scientific-and-technical-research-reports/energy-efficiency-and-ghg-emissions-prospective-scenarios-aluminium-industry

Musicco N, Gelfi M, Iora P, Venturelli M, Artioli N, Montorsi L, Milani M (2025) A review of hydrogen generation methods via aluminum-water reactions. Int J Thermofluids, 101152. https://doi.org/10.1016/j.ijft.2025.101152

Norsk Hydro ASA (2020) Hydro—second quarter 2020 invester presentation. https://www.hydro.com/Document/Index?name=Investor%20presentation%20Q2%202020.pdf&id=560591, July

Pelosi D, Longo M, Bidini G, Zaninelli D, Barelli L (2023) A new concept of highways infrastructure integrating energy storage devices for E-mobility transition. J Energy Storage 65:107364. https://doi.org/10.1016/j.est.2023.107364

Petrovic J, Thomas G (2011) Reaction of aluminum with water to produce hydrogen-2010 update. Washington DC.: Office of Energy Efficiency and Renewable Energy (EERE)

Puri P, Yang V (2010) Thermo-mechanical behavior of nano aluminum particles with oxide layers during melting. J Nanopart Res 12(8):2989–3002. https://doi.org/10.1007/s11051-010-9889-2

Reverdy M, Potocnik V (2020) History of inventions and innovations for aluminum production. In: TMS 2020 149th Annual meeting & exhibition supplemental proceedings, edited by the minerals, metals & materials society, the minerals, metals & materials series. Springer International Publishing, Cham, pp 1895–1910. https://doi.org/10.1007/978-3-030-36296-6_175

Roslan MF, Ramachandaramurthy VK, Mansor M, Mokhzani AS, Jern KP, Begum RA, Hannan MA (2024) Techno-economic impact analysis for renewable energy-based hydrogen storage integrated grid electric vehicle charging stations in different potential locations of Malaysia. Energ Strat Rev 54:101478. https://doi.org/10.1016/j.esr.2024.101478

Segatz M, Hop J, Reny P, Gikling H (2016) Hydro's cell technology path towards specific energy consumption below 12 kWh/Kg. In: Light metals 2016, edited by Edward Williams, Springer International Publishing, Cham, pp 301–5 https://doi.org/10.1007/978-3-319-48251-4_50

Setiani P, Watanabe N, Sondari RR, Tsuchiya N (2018) Mechanisms and kinetic model of hydrogen production in the hydrothermal treatment of waste aluminum. Mater Renew Sustain Energy 7(2):10. https://doi.org/10.1007/s40243-018-0118-8

Shkolnikov EI, Zhuk AZ, Vlaskin MS (2011) Aluminum as energy carrier: feasibility analysis and current technologies overview. Renew Sustain Energy Rev 15(9):4611–4623. https://doi.org/10.1016/j.rser.2011.07.091

Statistics Norway (2021) 09364: electricity prices in the end-user market, by type of contract, quarter and contents. Statbank Norway. SSB. September 30, 2021. https://www.ssb.no/en/system/

Testa V, Gerardi M, Zannini L, Romagnoli M, Santangelo PE (2024) Hydrogen production from aluminum reaction with NaOH/H2O solution: experiments and insight into reaction kinetics. Int J Hydrogen Energy 83:589–603. https://doi.org/10.1016/j.ijhydene.2024.08.152

Trowell K, Goroshin S, Frost DL, Bergthorson JM (2020) Aluminum and its role as a recyclable, sustainable carrier of renewable energy. Appl Energy 275:115112. https://doi.org/10.1016/j.apenergy.2020.115112

Trowell K, Goroshin S, Frost D, Bergthorson J (2022) Hydrogen production rates of aluminum reacting with varying densities of supercritical water. RSC Adv 12(20):12335–12343. https://doi.org/10.1039/D2RA01231F

Ujaczki É, Feigl V, Molnár M, Cusack P, Curtin T, Courtney R, O'Donoghue L et al (2018) Re-using bauxite residues: benefits beyond (critical raw) material recovery. J Chem Technol Biotechnol 93(9):2498–2510. https://doi.org/10.1002/jctb.5687

Vlaskin MS, Shkolnikov EI, Bersh AV, Zhuk AZ, Lisicyn AV, Sorokovikov AI, Pankina YuV (2011) An experimental aluminum-fueled power plant. J Power Sources 196(20):8828–8835. https://doi.org/10.1016/j.jpowsour.2011.06.013

Wang C, Liu Y, Liu H, Yang T, Chen X, Yang S, Liu X (2015) A novel self-assembling al-based composite powder with high hydrogen generation efficiency. Sci Rep 5(1):17428. https://doi.org/10.1038/srep17428

World Aluminium (2021) Primary aluminium production. 2021. https://www.world-aluminium.org/statistics/

Xu C, Niklas Herrmann X, Liu BH, Passerini S (2023) Addressing the voltage and energy fading of Al-air batteries to enable seasonal/annual energy storage. J Power Sources 574:233172. https://doi.org/10.1016/j.jpowsour.2023.233172

Xu C, Liu X, Sumińska-Ebersoldt O, Passerini S (2024) Al−air batteries for seasonal/annual energy storage: progress beyond materials. Batteries Supercaps, e202300590. https://doi.org/10.1002/batt.202300590

Yadav AK, Bharatee A, Ray PK (2023) Solar powered grid integrated charging station with hybrid energy storage system. J Power Sources 582:233545. https://doi.org/10.1016/j.jpowsour.2023.233545

Yang H, Zhang H, Peng R, Zhang S, Huang X, Zhao Z (2019) Highly efficient hydrolysis of magnetic milled powder from waste aluminum (Al) cans with low-concentrated alkaline solution for hydrogen generation. Int J Energy Res 43(9):4797–4806. https://doi.org/10.1002/er.4621

Zhang H, Sun Y, He F, Xianjin Y, Zhao Z (2014) Preparation and characterization of activated aluminum powder by magnetic grinding method for hydrogen generation: preparation and characterization of activated Al powder for hydrogen. Int J Energy Res 38(8):1016–1023. https://doi.org/10.1002/er.3123

Zhu B, Li F, Sun Y, Yuxin W, Shi W, Han W, Wang Q, Wang Q (2019) Enhancing ignition and combustion characteristics of micron-sized aluminum powder in steam by adding sodium fluoride. Combust Flame 205:68–79. https://doi.org/10.1016/j.combustflame.2019.02.007

Waterborne Transport. Hybrid Power Supply for Electrification of Port Infrastructures, Shore-to-Ship Power, and Ship Power and Propulsion

Juan Camilo Gomez Trillos and Urte Brand-Daniels

Abstract Hybrid power supply systems integrate the use of different power sources, converters and/or storages for the electrification of maritime applications in order to reduce environmental impacts and contribute to resilience. In the area of port infrastructure, there have been approaches to the hybridisation of RTG cranes, tugs and drayage trucks, which have been equipped with batteries in addition to combustion engines. These approaches can contribute to a reduction in fuel consumption of up to 40% and thus to a reduction in greenhouse gas emissions and air pollutants. With regard to shore-to ship power, there are promising solutions despite the lack of harmonised standards, such as charging with lithium-ion batteries due their high efficiency and power density, as well as battery swapping due to short recharging times and the flexible use of energy from the grid. In the area of hybrid power and propulsion systems, the comparison of diesel engines, fuel cells and batteries shows that diesel engines have lower investment costs, but also lower efficiencies of up to 52% compared to fuel cells with 60%. However, according to the EU HySeasIII project, fuel cells combined with batteries can contribute to a significant reduction in GHG emissions of up to 80%.

Keyword Battery storage · Decarbonization · Energy efficiency · Fuel cells · Maritime sustainability · Resilience

List of Abbreviations

AC	Alternative current
APS	Automated plug-in system
BC	Black carbon
BE	Battery electric
BESS	Battery energy storage system

J. C. G. Trillos (✉) · U. Brand-Daniels
Institute of Networked Energy Systems, German Aerospace Center (DLR), Oldenburg, Germany
e-mail: Juan.GomezTrillos@dlr.de

R. Scipioni et al. (eds.), *Hybrid Energy Storage*, Lecture Notes in Energy 47,
https://doi.org/10.1007/978-3-031-97755-8_11

CAPEX	Capital expenditures
CCCC	China communication construction co.
CP	Controllable pitch
DBE	Diesel battery electric
DC	Direct current
DCS	Data collection system
DE	Diesel electric
DoD	Depth of discharge
DWT	Deadweight tonnage
EGR	Exhaust gas recirculation
EMSA	European maritime safety agency
ESS	Energy storage system
EU	European Union
FP	Fixed pitch
GHG	Greenhouse gas
GM	Generator/motor
GT	Gross tonnage
HC	Hydrocarbons
HFCBE	Hydrogen fuel cell and battery electric
HFO	Heavy fuel oil
hp	Horsepower
HT-PEMFC	High temperature proton exchange membrane fuel cell
HVSC	High voltage shore connection
IAPH	International association of ports and harbours
IMO	International maritime organization
km	Kilometer
KPI	Key performance indicator
kV	Kilovolts
kW	Kilowatt
LCA	Life cycle assessment
LCC	Life cycle costing
Li-ion	Lithium-Ion batteries
lm	Lane meters
LT-PEMFC	Low temperature proton exchange membrane fuel cell
LVSC	Low voltage shore connection
MARPOL	International convention for the prevention of marine pollution from ships
MDO	Marine diesel oil
MGO	Marine gasoil
MW	Megawatt
MWh	Megawatt-hour
NMC	Nickel-manganese-cobalt Li-Ion battery
NMCA	Nickel-manganese-cobalt-aluminium
NOx	Nitrogen oxides
OPEX	Operational expenditures

OPS	Onshore power supply
PGM	Platinum-group metals
$PM_{2.5}$	Particulate matter 2.5 μm
PMS	Power management system
RoPax	Roll on/roll off and passenger ferry
RoRo	Roll on/roll off ferry
RTG	Rubber tired gantry crane
SBC	Shore-side battery charging
SBC-BS	Shore-side battery charging—battery swapping
SFOC	Specific fuel oil consumption
SPB	Shore-side power banks
SOFC	Solid oxide fuel cell
SOx	Sulphur oxides
SSE	Shore-side electricity
STS	Shore-to-ship
TEN-T	Trans-European transport network
TEU	Twenty-foot unit container

1 Introduction

Worldwide seaborne trade totaled 12,027 million tons in 2022, declining slightly by 0.4% in that year. However, a growth rate of 1.2% was recorded for 2023 and an expansion beyond 3% between 2024 and 2028 is expected (United Nations Conference on Trade and Development 2023). A growth trend will likely continue in the future. In addition to freight transport, maritime passenger transport is also growing, although it was significantly affected by the COVID pandemic. The passenger volume reached 31.7 million passengers in 2023, surpassing by 7% the pre-pandemic level in 2019. Moreover, a 10% increase in capacity is forecasted for the years 2024 through 2028 (Cruise Lines International Association 2024).

The Fourth Greenhouse Gas (GHG) Study by the International Maritime Organization (IMO) estimates that maritime shipping contributed approximately 2.8% to the global anthropogenic CO_2 emissions in 2018 (Faber et al. 2020). More recent statistics from the IMO Data Collection System (DCS) platform gives account of a fuel consumption totaling 212.3 million tons in 2021, resulting in estimated emissions of 660 million tons CO_2 after considering emission factors for different fuels (Marine Environmental Protection Comittee 2022). In comparison, worldwide anthropogenic CO_2 emissions were estimated at 36,816 million tons (Ritchie and Roser 2024). While operational and technical measures have led to decrease in energy intensity and higher efficiency, these gains have been outpaced by the growth of the sector, therefore leading to increase of the of the total emissions produced by this industry, what could lead to a higher impact in the future and a higher share of the total CO_2 emissions. In addition, maritime transportation has considerable challenges for its transition and a complete fleet renovation or retrofitting is expected to take several decades.

Although ships and their propulsion systems are often cited as the main source of emissions, the issue is much more complex. The maritime sector involves not only the water transportation activity itself, but also the activities occurring while ships are at berth as well as the handling of cargo and passengers at ports which can contribute massively to the environmental impact (Park 2022). On one side, the IMO is responsible for developing and maintaining international regulations for maritime shipping at international level. On the other side, port infrastructure and their regulations tend to be managed locally by each country. This leads to different regulations in force both on board or on shore.

In addition to the considerable pollution caused by cargo handling and passenger activities at ports, ships at berth have also considerable emissions. At berth, although the ship propulsion is not operating, other systems onboard require power. Often these systems are lighting, air conditioning, electronic systems and any load due to cargo handling. Thus, it is estimated that ships at berth emitted between 7.9 and 10.7 million tonnes of CO_2eq emissions per year between 2018 and 2023 in the EU, which represent a share of between 6.1 and 7.2% of the total emissions of ships bigger than 5000 gross tonnes in the EU (EMSA 2024). However, although ocean-going vessels at port are important contributors to local air pollutants and GHG emissions, some research also has shown that harbor crafts, cargo-handling equipment and heavy-duty diesel vehicles have also relevant emissions, comparable or even higher than those of the ships berthed at port, depending on the analyzed pollutant (Park 2022). Against this background, several of these technological options are being tested, developed and adopted both on as well as offshore to reduce local pollution and in addition GHG emissions of maritime transportation, including electrification, hybridization, and the use of renewable energy sources.

Hybrid power systems, which integrate the usage of multiple power sources, converters and/or storages, offer a promising solution for reducing local pollution and GHG emissions while also improving efficiency. In addition, electrification has been seen as one alternative to integrate renewable energies into the energy supply for waterborne transportation. Typically, hybrid power systems consist of internal combustion engines and energy storage systems including technologies like batteries, supercapacitors or flywheels. Other solutions may include fuel cells or other power-generation technologies (Geertsma et al. 2017; Damian et al. 2022; Inal et al. 2022). On the other hand, hybridization leads to the electrification of operations by the use of multiple power sources that can be integrated as required. It also enables the integration of energy storage systems which, depending on the system specifications, can be used as devices for peak shaving, load balancing and, in some cases, fully electric operation (Kalikatzarakis et al. 2018). In addition to a possible reduction in environmental impacts, this can also increase resilience, e.g. by enabling flexibilization, redundancy, storage options and, to a certain extent, independent operation.

The next subchapters will explore the state of the art of possible hybrid solutions in three key areas going from land to the shore by encompassing freight-handling infrastructure and vehicles/vessels at ports, shore-to-ship power technologies and onboard ship power and propulsion systems. In the latter case, selected key performance indicators (KPIs) for technologies such as traditional internal combustion engines, fuel cells and Li-ion batteries are explained and their possible developments partly up to 2050 are shown. The description of the state of the art will be finally exemplified with a case study of a hydrogen fuel cell and battery hybrid ferry concept previously developed in the project HySeas III.

2 Hybrid Power Solutions at Ports

Ports require extensive freight-handling infrastructure and vehicles. The construction and operation of both infrastructure and vehicles demand significant resources and can lead to considerable air pollution and in general environmental impacts. Traditionally, ports rely on local electrical grids for lighting, and to certain degree for equipment operation. On the other hand, Fossil fuels are used for heavy machinery like cranes and container handling equipment. With the need to reduce the emissions, minimize the local pollution and move towards sustainability, ports are increasingly electrifying their operations and integrating renewable energy sources like photovoltaic panels and wind turbines into their energy supply.

According to a survey of the International Association of Ports and Harbours (IAPH), 45% of the port investments in solar and wind energy are being executed timely, although also 46% of the surveyed replied that they currently do not have any planned investments in this subject (Notteboom and Pallis 2023). Energy Storage Systems (ESS) such as electrochemical batteries are also being used for the storage of fluctuating renewable energies and ensure a constant power supply and peak shaving, both for stationary as well as for mobile applications (Kermani et al. 2021). Alternative fuels or complete electrification are also envisioned to replace fossil fuels used in the operation.

In this context, hybrid solutions that combine multiple generation and storage technologies are becoming increasingly relevant as a potential alternative to systems powered only by fossil fuels. In the next section, the hybridization trends of three vehicles types commonly used in freight ports, rubber-tired gantry cranes, tugs and drayage trucks, will be analyzed exemplarily.

2.1 Rubber Tired Gantry (RTG) Cranes

Rubber Tired Gantry (RTG) cranes are used to move and store containers at ports. RTG cranes are mainly powered by diesel engines coupled with electric generators to provide the power to lift the containers (Starcrest Consulting Group LLC 2012; Antonelli et al. 2017). A typical RTG Crane is shown in Fig. 1, where the exhaust of the diesel engine can be seen as a vertical rising pipe to the right of the crane. The potential energy of the containers lifted is later dissipated during descent in resistive loads (Antonelli et al. 2017). Adding an energy storage allows the storage of the—otherwise—dissipated energy for further use later as lifting power. As the power for lifting containers can be partially supplied by the energy storage, a smaller generator can power the RTG crane and operate at an optimal point. This contrasts with the traditional operation for which the generator has to operate at multiple operational points. Statistics on the emissions of RTG cranes at the Ports of Long Beach and Port of Los Angeles, where some hybridization was tested more than a decade ago, show that approximately 11% of nitrogen oxides (NO_x) emissions and 9% of particulate matters (PM) from cargo handling equipment is emitted by RTG cranes (Starcrest Consulting Group LLC 2012).

Several projects have tested hybridization as one alternative to decrease the emissions and fuel consumption of these cargo-handling equipment. Recent research carried out by Antonelli et al. analyzed the typical operational loads of RTG cranes in the port of Livorno in Italy and potential savings that a hybrid system including electrochemical storage or supercapacitors combined with traditional diesel internal combustion engines could offer (Antonelli et al. 2017). By means of the collected data and simulations, this study identified the possibility of downsizing the internal combustion engine installed in the RTG crane from the original 414 kW. The main advantage of this is the operation of smaller engines for longer time at the design point. However, smaller engines have often poorer fuel specific fuel consumption

Fig. 1 Typical RTG crane at the Port of Kiel, Germany. The crane shown is not hybrid but has a similar function as the one shown in the figure (own figure)

compared to the bigger engines, what partially offsets the savings of operating the engines at their design point. The authors also analyzed the integration of 83.2 or 44.4 kWh Li-ion batteries with different cell chemistries. With the integration of lithium batteries in hybrid systems compared with systems only including internal combustion engines, the authors identified possible fuel savings from 30 to 60%, depending on the ICE engine sizing (Antonelli et al. 2017). Some costs analyses were carried out by the same authors by comparing a typically 414 kW ICE powered RTG crane with hybrid versions including 414, 165 kW or 36 kW engines in combination with batteries or supercapacitors. The main findings showed that the RTG versions with unmodified engine size have an estimated decrease in fuel costs of around 30%, whereas the engine-downsized versions manage to decrease the fuel costs down to 60%. Although the hybrid RTG versions with unchanged engine size result in a higher purchase cost, hybrid RTG cranes break even within 3–6 years due to the fuel savings (Antonelli et al. 2017). Other analyses carried out by Vlahopoulos and Bouhouras in 2022 confirm the aforementioned results with reductions of up to 33% of fuel consumption and payback times between 1.3 and 3.42 years (Vlahopoulos and Bouhouras 2022).

The implementation of hybrid RTG cranes is transitioning from a research topic into industrial application. Industrial equipment manufacturers already offer hybrid options as part of their standard RTG crane portfolio, claiming fuel savings of up to 40% depending on the operation (Liebherr 2019; Konecranes 2024). Some specifications of the Energy Storage Systems (ESS) installed on hybrid RTG cranes powered by battery systems are summarized in Table 1, with Battery Energy Storage Systems (BESS) systems of comparable size to those mentioned by Antonelli et al. (Corvus Energy 2024a).

Table 1 Reference projects of hybrid RTG cranes of the company Corvus Energy (2024a)

Year	Country	ESS capacity [kWh]	Number of RTG cranes	Location
2015–2016	China	94	20	Hybrid RTG Cranes of China Communication Construction Co. (CCCC) in the ports of Yangshang, Yidong and Waigaoqiao
2018–2020	China	79	48	Hybrid RTG Cranes of China Communication Construction Co. (CCCC) in China
2020	United States of America	170	25	25 hybrid RTG cranes in the of South Carolina

2.2 Tugs

Tugboats or simply tugs are vessels assisting other ships during maneuvers by pulling or pushing them (Wärtsilä 2024). Often these ships have a high power-tonnage ratio allowing them to generate significant thrust at low speed, what is commonly named bollard pull. A typical load profile of a tugboat shows that the engine load is much higher (ten-fold) when assisting another ship than during loitering or waiting (Vu et al. 2015). In efforts to reduce the air pollutants and GHG emissions, tugboats are being progressively electrified or hybridized.

Some industrial electric and hybrid solutions are already available in the market. The company DAMEN offers an electric tug fitted with battery packs with a storage capacity of around 2800 kWh (DAMEN 2024). Wärtsilä has also disclosed the production of thrusters for a 5000 hp (3728 kW) hybrid tug to be used in the Chinese market (Wärtsilä 2022). In addition, the company Rolls-Royce also disclosed the production of an LNG tugboat powered by two 1492 kW MTU gas engines, 2 × 500 kW electric motors and additionally onboard batteries with 904 kWh capacity for peak shaving during acceleration, maneuver or for electricity supply onboard (Rolls-Royce 2024). The specifications of the tugboats depend on operational aspects and particularly on the bollard pull required for the particular operations and cannot be generalized. More information on ship propulsion and power systems for ships will be explored in subchapter 11.4 (On Board Hybrid Ship Power and Propulsion).

In a study carried in 2010 and in the port of Los Angeles and Long Beach, reductions of emissions and fuel consumption of a hybrid power electric propulsion tugboat compared to a conventional were analyzed (Jayaram et al. 2010). The emissions of PM2.5, NOx and CO_2 emissions for the hybrid tug compared to a conventional tug were found to be 73%, 51% and 27%, respectively. In addition, fuel savings of about 25–28% in favor of the hybrid alternative were described. However, most of the impact was attributed to the use of gensets for propulsion rather than the onboard BESS. Moreover, according to a document published by Siraichi et al. in 2015, the implementation of the hybrid tugboat Tsubasa including 300 kWh lithium iron phosphate BESS onboard led to reductions of 20% in the fuel consumptions as well as in the CO_2 emissions for the operation of the ship compared to a conventional tugboat (Siraichi et al. 2015). According to the commercial producer of Tugboats DAMEN, it's model ASD TUG 2810 Hybrid can provide fuel savings of up to 30% and reduce the NOx, PM, HC and CO emissions by up to 42, 39, 44 and 46%, respectively (DAMEN 2014).

2.3 Drayage Trucks

Drayage refers to short distance transportation of goods, typically containers, within ports or surrounding areas. Containers are often unloaded from or loaded onto a vessel by one or multiple quayside cranes, with a frequency of around one to three minutes. The step either prior or subsequent to this process involves large fleets of trucks, conveying containers from or to the quayside cranes. For example, heavy-duty diesel trucks including drayage trucks accounted for 32, 43, 37 and 49% of the total NO_x, particulate matter ($PM_{2.5}$), carbon monoxide (CO), sulphur oxide (SO_2) and CO_2eq emissions by pollutant at the Port of New York and New Jersey in 2019, respectively (Park 2022).

As in the other cases presented, hybrid solutions are being explored, tested and implemented. Within a demonstration project in the port of Los Angeles in 2017, the company Kenworth developed a hydrogen fuel cell and lithium-ion battery class 8 truck with a 100kWh battery pack and 85 kW fuel cell from the company Ballard. The power of the entire system was rated at 360 kW/480 hp and the range of operation was estimated at 320 km (WaterstofNet 2018a). The company US Hybrid also tested a drayage truck with an 80 kW fuel cell and 30 kWh battery achieving a system power of 367 kW and a torque of 3930 Nm. The estimated range for the truck is 320 km with onboard hydrogen capacity of 25 kg (WaterstofNet 2018b). Furthermore, US Hybrid has also launched a hybrid natural gas-powered drayage truck with over 85 kWh battery capacity and a 8.9-L gas engine and an on-board battery charger of up to 20 kW (US Hybrid 2022). This engine is smaller than the 15-L used in the non-hybrid versions. The range of the truck is around 1000 miles (1600 km) with both compressed natural gas and battery and up to 35 miles (56 km) only with battery. The truck traction power is 340 hp (253 kW). According to the press releases of US Hybrid, the fuel economy of the hybrid drayage truck is as much as double as that of the original 15-L gas engine. However, no absolute numbers have been disclosed.

2.4 Section Concluding Remarks

Hybrid power technologies are being used to mitigate pollution and GHG emissions at port. Solutions for RTG cranes, tugs and drayage trucks were introduced in the former section. Hybridization contributed to the reduction of typically between 20 and 40% of the fuel consumption for RTG and tugboats and drayage trucks. At the same time significant reductions in air pollutants were also described. While specifications vary for each of these applications, a growing trend in the implementation of hybrid systems emerges, as these can decrease fuel consumption, air pollution and GHG emissions. Moreover, established producers of industrial equipment are already offering hybrid options in response to the demand of more sustainable port operations.

3 Hybrid Shore-to-Ship Power Solutions

When ships are berthed at port, they require energy to power their electrical and heat (or cooling) loads. These loads are often referred as hotel loads and frequently not bound to the propulsion system. Electrical loads are typically supplied by onboard diesel generators. Depending on the ship type and size, the hotel loads can fall in a range in the order of hundreds of kW for small ships to the order of tenths of MW for big cruise ships, the latter consuming considerable amounts of electricity for onboard equipment like air conditioning, lighting and other electrical loads. In fact, 25% of the available energy on board of cruise vessels is used for hotel loads (Marzi and Broglia 2019). Typically, diesel generators run on fossil fuels like heavy fuel oil (HFO), marine gasoil (MGO), marine diesel oil (MDO), hydrotreated vegetable oil or even natural gas. Depending on the fuel they burn, these generators can produce considerable amounts of air pollutants, particulate matter and GHG emissions. Although influenced by the manufacturer, engine size, operational point and fuel, typical specific fuel oil consumption (SFOC), falls into the range 180–210 g fuel oil/kWh of electricity generated, which leads to CO_2 emissions of between 575 and 670 g CO_2/kWh of electricity produced onboard (MAN Energy Solutions 2024). In addition, assuming an engine is Tier III—MARPOL Annex VI compliant, the NO_x emissions would be between 2.0 and 3.4 g/kWh (Marine Environmental Protection Comittee 2021). Particulate matter (PM) and black carbon (BC) emissions are in practice below 100 mg/kWh and 50 mg/kWh. However, NO_x, PM and BC emissions are highly dependent on the engine load, being often higher at partial load, which is often the case when ships are berthed (Kuittinen et al. 2024). Therefore, ships at berth are also considerable sources of air pollution and GHG emissions. For comparison, the GHG emissions of PV electricity ranges from 25.2 to 43.6 g CO_2eq/kWh and emissions for the production of electricity from fossil fuel generators can emit up to 1 kg of CO_2eq/kWh of electricity (Stucki et al. 2023).

Shore-Side Electricity (SSE), also known as Shore-to-Ship (STS) power, Onshore Power Supply (OPS) or—in a more traditional way—cold ironing, refers to the supply of electricity from shore to ships while they are berthed. The shore electricity supply allows to turn off the onboard generators, thus reducing the local air pollution. An additional benefit of SSE is the reduction of the net GHG emissions, as long as the electricity supplied from the shore grid or shore-based power supply has lower emissions than that generated onboard. This is particularly the case if renewable energies make an important part of the electricity generation matrix onshore.

According to Article 9 of the EU Regulation 2023/1804, SSE should be available for all passenger seagoing container and passenger ships in Trans-European Transport Network (TEN-T) maritime core ports by the beginning of 2030. This regulation targets container ships, RoRo ferries and passenger ships with a Gross Tonnage (GT) beyond 5000 GT, for which SSE should be used in at least 90% of the port calls (European Parliament and the Council of the European Union 2023). As a result, new solutions for SSE and practical use cases are expected to be developed or rather implemented in the following years.

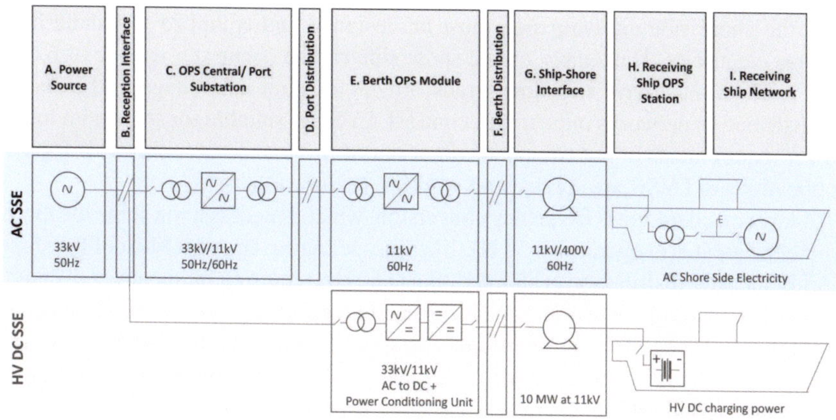

Fig. 2 Architecture of SSE systems considering main component blocks considering AC or DC electricity supply. Adapted from European Maritime Safety Agency (EMSA), 2022ab

The European Maritime Safety Agency (EMSA) produced already guidelines on the subject of SSE equipment, technology, planning and safety (European Maritime Safety Agency (EMSA) 2022a, b). EMSA so far considers different options of electricity supply to ships, including Onshore Power Supply (OPS), Shore-Side Battery Charging (SBC), SBC Battery Swapping (SBC-BS), Shore-Side Power Banks (SPB) and Port Generators.

Figure 2 shows the most important infrastructure components for SSE, focusing on the options OPS and SBC. Additional options using AC and DC SSE supply are shown. SSE consist systems of different components both onshore as well as onboard. Other configurations are also possible.

The different options will be explained to the light of hybrid power systems in the following sections and are described in detail in the current guidelines developed by EMSA (European Maritime Safety Agency (EMSA) 2022a, b). Although the options OPS, SBC and SBC-BS are not necessarily hybrid in nature, they will be explained to understand their differences compared to the other hybrid alternatives.

3.1 *Onshore Power Supply (OPS)*

OPS consist of the supply of electrical power to ships at berth from a shore-side source, such as the national grid or a local generation system. With this solution, the electricity generation from auxiliary engines onboard can be replaced and the generators turned off while at berth. Two types of OPS are existing; namely, High Voltage Shore Connection (HVSC) and Low Voltage Shore Connection (LVSC). HVSC is suited for ships with high power demand, such as container or cruise ships. It includes a centralized frequency conversion at port substation. However, both the shore side and ship side must be designed and equipped to handle high voltage electricity. The supply on the shore side can be occur at a voltage of 6.6 or 11 kV. Ships must have step-down transformers to adapt the voltage to the one of the distribution networks onboard. In contrast, LVSC is suitable for ships with lower power requirements or not equipped with a step-down transformer onboard. Typical applications of LVSC are service and working vessels or tankers.

A key consideration is frequency conversion, which is necessary in some locations where the local grid operates at 50 Hz, like Europe, Africa or Asia. Most of the ships worldwide have distribution grids operating at 60 Hz, requiring frequency conversion due to incompatibility when ship and local electricity grids have different frequencies. Frequency conversion entails additional costs for SSE infrastructure. The use of OPS could effectively reduce CO_2eq emissions and local air pollutants emissions at ports like SO_x, NO_x and particulate matter emissions, especially if the shore-side electricity is generated with renewable sources.

OPS can be hybrid or non-hybrid depending on the power source (Fig. 2a), which may include the national grid, port generators, renewable energy plants, electrical energy storage or emergency back-up units.

3.2 Shore-Side Battery Charging (SBC)

Shore-Side Battery Charging (SBC) builds on OPS to charge onboard BESS using AC or DC shore power supply. This option is increasingly relevant, as the fleet of hybrid and electric ship fleets is growing. Key features of SBC include fast charging capabilities, in some cases using high-voltage direct current (HVDC), for instance 2 MW at 1 kV DC. Moreover, greater flexibility in the internal ship arrangement, as components like transformers can be on land, an important feature for ships which internal spaces are occupied by battery compartments. Finally, depending on the shore-side electrical power source, these systems can provide reduced GHG emissions and reduce local pollution. Among the challenges are fire hazards due to battery overheating and thermal runaway, interconnectivity and interoperability between the onboard management system and shore-side charger and lack of standardized solutions.

The electrical connection in SBC can be either wired (AC or DC) or wireless (capacitive or inductive power transfer). Wireless charging offers some advantages like reduced exposure to mechanical wear and corrosion and simplified docking procedures, which are convenient for ships with opportunity charging like ferries, for which charging time is limited and connection and disconnection times can make charging unfeasible. However, the lower efficiency compared to wired charging options is a major obstacle to the widespread use of wireless charging systems (Khan et al. 2022).

One example of SBC can be found in the port of Oslo, where the company Cavotec installed in 2020 an Automated Plug-in System (APS) for e-ferry charging (see Fig. 3). The system is suitable for ferries with a charging connection in the bow,

Fig. 3 Automatic plug-in system of the company Cavotec for charging of e-ferries. *Source* Courtesy Cavotec (2024)

minimizing the space for onboard and shoreside equipment and maximizing the possible charging time during passenger boarding and offboarding. This system has been implemented for the electrification of the e-ferry service between Oslo City Hall Pier to Nesodden in Oslofjord providing charge service to the ferries MS Dronningen, MS Kongen and MS Prinsen (Cavotec 2020).

3.3 *Shore-Side Battery Charging—Battery Swapping (SBC-BS)*

In the case of SBC-BS, the ships' BESS are not recharged on board, but completely replaced (swapped). This reduces the turnaround times of electric/plug-in vessels at berth, eliminating the need of wait for recharging and enabling greater flexibility. For smooth operation, modularity and standardization are important prerequisites. In addition, the infrastructure of the interface between ship and shore should ensure fast and safe handling of the battery module units (European Maritime Safety Agency (EMSA) 2022a, b).

Some examples of this solution are already available. In 2023, Singapore has launched their first fully-electric cargo vessel called Hydromover with swappable batteries (Offshore Energy 2023). A photo of this vessel is shown in Fig. 4. The battery system called PrwSwäp includes 70 kWh $\times$ 6 NMC Li-ion batteries and is connected through cloud-based service and management provided by the company of

Fig. 4 A photo of the Hydromover, a fully-electric cargo vessel with interoperable swappable battery solution. *Source* Courtesy Yinson Green Tech (Offshore Energy 2023)

Shift Clean Energy from Vancouver. The lightweight, 18.5 m long vessel transports up to 25 tons of cargo. The improved energy efficiency and lower maintenance costs should save up to 50% of operating costs. The battery can be replaced within minutes to minimize downtime. The vessel also had to undergo a comprehensive risk assessment in relation to the operation and swapping of batteries to ensure compliance with international safety standards in the maritime industry.

A swappable battery container for inland shipping was also announced in 2021 by the company Wärtsilä. The battery containers are installed on the 104 TEU inland container vessel Alphenaar, which is powered simultaneously by two 20-foot containerized battery banks onboard (Wärtsilä 2021). According to representatives of the company developing these systems, this concept can allow the reduction of the emission of 1000 tons of CO_2 and 7 tons of NO_x per year. Each of the containers include 45 battery modules totaling 2 MWh and are charged with certified green energy at dedicated charging stations (Port of Rotterdam 2021; Zero Emission Services 2021).

3.4 Shore-Side Power Banks (SPB)

OPS, SBC and SSB-BS are not necessarily hybrid. An extension of the SBC and a hybrid solution are Shore-Side Power Banks (SPB). Power banks or shore side Electrical Energy Storage (ESS) units in containers are used to temporally store on-site electricity, which in some cases comes from renewable sources (European Maritime Safety Agency (EMSA) 2022a, b). This can push simultaneously the deployment of renewable energy sources in the port area and handle fluctuating/irregular electricity production (e.g., renewable energies). For instance, overnight charging or charging during off-peak hours can decrease the stress on the local grid and increase the use of cheaper electricity. Moreover, other services in the port area could be provided by power bank energy as the containerized storage units can be deployed anywhere and moved around the port. Given the challenges, the maintenance and safety care of a large number of containers that are associated with particular fire hazards, are mobile and have complex interoperability, can be costly (European Maritime Safety Agency (EMSA) 2022a, b). Finally, due to the low energy density of batteries, container energy banks require many units per MWh, which at the same time means a high space requirement that can be very expensive in port areas due to space scarcity. One example of SPB is the charging system implemented for the Amherst Islander II (fully electric) and Wolfe Islander IV (hybrid) passenger and car ferries operating in the Ontario lake in Canada. Due to the limited charging times, the ferries require fast charging. However, the local grids of Millhaven and Stella harbors have limited capacity. For this reason, each of the SBC stations servicing these ferries are equipped with a dedicated 3.0 MWh BESS (Leclanché 2022). These SBC systems charge the onboard 1.9 MWh (Amherst Islander II) and 4.6 MWh (Wolfe Islander IV) batteries. The BESS onshore are charged by the harbor grid and are connected to the ferry through DC-DC converters with a capacity of 1800 kW. In this way, the power drained from the local grid decreases from 1800 to around 1000 kW. Additionally, the BESS can also provide peak-shaving services during high consumption periods in the harbor area.

Another example is provided by a research project in the ports of Kiel in Germany and Gothenburg in Sweden. The EU-funded Sea Li-ion research project led by the shipping company Stena and BatteryLoop explored the possibility of reusing lithium-ion batteries from the automotive sector for a stationary ESS that supplies electricity for electric ferries in the ports of Kiel and Gothenburg (Powertrain International Web 2022). The classification society DNV also supported this project. So far, only the design of the ESS, the evaluation of the recycling potential of lithium-ion batteries, the impact of the ESS on the electricity grid and a business case for electric ferries and ESS in the Port of Gothenburg have been analyzed, but the aim is to make the system a reality by 2030 and to operate the RoPax ferry "Stena Elektra" between Gothenburg and Frederikshavn with it. The business case in the port of Gothenburg has shown that there is currently sufficient electricity in the port. However, placing an ESS in the port creates opportunities to free up electricity supply capacity for other system services and support the electricity grid. According to estimations of the project, 5% of the total electricity consumption in Gothenburg would be required to charge a ferry (Powertrain International Web 2022). In other ports, however, an ESS might be necessary to enable the charging of electric ferries without compromising grid stability or building additional infrastructure.

3.5 Port Generators

A final hybrid option discussed in the existing literature is port generators that utilize micro-generation (European Maritime Safety Agency (EMSA) 2022a). These systems aim to provide mobile and flexible electrical energy "on site", especially in the case of ports with limited access to the electricity grid or insufficient power for the demand placed by berthing ships. Some examples include hydrogen fuel cells or LNG hybrid power production barges, which can serve as port generators. A relevant consideration is the required space for the installation of the power generation units and safety aspects such as hazardous areas, low flash point or toxic fuels. Furthermore, the sustainability of such solutions can only be guaranteed depending on the energy sources used (e.g., green hydrogen) (European Maritime Safety Agency (EMSA) 2022a, b).

One example can be found at the Kirkwall pier in Scotland, where a 75 kW hydrogen fuel cell supplied by Arcola Energy was placed to provide electricity for ships and other activities. Moreover, the heat as by-product of the fuel cell is used for nearby buildings. The hydrogen is produced by electricity from 900 kW wind turbines and tidal turbines on the island of Eday and shipped to Kirkwall (BIG HIT Project 2024; Surf 'n' Turf Project 2024). The initiative was part of Surf 'n' Turf's community renewable energy project, funded by Local Energy Scotland and the Scottish Government's Local Energy Challenge Fund, managed by Community Energy Scotland and supported by partners EMEC, Orkney Islands Council, Eday Renewable Energy and ITM Power.

3.6 Concluding Remarks

In the previous section, several options for the shore-side supply of electricity to ships were discussed, including OPS, SBC, SBC-BS, SPB and port generators. Some publications have analyzed these alternatives with regard to different criteria (Khan et al. 2022; Mutarraf et al. 2022). For instance, Khan et al. (2022) evaluated different SSE options, including SBC, SBC-BS and power banks in the sense of hydrogen production, based on six parameters: cost, efficiency, environmental impact, recharge time, durability and reliability. Based on their assessment, Khan et al. could show that SBC with lithium-ion batteries performed best in terms of their high efficiency and power density. Battery swapping is a promising solution to save time (low recharging time) and utilize energy from the grid when demand is lower, but further improvements are needed to simplify the process. Finally, the hydrogen production system enables an external supply, which leads to a flexibility of the port and increases the continuity of the service.

Even though various SSE options are available and first examples of implementation exist, the power supply is still primarily provided by the conventional variant of the power systems on board. On the one hand, this has to do with the fact that there are no uniform standards especially for hybrid solutions like SBC as well as for electric vehicles (EV) and DC charging in the marine context. Moreover, the diffusion of electrically powered ships (especially larger vessels) still does not seem to be progressing as quickly as originally thought due to high costs, long charging times, system complexity, local grid conditions and a lack of charging infrastructure, which suffers the common chicken and egg problem. In particular, the low cost of MDO and HFO seems to be an important argument in favor of generating electricity on board instead of building infrastructure for SSE. However, an improved energy management system, e.g. in the form of energy storage systems that can absorb peak loads and use cheaper electricity at times of higher load, could help to improve the efficiency and reduce the ship's overall electricity costs (Tang et al. 2018; Kumar et al. 2019).

4 On Board Hybrid Ship Power and Propulsion

In the past sections, different hybrid solutions for ports and shore-side electricity were described. This section will be focused on the power and propulsion systems of ships.

Internal combustion engines consuming HFO and MDO are ubiquitous in most of the ships for propulsion and for the supply of electricity onboard. Nevertheless, concerns about air pollution and global warming potential have prompted the industry to explore alternatives to increase efficiency, reduce the environmental impacts and decrease operating costs, while maintaining safety and performance in the operations. Three not mutually exclusive approaches to minimizing these problems are (Geertsma et al. 2017):

- Abatement technologies: Engine modifications, exhaust gas recirculation (EGR), fuel water emissions, waste heat recovery systems, sulfur scrubbers and selective catalytic reduction (SCR) to combat air pollution. These technologies have largely been used to tackle NO_x emissions, which are regulated by the MARPOL VI annex and classified under different "tiers". Used so far to combat air pollution, these systems offer only a limited effect when it comes to reducing GHG emissions because gases like CO_2 are not captured or transformed in the process. Often, these systems increase the CAPEX and OPEX of ships.
- Alternative fuels: Replacing the current fossil fuels by other fuels like methanol, biodiesel, ammonia, hydrogen, among others. Depending on the pathway used to produce the alternative fuels, the net effect of these can be favorable or not in terms of air pollution and GHG emissions compared to the use of their fossil-based fuels counterparts. Additionally, these substances are at the moment mainly derived from fossil fuels and availability of renewable-based alternatives is still limited.

- Hybrid power systems: Design that integrates multiple power sources to provide efficient and innovative propulsive power. It can typically consist of a combination of internal combustion engines (such as marine diesel engines) and often ESS including batteries, supercapacitors, or flywheels. The designs are not limited to use diesel engines, but can also utilize fuel cells or any other generation technologies as power sources. One benefit of electrical propulsion is its higher efficient at low speed. Nevertheless, additional electrical components can introduce losses of between 5 and 15% of the propulsive power.

In the remainder of this subchapter, hybrid power and hybrid propulsion systems will be explored. The typical propulsion and power systems will be first introduced, followed by the hybrid power and hybrid propulsion systems.

4.1 Propulsion and Power Systems Options

Various designs can be categorized based on the type of propulsion and how it is powered. This situation is illustrated in Fig. 5, where the different options are shown. The option for shore-side electricity discussed in the previous section are also depicted for the different alternatives. Notice that two types of hybridization are possible: either the propulsion can be hybrid by using more than one driver (Fig. 5c and e) or the power supply can be hybrid obtaining power from different generation options (Fig. 5d–f). These two options can be as well combined (Fig. 5e).

The following sections describe different variations as presented in Fig. 5.

4.1.1 Mechanical Propulsion

The most typical option consists of a prime mover like a diesel engine or a gas turbine which is mechanically coupled in a direct way or through a gearbox to a propulsor (often a propeller). A separate electrical AC network supplied often by diesel generators provides electrical power for onboard electrical loads. If both ship and port have the necessary interfaces and infrastructure, shore-side electricity (SSE) may supply electricity for the onboard loads while at berth. A schematic representation of this option can be seen in Fig. 5a, where propulsion and the electrical loads are not coupled. Hybridization is not present in the propulsion or power system of this configuration.

This type of propulsion system reaches its maximum efficiency at design speed between 80 and 100% of the top speed. The advantages of this configuration include low conversion losses due to fewer conversion steps (main engine, gearbox and propeller), low complexity and low purchase cost compared to other options. However, some disadvantages are limited maneuverability due to the operational envelope of the propulsion engine, higher maintenance requirements under dynamic loads, poor fuel efficiency and high emissions at speeds below 70% of maximum

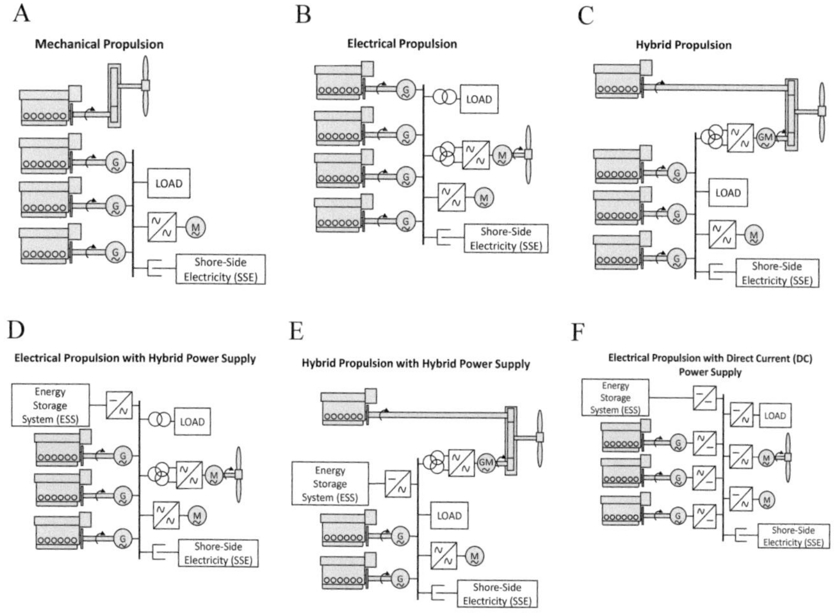

Fig. 5 Schematic comparison of different ship propulsion and power systems. Adapted from Geertsma et al. (2017) by adding Shore-Side Electricity (SSE)

speed, reduced availability in case of drive train failure leading to loss of propulsion, high NO_x emissions under certain operating conditions as well as noise and due to the transmission system.

Mechanical propulsion is the preferred application for ships operating at a single speed, typically cargo ships and fast crew suppliers. For other ship types that operate at low power in their operational envelope, like tugs or offshore vessels with dynamic positioning (DP), this type of propulsion result in a poor performance in terms of fuel consumption and emissions. In these cases, electric or hybrid solutions may be more suitable.

4.1.2 Electrical Propulsion

Figure 5b illustrates a typical electrical propulsion, which consists of multiple diesel engines coupled to electrical generators and integrated through a high voltage electrical bus. Propulsion motor drives and hotel loads are electrically fed from this bus, often requiring transformers and power electronic converters (Geertsma et al. 2017). As in the former case, if both ship and port have compatible interfaces and infrastructure, shore-side electricity (SSE) may supply electricity while at berth and the onboard generators can be shut down.

Electric propulsion is a fuel-efficient solution when the hotel load is significant compared to the propulsion load and the operating profile is diverse. The number of running engines and their operation is controlled by a power management system (PMS) matching the power required by propulsion and hotel load with the power produced by the generators. Important benefits of these systems include:

- Lower NO_x emissions, as these systems make use of more engines operating often at the design point, at which they typically produce less NO_x emissions.
- Reduced maintenance loads are reduced as the engines are shared by propulsion and hotel load.
- Lower noise and vibration are produced in the absence of transmission systems.
- High system availability, due to redundancy of engines and the possibility to operate in a broad operational envelope.
- Mechanical transmission requires a shaft to transfer the mechanical power produced by the engine to propellers. Electrical propulsion does not require a shaft connecting engine and propeller, so the absence of shaft-line along the ships allows more design-freedom.

In contrast, the main disadvantages of these systems are:

- Increased losses due to more power conversion stages, which leads to higher specific fuel consumption (SFC) near top speed.
- Poor fuel consumption and high emissions due to engines running at low part load in applications like Dynamic Positioning (DP) to achieve high availability.
- Susceptibility to voltage and frequency swings that can occur due to changing loads, which in turn can switch off electrical systems with the consequence of reduced reliability and availability.

Electrical propulsion is broadly used on cruise ships, ferries, drilling or offshore vessels with dynamic positioning, cable layers, icebreakers and naval vessels. Aspects like the redundancy of the engines, which is offered by these systems, have been particularly relevant for applications like DP, where maintaining the position even in fault conditions is necessary. Redundancy provides spinning reserve that guarantees the availability of power.

4.1.3 Hybrid Propulsion

Figure 5c illustrates a hybrid propulsion system, which combines a mechanical drive with an electrical motor. The electrical motor provides the power to drive the propulsion system at low speeds. At high speeds, the mechanical drive takes over and provides primary propulsion. In addition, the electrical motor can also act as generator to supply electricity to the loads (Geertsma et al. 2017). Therefore, this item is tagged as GM (Generator/motor). As in the two former cases, shore-side electricity (SSE) can supply electricity while at berth, provided that the corresponding interfaces and infrastructure is present onboard as well as onshore.

The main advantage of hybrid propulsion is the combined benefits of electrical and mechanical propulsion, often requiring a trade-off between their features, like efficiency, noise reduction and fuel consumption. Main applications for hybrid power systems are naval ships, towing vessels and offshore vessels. Hybrid propulsion can be provided by non-hybrid power systems, as in many cases the prime movers of the systems are diesel engines.

4.1.4 Electrical Propulsion with Hybrid Power Supply

Illustrated in Fig. 5d, a distinctive aspect in this case is the connection of all the power sources or ESS through an electrical bus bar. This contrasts with hybrid propulsion or purely mechanical propulsion, for which the propulsion engines directly provide mechanical power to the propulsion system. The onboard ESS allows the storage of electricity produced by the generators or even by shore-side electricity supply. By storing electricity from the grid onboard for later use in propulsion, the fuel consumption of the engines can be reduced (Geertsma et al. 2017). If the shore-side electricity charging the ESS is from low-GHG sources, such as renewables, the net GHG emissions of the ship can be reduced, not only while it is berthed but also during propulsion, this being an advantage over mechanical or electrical propulsion. Moreover, a combination of two or more types of power sources can be achieved, including:

- Combustion-based power supply based on diesel engines, gas turbines or steam turbines connected to an electrical generator.
- Electrochemical power supply from fuel cells
- Stored electrical power supply from energy storage systems, such as batteries, flywheels or super capacitors.

While the application of hydrogen fuel cells in ships has been so far been limited mainly to submarines and demonstration projects, the usage of battery energy storage for propulsion systems is becoming increasingly widespread. Different variants are available, depending on how the storage system is connected to the electrical system:

- At the main high voltage bus bar through an AC/DC converter (Option shown in Fig. 5d)
- At the low voltage bar through an AC/DC converter
- Directly using a DC/DC converter to the direct current link of the propulsion converter.

One of the main benefits is that the ESS can provide power when the operation of other generators at partial load is inefficient and can then be recharged by operating any generation source at its optimum operating point. Additionally, load balancing and peak shaving are possible with ESS, so that efficient operating points are maintained. Onboard storages enable recharging from shore-side reducing fuel consumption and local emissions. The integration of renewable sources can decrease the global warming potential of the ship's energy supply.

ESS can store the energy generated by braking motors, as it is the case with heavy crane installations and offshore vessels with heave compensation. Furthermore, ESS can provide backup power in case of failure of other generators, replace spinning reserves and enhance the availability of the propulsion, which is crucial for applications using DP.

Challenges include the complex control strategies compared to mechanical or electrical options and the additional maintenance and purchase costs. The higher costs can be partially offset by the ability to downsize some components due to redundancies and reducing component size through peak shaving.

The main applications so far are tugs, ferries and offshore vessels. An interesting example is the MV Hallaig, which demonstrated 35% fuel savings in the trial phase compared to a mechanical propulsion system. Around 24% of the fuel savings were due to overnight charging using electricity from the grid and 11% to optimizing operation with the energy management system (Geertsma et al. 2017). Some studies have found fuel savings between 7.9 and 17.6% for different types of ships (Karvounis et al. 2022).

4.1.5 Hybrid Propulsion with Hybrid Power Supply

This option combines the exceptional efficiency of direct mechanical drive and the flexibility of ESS for electrical supply, as found in Fig. 5e. By storing electricity from onboard generators, shore-side electricity or both, this option can reduce the net GHG emissions if the GHG emissions of the shore-side electricity are low. At low propulsion power, an electric drive propels the ship conveying power from electrical generators or the ESS. At high speeds, the mechanical drive takes over as prime mover. The electric drive can also work as generator supplying electricity to onboard electrical loads or for storage in the ESS. Therefore, the generator/motor is labelled "GM" in Fig. 5e.

Therefore, this alternative aims at combining the high efficiency of mechanical propulsion at high speed with those of the electrical propulsion with hybrid power supply at low speed. Main challenges are complicated control strategies and costs. Hybrid propulsion with hybrid power supply is being researched for tugs and yachts.

4.1.6 Electrical Propulsion with DC Hybrid Power Supply

DC grids offer several benefits over traditional AC grids, including lower fuel consumption and emissions at partial loads, lower noise levels and improved resilience to interference as the frequency does not play a role in the stability of the grid. This is because engines can operate at variable speeds in DC grids, reducing mechanical and thermal loads, in contrast to AC grids in which engines have to operate at a constant speed despite changes in the loads. Additionally, DC architectures require fewer and smaller switchgears and are less prone to faults spreading throughout the onboard grid. In turn, implementing a DC grid has the disadvantages of requiring extensive power electronic converters with high costs, fault protection systems and complex control strategies (Geertsma et al. 2017). This particular type of propulsion system together with DC hybrid power supply is demonstrated in Fig. 5f, where all the electricity is converted to direct current and fed to the busbar, before being reconverted back to AC for onboard electrical loads and propulsion.

DC power supplies are commonly used in submarines in combination with battery packs for air-independent propulsion. The technology has also been tested in ferries, offshore vessels, drilling ships, research vessels and wind farm support vessels. DC is also being used in the first completely electrical vessel, MF Ampere in Norway (Geertsma et al. 2017).

4.2 Trends of Hybrid Power and Propulsion Systems

A top-down picture of the topic of hybridization of power supply for ships will be presented in this section. The statistics for this section were inferred indirectly from data on the electrification of ships. Therefore, the focus here is mainly on ships with onboard ESS and in line with hybrid power supply (Fig. 5d and e). As of August 2024, and according to data of the class society DNV, 944 ships fitted with batteries are in operation and 451 on order (DNV 2024). The ships are categorized as hybrid, plug-in hybrid and pure electric, with a slightly different categorization in comparison with the previous section. Hybrid and plug-in hybrid refer to ships combining batteries with other power systems, having in the latter case the possibility of charge from shore. Pure electric refers to operation relying entirely on onboard batteries. Figure 6a shows that an important share (37%) of the ships fitted with onboard batteries are car/passenger ferries, followed by ships for other activities (20%), offshore supply vessels (11%) and fishing vessels. Moreover, Fig. 6b shows that most of the ships operate either in Europe excluding Norway (35%) or specifically in Norway (33%) with the rest elsewhere. Norway is a remarkable case for electrification of car/passenger ferries and the biggest market of this type of ship in Europe. In terms of different applications and as shown in Fig. 6c, 64% of the ships are hybrid, 17% are plug-in hybrid and 19% are fully electric. Finally, Fig. 6d shows that around 50% of the fleet operating with batteries is not more than four years old or was fitted in the last four years, with the ordered fleet being around one third of the current fleet by ship number.

The statistics reveal that hybrid, plug-in hybrid and pure-electric ships are gaining relevance. While, car/passenger ferries dominate the trend, similar solutions are being adopted for other ships. A big share of this trend is taking place in Europe. Finally, the orders of future ships fitted with hybrid, plug-hybrid and pure electric onboard power systems for the following years are comparable to the existing ships fitted with this solution, showing an increasing adoption.

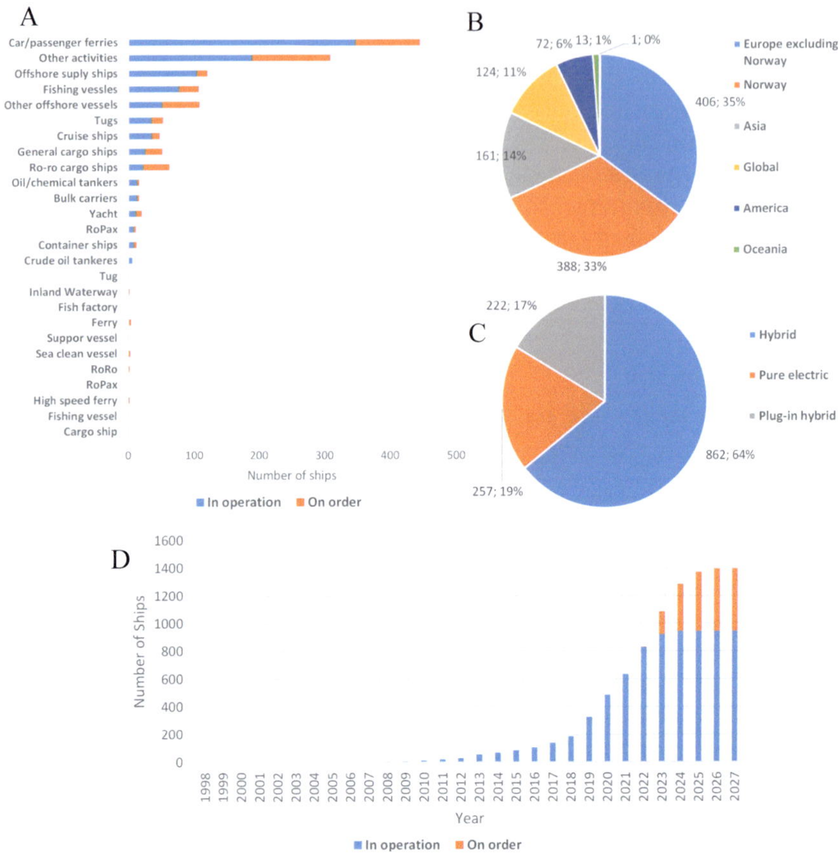

Fig. 6 Statistics on ships fitted with onboard batteries. The statistics are shown by ship type (**a**), location (**b**), type of power system (**c**) and number of ships per year (**d**). Own plots with data extracted from the Alternative Fuels Insight Platform of the classification society DNV (2024). Status 28.08.2024

4.3 Key Performance Indicators (KPIs) of the Components of Hybrid Power Supply Systems

Following the classification of the different propulsion and power systems for ships, a description of the KPIs will be presented in the following section. The section will start by describing the typical features of propulsion systems for ships. Later, the technical and economic features of diesel engines, fuel cells and batteries as important components of most of the hybrid power systems will be described and contrasted.

4.3.1 Ship Propulsion Requirements

Ships and their propulsion plants are traditionally optimized for their operation under calm water conditions. The power required by the propulsion system at constant speed depends on ship's frictional, residual and air resistance. Additionally, the acceleration required by the operational conditions may also drive the selection of the propulsion and power system. Ships requiring more acceleration demand a higher power-to-displacement ratio. However—and specially for cargo ships—lowering the fuel consumption has resulted in low power-to-displacement ratios, accelerating slower than previous designs (MAN Energy Solutions 2018). Off-design conditions related to rough weather or waves are handled in the design phase by adding a sea margin typically comprised in the range of 15–25% of the power required in calm water conditions (Taskar et al. 2016).

Table 2 highlights the diversity of ship types, engine speeds, and main engine power. The broad range of typical main engine power is largely due to variations in speed, size and engine types, even within each ship category.

4.3.2 Internal Combustion Engines, Fuel Cells and BESS

Most of the hybrid solutions so far envisioned for ships include either the traditional diesel engines combined with BESS or fuel cells. The main features of these systems will be introduced in the following sections.

Internal Combustion Engines

Both Diesel and Otto cycles can in general be used to produce mechanical power for propulsion. However, the Diesel cycle offers some advantages related to higher efficiency, fuel quality and higher compression ratios (Hannemann 2024). Being more common, the following section will describe some KPIs for diesel engines. KPIs for diesel engines depend heavily on the size and type of engine. Medium speed four-stroke engines have a higher specific power and a higher power density than low-speed two stroke engines. However, the latter have a superior fuel economy and therefore a higher efficiency. Exemplary values for two- and four-stroke diesel engines are summarized and compared in Table 3 (MAN Energy Solutions 2018).

Fuel Cells

Although hydrogen fuel cells have found so far limited applications for ships, these devices are being increasingly studied and tested. A summary of KPIs for different fuel cell (FC) technologies can be found in Table 4.

Future technology targets as defined by the Clean Hydrogen Joint Undertaking in the EU can be found in Table 5. It is noticeable that the FC module CAPEX for heavy duty vehicles are estimated to be lower than those for maritime applications. CAPEX for maritime applications are expected to be around 1000 EUR/kW by 2030 in comparison with heavy duty vehicles, for which CAPEX are expected to be below 100 EUR/k. In addition, the system lifetime is expected to be longer for maritime

Table 2 Typical characteristics of different ship types.

Category	Type	Propeller	Main engine type	Size factor	Speed (kn)	Typical main engine power (MW)
Tanker	Crude oil	1 FP	2-stroke	dwt	13–17	10–35
	Gas tanker/ LNG carrier	1 FP	2-stroke, steam turbine	dwt/cubic meter (cbm)	16–20	2–45
	Product	1 FP	2-stroke	dwt	13–16	2–12
	Chemical	1 FP	2-stroke	dwt	15–18	3–20
Bulk carrier	Ore carrier	1 FP	2-stroke	dwt	14–15	3–32
	Regular	1 FP	2-stroke	dwt	12–15	5–64
Container ship	Liner carrier	1 FP or 2 FP	2-stroke	teu	20–23	10–80
	Feeder	1 FP or 1 CP	2 or 4-stroke	teu	18–21	< 12
General cargo ships	General cargo	1 FP	2 or 4-stroke	dwt/nt	14–20	2–28
	Coaster	1 FP or 1 CP	2 or 4-stroke	dwt/nt	13–16	< 12
RoRo cargo ship	-	1 CP or 2 CP	2 or 4-stroke	Lane meters (lm)	18–23	0.7–32
Passenger cargo ship		2 CP	2 or 4-stroke	Passengers/ lm	18–23	0.7–32
Passenger ship	Cruise Ship	2 CP	4-stroke	Passengers/ gt	20–23	0.8–97
	Ferry	2 CP	4-stroke	Passengers/ gt	16–23	0.8–68

[a] *FP* fixed pitch, *CP* controllable pitch, *DWT* deadweight tonnage, *TEU* twenty-foot unit, *GT* gross tonnage, lm lane meters
Source MAN Energy solutions (MAN Energy Solutions 2018) and data from ship-db.de (Hannemann 2024)

applications by 2030 (see KPI Fuel cell system lifetime) than for heavy duty vehicles (see KPI FC stack durability). Since maritime applications require high power (see Table 2), the fuel cell power rating is expected to increase from around 500 kW in 2020 to around 20 MW in 2030. Finally, topics related to the loading of platinum-group metals (PGM) and power density at cell level are also considered for the development of stacks for heavy duty vehicles.

BESS

BESS are rapidly evolving and being adopted for maritime applications. Table 6 summarizes the features of some products used for maritime applications according to specifications made public by their manufacturers.

Table 3 KPIs of diesel engines

KPI	Unit	Two-stroke diesel engines	Four-stroke diesel engines
Power range	kW	3200–82,440	1290–19,200
Specific power	W/kg	32–50	57–336
Power density	W/L	26–44	44–295
Specific fuel oil consumption at 75%	g/kWh HFO	151.5–175.5	174.4–202[a]
Specific gas consumption + specific pilot oil consumption (1.5%) at 75%	g/kWh methane + g/kWh HFO	(126.3 + 2.9)–(133.9 + 3.1)	N/A
Specific gas consumption + specific pilot oil consumption (5%) at 75%	g/kWh methanol + g/kWh HFO	(306.9 + 9.8)–(322.1 + 10.2)	N/A
Capital costs	$/kW	240–420 Korberg et al. (2021), Dotto and Satta (2023)	238–493 Talluri et al. (2016), Wärtsilä (2016), Korberg et al. (2021), Karvounis et al. (2022), Dotto and Satta (2023)

Based on the Marine Engine Program of MAN Energy Solutions (2024). Capital costs are taken from literature references

[a] Diesel generators for maritime applications typically consist of a four-stroke diesel engine which and an electrical generator assumed with an efficiency in the range 95–97%; therefore, the SFOC for diesel generators is the range 180–213 g/kWh_{el}

As a result of the highly dynamic market of BESS and their parallel development for massive markets like land transportation, their prices and particularly those of Li-ion-based BESS are decreasing rapidly due to technological progress, upscaling of production and economies of scale. Figure 7 shows predicted price trends for maritime batteries from the E-Ferry project. Industry sources report current prices ranging from 400 to 420 EUR/kWh in 2024. However, the prices vary depending on system scope and size. These prices are considerably higher than those for automotive applications, for which forecasts predict a price at around 80 USD/kWh (approx. 76 EUR/kWh) by 2026 and faster than original predictions to higher energy densities related with technological development and decreases in metal prices like lithium and cobalt, which contribute considerably to the cost of producing Li-ion batteries (GoldmanSachs 2024).

In general, KPIs for automotive applications are more developed and show the trends that the industry is moving toward. A summary of present and future KPIs for lithium-ion batteries is presented in Table 7.

Table 4 KPIs of fuel cells

KPI	Unit	LT-PEMFC	HT-PEMFC	SOFC
Operating temperature	°C	65–85	140–180	500–1000
Electrical efficiency	%LHV	40–60	40–50	50–65
Hydrogen purity	–	> 99.98% H_2	< 3% CO	< 20 ppm S
Cooling medium	–	Water mixture	Thermal oil	Air
Specific power	W/kg	125–750	25–150	20–80
Power density	W/L	50–400	10–100	10–40
Stack life time	Thousand hours	5–35	5–20	20–90
Start-up time (cold)	Cold	< 10 s	10–60 min	> 30 min
Load transients	Idle-rated power	< 10 s	< 5 min	< 60 min
Capital costs 2021 *(2030)*[*]	$/kW	1000–2500 *(60–600)*	3000–5000 *(150–1500)*	3500–8000 *(500–2000)*

[a] Future expected values italic under current values
Taken from van Biert and Visser (2022)

Table 5 KPIs and future targets for fuel cell technology for heavy duty vehicles and for maritime applications

	KPI	Unit	2020	2024	2030
Heavy duty vehicles	FC module CAPEX	EUR/kW	1500	< 480	< 100
	FC module availability	%	85%	95%	98%
	FC stack durability	Hours	15,000	20,000	30,000
	FC stack cost	EUR/kW	> 100	< 75	< 50
	Power density	W/cm^2	1 at 0.650 V	High TRL: 1.0 at 0.675 V Low TRL: 1.2 at 0.650 V	High TRL: 1.2 at 0.675 V Low TRL: 1.5 at 0.650 V
	PGM loading	g/kW	0.4	High TRL: 0.35 Low TRL: < 0.30	High TRL: 0.30 Low TRL: < 0.25
Maritime applications	FC power rating	MW	0.5	3	10
	Hydrogen bunkering rate	Ton H_2/h	0	2	20
	Fuel cell system lifetime	Hours	20,000	40,000	80,000
	PEMFC CAPEX	EUR/kW	2000	1500	1000

Based on Tables 16 and 17 of the Strategic Research Agenda 2021–2027 of the Clean Hydrogen Joint Undertaking (2022)

Table 6 KPIs of BESS for maritime applications Corvus Energy (2024b), Leclanché Energy Storage Solutions (2024)

KPI	Unit	Leclanché MRS-3/65	Leclanché MR-3/72	Corvus energy Orca ESS
Battery cell chemistry		G/NMC, 65 Ah	G/NMCA, 72 Ah	Li-ion NMC/ Graphite
Operating temperature		Charge: 0 to + 45 °C; Discharge: − 20 to + 55 °C		
Electrical efficiency		Typically, 85–90%, round	Typically, 85–90%, round	Typically, 85–90%, round
C-Rate discharge, peak, 20 s	C-Rate	4.6C	TBC	3C
C-Rate charge, peak, 20 s	C-Rate	2.8C	TBC	3C
C-Rate discharge, continuous	C-Rate	2.7C	2.4C	
C-Rate charge, continuous	C-Rate	1.8C	1.5C	
Cooling medium		Liquid cooled	Liquid cooled	
Specific energy (pack)	Wh/kg	152–157	N/A	77
Energy density (pack)	Wh/L	249–270	N/A	88
Specific energy (system)	Wh/kg	101	112	76
Energy density (system)	Wh/L	108	120	76
Life time	Cycles	7000 at 80%DoD 4000 at 100%DoD	6000 at 80%DoD	–

Alternative cell chemistries are under development. This implies developing new cathodes, anodes, electrolytes in lithium-based batteries or even substituting lithium completely for other elements like sodium (Cai et al. 2024), potassium (Xu et al. 2023), calcium (Stievano et al. 2021), magnesium (Dominko et al. 2020), or aluminum (Elia et al. 2021). Expected KPIs for some of these emerging battery technologies at cell level are currently collected on a regular basis by Batteries Europe (Batteries Europe 2023).

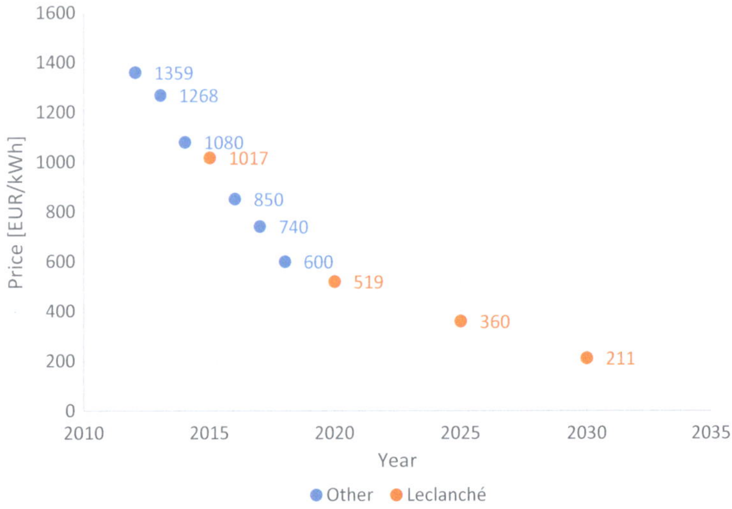

Fig. 7 Battery pack prices for maritime applications. Data gathered by Marstal Navigationsskole combined with estimated and realized prices from the company Leclanché. *Source* Own plot with data from Kortsari et al. (2022)

Table 7 KPIs for lithium-ion batteries

KPI	Unit	Level	2020	2030	2050
Gravimetric energy density	Wh/kg	Pack	90–180	190–2320	> 250
		Cell	160–260	275–320	> 350
Volumetric energy density	Wh/L	Pack	250–400	450–550	> 600
		Cell	450–730	750–900	> 1000
Gravimetric power density	W/kg (100–20% SOC)	Cell	340–500	800–1100	> 1200
Fast charging time	min (20–80% SOC, 25 °C)	–	~ 1000	~ 2000	> 3000
Battery lifetime to 80% end-of-life capacity for BEV	Cycles (25 °C)	–	15–30	10–15	< 10
Battery lifetime to 80% end-of-life capacity for stationary applications	Cycles (40–50 °C)	–	~ 1000	Up to 2000	2500–5000
Calendar life	Years (80% energy)	–	10	10–15	15–20
Cost targets	EUR/kWh	Cell	60–100	40–60	< 50
	EUR/kWh	Pack	90–140	65–110	40–70
Collection/take back rate	–	–	–	> 50%	> 90%
Recycling efficiency by weight	–	–	–	> 40%	> 90%
Economy of recycling	–	–	–	~ 150%	~ 50%

Taken from Armand et al. (2020)

4.3.3 Comparison of Different Power Generation and ESS Technologies

In the previous section, a few key indicators about the three key technologies were summarized. Combustion engines (diesel) as prime movers have a lower upfront cost compared to fuel cells, especially in its LT-PEMFC and SOFC variations. Batteries are not directly comparable because they are energy storage technologies and not a mechanical or electrical power generation technology. Nevertheless, as their prices are decreasing rapidly, they are becoming an option to store and provide power for automotive applications and further onboard ships (GoldmanSachs 2024).

Considering the specific fuel consumption of the diesel engines and the heating value of the fuels they can burn, their efficiency is in the range between 42 and 52%, depending on the size and type of engine. The addition of one electrical generator decreases the efficiency to around 46% in the best cases considering the production of electricity. This is lower than the efficiency of fuel cells which is in the best cases around 60%. Therefore, fuel cells promise a higher efficiency than the internal combustion engines, with the tradeoff so far of a higher upfront cost. Specific power and power density are relevant aspect for transportation equipment and those of fuel cells and diesel engines are comparable.

Scalability is still an issue for fuel cells. The power demands for ships can go up to two-digit MW, whereas currently the fuel cell modules offered in the market have rated power around hundreds of kW. Therefore, power upscaling is one of the KPIs for the upcoming years for fuel cells (see Table 7). Although in principle modularity allows to put several modules together to meet the power demands, in reality this becomes unpractical. Diesel engines as the incumbent technology come in power outputs, often with smaller power in their four-stroke variation and higher in the two-stroke.

The integration of these elements into hybrid power systems also requires a considerable amount of equipment for the control of the different devices, fault protection, voltage conversion, rectification, among others. This equipment is not mentioned here, but can certainly have considerable additional costs and conversion losses during the operation.

The following section will show an example of a hybrid hydrogen fuel cell battery electric ferry, encompassing the motivations, the specifications of the system and how this system compares to other alternatives in environmental and economic terms.

5 Case Study: Electric Propulsion Hybrid Power System Ferry Developed in the Project HySeas III

Building on previous sections, this section showcases an example vessel with an electrical propulsion system and hybrid power supply (see Sect. 4.1.4). The EU Horizon 2020 project HySeas III (Grant Agreement 769,417, 2018–2022) conceptualized this hybrid Hydrogen Fuel Cell and Battery Electric (HFCBE) passenger/car

(RoPax) ferry, for which approval in principle was granted by the end of the project (European Commission 2022).

5.1 Motivation for the Concept Development

The HySeas III project aimed at developing and testing a hybrid Hydrogen Fuel Cell and Battery Electric (HFCBE) RoPax ferry concept, with the final goal of creating the world's first sea-going hydrogen-powered vessel for the Orkney Islands in Scotland. The project built on the previous feasibility studies (HySeas I and HySeas II) and tested the effectiveness of hydrogen and fuel cells in seagoing vessels through testing a real size power system on land. The selection of the route and the operation site aimed also at the hydrogen availability and utilization in the Orkney Islands, a location where hydrogen infrastructure is already existing and has been installed by other parallel testing projects. The outcomes contributed to paving the way for future ship construction by conceptualizing the ferry and its hybrid propulsion and power system, designing and testing the power systems at real scale, conceptualizing the refueling infrastructure and conducting market-potential and sustainability assessments. A rendering of the concept developed in the project can be seen in Fig. 8, where the compressed hydrogen tanks can be seen over deck.

Fig. 8 Rendering of the HySeas III RoPax hydrogen fuel cell RoPax ferry concept. *Source* Gomez Trillos and Draheim (2022), created by Courtesy of ABL Group/Caledonian Maritime Assets Limited

Table 8 Ship general specifications

Ship Specification	Value	Unit
Route	Kirkwall-Shapinsay	–
Number of crossings per day	12	–
Operation speed	9.5	kn
Maximum speed	11.0	kn
Length overall	40	m
Breadth (moulded)	11.5	m
Depth (main deck)	2.5	m
Type of ferry	Double-ended	–
HGV	2	Units
Cars	16	Units
Deadweight	115	Ton
Lightship	315	Ton
Crew	4	Person

Source HySeas Preliminary General Arrangement Caledonian Maritime Assets Limited (2021)

5.2 *Technology Neutral Requirements*

The carrying capacity, number of crossings and speed were specified according to local transportation demands between the ports of Kirkwall and Shapinsay. Based on these specifications, a design was drafted and its corresponding power demands were estimated. Table 8 outlines the key specifications for the concept developed in the project.

An electrical propulsion with hybrid power supply including fuel cells and batteries was designed to meet the calculated power demand, based on the specifications. Key propulsion and power system specifications are summarized in Table 9. The fuel cell and battery capacity on board was estimated according to the operational profile of the ship and to meet the power requirements onboard, with margins for weather conditions, redundancy in case of failure and considering emergency operation with system limitations.

5.3 *Operational Profile and Hybrid Power System Solution*

The daily operation of the ship consists of 12 crossings between Kirkwall and Shapinsay, covering a distance of 4 nm (7.2 km) at a service speed of 9.5 kn (17.6 km/h), each crossing taking approximately 25 min. The loads considered for the propulsion design are depicted in Fig. 9, which indicates the highest loads at around 900 kW, and around 400 kW during steady crossing. When the ship is docked at port, the loads

Table 9 Propulsion and power system specifications Gomez Trillos and Draheim (2022)

Feature	Value	Unit	Note
Propulsion load	850	kW_{el}	Estimated
Hotel load	50	kW	Estimated
Maximum load	900	kW_{el}	Propulsion + Hotel load
Fuel cells	$100 \times 6 = 600$	kW_{el}	Assumed as FCMove® 100 kW fuel cell of the company Ballard
Battery capacity	740	kWh	Charging overnight and from fuel cells during operation
Hydrogen consumption	~ 120	kg/day	Estimated
Hydrogen storage onboard	320	kg	Per design and according to fueling frequency specifications

are just under 200 kW, accounting for minimum operation of thrusters to maintain the position of the ship hotel loads.

To meet the power demands, various strategies with the specified hybrid power system comprised of fuel cells and batteries were analyzed. A strategy that maximizes the battery usage with overnight charging was deemed optimal and will be considered hereafter. This approach is exemplified in Fig. 10, where the fuel cells operate at around 45% of their rated power, near their optimal range (10–40%), minimizing therefore the specific hydrogen consumption per unit of electricity produced. The batteries absorb the demand peaks during acceleration and maneuvering, while negative values represent battery charging from the fuel cells when docked. During

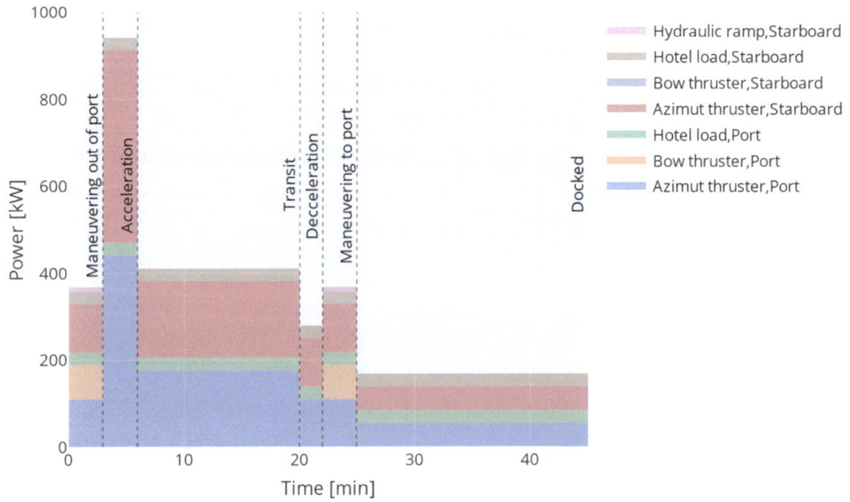

Fig. 9 Load profile as assumed for the HySeas III concept Gomez Trillos and Draheim (2022)

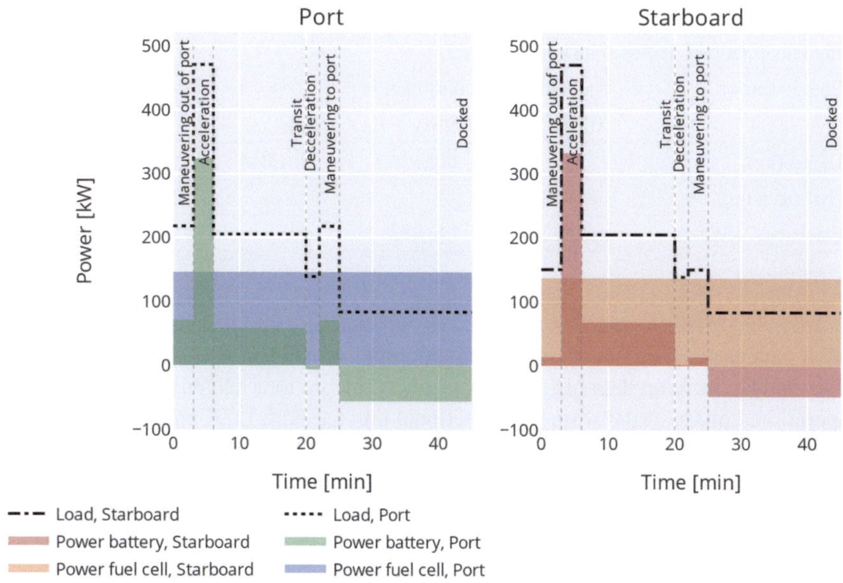

Fig. 10 Operational strategy using maximum battery and maintaining the operational point of the fuel cell in their highest efficiency Gomez Trillos and Draheim (2022)

each crossing, the net charge in the battery decreases slightly, so overnight charging is necessary to compensate for the net energy discharged during the entire day. In addition as shown in Fig. 10, the loads are divided into port and starboard sides more or less evenly and therefore reach around 450 kW for each side (Compare with Fig. 9). The total net power supply for each of the sides is shown in dotted black lines.

5.4 *Sustainability: Environmental and Economic Assessments*

Environmental and economic assessments of the HFCBE RoPax ferry were carried out and compared against other alternatives like Diesel Electric (DE), hybrid Diesel Battery Electric (DBE) and Battery Electric (BE), the latter based completely on batteries. The HFCBE, DBE and BE were considered as plug-in alternatives with charging overnight. Life cycle assessment (LCA) was used to assess the environmental impacts according to the ISO 14040/14044 standards and using the ILCD2.0—2018 impact assessment method. Construction, operation and end-of-life management were considered. The software Brightway2, its user interface ActivityBrowser and the database ecoinvent 3.7 were used to carry out the calculations (Wernet et al. 2016; Mutel 2017; Steubing et al. 2020). The economic assessment was

Table 10 Yearly energy carrier and fuel/electricity consumption of the different considered alternatives Gomez Trillos and Draheim (2022)

Energy carrier	Fuel/electricity consumption			
	HFCBE	DBE	DE	BE
Diesel (kg)	–	174,625	201,044	–
Hydrogen (kg)	49,988	–	–	–
Electricity (kWh)	146,482	146,482	–	1,095,368

carried out via Life Cycle Costing (LCC) by considering the construction costs, operation expenses, energy carriers, component replacement and end-of-life processing. The analysis built on data of fuel consumption calculations and energy flows for the operational phase of the ship. The functional unit for both LCA and LCC was 1 km of crossing over a 30-year ship lifetime. For the HFCBE alternative, hydrogen was assumed as produced in the UK and electricity from UK's grid was assumed for battery charging overnight. Some results of this work can be found in other documents (Trillos et al. 2021; Gomez Trillos and Draheim 2022; Kazemi Esfeh et al. 2022).

As the project focused on developing an innovative hybrid power system based on hydrogen fuel cells and Li-ion batteries, which impacts directly the energy carriers used by the ship and the emissions, the operational profile was therefore emphasized. A summary of the yearly energy carrier and electricity consumption according to the modelled considerations can be seen in Table 10. According to the estimations carried out in the project, the operation of the ship requires approximately 50 tons of hydrogen per year and it is additionally supported with the charging of around 146 MWh of electricity overnight from the grid for battery charging. For the diesel-based alternatives, around 174 tons of diesel are consumed by the hybrid DBE alternative and 201 tons are consumed by the DE alternative on a yearly basis. The latter had a higher fuel consumption because of not having the possibility of storing electricity onboard. Finally, the BE alternative consumes around 1095 MWh per year and no fuel, as this alternative is completely based on electricity charged from shore.

5.5 *Environmental Assessment*

The results of the GHG emissions per km of crossing are shown in Fig. 11. The operation phase has the higher share among the three considered phases and for all the alternatives. This increases from 76.3% for the HFCBE up to 96.1% for the DE alternative. Moreover, the construction has higher relative impact in the case of the HFCBE (22.9%) than in the case of the DE alternative (3.8%). The end-of-life was found to have a limited impact in the life cycle, although the data for the scrapping process was limited and scrapping on site was assumed for this phase.

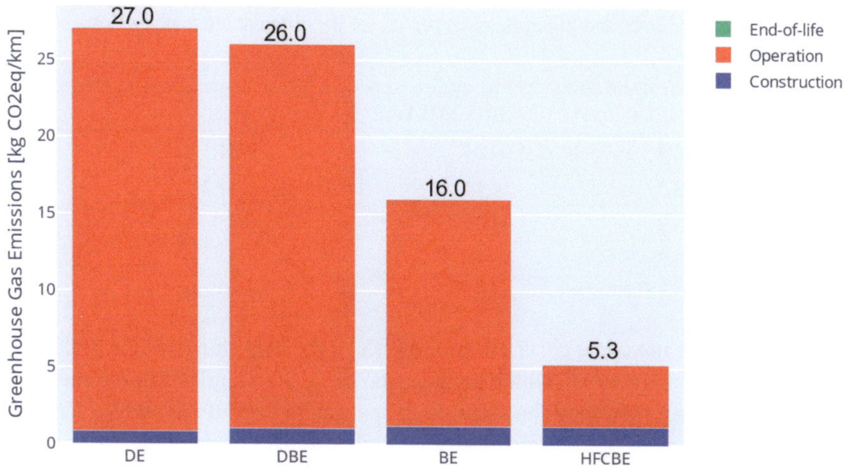

Fig. 11 Complete life cycle GHG emissions for the different alternatives considered for the analysis of the concept in the project HySeas III Gomez Trillos and Draheim (2022)

When comparing the impact per km of crossing, the DE alternative was found as the one with the highest GHG emissions with 27.0 kg CO_2eq/km. The hybrid DBE alternative achieve certain reductions with an impact of 26.0 kg CO_2eq/km mainly linked to the lower diesel fuel consumption but in turn with higher emissions for the construction of the ship due to the onboard batteries, despite considering replacement of the batteries during the lifetime, the impact of this was relatively minor in relation to the impacts of fuel combustion on board. Therefore, the DBE alternative achieves a reduction of the GHG emissions by approximately 3.7% compared to the DE alternative when all the life cycle is considered. The reduction achieves 5.0% when only the operation phase is considered. For the BE alternative, for which only electricity is considered for the operation, the reduction compared to the DE alternative is even higher. In this case, the impact per km of crossing reduces to 16.0 kg CO_2eq/km, therefore resulting a reduction of 40.7% compared to the DE alternative. Finally, the hybrid HFCBE alternative developed in HySeas III had estimated GHG emissions of 5.3 kg CO_2eq/km, thus a reduction of 80.4% compared to the DE alternative, considering the assumption that the hydrogen consumed by the power system onboard is produced via electrolysis with electricity sourced from wind power.

5.6 *Economic Assessment*

Table 11 summarizes the total life cycle costs as well as the life cycle costs per km estimated for the different alternatives. The HFCBE alternative is estimated to have costs per km and for the entire life cycle costs 51.1% higher than those of the DE alternative. The contribution of different cost items can be seen in Fig. 12. The main

Table 11 Life cycle costs and life cycle costs per km for the different alternatives Gomez Trillos and Draheim (2022)

Alternative	Life cycle cost (MEUR2021)	Life cycle cost per km (EUR2021/km)	Comparison with DE (%)
HFCBE	36.8	69.83	+ 51.1
BE	28.5	54.14	+ 17.2
DBE	25.3	48.06	+ 4.0
DE	24.3	46.21	–

contributor to the life cycle costs are the CAPEX of the ship shown in the plot under the item "ship construction", which includes the metal work and ship equipment but excludes the power system of the ships. In the particular case of the HFCBE and BE alternatives, these costs are higher due to a higher construction costs assumed for these alternatives. Next to the CAPEX, the personnel costs for the operation of the ship also contribute a considerable share to the total life cycle costs and are the ones with the second highest share for the BE, DBE and DE alternatives. The personnel costs were assumed equal for all the alternatives. Following this, the costs of the fuels and electricity are the third contributor in all the cases, with a marked contribution in the case of the HFCBE alternative, for which the hydrogen costs are higher than the personnel costs. Interestingly, the power train and the replacements of batteries (and in the case of the HFCBE alternative fuel cells) contribute in lesser degree to the life cycle cost of the ship, being more relevant for the HFCBE alternative which includes both batteries and fuel cells.

5.7 Conclusions of the Case Study and Outlook

This chapter described the requirements necessary for the operation of a passenger/car ferry. Built on these specifications, a solution based on hybrid hydrogen fuel cell and battery power system was developed, considering the power demands according to estimations made in the project HySeas III. Following this, sustainability assessments were carried out, showing that the proposed solution allows to reduce the GHG emissions considering all the life cycle phases from 27.0 for a diesel electric alternative to 5.3 $kgCO_2eq/km$ for the hybrid hydrogen fuel cell and battery alternative, therefore allowing a reduction of 80.4% of the GHG emissions per km. This result was obtained considering hydrogen produced via electrolysis fed with onshore wind electricity. Even some reductions estimated at 3.7% were obtained for a hybrid diesel electric alternative also assessed as comparison in the project. Nevertheless, the GHG reductions come along with higher expected life cycle costs, which were estimated as 51.1% higher than those of a diesel electric alternative due to the high CAPEX and hydrogen costs for operating the ship. In general, this study case shows that the hybrid power solutions for ships may provide environmental benefits in terms of the

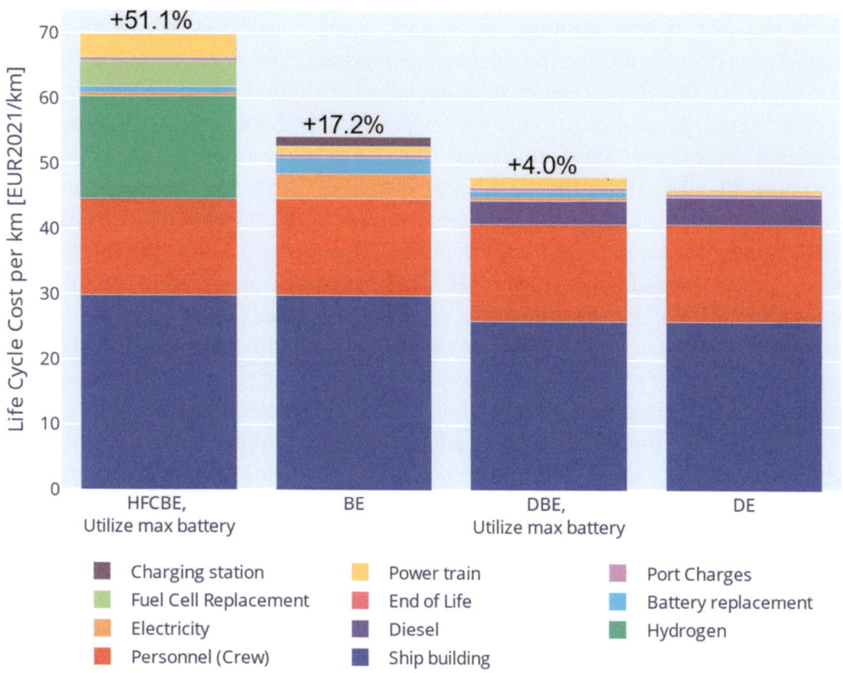

Fig. 12 Contribution of different cost items to the life cycle cost per km for the different alternatives considered in the study. Discount rate of 3.5% assumed Gomez Trillos and Draheim (2022)

reduction of potential GHG emissions. However, some tradeoffs between the reduction of GHG emissions and the upfront and operational costs exist, as the hybrid solutions were found to be more expensive than more traditional and non-hybrid solutions.

Although the construction of the HySeas III concept has not yet been realized, in this dynamic market, ships using fuel cells for their power systems have already been built in parallel projects. These include, for example, the MF Hydra in Norway, the MV Sea Change in the USA and the Suchetha ferry in India, which are already testing these technologies under operating conditions.

6 Conclusions

Waterborne transportation is under increasing pressure to minimize its contributions to climate change and pollution. The introduction of solutions for ships is a relevant part of the measures being taken in this direction, but other equipment at the port as well as the energy supply when ships are at berth are also seen as part of the solution.

To this end, hybrid solutions are already being considered, which have been discussed in detail in the previous chapters. For instance, for ports, the use of Rubber-Tire Gantry (RTG) Cranes has been presented, where hybridization makes it possible to store the energy used to lift containers, which would otherwise be dissipated during container descending. The fuel savings from these solutions are estimated to be in the range 30–60%. Hybrid tugs, ships used to assist other ships during maneuver, are also being considered as a solution to decrease pollution and GHG emissions at ports, already with some offers from commercial producers of these pieces of equipment in the market. Fuel savings of 20–28% have been described in the literature for hybrid tugs. Finally, drayage trucks carrying containers ashore within ports were also portraited as a means of transport where hybridization could also bring benefits such as reducing air pollution with the added possibility of reducing GHG emissions due to higher efficiency and the possibility of use green electricity from the grid.

Shore-side electricity (SSE) applications aim at the supply of electricity to ships while at berth were shown. Hybrid solutions in this regard were discussed and are promising, especially when the electricity supply at port is limited. Conventional solutions such as diesel generators, especially for large ships at berth, have a non-negligible share of total CO_2 emissions and air pollutants like NO_x, SO_x and $PM_{2.5}$. Alternative variants such as shore-side power supply, e.g., with renewable energies (OPS) via the local grid and the combination with storage on land (power banks), own port energy generators like fuel cells and the use of battery-electric ships (e.g., charging with the help of battery swapping) can reduce pollution and CO_2 emissions in the future and increase the resilience of the supply in the ports. If hybrid power systems onboard ships are increasingly adopted in the future and the energy stored on board is used for ship propulsion, other operational phases of the ships can also be impacted and additional benefits in terms of reduction of air pollution and GHG emissions are expected. Even if various SSE options are available, the lack of uniform standards, e.g., in the area of interchangeable batteries, eclectically operated ships and direct current charging in a maritime context are preventing implementation. In addition, the high costs of electricity and hydrogen and the associated propulsion systems compared to the low costs of MDO and HFO are also a deterrent to developing an infrastructure for SSE. However, an improved energy management system to absorb peak loads and use cheaper electricity at times of higher load could help to improve efficiency and reduce the ship's overall electricity costs.

Hybrid power and propulsion systems on ships were also highlighted in comparison to traditional mechanical propulsion systems. The current increasing trends of the adoption of hybrid power for ships based on battery energy storage systems were briefly explained. Current key performance indicators of the power demands for ships as well as the current performance of some key equipment like internal combustion engines, fuel cells and batteries were briefly described and compared. In addition, statistics of the hybridization of ships based on battery energy storage on board were shown, from which it can be concluded that Europe and Norway are the main locations where these solutions are being implemented. Moreover, car/passenger ferries are the main type of ship for which these systems are being adopted, hybrid and plug-in hybrid are the main ship types in terms of the number of active or ordered vessels, and the number of ships fitted with batteries is generally increasing.

Finally, the particular case of the project HySeas III was presented showing that the hybrid hydrogen fuel cell and battery electric concept allows a reduction of 80.4% of the GHG emissions during its entire life cycle compared to a diesel electric alternative. However, the HFCBE alternative was estimated with a life cycle cost of 69.83 EUR2021/km, exceeding by 51.1% that of a comparable diesel electric alternative. One prerequisite for this is that the hydrogen consumed by the ship is produced using wind power and assuming the average GHG emissions from the grid in the UK for the overnight electricity supply to charge the batteries onboard. Therefore, tradeoffs between the reduction of environmental impacts and the costs remain to be one of the barriers to overcome for these innovative hybrid power alternatives.

References

Antonelli M, Ceraolo M, Desideri U, Lutzemberger G, Sani L (2017) Hybridization of rubber tired gantry (RTG) cranes. J Energy Storage 12:186–195. https://doi.org/10.1016/j.est.2017.05.004

Armand M, Axmann P, Bresser D, Copley M, Edström K, Ekberg C, Guyomard D, Lestriez B, Novák P, Petranikova M, Porcher W, Trabesinger S, Wohlfahrt-Mehrens M, Zhang H (2020) Lithium-ion batteries—current state of the art and anticipated developments. J Power Sources 479:228708. https://doi.org/10.1016/j.jpowsour.2020.228708

Batteries Europe (2023) KPIs benchmarking II—October 2023. https://batterieseurope.eu/results/kpis-benchmarking-2/kpis-benchmarking-2-october-2023/

BIG HIT Project (2024) BIG HIT project—project elements—hydrogen fuel cell. https://www.bighit.eu/hydrogen-fuel-cell

Cai X, Yue Y, Yi Z, Liu J, Sheng Y, Lu Y (2024) Challenges and industrial perspectives on the development of sodium ion batteries. Nano Energy 129:110052. https://doi.org/10.1016/j.nanoen.2024.110052

Caledonian Maritime Assets Limited (2021) HySeas preliminary general arrangement

Cavotec (2020) Cavotec's next generation e-ferry charging solution enters service in Norway. https://www.mynewsdesk.com/cavotec/pressreleases/cavotecs-next-generation-e-ferry-charging-solution-enters-service-in-norway-2985750

Cavotec (2024) MoorMaster: electric vessels. https://www.cavotec.com/en/your-applications/ports-maritime/automated-mooring/electric-vessels

Clean Hydrogen Joint Undertaking (2022) Strategic research and innovation agenda 2021–2027. https://www.clean-hydrogen.europa.eu/document/download/8a35a59b-a689-4887-a25a-6607757bbd43_en

Corvus Energy (2024a) Corvus energy references. https://corvusenergy.com/projects/

Corvus Energy (2024b) Corvus Orca ESS. https://corvusenergy.com/products/energy-storage-solutions/corvus-orca-energy/

Cruise Lines International Association (2024) State of the Cruise Industry Report 2024. https://cruising.org/-/media/clia-media/research/2024/2024-state-of-the-cruise-industry-report_updated-050824_web.ashx

DAMEN (2014) Broshure DAMEN ASD TUG 2810. https://joules-project.eu/Joules/resources/Damen%20ASD%20Tug%202810.pdf

DAMEN (2024) Electric tugs. https://www.damen.com/vessels/tugs/electric-tugs

Damian SE, Wong LA, Shareef H, Ramachandaramurthy VK, Chan CK, Moh T, Tiong MC (2022) Review on the challenges of hybrid propulsion system in marine transport system. J Energy Storage 56:105983. https://doi.org/10.1016/j.est.2022.105983

DNV (2024) Alternative Fuels Insight (AFI). https://www.dnv.com/services/alternative-fuels-insights-afi--128171/. Accessed 28 Aug 2024

Dominko R, Bitenc J, Berthelot R, Gauthier M, Pagot G, Di Noto V (2020) Magnesium batteries: current picture and missing pieces of the puzzle. J Power Sources 478:229027. https://doi.org/10.1016/j.jpowsour.2020.229027

Dotto A, Satta F (2023) Techno-economic optimization of hybrid-electric power plants onboard cruise ships. Energy Convers Manag: X 20:100436. https://doi.org/10.1016/j.ecmx.2023.100436

Elia GA, Kravchyk KV, Kovalenko MV, Chacón J, Holland A, Wills RG (2021) An overview and prospective on Al and Al-ion battery technologies. J Power Sources 481:228870. https://doi.org/10.1016/j.jpowsour.2020.228870

EMSA (2024) EU-MRV reporting system. https://mrv.emsa.europa.eu/#public/eumrv. Accessed 27 June 2024

European Commission (2022) Realising the world's first sea-going hydrogen-powered RoPax ferry and a business model for European islands. https://cordis.europa.eu/project/id/769417/results/de

European Maritime Safety Agency (EMSA) (2022a) Shore-side electricity—guidance to port authorities and administrations: part 2—planning operations and safety. https://emsa.europa.eu/electrification/sse.html

European Maritime Safety Agency (EMSA) (2022b) Shore-side electricity—guidance to port authorities and administrations: part 1—equipment and technology. https://emsa.europa.eu/electrification/sse.html

European Parliament and the Council of the European Union (2023) Regulation (EU) 2023/1804 OF The European Parliament and of the Council of 13 September 2023 on the deployment of alternative fuels infrastructure, and repealing Directive 2014/94/EU

Faber J, Hanayama S, Zhang S, Pereda P, Comer B, Hauerhof E, van der Loeff WS, Smith T, Zhang Y, Kosaka H, Adachi M, Bonello J-M, Galbraith C, Gong Z, Hummels D, Kleijn A, Lee DS, Liu Y, Lucchesi A, Mao X, Muraoka E, Osipova L, Qian H, Rutherford D, Suárez de la Fuente S, Yuan H, Velandia Perico C, Wu L, Sun D, Yoo D-H, Xing H (2020) Fourth IMO GHG study 2020. https://www.imo.org/en/OurWork/Environment/Pages/Fourth-IMO-Greenhouse-Gas-Study-2020.aspx

Geertsma RD, Negenborn RR, Visser K, Hopman JJ (2017) Design and control of hybrid power and propulsion systems for smart ships: a review of developments. Appl Energy 194:30–54. https://doi.org/10.1016/j.apenergy.2017.02.060

GoldmanSachs (2024) Electric vehicle battery prices are expected to fall almost 50% by 2026. https://www.goldmansachs.com/insights/articles/electric-vehicle-battery-prices-are-expected-to-fall-almost-50-percent-by-2025

Gomez Trillos JC, Draheim P (2022) HySeas III: HySeas concept vs conventional—final report—deliverable 6.4

Hannemann R (2024) Ship-DB. www.ship-db.de. Accessed 13 Feb 2024

Inal OB, Charpentier J-F, Deniz C (2022) Hybrid power and propulsion systems for ships: Current status and future challenges. Renew Sustain Energy Rev 156:111965. https://doi.org/10.1016/j.rser.2021.111965

Jayaram V, Khan Y, Miller W, Welch WA, Johnson K, Cocker DR (2010) Evaluating emission benefits of a hybrid tug boat. https://ww2.arb.ca.gov/sites/default/files/2020-12/hybridreport1010_remediated.pdf

Kalikatzarakis M, Geertsma RD, Boonen EJ, Visser K, Negenborn RR (2018) Ship energy management for hybrid propulsion and power supply with shore charging. Control Eng Pract 76:133–154. https://doi.org/10.1016/j.conengprac.2018.04.009

Karvounis P, Dantas JLD, Tsoumpris C, Theotokatos G (2022) Ship power plant decarbonisation using hybrid systems and ammonia fuel—a techno-economic–environmental analysis. JMSE 10:1675. https://doi.org/10.3390/jmse10111675
Kazemi Esfeh S, Monnerie N, Mascher S, Baumstark D, Kriechbaumer D, Neumann N, Eschmann J, Jochem P, O'Sullivan M, Gomez Trillos JC, Vogt T, Brand U, Asif Ansar S (2022) Zukünftige maritime Treibstoffe und deren mögliche Importkonzepte: Kurzstudie. https://elib.dlr.de/186857/2/kurzstudie-maritime-treibstoffe.pdf
Kermani M, Shirdare E, Parise G, Martirano L (2021) Integrated system of energy storage technologies for demand control and energy saving in ports. In: 2021 IEEE industry applications society annual meeting (IAS). IEEE, pp 1–5. https://doi.org/10.1109/IAS48185.2021.9677039
Khan HH, Foti S, Mumtaz F, Testa A (2022) A review of shore infrastructures for electric ferries. In: 2022 International symposium on power electronics, electrical drives, automation and motion (SPEEDAM). IEEE, pp 430–435. https://doi.org/10.1109/SPEEDAM53979.2022.9842000
Konecranes (2024) Rubber-tired gantry cranes. https://www.konecranes.com/port-equipment-services/container-handling-equipment/rubber-tired-gantry-cranes
Korberg AD, Brynolf S, Grahn M, Skov IR (2021) Techno-economic assessment of advanced fuels and propulsion systems in future fossil-free ships. Renew Sustain Energy Rev 142:110861. https://doi.org/10.1016/j.rser.2021.110861
Kortsari A, Mitropoulos L, Heinemann T, Mikkelsen H, Aifadopoulou G (2022) Evaluating the economic performance of a pure electric and diesel vessel: the case of E-ferry in Denmark. Trans. Marit. Sci. 11:95–109. https://doi.org/10.7225/toms.v11.n01.008
Kuittinen N, Koponen P, Vesala H, Lehtoranta K (2024) Methane slip and other emissions from newbuild LNG engine under real-world operation of a state-of-the art cruise ship. Atmos Environ: X 23:100285. https://doi.org/10.1016/j.aeaoa.2024.100285
Kumar J, Kumpulainen L, Kauhaniemi K (2019) Technical design aspects of harbour area grid for shore to ship power: state of the art and future solutions. Int J Electr Power Energy Syst 104:840–852. https://doi.org/10.1016/j.ijepes.2018.07.051
Leclanché (2022) Zero emission ferry and onshore battery energy storage system—Amherst Islander II & Wolfe Islander IV. https://www.leclanche.com/wp-content/uploads/2021/10/Zero-Emission-Ferry-and-Onshore-Battery-Energy-Storage-System_04.2022-1.pdf
Leclanché Energy Storage Solutions (2024) Navius MRS-3: marine battery system. https://www.leclanche.com/wp-content/uploads/2020/04/LECLANCHE-Marine-Battery-Systems-brochure-2.pdf
Liebherr (2019) Liebherr container cranes—rubber tyre gantry cranes. https://www.liebherr.com/shared/media/maritime-cranes/downloads-and-brochures/brochures/lcc/liebherr-rtg-cranes-brochure.pdf
MAN Energy Solutions (2018) Basic principles of ship propulsion. https://www.man-es.com/docs/default-source/marine/5510-0004-04_18-1021-basic-principles-of-ship-propulsion_web.pdf
MAN Energy Solutions (2024) Marine engine programme. https://man-es.com/docs/default-source/marine/marine-engine-programme-20205656db69fafa42b991f030191bb3bbb4.pdf?sfvrsn=9cac9964_114
Marine Environmental Protection Comittee (2021) Resolution MEPC.328(76)—amendments to the annex of the protocol of 1997 to amend the international convention for the prevention of pollution from ships, 1973, as modified by the protocol of 1978 relating thereto—2021 Revised MARPOL Annex VI. https://wwwcdn.imo.org/localresources/en/OurWork/Environment/Documents/Air%20pollution/MEPC.328(76).pdf
Marine Environmental Protection Comittee (2022) Energy efficiency of ships: report of fuel oil consumption data submitted to the IMO ship fuel oil consumption—database in GISIS (reporting year 2021). https://wwwcdn.imo.org/localresources/en/OurWork/Environment/Documents/Air%20pollution/MEPC%2079-6-1%20-%20Report%20of%20fuel%20oil%20consumption%20data%20submitted%20to%20the%20IMO%20Ship%20Fuel%20Oil%20ConsumptionDatabase…%20(Secretariat).pdf

Marzi J, Broglia R (2019) Hydrodynamic tools in ship design. In: Papanikolaou A (ed) A holistic approach to ship design. Springer, Cham, pp 139–207. https://doi.org/10.1007/978-3-030-02810-7_6
Mutarraf MU, Guan Y, Xu L, Su C-L, Vasquez JC, Guerrero JM (2022) Electric cars, ships, and their charging infrastructure—a comprehensive review. Sustain Energy Technol Assess 52:102177. https://doi.org/10.1016/j.seta.2022.102177
Mutel C (2017) Brightway: an open source framework for life cycle assessment. JOSS 2:236. https://doi.org/10.21105/joss.00236
Notteboom T, Pallis T (2023) World ports tracker. https://sustainableworldports.org/wp-content/uploads/IAPH-World-Ports-Tracker-2023.pdf
Offshore Energy (2023) Singapore's 1st fully electric cargo vessel, the Hydromover, hits the water. https://www.offshore-energy.biz/singapores-first-fully-electric-cargo-vessel-the-hydromover-hits-the-water/
Park GY (2022) Emissions analysis of the port drayage truck replacement program and local air quality: the case of the Port of New York and New Jersey. Case Stud Transp Policy 10:1407–1416. https://doi.org/10.1016/j.cstp.2022.05.004
Port of Rotterdam (2021) First emission-free inland shipping vessel on energy containers in service. https://www.portofrotterdam.com/en/news-and-press-releases/first-emission-free-inland-shipping-vessel-on-energy-containers-in-service
Powertrain International Web (2022) Sea Li-ion to electrify large ships. https://www.powertraininternationalweb.com/sustainability/sea-li-ion-to-electrify-large-ships/
Ritchie H, Roser M (2024) CO_2 emissions. Our World in Data
Rolls-Royce (2024) First LNG tugboat with hybrid system goes into operation in Singapore with mtu gas engines from Rolls-Royce. https://www.rolls-royce.com/media/press-releases/2024/24-05-2024-first-lng-tugboat-with-hybrid-system-goes-into-operation-in-singapore-with-mtu.aspx
Siraichi K, Minami S, Kodera M (2015) Development of the hybrid tugboat system. Transit 80:100
Starcrest Consulting Group LLC (2012) Rubber-tired gantry crane hybridization demonstration project—final report. https://sustainableworldports.org/wp-content/uploads/CAAP-lbct-ecocrane-final-report-january-2012-1.pdf
Steubing B, de Koning D, Haas A, Mutel CL (2020) The activity browser—an open source LCA software building on top of the brightway framework. Softw Impacts 3:100012. https://doi.org/10.1016/j.simpa.2019.100012
Stievano L, de Meatza I, Bitenc J, Cavallo C, Brutti S, Navarra MA (2021) Emerging calcium batteries. J Power Sources 482:228875. https://doi.org/10.1016/j.jpowsour.2020.228875
Stucki M, Götz M, de Wild-Scholten M, Frischknecht R (2023) Fact sheet: environmental life cycle assessment of electricity from PV systems. https://iea-pvps.org/wp-content/uploads/2024/05/Task-12-Fact-Sheet-v2-1.pdf
Surf 'n' Turf Project (2024) Surf 'N' Turf. https://www.surfnturf.org.uk/
Talluri L, Nalianda DK, Kyprianidis KG, Nikolaidis T, Pilidis P (2016) Techno economic and environmental assessment of wind assisted marine propulsion systems. Ocean Eng 121:301–311. https://doi.org/10.1016/j.oceaneng.2016.05.047
Tang R, Wu Z, Li X (2018) Optimal operation of photovoltaic/battery/diesel/cold-ironing hybrid energy system for maritime application. Energy 162:697–714. https://doi.org/10.1016/j.energy.2018.08.048
Taskar B, Yum KK, Steen S, Pedersen E (2016) The effect of waves on engine-propeller dynamics and propulsion performance of ships. Ocean Eng 122:262–277. https://doi.org/10.1016/j.oceaneng.2016.06.034
Trillos JCG, Wilken D, Brand U, Vogt T (2021) Life cycle assessment of a hydrogen and fuel cell Ropax Ferry prototype. In: Albrecht S, Fischer M, Leistner P, Schebek L (eds) Progress in life cycle assessment 2019. Springer, Cham, pp 5–23. https://doi.org/10.1007/978-3-030-50519-6_2

United Nations Conference on Trade and Development (2023) Review of maritime transport—2023: towards a green and just transition. unctad.org/system/files/official-document/rmt2023_en.pdf

US Hybrid (2022) CNG/RNG parallel hybrid powertrain technology. https://ushybrid.com/wp-content/uploads/2022/05/USH_CNGPHET_Brochure_V7_2022_digital.pdf

van Biert L, Visser K (2022) Fuel cells systems for sustainable ships. In: Sustainable energy systems on ships. Elsevier, pp 81–121. https://doi.org/10.1016/B978-0-12-824471-5.00010-4

Vlahopoulos D, Bouhouras AS (2022) Solution for RTG crane power supply with the use of a hybrid energy storage system based on literature review. Sustain Energy Technol Assess 52:102351. https://doi.org/10.1016/j.seta.2022.102351

Vu TL, Ayu AA, Dhupia JS, Kennedy L, Adnanes AK (2015) Power management for electric tugboats through operating load estimation. IEEE Trans Contr Syst Technol 23:2375–2382. https://doi.org/10.1109/TCST.2015.2399440

Wärtsilä (2016) LNG as a marine fuel boosts profitability while ensuring compliance. https://cdn.wartsila.com/docs/default-source/services-documents/white-papers/wartsila---bwp-lng-as-a-marine-fuel-boosts-profitability-while-ensuring-compliance.pdf?sfvrsn=26f78b45_10

Wärtsilä (2021) Wärtsilä swappable battery containers enabling inland waterway vessels to operate with zero emissions. https://www.wartsila.com/media/news/07-09-2021-wartsila-swappable-battery-containers-enabling-inland-waterway-vessels-to-operate-with-zero-emissions-2971607

Wärtsilä (2022) Wärtsilä propulsion solution selected for sustainable new hybrid tug. https://www.wartsila.com/media/news/05-09-2022-wartsila-propulsion-solution-selected-for-sustainable-new-hybrid-tug-3151001

Wärtsilä (2024) Wärtsilä Encyclopedia of marine and energy technology: tugs. https://www.wartsila.com/encyclopedia/term/tugs

WaterstofNet (2018a) KENWORTH: fuel cell hybrid drayage truck. https://fuelcelltrucks.eu/project/kenworth-fuel-cell-hybrid-drayage-truck/

WaterstofNet (2018b) US HYBRID: fuel cell Class 8 drayage truck. https://fuelcelltrucks.eu/project/us-hybrid-fuel-cell-class-8-drayage-truck/

Wernet G, Bauer C, Steubing B, Reinhard J, Moreno-Ruiz E, Weidema B (2016) The ecoinvent database version 3 (part I): overview and methodology. Int J Life Cycle Assess 21:1218–1230. https://doi.org/10.1007/s11367-016-1087-8

Xu Y, Titirici M, Chen J, Cora F, Cullen PL, Edge JS, Fan K, Fan L, Feng J, Hosaka T, Hu J, Huang W, Hyde TI, Imtiaz S, Kang F, Kennedy T, Kim EJ, Komaba S, Lander L, Le Pham PN, Liu P, Lu B, Meng F, Mitlin D, Monconduit L, Palgrave RG, Qin L, Ryan KM, Sankar G, Scanlon DO, Shi T, Stievano L, Tinker HR, Wang C, Wang H, Wang H, Wu Y, Zhai D, Zhang Q, Zhou M, Zou J (2023) 2023 roadmap for potassium-ion batteries. J Phys Energy 5:21502. https://doi.org/10.1088/2515-7655/acbf76

Zero Emission Services (2021) First emission-free inland shipping vessel on energy containers in service: Zero Emission Services commences operation. https://zeroemissionservices.nl/en/zero-emission-services-commences-operation/

Hybrid Energy Storage Systems in Rail Transport

Michela Longo, Linda Barelli, and Dario Zaninelli

Abstract As of now, decarbonization is a central theme. In the European context, 40% of railway lines are operated by diesel trains. Among countries such as Italy, Germany or UK, even values as high as 60% are reached. In order to successfully achieve the 2030 and 2050 targets, in the past it has been considered to electrify all remaining lines. Hybrid trains, however, are an alternative to this. In this chapter, solutions for Hybrid Energy Storage Systems in rail transport will be discussed.

Keywords Transportation systems · Hydrogen · Railway systems · Battery electrical multiple unit · Techno-economic assessment

1 Introduction

Environmental factors related to rail transport are Green House Gases (GHG) and local pollutant emissions. GHGs evaluation must be computed with a well-to-wheel approach (WTW), which considers both emissions from the conversion and use of fuel to operate the vehicle, and emissions from fuel production and fuel transport (Choi and Song 2018; da Fonseca-Soares et al. 2024; Kwakwa et al. 2023). The level of Carbon depends on the type of rolling stock supply. To illustrate the advantages of one traction system over another, it is necessary to analyze aspects both from the point of view of the emissions that this traction system entails and the cost of implementing a traction system. Diesel traction has higher CO_2 emissions than electric traction, although on low-traffic lines it can be more economical than electric traction. It must be kept in mind that electric traction has a fixed initial cost, so the construction of an electrified line does not depend on the number of trains running on it. Still, the economic return on construction depends on the useful traffic on the line (Ogunkunbi

M. Longo (✉) · D. Zaninelli
Department of Energy, Politecnico di Milano, Milano, Italy
e-mail: michela.longo@polimi.it

L. Barelli
Department of Engineering, University of Perugia, Perugia, Italy

R. Scipioni et al. (eds.), *Hybrid Energy Storage*, Lecture Notes in Energy 47,
https://doi.org/10.1007/978-3-031-97755-8_12

and Meszaros 2023; Sasse and Trutnevyte 2023). The useful traffic is the number of people carried for passenger trains and tons of goods for freight trains. Therefore, diesel traction is used today for complementary lines with low traffic density with single track, while lines with high traffic density prefer electrified lines. Analyzing more in detail, diesel traction has the energy source on board and has torque at low speeds, which makes it perform better with higher acceleration, but as mentioned earlier, it has higher emissions than electric traction (Praticò and Fedele 2023). The introduction of the Battery Electric Multiple Unit (BEMU) would result not only in reduced emissions but also in the introduction of the electric energy source on the trainset represented by the traction battery. Moreover, also the emissions generated by fuel consumption in the production of electricity for traction should be accounted for (Kapetanović et al. 2024; Klebsch et al. 2018, 2019). For the electric rolling stock, the increase of renewable sources for electricity production must be considered in GHG emissions calculation (see Fig. 1). Since 2017, many railway companies in Europe have purchased electricity from renewable sources, reducing CO_2 emissions per passenger by 15% compared to purchasing electricity directly from national operators. As far as freight transport is concerned, there has been a reduction of around 6%. The European Union announced on 16th December 2020 that 2021 would be the European Year of Railways (IEA, 2020).

This initiative goes towards the promotion of environmentally friendly public transport for both companies and citizens and is part of the goal of achieving climate neutrality by 2050 sought with the Green Deal. Transport accounts for 25% of greenhouse gas emissions in Europe. However, rail only accounts for 0.4% of these emissions and is the only sector that has drastically reduced emissions since 1990. Only 11% of passengers and 7% of goods travel by rail. In addition to this, the European Union wants to increase these percentages to reduce greenhouse gas emissions, especially from the transport of goods by land and sea. During the Covid pandemic, the

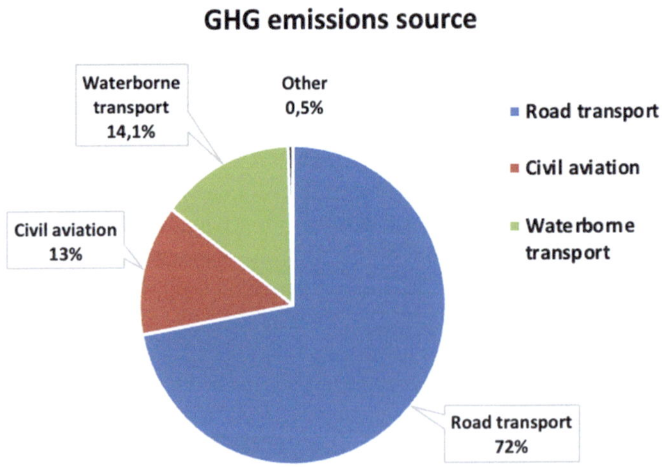

Fig. 1 Greenhouse gas emissions from transport

railway was a means of transport for essential goods, including medicines and basic services. This sector was also severely affected by a decrease in passenger numbers due to health restrictions imposed by European states in response to the pandemic. A supplementary package that created a united railway area without institutional barriers in favor of economic growth has also been designed (European Parliament 2022).

As it was reported earlier with the Green Deal, Europe has set a target to reduce greenhouse gas emissions by 50% by 2030 and 90% by 2050 compared to 1990. In Europe, most railway lines are electrified, although there is a 20% use of diesel rolling stock. On lines with low passenger density, the electrification process is not economically sustainable. Alternatives to diesel propulsion are being investigated, such as the introduction of hydrogen and the use of battery-powered trains (BEMU) that can reduce the level of greenhouse gases (G. and I. T. Directorate-General 2021).

Over the decades, the technology of a Diesel Multiple Unit (DMU) has improved considerably in terms of both emission and noise reduction and energy absorption during operation. Many DMUs feature a battery to ensure hybrid operation (Praticò and Fedele 2023). However, European Union has decided to stop the production of diesel vehicles; this decision makes clear that the objective of decarbonization in rail transport is to be achieved by moving in a different direction, namely through the introduction of new technologies (Ansa 2022).

In the classification of railway lines three classes are distinguished:

- Fully electrified lines which are subdivided into regional transport lines and high-speed lines, where run Electrical Multiple Unit (EMU). The rolling stock operation consists of the absorption of electrical power through the catenary by means of a pantograph. Electrified lines in the European Union have different traction systems that differ in terms of catenary voltage. For historical reasons too, most of the European network has an electrical voltage of 15 kV AC single-phase with a frequency of 16.7 Hz or 25 kV AC single-phase with a frequency of 50 Hz for ordinary regional rail service. In Italy, there is a 3 kV DC power supply for regional rail transport (Arboleya et al. 2020). Furthermore, the Technical Specifications for Interoperability (TSI) for Energy subsystem establish in Sects. 7.2.1 and 7.2.2 that the railway lines of the various countries must be interoperable, although it is left up to each country to arbitrarily choose the traction system (Commission Regulation 2014).
- Non-electrified lines that do not have any power supply system are travelled by DMU and by diesel-electric rolling stock.
- Mixed or hybrid lines are those railway lines with partial electrification. Partial electrification is due to the presence of electrified line sections and stations since these are shared with other lines that are fully electrified. Such lines are run by DMU or Battery Electrical Multiple Unit (BEMU).

The railway represents a transport sector that is widely used to connect isolated areas of Europe and thus represents a link between internal and cross-border European regions. In Italy, this situation is not the same case as European contest because many areas of Italy are not connected by any railway lines.

In the Italian railway sector, a distinction must be made between basic lines and complementary lines. Fundamental lines are those lines with a high level of traffic and an electrified infrastructure connecting the main cities, while complementary lines are single-track lines with medium to low traffic and a non-electrified infrastructure. From an infrastructural point of view, mixed lines are complementary lines with a low level of electrification compared to the entire length of the line. The three lines here investigated, which run from the city of Pavia, are medium-low traffic lines and have a medium-low electrification level concerning the length of the line. Diesel traction is the solution used on these sections because any electrification of the line would entail an unsustainable cost with respect to the traffic on these lines. Traditional diesel traction is flanked by diesel-electric traction to reduce the environmental impact of rolling stock. BEMU is an economical and environmentally sustainable solution for these lines. BEMUs are rolling stock with a traction battery that can be used on sections of line where there is no catenary, while where there is a catenary, the rolling stock is powered using a pantograph. The pantograph allows the battery to be recharged both while running and when the rolling stock is at a standstill at the station. BEMU is used in some European countries such as Germany, France, Austria, Netherlands, England, and Norway (CHAMARET 2019; IEEE Industrial Electronics Society 2019; Thorne et al. 2019; Heckele et al. 2022), while it is not yet used in Italy.

2 Main Trends in Rail Transport Sector

Mobility as a resource for environment and community: public transit, specifically the one on two wheels and rail, needs to be decarbonized in order not to further impact the Planet's health and, at the same time, allow to move around in an efficient and sustainable way.

- *On a global scale*, rail industry is undergoing a transformation, with hybrid trains emerging as the next key technology, towards a sustainable and efficient rail transport. Such trains, combining traditional diesel motors with electric traction systems, offer a practicable solution in reducing GHG emissions and fuel consumption. In the following, the five main trends that shape hybrid train market sales are presented:
- *Growing emphasis on environmental sustainability*: Environmental sustainability is the leading aspect of hybrid train markets. Governments and regulatory bodies all over the world are setting stricter and stricter targets for emissions, to fight against climate change and reduce atmospheric pollution. Hybrid trains are seen as a crucial part of this strategy since they can significantly reduce carbon emissions concerning conventional diesel trains. These trains are able to circulate on non-electrified lines, which compose the largest share of global railway lines, using energy coming from batteries in urban areas to reduce to the maximum extent possible local pollution. The push towards more ecological transport solutions is

stimulating the demand for hybrid trains, with producers focusing on developing more efficient and more ecological models every year.

- *Technological progress in batteries and energy storage systems*: one of the most relevant tendencies in the hybrid train market is the rapid evolution of batteries' technology and energy storage systems. Innovations, such as the ones in the field of Lithium-Ion and solid-state batteries, as well as other storage solutions, are improving the performances, autonomy and efficiency of hybrid trains. These progresses allow for longer working periods in electrical-only mode, reducing the dependence on diesel motors. In addition to this, regenerative braking systems, able to capture and store energy during braking, are becoming more and more diffused. These technologies not only play in favor of making hybrid trains more ecological and sustainable but also reduce operative costs, acting on fuel consumption and maintenance requirements.
- *Government incentives and funding programs*: Governments all over the world are introducing incentives and funding programs to promote hybrid trains' adoption. These initiatives include grants, subsidies, and tax relief projects for railway operators and producers that invest in hybrid technology. As an example, in Europe, the European Green Deal is aimed at making railway transport more sustainable, distributing significant funding for the development and diffusion of hybrid trains. Parallel to this, in Asia and North America, governments are prioritizing projects of railway infrastructures that contemplate hybrid technologies. These incentives are accelerating the market's growth, making hybrid trains more appealing and financially feasible for operators.
- *Expansion of hybrid trains' applications*: The possible applications of hybrid trains are growing beyond passenger service, including the transport of goods and high-speed service. Freight transport operators are adopting an ever-increasing number of hybrid locomotives in order to keep up with the targets set for emissions reductions and raising fuel efficiency. In addition to this, high speed hybrid trains able to guarantee faster long-haul routes, while being cleaner and more efficient are currently being developed. This diversification is enlarging the hybrid trains' market, because operators, in various fields, recognize the advantages related to lower emissions, lower operating costs, and better performances. The versatility of hybrid trains makes them suitable for a wide range of rail services, further pushing the possibilities of adoption.
- *Integration with intelligent and connected technologies*: Hybrid trains' market is being revolutionized by the integration of intelligent and connected technologies. Advanced monitoring systems, IoT devices and data analysis are being used towards optimizing rail operations, increasing energetic efficiency and bettering user experience. As an example, predictive maintenance technologies allow operators to monitor working conditions of hybrid trains in real time, avoiding mechanical faults and reducing the downtimes. Furthermore, intelligent energy management systems allow for a more efficient battery recharge and refueling. These technological progresses are making hybrid trains more reliable, convenient and easy to use, contributing to their ever-increasing popularity.

3 Example Technologies to be Adopted

3.1 *Battery Electric Multiple Unit (BEMU)*

An important goal for the freight and passenger transport sector is the reduction of CO_2 emissions into the air. In Europe, transport accounts for 25% of greenhouse gas emissions and is the main cause of urban pollution gases (G. and I. T. Directorate-General 2021).

These emissions must also be reduced to ensure a slowdown in climate change that has been affecting the world in recent years. This climate change is due to the greenhouse effect caused by the CO_2 emissions mentioned above.

Europe's response to this problem is the introduction of increasingly low-emission transport. Rail transport is a sector with very low carbon emissions for both passenger and freight transport. However, diesel-powered rolling stock is currently used in this sector for low and medium-traffic density routes.

Through the Green Deal signed by the European Commission, the European Union has set the goal of reducing net greenhouse gas emissions by at least 55% by 2030 compared to 1990 levels, combined to have no more diesel vehicles on the market by 2050, achieve carbon-free transport (European Commission 2022). In the railway sector, substitutes for diesel traction are being considered with the aim of reducing both environmental and noise pollution. Possible solutions are rolling stock with batteries, hydrogen fuel cells, and other low-carbon energy sources. In diesel rolling stock, hybrid rolling stock is considered. Such rolling stock has a diesel-electrified traction that reduces CO_2 emissions into the air (Praticò and Fedele 2023). Complete electrification of the lines is not applied on these railway sections because the cost of electrification is high compared to the useful passenger traffic, and this investment would not be recovered.

In recent decades, special attention has been paid to vehicles with an onboard energy storage system for use in railway sections where there is no catenary. Such OESS (Onboard Energy Storage System) represents a very efficient electrical traction with the use of a reasonable amount of regenerative braking. Such vehicles represent a solution as they minimize costs by reducing maintenance and installation of electrical infrastructure requirements.

One problem with the use of these rolling stock materials is the materials used to manufacture the batteries. Rail represents the lowest carbon-emitting transport sector today and the most efficient, although a 9% share of total passenger and freight transport railways account for less than 2% well-to-wheel GHG, and it is about 3% of total final energy use (IEA 2022). Moreover, the low energy demand per passenger per kilometer is because there are lower losses caused by friction and drag. The higher energy efficiency of electric motors compared to diesel engines is due to regenerative braking and the high storage capacity; however, in the account of GHG emissions related to electrification in railways, emissions of fuels that are used to produce the electric energy are considered. In electricity production, renewable sources are also considered. In Europe, approximately 40% of electricity is produced with minimal coal, with an average of 20% produced directly from renewable sources (IEA 2022).

All over the world, solutions are being developed to reduce the environmental impact of railway companies' rolling stock by increasing their production and power plants to promote energy demand interaction. Production of energy from renewable sources has increased, including nuclear energy, and this production contributes to supply railway traction. In the global railway world, the mix of renewables and nuclear energy accounted for 13.5% as of 2015, and it has doubled since 1995.

BEMU technology

Bimodal battery trains have replaced diesel-powered rolling stock in complementary hybrid lines in some European countries, including Germany, England, France, Austria, and Poland. Worldwide, in some states such as Japan and even Russia, bimodal battery trains are used, while in North and South America, diesel traction is still widely used (IEA 2022).

BEMU is a train with battery traction where there are no electrified sections on a line, while there is a catenary. The train through the pantograph absorbs power to recharge the battery. The problem of the current absorption limit that occurs when the train is stationary in the station to recharge does not appear in the case of recharging in a running train. While running, the train with the pantograph raised in the electrified sections can absorb currents that are in the order of magnitude larger than the 200 A imposed by the TSI Energy standards (RFI 2022). The use of a BEMU over a diesel drive also has the beneficial effect of reducing particulate matter and nitrogen oxides NO_x (NASA 2022).

Returning to the discussion of the different railway infrastructures for regional lines in European states, the presence of a higher voltage value for the same absorbed current results in greater power absorbed by the rolling stock for charging the catenary batteries. Higher power results in a shorter charging time for the trainset battery so that the use of the BEMU in the railway company commercial services is feasible.

In Italy, as there is a lower voltage value, charging times are longer and are not compatible with the railway companies' dwell times, thus limiting the use of such trainsets for passenger transport on mixed lines. The lines, where BEMU can operate, are lines with a low to medium useful traffic index and have electrified sections at least in the line head stations for charging the rolling stock.

BEMU has the traction system on board, so on the lines where it is applied, it does not need any major infrastructure intervention. On railway lines with a low average useful traffic index, BEMU represents an opportunity and a solution that minimizes infrastructure maintenance costs. However, the use of on-board batteries has some limits because battery trains are used on lines that are 60 to 100 km long, while this parameter also varies depending on the rail line and schedule characteristics (Streuling et al. 2020). Another limitation of the battery-powered train is that on lines where there is a very steep gradient, the status charge of the battery drops more quickly. The batteries that are mainly used as traction batteries for bi-modal battery trains are nickel metal hydride (Ni-MH) and lithium-ion (Li-ion) batteries. Ni-MH batteries have a very high robustness with low maintenance, although they have a low efficiency and a high self-discharge rate. However, the batteries that are being researched the most are lithium batteries, which have higher energy densities with higher efficiency and a lower self-discharge rate. Li-ion batteries are often considered to be homogeneous groups and so the batteries with high energy density include nickel-metal cobalt and lithium manganese oxide, but they have short lifespan (Ghaviha et al. 2017). BEMU trains are similar in size to EMU trains. The only difference is the greater weight of the rolling stock due to the presence of the traction battery. Through roof-mounted pantographs, BEMU operates like a normal EMU on an electrified route. Logically, the battery is used for the non-electrified sections while the pantograph is used as a charging system for the batteries both when the train is moving and when it is at a standstill at the terminus station.

The catenary is a very fast charging system for the batteries as it can provide the battery with power in the order of MW that can recharge the traction batteries quickly (Heckele et al. 2022). Nowadays, the performance of a BEMU depends on the distance of the railway line without a catenary and the capacity of the battery. One parameter that must be considered is the degradation of battery capacity due to load utilization. Other factors that determine a battery utilization limit are stringent running conditions and the climatic peculiarities of the line. These factors have an impact on the utilization range of the traction battery. Manufacturers still lack the capabilities and data to determine the long-term impact of these factors on battery performance. This leads manufacturers to force the use of traction batteries for specific lines where simulations of driving profiles have been made (Heckele et al. 2022).

The useful capacity of the battery does not correspond to the nominal capacity of the battery specified in the battery datasheet. Normally, an attempt is made to have a state of charge (SoC) of the battery in a range between 20 and 80% to preserve the capacity of the battery and thus extend its service life as much as possible, so the capacity is also designed to work in a state of charge (SoC) between 40 and 80%. In the battery design, the cases of failures and extreme weather conditions are also considered to establish a buffer that for a defined short time can guarantee more power without damage to the battery. Such situations are considered to avoid the case where the BEMU stops due to battery depletion.

One aspect that greatly increases the efficiency of the BEMU is regenerative braking because if there is no other train in the vicinity that can absorb this energy to accelerate, the train uses this energy to recharge the traction battery (Heckele et al. 2022).

After presenting all the main advantages of using a BEMU train, the negative aspects of adopting such a means of transport are also discussed. One disadvantage is that the BEMU is 10% heavier than a fully electric train due to the weight of the traction batteries.

Another disadvantage is the construction of charging points for the batteries and the organization of charging times throughout the day. These times must be compatible with the dwell times required by the railway companies. The addition of recharging times for battery trains can lead to problems with the smooth running of the train service provided by the railway undertakings and require changes to the infrastructure, as such high recharging capacities require ad hoc recharging solutions for possible stations (Heckele et al. 2022).

The main problem for eventual use is the charging time of battery trains in stations. Energy TSI stipulates that a standstill train can draw a maximum current of 200 A from the catenary. This current limit is set because the elastic catenary, being under tension, heats up at the point of contact between the pantograph creep and the contact wire and the contact wire can break, causing damage to the infrastructure (European Commission 2014).

This maximum current limit is one of the most critical points for the use of BEMU in some European countries. In Europe, the infrastructure for regional railway lines is supplied with a 15 kV AC single-phase power supply, whereas in Italy these lines are supplied with a 3 kV DC power supply. This difference in power supply systems for the traction of electric or battery-powered rolling stock results in a substantial difference in the recharging time of BEMU at the end of the line (Arboleya et al. 2020). The difference from a rolling stock with diesel traction is represented by the fact that BEMU has a traction system on board the train, however, BEMU requires a charging infrastructure for the battery, which entails a modification to the existing railway infrastructure.

Charging and filling infrastructure

Battery trains can use the catenary in electrified sections and the battery in non-electrified sections as the traction system. This means that the battery train in electrified sections uses the pantograph, which, when placed in contact with the catenary, draws electric power. In this operating regime, BEMU behaves like an EMU train; in addition to that, BEMU can use the catenary contact line to recharge the battery when it has a low state of charge, and the rolling stock is in motion. Based on the rolling stock running in Europe and Japan, the various charging systems for each BEMU are listed. In Europe AC railway power supply system's generally operate at 15 kV at 16, 7 Hz or 25 kV at 50 Hz. These voltage values are standardized by EN 50,163, and to ensure maximum interoperability between the various European railway systems, the technical specifications for energy interoperability consider these two standards.

EN 50,163 stipulates the use of a frequency of 50 Hz for recharging infrastructure in the case of a nominal voltage of 25 kV AC. However, battery-powered trains in states with a power supply of 15 kV AC with a frequency of 16.7 Hz must recharge with a frequency of 50 Hz. To recharge a multisystem vehicle for both AC traction systems, a single transformer with two voltage taps on the primary side can be used to change the number of transformer turns to change the transformer turn ratio.

Unlike vehicles that have a single traction system, multisystem vehicles need an additional switch in the primary side of the transformer, and components that are subject to a voltage of 25 kV must be sized for a higher voltage value. Consequently, standard EN 501 63 indicates a maximum voltage value of 29 kV for the nominal value of 25 kV.

A reason why a higher voltage is used at the same frequency, which in this case is the frequency of 50 Hz, is that the inductive interference of voltages is proportional to the frequency of the voltage, so a higher voltage value compensates these effects.

An alternative solution to this problem may be to reduce the distance between substations (Dschung and Ludolf 2021). In isolated lines, this problem is irrelevant as the reactance of the line is negligible.

In Germany, Talent 3 can recharge through the 15 kV overhead line at a frequency of 16.7 Hz or use regenerative braking like Electrostar and D-Train in the UK (Thorne et al. 2019). In Japan, the EV-E301 V uses a DC/DC converter to lower the voltage from the 1500 V DC of the overhead line to the 630 V of the battery voltage. In non-electrified lowered pantograph sections, BEMU can recharge its battery pack through regenerative braking (Thorne et al. 2019).

3.2 Hydrogen Trains

An alternative, clean and with great potential, that is currently acting towards reducing the emission of large amounts of carbon dioxide in the atmosphere is Hydrogen.

In Europe, hydrogen train experimentation has already begun, as a new frontier of sustainable mobility. Hydrogen trains represent the ecological evolution of rail transport modes, that in these years have traditionally been powered by fossil fuels. It is a new transport technology that allows carriages to move using entirely clean energy, fully respecting the environment. Green hydrogen is a zero emission energy vector, renewable, and very efficient: in the specific, it allows for high performances thanks to the capacity to carry a large amount of energy per kg of fuel. According to estimates, the employment of a hydrogen train can avoid over 4.400 tons of CO_2 emissions in a single year. In this light, it is possible to understand why they represent a green solution to pollution and global warming.

In terms of structure, on top of the carriage a fuel cell is installed, acting as the nucleus of the whole system, and allowing to mix the hydrogen located in the tanks with oxygen naturally present in the environment. The only emissions that are produced by this green train are vapor and condensate water, two of the components generated by the meeting of hydrogen with oxygen in the fuel cell: a 100% green and eco-friendly mobility.

A completely new scenario is being opened by hydrogen trains in transport systems. This paradigm shift is registering as a first improvement the cutting of a notable amount of GHG emissions, around 40%. Moreover, if on top of green hydrogen also grey hydrogen is used (the nomenclature derives from the employment of fossil sources for hydrogen production, typically natural gas), an overall lessening of pollution concerning the oil engines currently operating is to be expected anyway.

In a not-so-distant future, in which an increase of hydrogen production is estimated, the cost of the green energy vector will be more competitive with respect to diesel.

Another benefit is in terms of the refuelling process. In the future it will become a fast operation, in the order of minutes, allowing for a drastic reduction of stopping times for trains. This, combined with the trains being able to run for 18 consecutive hours between refuelling stops, will provide an unprecedented level of service.

Table 1 Hydrogen SWOT analysis

Strengths	Weakness
• Available technology • Does not produce direct emissions • Fuel efficient • Renewable technology (depending on the type of hydrogen)	• Efficiency of hydrogen supply chain • Expensive • Highly flammable • Difficult to revamp • Storage challenge
Opportunities	**Threats**
• Government and European Union support • Become and innovative company	• Suppliers of key components not available • Type of hydrogen sustainable (Blue, Green or Grey) • Can be produced with fossil fuels (not really sustainable in this case)

In some European countries, such as Germany, hydrogen passenger trains are already fully operational and regularly used by travellers. In UK and France, some proposals have been brought forward in order to fully substitute diesel trains with hydrogen ones, in the next twenty years, in those lines that are difficult to electrify. In Italy, as it's possible to read in the Strategical National Guidelines for Hydrogen, as much as half of non-electrifiable national lines could be converted to hydrogen before 2030. A more sustainable and not-too distant-tomorrow.

In Table 1 it is possible to see a hydrogen SWOT analysis, identifying points of strength, weakness, opportunities and threats for this technology.

4 Guideline for the Electrification of Rail Infrastructure

Climate change is the challenge of our time that will determine our future. Climate change is accelerated by the presence of greenhouse gases in the atmosphere, of which carbon dioxide (CO_2) is the most significant. Transport is one of the sectors with the greatest responsibility for CO_2 emissions (IEA2023a, 2023b). In transportation, the road infrastructure segment needs major decarbonization measures, but it is not the only one. Non-electrified railway lines represent a major sustainability challenge, and currently, about 33% of lines in Italy use diesel traction, contributing significantly to greenhouse gas emissions (García-Olivares et al. 2020). Electrification through the implementation of new catenary infrastructure is not always easy to implement due to geographical and structural constraints, such as the presence of tunnels and bridges. At the same time, the implementation of new catenary electrification systems may not be economically viable for short line lengths or short sections where catenary electrification does not exist. Catenary electrification is the traditional method of reducing emissions in rail transport. It offers high energy efficiency but involves high infrastructure costs (around €1 million/km) and it is more suitable for high-frequency lines (Kilsby et al. 2017). To address these requirements, this chapter provides guidance on how to make such lines more sustainable by analyzing two technological solutions:

- Battery-powered trains;
- Hydrogen fuel cell trains.

4.1 Technical Solutions for Railway Electrification

Battery Trains

Battery-powered trains are a promising and adopted solution for short to medium-distance non-electrified railways. The technology is based on the use of lithium batteries, which are known for their high energy density and efficiency. Battery trains can also operate in bimodal mode, using the energy stored in the batteries and using electrified sections of track. This technology involves the use of lithium batteries, which are already used in vehicles because of their energy efficiency and the possibility of recharging them at dedicated stations or sections via the overhead contact line. Table 2 summarizes the advantages and limitations of battery trains (Pugi et al. 2024; Abiko 2013).

Many applications require the use of batteries sized for the type of service to be performed. For correct sizing, it is advisable to first simulate the total energy consumption to be combined with the required range and operating conditions. Battery trains are a viable solution for regional lines of medium length and moderate frequency. However, a detailed planning of the charging phase should be defined and

Table 2 Advantages and limitations of battery trains

Advantages and limitations	Description
Emission reduction	No exhaust gases during operation
Noise reduction	Silent operation in motion, leading to a more comfortable service for users
Modularity	Ability to add or reduce batteries, improving vehicle flexibility
Energy efficiency	Batteries allow optimal energy use, supported by regenerative braking
Lower operating costs	Cost per kilometer run is lower than the diesel equivalent, due to lower electricity prices and lower maintenance requirements
Reduced autonomy	Batteries still have a limited range for vehicles of this size and require frequent recharging
Storage system footprint	Traction batteries have a significant weight, which reduces the weight of transportable passengers
Dedicated infrastructure	A charging station is still required
Geographical limitations	High slopes significantly reduce battery life

adapted to the service to be provided. The use of regenerative braking in this scenario is a very useful tool to increase the available energy of the vehicles. Finally, the rapid development of traction batteries in terms of energy efficiency could provide a further boost to the use of these vehicles, enabling them to travel ever longer distances.

Hydrogen Trains

Hydrogen cell trains are an innovative solution for decarbonizing non-electrified railways, especially in regions with long distances between stations or complex geographical conditions. This technology uses hydrogen as the primary fuel for the cells to generate electricity through electrochemical oxidation, producing water vapor and avoiding CO_2 emissions. Although fuel cells have significantly lower efficiency than lithium batteries, their energy efficiency is much higher than that of an internal combustion engine. Today, fuel cells with efficiencies over 60% are being investigated for transport applications. Table 3 summarizes the advantages and limitations of hydrogen trains (Ding and Wu 2024; Nqodi et al. 2023).

The most widely used fuel cell technology for traction is PEM (polymer electrolytic membrane). Hydrogen trains are an optimal solution for long non-electrified lines due to their autonomy and ability to operate without extensive electrical infrastructure. However, the initial cost of refuelling infrastructure and the production of green hydrogen remain the main barriers to large-scale deployment. The choice of this technology depends heavily on local conditions and sustainability priorities.

Table 3 Advantages and limitations of hydrogen trains

Advantages and limitations	Description
Emission reduction	No emissions during operation
Long haul autonomy	H_2 trains have a high autonomy (more than 800 km)
Noise reduction	Silent operation in motion, leading to a more comfortable service for users
High adaptability	Only the track and a refuelling station are required for operation
Modularity	Fuel cells can be designed at different power levels
High initial costs	Charging infrastructure, vehicle and fuel cell are still unaffordable
Reliability	Fuel cells have made progress in recent years. Further improvements are needed to ensure the longevity and stability required for rail applications
Power supply	Hydrogen production is not necessarily from renewable resources
Presence of tanks	Hydrogen tanks require significant volume and a dedicated propulsion system, affecting train design

4.2 Technical Analysis Implementation

Battery Trains

As explained above, in order to implement a service based on battery trains, it is necessary to know the energy demand on the line through a simulation. To do this, it is necessary to select a type of rolling stock and build a consumption model (Colombo et al. 2023). The next step is to start the battery sizing process as it follows. The limitations introduced for the battery sizing were described in (1, 2):

$$\mathrm{P}_{\mathrm{disch}} = \mathrm{P}_{\mathrm{train}} + \mathrm{P}_{\mathrm{aux}} + \mathrm{P}_{\mathrm{add}} \tag{1}$$

$$E_{tot} = \eta_{disch} \cdot \frac{P_{disch}}{C_{disch}} \tag{2}$$

where:

- P_{disch} is the discharging power, the useful power supplied by the battery to ensure both traction and auxiliary services.
- P_{train} is traction power;
- P_{aux} is the power needed for the auxiliaries
- P_{add} is a set condition to be sure to cover all the requirements.
- E_{tot} is the total energy provided by the battery;
- η_{disch} is the battery efficiency;
- C_{disch} is the discharging rate of the battery.

For the lithium-ions battery it is possible to assume a complete charge/discharge cycle to 80%. C-rate, instead, identifies the discharge rate of the battery linked to its capacity. This highlights the discharging current at which the battery begins to discharge over its nominal capacity. Having to size the battery satisfying both the power and the energy requirements to complete the journey, it is necessary to have an estimation of the energy consumed (E_{cons}) with the addition of a margin (E_{add}) for non-regular operation or stress conditions so that the battery does not discharge completely (3).

$$E_{cons} = E_{tot} + E_{add} \tag{3}$$

Therefore, the maximum one needs to be selected between the two obtained values and this value must be used to estimate the total necessary capacity of the battery (C_{tot}) (4).

$$C_{tot} = \frac{\max(E_{cons}, E_{tot})}{V_{batt}} \tag{4}$$

The battery installed on-board reaches the desired voltage by connecting a number of cells in series (N_s), each of which increases its overall potential by its own voltage. The parallel connection instead increases the total capacity instead (N_p). It is important to use the same type of cell with the same voltage (V_{cell}) and capacity (C), since modules made from cells with different voltages, capacities and dimensions could cause imbalances (5, 6).

$$N_s = \frac{V_{batt}}{V_{cell}} \tag{5}$$

$$N_p = \frac{C_{tot}}{C} \tag{6}$$

The series–parallel configuration allows to reach the desired voltage and current values starting from a certain number of standard cells. According to the total number of cells and the number of cells connected in parallel, it is useful to compute more accurately the new value of energy needed to satisfy the highlighted specifications E_{tot_new} (7, 8) and the maximum absorbed current by the battery I_{max} (9).

$$C_{tot\_new} = C \cdot N_p \tag{7}$$

$$E_{tot\_new} = C_{tot\_new} \cdot V_{batt} \tag{8}$$

$$I_{max} = C \cdot C_{rate} \cdot N_p \tag{9}$$

The maximum absorbed current value is necessary to compute the battery charging power P_{ch} (10), so as to have an estimation of the recharging time necessary t_{ch} (11), taking into account also the energy consumed for the journey.

$$P_{ch} = I_{max} \cdot V_{batt} \tag{10}$$

$$t_{ch} = \frac{E_{cons}}{P_{ch}} \tag{11}$$

The charging time is an element that must be considered with the location and management of the charging stations since the recovable share of regenerative braking energy is not enough for recharging the battery. Furthermore, the good performance of the battery also depends on this parameter. The required performance directly affects the system's energy consumption. Therefore, knowing that the necessary energy depends on load, structural characteristics of the lines, speed profile and so on, an increase in the total mass of the train due to the addition of the battery on-board could have an impact on the overall system (12). This highlights the weight of the battery after the sizing process, combined with the specific energy of the storage system (e).

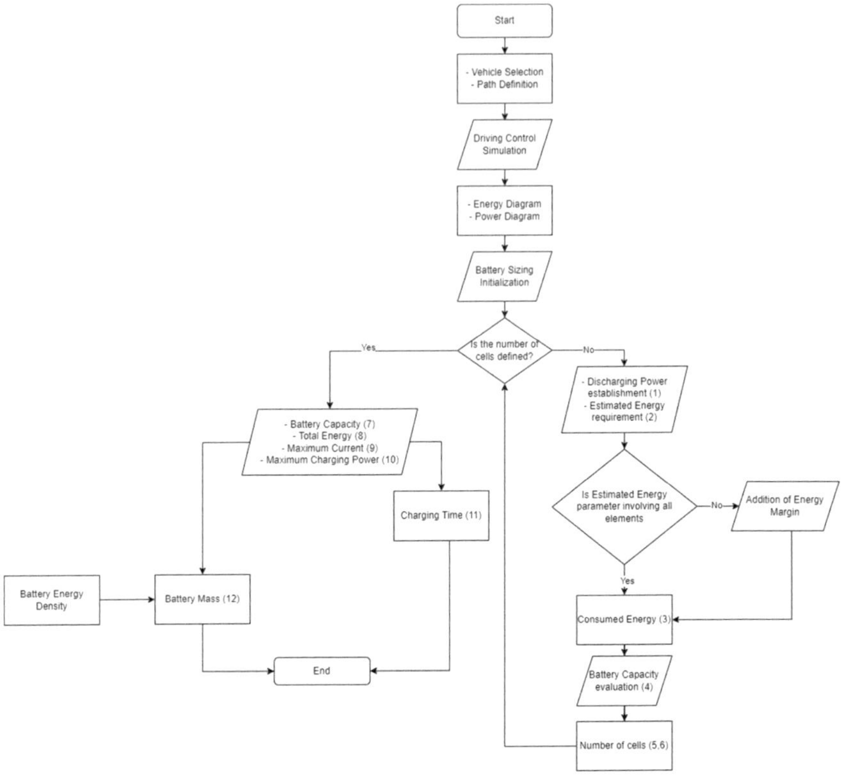

Fig. 2 Battery train implementation process

$$m_{batt} = \frac{E_{tot\_new}}{e} \tag{12}$$

Figure 2 highlights the process to be followed for the battery train implementation.

Hydrogen Trains

Similarly, in the case of hydrogen electrification, a test drive is required to determine consumption and, again, a suitable vehicle should be selected based on the characteristics of the route. If a vehicle with these characteristics does not already exist, it is advisable to refer to another vehicle with similar characteristics. Once the profile for energy and power is simulated, it is possible to provide the battery and fuel cell sizing. The fuel cell sizing is developed depending on the average power required to operate the line. The required energy will be provided by the fuel cell, which will supply the battery or the traction motors alternatively (Iskandar and Maheri 2022). Thus, the hydrogen on board will be enough to supply all the load, when the battery will satisfy the peak demand during traction. A coherent evaluation of the energy implies the use of (13).

$$E_{tot} = \frac{E_{el1}}{\eta_{FC}} + \frac{\left(\frac{E_{el2}}{\eta_{FC}}\right)}{\eta_{batt}} \tag{13}$$

where:

- E_{tot} is the total energy required for the path expressed in kWh;
- E_{el1} is the energy consumption in case of direct flux fuel cell-motors in kWh;
- E_{el2} is the energy consumption in case of fuel cell supplying both battery and motors in kWh;
- η_{FC} is the efficiency of fuel cell assumed constant and equal to 0.6;
- η_{batt} is the efficiency of battery assumed equal to 0.9 and constant.

Then, considering now a hydrogen energy density equal to 5.6 MJ/l (compressed hydrogen at 700 bar) [40], it is possible to obtain the volume of required H_2 (V_{H_2}) as a function of the total energy (E_{tot}) and the energy density (u) (14).

$$V_{H_2} = \frac{E_{tot}}{u} \tag{14}$$

Since hydrogen tanks can contain up to 0.039 kg/l at 25 °C and 700 bar, it is possible to estimate the H_2 mass (m_{H_2}), knowing the tank capacity (C_{tank}) (15).

$$m_{H_2} = V_{H_2} \cdot C_{tank} \tag{15}$$

However, it must be considered that the train is not operating while carrying the H_2 mass. The weights of the on-board devices, the cooling systems, the air management and the whole tank, must be considered, while considering the weight of the fuel cell. Then, to size the lithium batteries to side to the fuel cell, the maximum energy and power need to be evaluated. It is necessary to size in energy and power and after adopting the major value between the two (16, 17).

$$m_{Li\_p} = \frac{\Delta P_{max} \cdot 1000}{p \cdot \eta_{batt}} \tag{16}$$

$$m_{Li\_e} = \frac{\Delta E_{max} \cdot 1000}{u \cdot \eta_{batt}} \tag{17}$$

With m indicating the mass for the two different types of sizing. Specifically ΔP_{max} measured in kW is the difference between the providable power from tank and required power and ΔE_{max} in Wh represent the difference between the providable energy from tank and required energy. p represents power density [W/kg] and u energy density [Wh/kg]. Figure 3 reports the overall sizing process for hydrogen/ battery hybrid train.

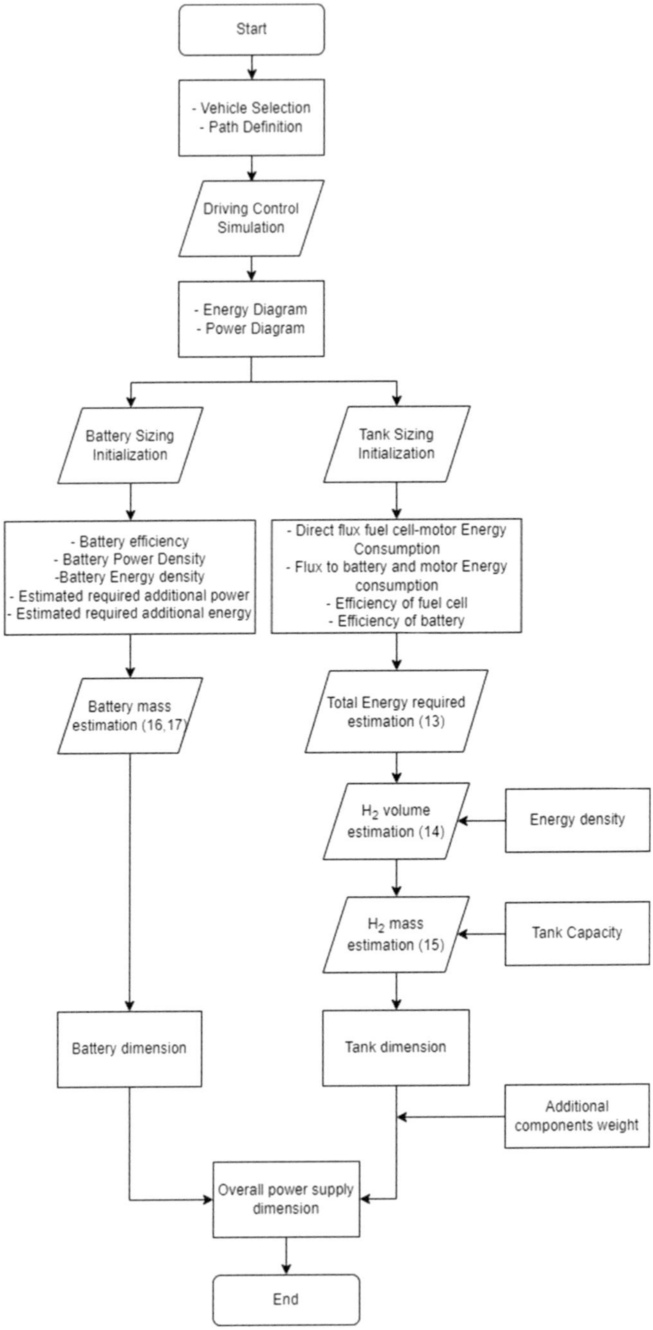

Fig. 3 Hydrogen train implementation process

Catenary-less Electrification Outcomes

After the analysis of the methodologies of catenary-less electrification, two relevant factors prevent the adoption of a new construction of electrified networks. The type of the lines is an impacting factor, secondary branches in regional network are non-interested by heavy traffic, meaning the electrification is not profitable. Typically, electrified lines have a considerable frequency, in fact usually capacious trains are employed. Thus, since relevant economic investments are needed to transform the paths, it would be safer saving efforts for higher density routes. Battery-traction trains through the preliminary sizing, under favorable assumptions, are solutions that could be effectively adopted, since they satisfy primary constraints. Starting from power and energy as requirements, a light train could be transformed to a battery-traction train with the state-of-art technology. In this sense, it was a confirmation that this is an enough-mature alternative for non-electrified tracks. On the other hand, longer pathways could exploit the features of hydrogen-trains. Again, from power and energy demand it was proved a preliminary and non-optimized sizing of a light train with fuel cell stack and battery pack. In this case, the adoption of equipment for hydrogen provision could result simpler and suitable for the line. In addition, other possibilities of optimization, also from a technological point of view, can be evaluated but this brief report can be intended as an initial backbone for further studies.

Operating Cost Estimation Process

Finally, it is worth highlighting the costs that would be incurred with each of the proposed solutions compared to the baseline scenario, which uses diesel for traction. Figure 4 proposes the model to be considered when assessing the costs of (a) diesel, (b) decarbonised with battery, (c) decarbonised with hydrogen.

As Fig. 4 shows, the processes for estimating operating costs are similar. The process shown in Fig. 4c is also similar to that shown in Fig. 4a and b, except that in the case of hydrogen trains, operating emissions costs are not considered. Instead, these costs need to be considered in the electrolysis hydrogen production phase, depending on the sources feeding the electrolyser.

5 Technical–economic and Sustainability Analysis

In order to evaluate possible investments and in general the uptake of such innovative technologies, Multi-Criteria Decision Analysis (MCDA) is considered. This method is a structured process for assessing options with conflicting criteria and choosing the best result. MCDA is analogous to a cost–benefit analysis but evaluates numerous criteria, apart from just cost. MCDA has operations in several fields, including business, government and everyday life. For illustration, we can use MCDA to decide

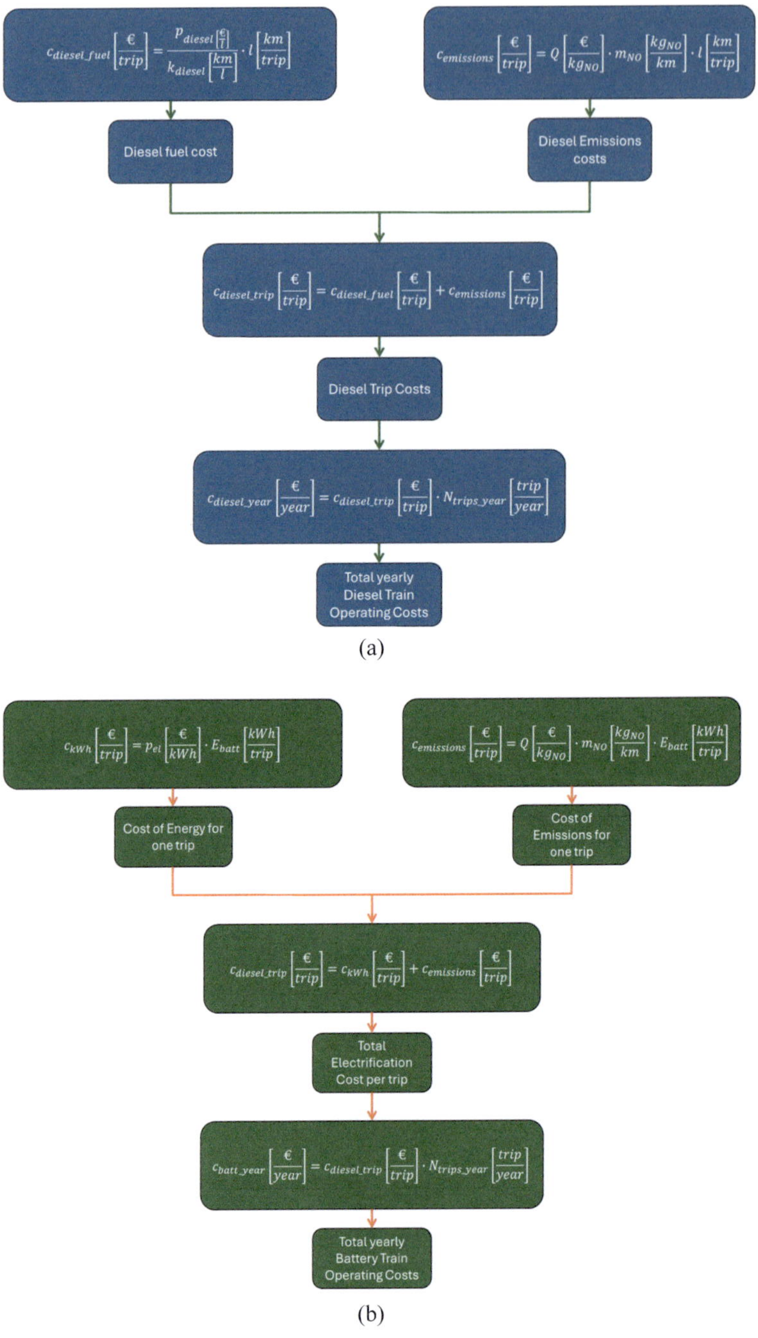

Fig. 4 Operating costs estimation process for: **a** diesel train, **b** battery train, **c** hydrogen train

Fig. 4 (continued)

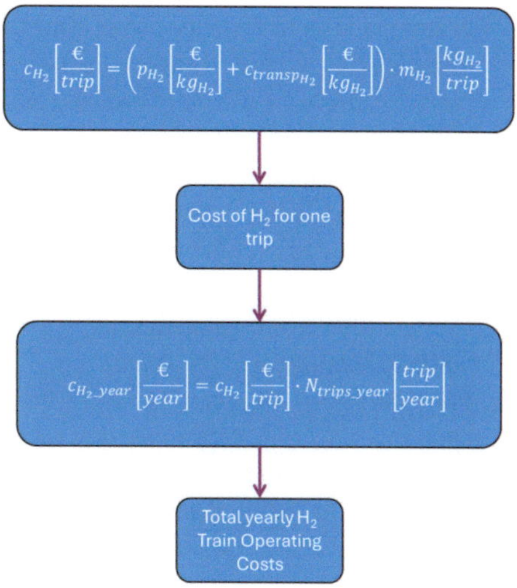

(c)

between vendors or to select equipment that meets all the institution/industry requirements. The steps, needed to perform a Multi-Criteria Decision Analysis effectively, consist in:

1. First, by defining the *objective*: To achieve an ideal mode of transport for the people with which considering the contextual factors for the analysis. In our case, it is (Railways) for Public Transportation.
2. Defining the *criteria*: By developing the criteria to represent the norms for different options. This relates to the measures we assume are important in determining the most valuable choice. We used the parameters such as price (or) cost for transport usage, time taken from origin to destination, distance covered, reliability of the transport and emission control.
3. *List of choices (or) alternatives*: This step is to find choices that meet our criteria to a certain degree. For illustration, we used the criteria to offer affordable and durable characteristics and a list of top three options namely, trains (electrified), buses and self-driven cars.
4. Determine the *performance values*: Each criterion is likely to have its own performance value or the measure to use for ranking it in comparison to other criteria. Price, time and distance use a numeric performance value. Further private criteria such as Reliability and Emission control use different measures, similar to 'high (or) excellent', 'above average,' 'average,' 'below average' and 'low (or) poor.'
5. *Rate the choices*: Rating our choices involves determining how each option compares to identified criteria. A criterion similar to price is a non-beneficial criterion, meaning that a lower value is preferable. For illustration, we probably prefer a mode of transport offering lower prices (in terms of cost only). In our standings, the lowest price would have the higher rank. In comparison,

criteria similar to distance cover, time taken, and reliability are beneficial criteria, meaning that advanced values are better and directly proportional to the rankings.
6. *Normalization of the performance values*: Normalization refers to the act of adapting the values so that they operate on a common scale. To do this, we can perform mathematical operations to convert the values. For non-beneficial criteria, divide the smallest value by the performance value. For beneficial criteria, you first convert your values using a conversion scale. The smallest rank would correspond to the numeric one and the highest would correspond to five. After converting the performance values, divide each by the maximum value to homogenize the scale.
7. *Substitution of Weights*: Multiply each by the weights assigned to the corresponding criterion, represented as a numerical value. The sum of the weightage must be equal to 100%. In our case, we assumed weights equally for all the parameters to obtain the Weighted Normalized Decision Matrix.
8. *Calculating the performance scores:* Eventually, after calculation with equal weights for each option, the performance scores (performance factor assumed: (0) Lowest, (1) Highest) are obtained and compared for the investigated alternative options. The option with the highest score would be with the most value, according to the criteria.

6 Conclusion

With the new European directives and the net-zero emissions target set for 2050, the electrification of transport has become an important research topic. In recent years, several studies have been published proposing decarbonisation methods for rail transport that rely on the use of diesel fuel. The most promising technologies are battery trains and hydrogen trains. This chapter analyses both technologies, highlighting their advantages, disadvantages and deployment situations. Moreover, it proposes a methodology for the electrification of railway lines using these technologies. Finally, methods for estimating operating costs were proposed, highlighting the dependence on fuel and energy costs. In conclusion, battery and hydrogen powered trains will be the optimal solution for railway lines that are difficult to electrify, due to their flexibility, energy efficiency and emission reduction, as the technologies progress.

Hybrid trains' market is heading towards a significant growth, guided by the ever-increasing attention to sustainability, technological progress, government incentives, expansion of applications and integration of intelligent systems. While the global railway industry keeps evolving, hybrid trains offer a valuable solution for a cleaner, more efficient and more versatile rail transport. These tendencies not only emphasize the potential of hybrid trains, but also highlight the innovative steps towards a greener and more connected future in rail transport.

References

Abiko H (2013) Development of catenary and storage battery hybrid train system. Jpn Railw Eng 179:6–9. Print

Ansa (2022) Ansa website: stop selling of diesel vehicles https://www.ansa.it/canale_motori/notizie/attualita/2022/06/08/stop-alla-vendita-di-auto-benzina-diesel-gpl-dal-2035-via-libera-dal-parlamento-europeo_32037239-8d4a-4a3c-98e9-0c933fb7e168.html. Accessed 28 Sep 2022

Arboleya P, Mayet C, Mohamed B, Aguado JA, de la Torre S (2020) A review of railway feeding infrastructures: mathematical models for planning and operation. eTransportation 5:100063. ISSN 2590-1168. https://doi.org/10.1016/j.etran.2020.100063

CHAMARET (2019) Sharing battery benchmark/experience/use cases to boost railway production (Replacement to closed diesel lines). SNCF Mobilités André-Philippe

Choi W, Song HH (2018) Well-to-wheel greenhouse gas emissions of battery electric vehicles in countries dependent on the import of fuels through maritime transportation: a South Korean case study. Appl Energy 230:135–147. ISSN 0306-2619. https://doi.org/10.1016/j.apenergy.2018.08.092

Colombo CG, Borghetti F, Foiadelli F, Longo M, Yaici W, Zaninelli D (2023) From diesel to electric bimodal train: case study in Italy for decarbonization of railway lines. In: 2023 13th international conference on power, energy and electrical engineering (CPEEE), Tokyo, Japan, pp 357–363. https://doi.org/10.1109/CPEEE56777.2023.10217645

Commission Regulation (2014) https://eur-lex.europa.eu/eli/reg/2014/452/oj/eng

da Fonseca-Soares D, Eliziário SA, Galvincio JD, Ramos-Ridao AF (2024) Greenhouse gas emissions in railways: systematic review of research progress. Buildings 14:539. https://doi.org/10.3390/buildings14020539

Ding D, Wu XY (2024) Hydrogen fuel cell electric trains: technologies, current status, and future. Appl Energy Combust Sci 17:100255. ISSN 2666-352X. https://doi.org/10.1016/j.jaecs.2024.100255

Directorate-General for Research and Innovation Smart (2021) Electrification of the transport system expert group report https://ec.europa.eu/programmes/horizon2020/en/news/electrification-transport-system-expert-group-report

Dschung Dr.-IF, Ludolf Dipl-IM (2021) 50 Hz train charging station for battery electric trains (BEMU)

European Commission (2022) An European green deal. https://www.ec.europa.eu/info/strategy/priorities-2019-2024/european-green-deal_en. Accessed 09 Nov 2022

European Parliament (2022) 2021: the European year of rail. https://www.europarl.europa.eu/news/it/headlines/eu-affairs/20210107STO95106/2021-l-anno-europeo-delle-ferrovie. Accessed 28 Sep 2022

Fedele E, Iannuzzi D, del Pizzo A (2021) Onboard energy storage in rail transport: review of real applications and techno-economic assessments. IET Electr Syst Transp 11(4):279–309. https://doi.org/10.1049/els2.12026

García-Olivares A, Solé J, Samsó R, Ballabrera-Poy J (2020) Sustainable European transport system in a 100% renewable economy. Sustainability 12:5091. https://doi.org/10.3390/su12125091

Ghaviha N, Campillo J, Bohlin M, Dahlquist E (2017) Review of application of energy storage devices in railway transportation. Energy Procedia 105:4561–4568. https://doi.org/10.1016/j.egypro.2017.03.980

González-Gil A, Palacin R, Batty P, Powell JP (2014) A systems approach to reduce urban rail energy consumption. Energy Convers Manag 80:509–524. https://doi.org/10.1016/j.enconman.2014.01.060

Heckele L, Tesar M, Igelspacher J, Brunner J, Gratzfeld P (2022) Steigerung der Leistungsdichte und des Wirkungsgrades von Straßenbahnantrieben durch den Einsatz hochdrehender Maschinen. e & i Elektrotechnik und Informationstechnik 139(2):186–194. https://doi.org/10.1007/s00502-022-01011-6

IEA (2020) https://www.iea.org/reports/energy-technology-perspectives-2020

IEA (2023) Tracking clean energy progress 2023, IEA, Paris https://www.iea.org/reports/tracking-clean-energy-progress-2023. Licence CC BY 4.0

IEA (2023) Global CO_2 emissions from transport by sub-sector in the net zero scenario 2000–2030 IEA, Paris. https://www.iea.org/data-and-statistics/charts/global-co2-emissions-from-transport-by-sub-sector-in-the-net-zero-scenario-2000-2030-2. Licence CC BY 4.0

IEEE Industrial Electronics Society (2019) Conference (45th: 2019: Lisbon, Universidade Nova de Lisboa, Institute of Electrical and Electronics Engineers, and IEEE Industrial Electronics Society, Proceedings, IECON 2019–45th Annual Conference of the IEEE)

International Energy Agency. The future of rail opportunities for energy and the environment in collaboration with (Online). Available www.iea.org/t&c/

Iskandar NA, Maheri A (2022) Techno-economic assessment of hydrogen refuelling station: case study of hydrogen train. In: 2022 7th international conference on environment friendly energies and applications (EFEA). Bagatelle Moka MU, Mauritius, pp 1–6. https://doi.org/10.1109/EFEA56675.2022.10063828

Kapetanović M, Núñez A, van Oort N, Goverde RMP (2024) Energy use and greenhouse gas emissions of traction alternatives for regional railways. Energy Convers Manag 303:118202. ISSN 0196-8904. https://doi.org/10.1016/j.enconman.2024.118202

Kilsby P, Remenyte-Prescott R, Andrews J (2017) A modelling approach for railway overhead line equipment asset management. Reliab Eng Syst Saf, Volume 168:326–337. ISSN 0951-8320. https://doi.org/10.1016/j.ress.2017.02.012

Klebsch W, Heininger P, Geder J, Hauser A (2018) Battery systems for multiple units: emission-free drives powered by lithium-ion cells. Frankfurt am Main

Klebsch W, Heininger P, Martin J (2019) Alternatives to diesel multiple units in regional passenger rail transport: assessment of systemic potential. Frankfurt am Main

Klebsch W, Guckes N, Heininger P, eV V Evaluation of climate-neutral alternatives to diesel multiple units: economic viability assessment based on the example of the ›Düren network‹ VDE study (Online). Available www.vde.com

Kwakwa PA, Adjei-Mantey K, Adusah-Poku F (2023) The effect of transport services and ICTs on carbon dioxide emissions in South Africa. Environ Sci Pollut Res 30:10457–10468

NASA (2022) https://www.nasa.gov/wp-content/uploads/2018/01/2022_nasa_strategic_plan_0.pdf

NASA Climate. Emission-free battery trains for Norway. https://climate.nasa.gov/resources/global-warming-vs-climate-change/ (Online). Available

Nqodi A, Mosetlhe TC, Yusuff AA (2023) Advances in hydrogen-powered trains: a brief report. Energies 16:6715. https://doi.org/10.3390/en16186715

Ogunkunbi GA, Meszaros F (2023) Preferences for policy measures to regulate urban vehicle access for climate change mitigation. Environ Sci Eur 35:42

Praticò FG, Fedele R (2023) Economic sustainability of high-speed and high-capacity railways. Sustainability 15:725. https://doi.org/10.3390/su15010725

Pugi L, di Carlo L, Kociu A, Berzi L, Delogu M (2024) A tool for design and simulation of battery operated trains. In: F. Bellotti et al. Applications in electronics pervading industry, environment and society. ApplePies 2023. Lecture notes in Electrical Engineering, vol. 1110. Springer, Cham. https://doi.org/10.1007/978-3-031-48121-5_62

Quantum Fuel Systems Technology Worldwide, Inc. High-pressure hydrogen storage systems. In: Hydrogen and fuel cell summit VIII.

Rail—Analysis—IEA https://www.iea.org/reports/rail. Accessed 09 Nov 2022

Reimann S, Gratzfeld P (2023) New light rail vehicle and drivetrain concepts for catenary free operation of branch lines. Transp Res Procedia 72:64-71. https://doi.org/10.1016/j.trpro.2023.11.323

RFI. RFI website: information about RFI and its staff. https://www.rfi.it/it/chi-siamo/le-nostre-persone-.html. Accessed 19 Sep 2022

Sasse JP, Trutnevyte E (2023) Cost-effective options and regional interdependencies of reaching a low-carbon European electricity system in 2035. Energy 282:128774

Streuling C, Pagenkopf J, Schenker M, Lakeit K (2021) Techno-economic assessment of battery electric trains and recharging infrastructure alternatives integrating adjacent renewable energy sources. Sustainability (Switzerland) 13(15)

Thorne R, Amundsen AH, Sundvor I (2019) With Institute of Transport Economics Oslo. Battery electric and fuel cell trains: maturity of technology and market status. TOI 2019

Union (2014) Commission Regulation (EU) No 1301/2014 of 18 Nov 2014 on the technical specifications for interoperability relating to the 'Energy' subsystem of the rail system

Battery Systems for Air Transport Climate Neutrality

Michele De Gennaro and Helmut Kuehnelt

Abstract The aviation industry faces significant challenges in achieving climate neutrality by 2050, requiring a transition to advanced propulsion technologies and energy storage systems. This chapter examines the role of battery technologies in enabling electrified aircraft, assessing the current state-of-the-art battery systems, certification standards, and key performance indicators. It explores full-electric, hybrid-electric, and more-electric aircraft architectures, evaluating their feasibility across market segments and projecting battery performance requirements, the airborne battery market, and emission reductions through electrification by 2050. Case studies from ongoing European projects, such as HERA and HECATE, alongside industrial programs like Heart Aerospace's ES-30 and ATR-EVO, are also discussed. Additionally, a structural battery concept is presented as a disruptive innovation for multifunctional energy storage, highlighting its potential to achieve higher energy densities compared to conventional integration concepts. The findings indicate that while battery technology advancements are critical for decarbonizing aviation, hybrid-electric and more-electric architectures will contribute approximately 3.5% to total emissions reductions. Though modest, this impact represents about 10% of the emissions reductions expected from technological developments by 2050. Future progress in safe, high-energy–density batteries and certification frameworks will determine the pace of electrification in commercial aviation, shaping the industry's long-term sustainability.

Keywords Airborne batteries · Hybrid-electric aircraft · More-electric aircraft · Structural batteries · Battery certification · Aircraft decarbonization · Electric propulsion · Climate-neutral aviation

M. De Gennaro (✉) · H. Kuehnelt
AIT Austrian Institute of Technology GmbH - Center for Transport Technologies, Vienna, Austria
e-mail: michele.degennaro@ait.ac.at; michele.degennaro.phd@gmail.com

H. Kuehnelt
e-mail: helmut.kuehnelt@ait.ac.at

R. Scipioni et al. (eds.), *Hybrid Energy Storage*, Lecture Notes in Energy 47,
https://doi.org/10.1007/978-3-031-97755-8_13

Abbreviations

ATM	Air Traffic Management
CF (RP)	Carbon Fiber (Reinforced Polymer)
CO_2	Carbon Dioxide
CORSIA	Carbon Offsetting and Reduction Scheme for International Aviation
CS	Certification Specification
CSMU	Cell and Structural Monitoring Unit
DEP	Distributed Electric Propulsion
DoD	Depth of Discharge
DoH	Degree of Hybridization
EASA	European Union Aviation Safety Agency
ECS	Environmental Control System
EiS	Entry into Service
ETS	Emissions Trading System
EU	European Union
FAA	Federal Aviation Administration
GED	Gravimetric Energy Density
HER	Hybrid Electric Regional
HPA	Hydrogen-Powered Aircraft
IPS	Ice Protection Systems
KPI	Key Performance Indicator
MEMS	Micro-Electro-Mechanical Systems
MRV	Monitoring, Reporting, and Verification
MOC	Means of Compliance
MTOW	Maximum Take-Off Weight
NMC	Nickel Manganese Cobalt
NOx	Nitrogen Oxides
RPK	Revenue Passenger Kilometers
SAF	Sustainable Aviation Fuel
SC-VTOL	Special Condition Vertical Take-Off and Landing
SB	Structural Battery
SHM	Structural Health Monitoring
SMR	Short-and-Medium Range
SOH	State of Health
SOC	State of Charge
TRL	Technology Readiness Level
VED	Volumetric Energy Density

1 Introduction to the Challenges of Air Transport Climate Neutrality

The aviation industry plays a crucial role in the European Union, contributing significantly to its economic growth, societal development, and cohesion. It supports 13.5 million jobs in Europe and generates one trillion € in economic activity, accounting for 3.6% of employment and 4.4% of the EU's GDP, (ATAG 2021). As with other transportation sectors in the European economy, the aviation industry is undergoing a technological revolution aimed at delivering aircraft technologies that offer improved environmental performance and lower fuel consumption, thus reducing operational costs and increasing profitability for airlines.

Focusing on the environmental aspect, particularly emissions from air transport at the global level, the technological advancements in the latest generation of aircraft are being outpaced by the increasing demand for air travel. Air transport continues to grow at a steady rate of 5–7% per year, depending on the region, with more than 4.6 billion passengers globally and over 1 billion passengers on scheduled flights to, from, and within the EU (based on pre-pandemic data, corroborated by consolidated data from 2023). This market is expected to double in size within the next three decades, potentially surpassing 20 trillion Revenue Passenger Kilometers (RPKs) by 2050. This growth will place enormous pressure on the industry's ability to achieve carbon neutrality by 2050, as outlined in the European Green Deal.

From a technological standpoint, moving away from energy-dense fossil fuels is a formidable challenge due to the significant power and energy requirements of aircraft. There is no "silver bullet" solution to this issue. Strategically, the EU has identified four synergistic pillars to address this challenge, (EASA 2022):

- Improving aircraft technologies by introducing more-electric, hybrid-electric, and hydrogen-based propulsion systems, with a target of reducing carbon emissions by 38% by 2050.
- Adopting (sustainable) Synthetic Aviation Fuels (SAF), targeting a 46% reduction in carbon emissions by 2050.
- Implementing economic measures, such as the Carbon Offsetting and Reduction Scheme for International Aviation (CORSIA), as an interim measure if the two targets above prove difficult to achieve, aiming for a 10% carbon reduction by 2050.
- Deploying improved operations and Air Traffic Management (ATM), with a goal of cutting carbon emissions by 6% by 2050.

The successful implementation of these measures could enable the aviation industry to achieve carbon neutrality by 2050. However, this will not directly translate into climate neutrality, as air travel also produces non-CO_2 emissions. These emissions include nitrogen oxides (NOx), aerosol particles (soot and sulfur-based), and water vapor, which contribute to radiative effects, cloud formation, and contrail cirrus formation. The impact of these non-CO_2 emissions on climate change remains largely undetermined. While the scientific community agrees that they have a detrimental effect on climate change, their impact relative to CO_2 is poorly understood, with opinions ranging from negligible to as significant as CO_2 emissions, (Lee et al. 2023). Despite these uncertainties, there is consensus that further research is necessary to better understand the effects of non-CO_2 emissions, and caution is advised in addressing them until they are fully comprehended. This caution also applies to quantifying the climate impact of alternative fuels, such as SAF and hydrogen, which may exacerbate non-CO_2 effects while reducing net CO_2 emissions. Therefore, reliable impact assessments are required within the framework of the Monitoring, Reporting, and Verification (MRV) of the EU Emissions Trading System (ETS), (Transport and Enviroment 2024).

In this complex political, scientific, and technological landscape, the European Union has launched the Clean Aviation Joint Undertaking in 2021, with the mission to execute the reference European funding programme aimed at developing the portfolio of new aircraft technologies needed to achieve the objective of climate neutrality in air travel by 2050. With a budget of approximately €1.7 billion in public funding and €2.1 billion in private investment, the Joint Undertaking focuses on maturing technologies capable to achieve a reduction in greenhouse gas emissions at the aircraft level of at least 30% compared to the 2020 baseline. This effort addresses the first pillar mentioned earlier, i.e. improvement of aircraft technologies.

The Clean Aviation program is structured around three main thrusts: the Hybrid Electric Regional (HER), the Short-and-Medium Range (SMR), and the Hydrogen-Powered Aircraft (HPA). The 2021 Strategic Research and Innovation Agenda (SRIA) (Clean Aviation 2021), along with its ongoing 2024 revision launched in September 2024 (Clean Aviation 2024), focuses on evolving technologies for the regional and short-to-medium range aircraft segments and on disruptive technologies in the hydrogen field. The HER and SMR segments are respectively responsible for 5% and 50% of air travel and associated emissions. Clean Aviation aims to reduce fuel consumption of these aircraft types by respectively 50% and 30% against the 2020 baseline. This reduction, combined with the deployment of SAF with an 80% lower carbon footprint, could lead to net CO_2 reductions of 90% and 86% by 2035 (the projected Entry-into-Service year), deploying its potential at fleet level by 2050. The HPA thrust involves, instead, the development and test-bench demonstration of hydrogen aircraft technology components, such as tanks, fuel cells, combustors, and onboard hydrogen distribution systems, targeting the commuter and short-range segments as possible early adopters for fuel cell and hydrogen combustion systems, should these technologies prove successful. Additionally, technology spillovers from the HER, SMR, and HPA thrusts are expected to impact long-range aircraft, which, while not explicitly addressed by the SRIA or Clean Aviation program, are responsible for 45% of global air travel and associated emissions.

Given these complexities, achieving carbon neutrality in aviation requires the development of advanced technologies and supporting infrastructures, and the achievement of this goal is all but trivial. Nevertheless, the technological path outlined above suggests that significant improvements over the current generation of aircraft are feasible, and that investing in these technologies is crucial for the European aeronautics industry to maintain and strengthen its dominant position in the civil aircraft market.

In this context, highly energy-dense batteries are recognized as a crucial element in various segments of aerial vehicles, with an impact on safety critical systems related to full electric, hybrid electric and more electric aircraft architectures. In this chapter, we provide an overview of the current state-of-the-art in the field of airborne battery technologies, focusing on an end-user perspective.

We begin by examining the airworthiness certification landscape as of 2024 and presenting the key performance indicators for state-of-the-art airborne battery systems. On this basis we provide a description of full electric, hybrid electric and more electric aircraft architectures, and of their high-level matching with different aircraft segments, deriving minimum and desirable key performance indicators for battery systems for Entry-into-Service by 2035, coherently with the time horizon given in the SRIA 2024, (Clean Aviation 2024). Additionally, we size the airborne battery market and its potential impact in 2050 based on fleet penetration forecasts. Finally, we present a concise overview of the on-going research and development efforts in the area of hybrid electric aircraft. Specifically, we present two research case studies based on the results of the research projects IMOTHEP and of the Clean Aviation projects HERA and HECATE, complemented by two additional on-going case studies from two aircraft development programmes, i.e. the Heart Aerospace ES-30 and the ATR-EVO. We conclude the chapter by presenting the results of the research projects SOLIFLY and MATISSE, aimed at developing multifunctional batteries with load-bearing capabilities to achieve "mass-less" energy storage within aircraft structures. Such technology, despite its relatively low maturity, has the potential of achieving significantly higher energy densities, unlocking new use cases beyond the case studies presented above.

2 State-of-the-Art for Airborne Battery Systems: Airworthiness Certification Specifications

Uncompromised safety is at the very heart of the aviation business, and airworthiness certification is the cornerstone to ensure it. Electrified propulsion and safety–critical battery energy storage systems represent a novel element in the aircraft industry, calling for new regulations and compliance methodologies. Concerning the batteries, they are expected, in future, to store a substantially higher amount of electrical energy at higher voltage levels than batteries utilized in aircraft so far.

The reference EASA regulatory framework on this subject has been under development since 2019, and it covers safety of propulsion batteries for *VTOL-capable aircraft*, i.e. SC-VTOL (EASA), and its Means of Compliance (MOCs). SC-VTOL stands for "*special conditions vertical take-off and landing*" and it comprises aircraft capable of carrying up to 9 passengers with a *maximum take-off weight* (MTOW) of

5,700 kg, operating over congested areas or in commercial air transport. The MOCs have published in four consecutive parts (MOC-1, Issue 2, May 2021; MOC-2, Issue 3, Dec. 2022; MOC-3, Issue 2, June 2023 and MOC-4, Issue 1, Dec. 2023) and will be kept up to date with latest technological improvements and insights. Moreover they have been identified by EASA as the baseline for the upcoming battery certification specifications for large passenger aircraft (i.e. CS-25, as stated in the "Electric/ Hybrid Propulsion System (EHPS)—Progress and Roadmap to means of compliance definition" event held by EASA in Dec. 2023).

In brief, the MOC-3 SC-VTOL outlines safety, performance, and certification standards for these aircraft, addressing both manned and unmanned platforms, while focusing on the aspects of safety objectives and system design, electric and hybrid propulsion, operational safety and performance, human factors and automation and energy management and battery systems. On the latter, EASA only considers commercially available battery technologies and chemistries, which, at present, show rather low safety levels. Battery fire due to cell thermal runaway is considered the most safety–critical ("*catastrophic*") event, which needs to be taken into account due to failure modes intrinsic to the cells that are inherently uncontrollable (e.g. issues of cell manufacturing defects, material impurities and usage-accumulated effects as dendrite growth). In the SC-VTOL, the battery needs to show compliance with respect to:

(a) Fire protection of the energy storage and its installation in the aircraft, specifically its crash resistance with mechanical load factors depending on the installation location (VTOL. 2325 (a) (4), MOC-1), in designated fire zones (VTOL. 2330, MOC-2, yet incomplete for hybrid-electric systems), and of the lift/thrust system installation (VTOL. 2440, MOC-3).
(b) The impact of various battery-related factors, e.g., ageing, thermal (management) limits or variation of accessible power on the aircraft performance (VTOL. 2105, MOC-2).
(c) Provision of reliable information of the accessible energy and power to the pilot (VTOL. 2430 (a) (3) and (a) (4), MOC-2).

Therefore, in addition to the general aircraft component design requirements, EASA requires the deployment of a multi-layer safety approach (cell-to-battery-to-aircraft) which accounts for: (i) *safety at cell level*, a full characterisation to identify the worst-case combinations of conditions for cell thermal runaway and degradation, e.g., due to ageing and environmental conditions during operation, also to be assessed on used/degraded batteries; (ii) *safety at battery level*, a design that inhibits thermal runaway propagation of a single cell to another and contains within the battery or module the effects of simultaneous thermal runaway of multiple cells (i.e. currently set by EASA to 20% of the cells in the battery or module) for the time needed for Continued Safe Flight and Landing; (iii) *safety at installation level*, crashworthiness, and the provision of reliable information of the battery operation and SoX (state of charge, energy, power, health, safety, etc.) to the pilot.

The design requirements at battery level (i.e. non-propagation and containment) have highest impact in the design choices for airborne batteries and achievable system performance, and they are considered by the aeronautic industry as particularly challenging. On the other hand, at present, there is a lack of long-term service experience on such systems compared to other energy storage and conversion systems, such as tanks and aero-engines, justifying in part the imposition of such very challeging safety requirements. This is also aknowledged by the MOC-3 SC-VTOL, which does recognise that these requirements have substantial impact on weight penalty, hence open up the possibility of a "*modularized [propulsion] battery system composed out of battery modules to comply [with the certification requirements] at battery module level, instead of at battery system level*". The module itself is defined as "*a group of electrically interconnected cells in series and/or parallel arrangement contained in a single enclosure that ensures that no fluids, flames, gasses, smoke, or fragments enter other modules, and that no thermal runaway is propagated from one module to the others during normal operation or failure conditions*". A battery system is an assembly of electrically interconnected battery modules (modularized battery) or cells in series and/or parallel, plus any protective, monitoring, alerting circuitry or hardware inside or outside of the battery, its packaging, and the designed venting provisions.

Therefore, at present, the most followed approach consists in satisfying the safety provisions described under "*(ii) safety at battery level*" at module level. This approach is expected to be extended from SC-VTOL to large aircraft, with the possibility of re-considering the criticality of specific failure modes, should future battery technologies demonstrate better performance under "*(i) safety at cell level*" or, in other words, demostrate significantly less probability to develop thermal runaway and degradation effects in operating conditions.

3 State-of-the-Art for Airborne Battery Systems: Key Performance Indicators

Table 1 provides an overview of the current state-of-the-art airborne battery systems, with a focus on products that are either already available on or close to the market. At present, propulsive batteries have been developed and certified for fully electric light sport aircraft only (specifically those within CS-LSA, a subcategory of general aviation CS-23, such as the Pipistrel Velis Electro), while batteries for eVTOL (electric Vertical Take-Off and Landing) vehicles are emerging. However, the level of publicly available information on existing CS-23 and eVTOL systems remains limited. Additionally, tear-down data for these systems is scarce, as few systems currently exist, and most are relatively new and produced in small series. In this analysis the authors have retrieved data for 6 systems, from 2020 onward: the Pipistrel Velis Electro (Textron Pipistrel), the EPS Epic (Electrical Power Systems), the H55 (H55), the Evolito (Evolito Ltd), the Joby S4 eVTOL (Joby Aviation), and the Cuberg (Cuberg). The data presented have been collected and compiled from various sources, including data sheets, test reports, company websites, and flight manuals. These consolidated data are presented in Table 1 using regular font. Where consolidated data are not available, estimates have been reconstructed from alternative sources, based on relevant conditions and performance metrics. These estimated data are also reported in Table 1 using *italic font*. Complementarily, the following observations are made:

Battery Cells: Commercial cylindrical cells (18,650 or 21,700 format, with a capacity ranging from 3 to 5 Ah) are, at present, most commonly used in the aeronautic industry (see Pipistrel, EPS, H55, Evolito). However, this format requires a high number of cells per module or pack, resulting in low packaging efficiency. Recent designs (e.g., Joby, Cuberg) are adopting larger pouch cells with capacities ranging from 20 to 50 Ah to achieve higher energy density and improve the packaging efficiency. The most common cell chemistries in use are belonging to the family of NMC|liquid electrolyte|graphite, providing gravimetric energy densities between 230 and 290 Wh/kg at the cell level. In one case, Cuberg uses their Li-metal anode technology, aiming at surpassing 300 Wh/kg energy density.

Charge/Discharge: Energy densities (gravimetric and volumetric) at the cell and module/pack level are typically measured at very low discharge rates (e.g., C/10 or C/20) to report high KPI figures. However, these rates are not representative of the typical airborne duty cycle, such as that for an eVTOL flight mission. In real world conditions, systems fully and hybrid electric fixed wing aircraft must operate within charge/discharge rates from 1 to 3C (with peaks exceeding 5C) while eVTOL and large aircraft that use batteries to boost at peak power demands will require much higher discharge rates (up to 10–12C for several minutes). While these rates meet aircraft mission requirements, they are expected to diminish the overall energy density performance of the system. Consolidated data on the correlation between energy density, lifetime, and C-rate are, at present, unavailable.

Battery System Thermal Operation and Conditioning: State-of-the-art battery modules and packs are designed to operate within a temperature window ranging from − 20 °C to 0 °C up to + 60 °C during discharge, and from 0 to 10 °C up to + 45–60 °C during charging. At the extremes of this range, the performance is likely reduced, and manufacturers typically recommend a narrower operating range to optimize performance. For instance, Pipistrel limits engine start temperatures to between 0 and 45 °C, recommending temperatures below 40 °C for high-power maneuvers such as prolonged maximum-rate climbs. Active liquid cooling systems are implemented as deemed necessary by regulators. However, gravimetric data on module-level performance often refers to dry weight (i.e., without coolant), and details of the integration penalty are generally not disclosed. Similarly, data on battery cell/module capacity degradation as a function of thermal constraints is not available, though it is likely that battery aging is influenced by factors such as power profile, state-of-charge (SOC) margins, and the thermal and mechanical operating envelope (temperature, vibration, and mechanical shock).

Safety Provisions: As outlined in Sect. 2, the airworthiness certification specifications for battery systems are evolving, with the SC-VTOL undergoing multiple revisions and updates over the past five years. The Pipistrel and EPS batteries have been

Table 1 SotA in 2024 of aeronautic batteries for propulsive and safety–critical applications (CS-23/eVTOL)

Manufacturer/aircraft (if any)			Pipistrel velis electro	EPS EPiC (e.g. diamond eDA40)		H55 (e.g. bristell B23)	Evolito	JOBY S4 eVTOL	Cuberg (prototype)
				1.0 Energy	2.0 Energy				
Year			2020	2022	2025 (Expected)	2022	2023/2024	2023	2024
Level			Pack	Module	Module	Sub-module	Module	Module	Module
Aircraft category			CS-23 LSA	CS-23 Level 1–2	CS-23	CS-23	eVTOL	eVTOL	n.a
Certification level			Aircraft, EASA LSA	TSO, FAA, exp. 2024	n.a	EASA & FAA TC pursued	n.a	n.a	n.a
Cell	GED [Wh/kg]		236	270	n.a	230–260	240/260	288	395^{*} [330^{**}]
	VED [Wh/l]		654	760	n.a	650–700	640/700	n.a	797.2^{*} [666^{**}]
Module	Configuration		96s12p	10s18p	12sXp	9 s	168s2p	30 cells	10s6p
	Energy [kWh]		$12.4^{\#}$	2.27	3.19	~ 0.15	5.44/6.00	~ 6	4.7^{*} [3.93^{**}]
	Max voltage [V]		394	44.5	50.4	37.8	706	low	42.8
	Max. cont. (peak) CH | DCH rate		1C | 2C (up to 3C)	3C | 2.5C (up to5 C)	3C | 2.5C (up to 5C)	n.a	2C | (up to 10C)	n.a	0.2C | (up to 5.4C)
	GED [Wh/kg]		$172^{\#}$	200	265	200	165/181	235	284.8^{*} [239^{**}]
	VED [Wh/l]		$230^{\#}$	250	350	n.a	230/250	?	320^{*} [274^{**}]
	Packaging efficiency	gravimetric	$73\%^{\#}$	74%	n.a	n.a	69%	81%	72%
		volumetric	$35\%^{\#}$	33%	n.a	n.a	36%	n.a	40%

(continued)

Table 1 (continued)

Manufacturer/aircraft (if any)		Pipistrel velis electro	EPS EPiC (e.g. diamond eDA40)		H55 (e.g. bristell B23)	Evolito	JOBY S4 eVTOL	Cuberg (prototype)
			1.0 Energy	2.0 Energy				
	Cycle life (SOH, typical DoD)	600, 2,000 Planned (80% SOH, 70% DoD)	2,000 (Unconfirmed) (80% SOH)		> 1,500 (Unconfirmed)	n.a	n.a	692 (90% SOH, 55% DoD)
	Cooling	liquid	air or liquid	liquid	air or liquid	liquid	liquid	passive

Legend *GED* (Gravimetric Energy Density); *VED* (Volumetric Energy Density); CH ∣ *DCH* (Charge ∣ Discharge); *SOH* (State-of-Health); *DoD* (Depth-of-Discharge); Packaging Efficiency (energy density at module level/energy density at cell level); regular font = consolidated data (i.e. data sheet, test report, company website, flight manual); *italic* = reconstructed data; n.a. = not available; # = values at mid of life; * = at C/20, 45 °C; ** = at 1C certified according to the RTCA DO311a standard, demonstrating thermal runaway containment at the full battery pack level (as opposed to the 20% of cells required under MOC-3 SC-VTOL). Most systems utilize conventional thermal insulation materials to meet the thermal runaway non-propagation and containment requirements, adopting cell technologies that do not inherently prevent its onset. This significantly impacts module design, ultimately limiting packaging efficiency and, as a result, the energy densities achievable at the system level.

Based on the data presented, the state-of-the-art for airborne battery systems can be summarized as follows:

- Only one fully electric aircraft has been type-certified yet under the light sport aircraft (LSA) category, with non-modularized battery packs. No battery has been certified under the new EASA SC-VTOL regulation. Meeting its safety requirements at the module level will have a clear impact on design.
- A shift from cylindrical to pouch cells to improve energy density and packaging efficiency at the system level is envisaged.
- Current cell technologies deliver gravimetric energy densities at the cell level ranging from 230 to 290 Wh/kg, resulting in values between 170 and 235 Wh/kg at the module level, with a gravimetric packaging efficiency of 73–81%. Volumetric energy density at the module level ranges from 230 to 350 Wh/liter (33–36% packaging efficiency). New cell technologies are expected to push gravimetric energy density above the 300 Wh/kg mark in the short-term future.
- Charge/discharge C-rates for most systems fall within the 1–3C range, balancing the need to meet energy and power duty cycles. Peak rates exceeding 5C are possible.
- The cyclability of the battery system is claimed to be in the range of 2,000 cycles at 55–70% depth of discharge, consistent with cell performance, allowing approximately 12 months of operation at 6 flight cycles per day in intensive aircraft utilization. However, the mean time between overhaul for the battery of the first type-certified electric light sport aircraft is currently limited by the regulator to 600 cycles.

Active liquid cooling is currently being considered, and in most cases will remain necessary for proper thermal conditioning of the battery system to deliver the expected performance and durability.

4 Battery Systems within Next Generation Aircraft: Aircraft Configurations, Performance Indicators and Market Projections by 2050

The state-of-the-art depicted in Sects. 2 and 3 highlight a scenario where airborne batteries are at the very beginning of their development cycle, with both airworthiness certification specifications and technologies that are rapidly evolving. The aim of this section is to project the expected development of battery systems technologies over the next decades, framing their implementation across different categories of aerial vehicles. This is done with the aim of quantyfing what is realistically achievable by aircraft electrification by 2050, and which demand of airborne batteries is expected in the next decades. However, before doing so, three main concepts for aircraft electrification need to be introduced, i.e. full electric, hybrid electric and more electric aircraft. They can be defined as follows.

A full electric aircraft is defined as an aircraft that operates solely on electrical energy, without reliance on combustion engines, whether powered by fossil or sustainable fuels. These aircraft typically use one or more electric motors to drive propellers or fans, drawing energy from an onboard battery or fuel cell system. This architecture results in zero emissions during flight, making it the most favorable option for minimizing both CO_2 and non-CO_2 effects associated with air travel. The viability of full electric aircraft depends on the energy storage and conversion system's ability to simultaneously meet the energy and power requirements of the aircraft. This includes considerations of gravimetric (weight-related) and volumetric (space-related) energy and power density, as well as the overall systems' capacity.

A hybrid electric aircraft, by contrast, combines conventional fuel-powered engines with electric propulsion systems, offering a balance between the efficiency of electric power and the higher energy density of traditional fuels. This configuration allows the aircraft to switch between, or simultaneously use, both power sources, optimizing energy consumption according to the specific requirements of the different flight phases. There are two main configurations of hybrid powertrains:

i. Parallel hybrid: in this configuration, both the combustion engine and the electric motor are connected to the propeller or fan driving shaft. The electric motor functions as a booster, supplementing power, and torque to the combustion engine during power-intensive phases like takeoff, climb, and go-around. During cruising, the combustion engine operates alone for maximum thermal efficiency. While this setup is more flexible, it requires a mechanical coupling between the shafts of the combustion and electric motors, introducing a non-negligible complexity in the gearbox assembly.
ii. Series hybrid: here, the combustion engine generates electricity to power the electric motor, which is the only one mechanically connected to the propeller or fan driving shaft. This configuration eliminates the need for mechanical coupling but limits total power output to the specifications of the electric motor, with the combustion engine functioning as a range extender.

In both configurations, the key performance metric is the Degree of Hybridization (DoH), defined as the ratio between the maximum continuous electric power and the total maximum power of the propulsion system at the shaft.

A more electric aircraft focuses on the electrification of non-propulsive systems while still relying on conventional jet engine technology for propulsion. This approach involves converting all subsystems to electric, including the environmental control systems (ECS, cabin temperature and pressure control), the ice protection systems (IPS, eliminating the need to bleed hot air from the engine compressor, enabling a "bleedless" engine design), and the actuation systems (replacing pneumatic and hydraulic systems for controlling flaps, slats, ailerons, spoilers, rudder, trim, and landing gear with electro-actuators). Although these subsystems typically account for only 1–5% of the total block energy required for the flight (depending on aircraft size and mission profile), their electrification offers several advantages. These include weight reduction, improved maintainability (with maintenance being the second-largest cost driver for airlines after fuel), and increased aircraft reliability. Given the safety–critical nature of these subsystems, i.e. malfunctions can lead to catastrophic failures and consequent loss of the aircraft, their certification is subject to the same safety standards as propulsive systems.

Aerial vehicles are classified according to their certification specifications category. Different categories are established by regulatory authorities based on their intended use, design, size, and technical capabilities. These categories guide the certification process, defining the specific requirements each aircraft must meet to be deemed airworthy. The primary regulatory bodies overseeing aircraft certification include the Federal Aviation Administration (FAA) in the United States and the European Union Aviation Safety Agency (EASA) in Europe. Focusing on fixed wing aircraft, two major categories can be identified:

Part-23. Aircraft certified under Part-23 (FAA) and CS-23 (EASA) include small to medium-sized aircraft designed for general aviation, personal use, instruction, and short-haul commercial flights. Part-23 regulations apply to airplanes with a maximum takeoff weight of up to 8,618 kg (19,000 lbs) and seating capacities ranging up to 19 passengers (pax), depending on the specific sub-category. The following sub-categories can be identified: (i) normal category, including aircraft for non-aerobatic operations, designed to carry up to 9 passengers and have a maximum takeoff weight of up to 5,670 kg (12,500 lbs); (ii) utility category, including planes with limited aerobatics capabilities (steep turns, spins), at the same passenger and weight threshold of the normal category; (iii) acrobatic category, including aircraft capable of unlimited aerobatic maneuvers (such as loops, rolls, and inverted flight) within the same weight limit of the normal category but being very often limited to single- or dual-seaters, and (iv) commuter category, which includes aircraft designed for short regional flights, capable of carrying up to 19 passengers, with a maximum takeoff weight of up to 8,618 kg (19,000 lbs). Being commuters fit for commercial operations, these must meet more stringent performance and safety requirements than those in the normal category, especially regarding engine-out performance, emergency systems, and passenger safety equipment. For all Part-23 aircraft, certification standards emphasize ensuring safe operation during different flight phases, including takeoff, climb, cruising, and landing. Performance tests also ensure that the aircraft can handle emergency situations like engine failure and bad weather, while maintaining minimal structural weight to optimize fuel efficiency and performance.

Part-25. Large aircraft are governed by the Part-25 (FAA) and CS-25 (EASA) regulations. These aircraft are typically designed for commercial air transport, including passenger and cargo operations and they can carry more than 19 passengers and have a maximum takeoff weight above 8,618 kg (19,000 lbs), with no upper limit identified. Due to their size, capacity, and operational demands, Part-25 aircraft face the strictest certification requirements to ensure safety, reliability, and performance during all phases of flight, as well as fuel efficiency standards, environmental regulations (such as noise and emissions limits), and specific operational requirements for high-altitude or transoceanic flights. Part-25 includes passenger aircraft belonging to the categories of regional turboprop, regional jet, short-and-medium range, and long-range aircraft, covering, de-facto, the vast majority of commercial aircraft operations. Additionally, cargo aircraft, often derived by the conversion of passenger aircrafts, and business jets are also certified under this category.

Focusing on the Part-25 passenger aircraft, the **regional aircraft segment** is commonly identified by a reference mission up to 1,000 km and passenger count up to 130. This is divided in the regional turboprop (lower sub-segment, later referred to as "regional low"), encompassing aircraft architectures capable of 30–70 passengers (e.g. ATR-42/72 airframes), reference mission of 500 km and power installed from 2-to-5 MW in open propeller configuration, and the regional jet (upper sub-segment, later referred to as "regional high"), with aircraft architectures capable of 100–130 passengers (e.g. A220/Embraer 190 airframes), reference mission up to 1,000 km and power installed in the range of 10 MW in ducted fan configuration. The regional segment is estimated to cover approximately 5% of the global air travel demand.

Moving up to the **short-and-medium range aircraft segment**, this is identified by the reference mission of 1,500–2,000 km, passenger count in the range of 150–250, and power installed of 15-to-25 MW in ducted fan configuration. The A320 and B737 airframe belongs to this category, covering approximately 50% of the global air travel demand and the near totality of the intra-European air traffic. This segment, moreover, constitutes the most profitable in the aviation industry, concentrating above 80% of the combined Airbus and Boeing aircraft yearly production, measured as aircraft units delivered per year.

Finally, the **long-range aircraft segment** is identified by the reference mission of 6,000–10,000 km, passenger count above 300, and power installed of approximately 40 MW, covering the remaining 45% of global air travel demand. The A350 and B777/B787 airframes belong to this category, that is characterized by a significant lower volume of aircraft delivered per year as well as a significantly different utilization pattern of the aircraft compared to the short-and-medium range segment. On the latter, long-range aircraft performs a significant lower number of rotations per year (take-off and landing cycles, i.e. 2 rotations/day) against the short-and-medium range (i.e. up to 6 rotations/day).

Part-23 and Part-25 solely address fixed wing aircraft, therefore not including rotorcraft, which fall under Part-27 and Part-29, and VTOL, which fall under EASA SC-VTOL in EU and under FAA Part-23 in the US. For this chapter, we mostly focus on Part-23 commuters and Part-25 passenger aircraft, according to the market segmentation described above.

Table 2 reports a summary of the fitness of the full electric, hybrid electric and more electric architectures to the different aircraft categories, with a focus on Entry-into-Service by 2035. Full electrification is doable for Part-23 light and commuter airframes, and in the lower bound of regional-low segment (up to 30 pax). This can be sustained by batteries, fuel cells or by systems which combine both technologies. Examples of full electric airframes here are the Pipistrel Velis Electro (Textron Pipistrel), the Pipistrel Miniliner (available so far only as a concept, developed in the European funded project UNIFIER19, (European Commission)) and the Hearth Aerospace ES-30 (first demonstrator unveiled in Sept. 2024, (Heart Aerospace)). The hybrid electric domain shall cover the segments from the commuter to the regional-low, starting with a DoH of 50% for the smallest aircraft down to 20–30% for the larger aircraft. Instead, from the regional high to the short-and-medium and long-range segment, the propulsion system is not considered an immediate candidate for electrification given its power and weight requirements, hence they are considered fit for implementing the more electric aircraft architecture to reduce the fuel consumption.

Note that this fitness analysis should not be considered final, and several concepts which deviates from it can be retrieved in literature. For example, Elysian aircraft is studying the feasibility of a full electric regional in the 90 pax range (Elysian Aircraft), and DLR is showcasing a hybrid electric configuration at high passenger density suitable for short-range, (DLR). Furthermore, a press release of Airbus from March 2025, (Airbus 2025), highlights the increasing role of hybridization and electrification technologies in complementing fuel efficiency gains achieved through engine advancements. Although no technical specifications have been disclosed, the statement supports the applicability of more-electric aircraft technologies in the SMR segment and suggests the potential for mild hybridization using batteries or fuel cells, possibly enabling up to a 5% reduction in carbon emissions.

However, due to the limited industrial backing of the Elysian and DLR concepts, as well as the lack of detailed information provided by Airbus in the press release, hybrid-electric propulsion, even in its mildest form, has not been deemed fit for Entry-into-Service by 2035 in the regional-high, SMR and LR segments. This decision reflects the authors' intention to adopt a conservative approach, focusing only on

Table 2 Fitness of the full electric, hybrid electric and more electric architectures to the main Part-23/25 aircraft segments with focus on Entry-into-Service by 2035

		Full electric	Hybrid electric	More electric
Part-23	Light (2–4 pax)	✓	n.a	n.a
	Commuter (up to 19 pax)	✓	✓	n.a
Part-25	Regional low (30–70 pax)	✓	✓	n.a
	Regional high (100–130 pax)	n.a	n.a	✓
	Short-medium range (150–250 pax)	n.a	n.a	✓
	Long-range (300 + pax)	n.a	n.a	✓

configurations with a higher degree of technological maturity and industry support. As a result, this option is not included in Table 1.

As per today, the only electric aircraft technology that has reached full maturity (Technology Readiness Level 9) is the CS-23 light aircraft. However, this segment has minimal impact on commercial air travel and its influence on global aviation markets is negligible.

Full-electric and hybrid-electric architectures, particularly within the commuter and regional segments, are in an advanced stage of development, with early ground demonstrators expected to be unveiled in the coming years (2025/26, reaching TRL 5). Based on the outcomes of these demonstrators, flight testing could be feasible by 2030 (TRL 7), with the first commercial Entry-into-Service by 2035 (TRL 9). From that point, it will likely take at least 20 years to achieve full fleet penetration in the European market. Therefore, it is realistic that no more than 75% of the European regional fleet can be replaced with hybrid-electric aircraft by 2050.

In the case of more electric aircraft architectures, even though this architecture has not yet been deployed in full, few key components have already reached full maturity. For example, the B787 is the first large scale aircraft adopting ad electrical Environmental Control System (ECS), while the A380 adopts electro-hydrostatic actuators for the flight controls. In general, several key components such as the electrical ECS, the ice protection systems (IPS), and electro-actuation systems have already been demonstrated at TRL 5/6 within the CleanSky2 Integrated Technology Demonstrator Systems. These systems, primarily targeting the SMR market, are expected to undergo flight testing by 2030, leading to integration into SMR aircraft with an anticipated Entry-into-Service around 2035 (sometimes referred to as "SMR + "). Later, these technologies can be scaled for long-range aircraft ("LR + "), with an Entry-into-Service target by 2040. Like the regional segment, a 20-year timeline is projected for full fleet penetration in Europe. Therefore, no more than 75% of short-medium range aircraft can be realistically replaced by the more electric SMR + by 2050, down to 50% for the more electric LR + by the same reference year.

Assuming the hybrid electric architecture capable of reducing fuel consumption and CO_2 emissions by 40% and more electric architecture capable of reducing CO_2 emissions by 4% in the short and medium range and by 2% in the long range, the expected impact of electrification in commercial aviation will allow for reducing air travel emissions by 3–3.5% by 2050, and up to 5% in the longer term, when full fleet replacement will be achieved, with a possibility to exceed this mark, should mild hybridization materialize in the regional high, SMR and LR segments. In other words, electrification has the potential to cover approximately 10% of the carbon emissions reduction target from technological improvements reported in Sect. 1, with the remaining part covered by aerodynamic improvements, structural improvements and, most prominently, engine technology improvements.

Based on the presented data, and solely focusing on commercial operations, Table 3 reports a projection of required airborne battery performance to enable full, hybrid and more electric aircraft across the categories above. Additionally, it provides an overall sizing of the market for airborne batteries in 2050.

Table 3 Reference airframe and powertrain data, battery systems key-performance indicators and estimation of the airborne battery market in the year 2050 per aircraft category (commuter, regional, short-medium range, and long-range)

		Part-23	Part-25			
		Commuter	Regional low	Regional high	Short-medium range	Long range
Pax		≤ 19	30–70	100–130	150–250	≥ 300
Reference mission	Range [km]	200	200–500	1,000	1,500–2,000	6,000–10,000
	Duration [mins]	60	60–120	60–120	60–150	360 +
Installed power [MW]		0.5–1.0	2.0–6.0	10	20	40
Conversion type		Full electric	Full/hybrid electric	More electric		
Installed electrical power [MW] ∣ DoH		0.5–1.0 ∣ 100%	2.0 ∣ 100%–20%	0.5 ∣ 5%	1.0 ∣ 5%	2.0 ∣ 5%
Expected on-board energy storage capacity [MWh]		1.0	1.0–2.0	0.5–1.0	1.0–2.0	1.0–2.0
Gravimetric energy density (pack-level) [Wh/kg]	Minimum	350	400 +			
	Desirable	500	600 +			
Cyclability [#]		2,000 +				
2023 Global production [units/year]		100	40	140	1,300	150
2050 Market projections [units/year]		200	120–150	300–400	3,000	300
2050 Airborne battery market size [GWh]	New deliveries	0.2	0.08–0.24	0.15–0.4	3.0–6.0	0.3–0.6
	Flight rotations per day [#]	2	6	6	6	2
	Refurbishment legacy fleet	1.0	1.2–3.6	2.2–6	45.0–90.0	4.5–9.0
	Total	1.2	1.28–3.84	1.35–6.4	48.0–96.0	4.8–9.6
	Market share	~ 1.5%	~ 3.5%	~ 5.0%	~ 80%	~ 10%
Emission reduction by electrification		~ 1.5%			~ 1.5%	~ 0.5%

In this projection, the five considered categories for aircraft commercial operations, i.e. commuter, regional (low/high), short-medium and long-range, are converted to full, hybrid and more electric architectures, with an electric power installation from 0.5 to 2.0 MW, depending on the aircraft and related conversion architecture. Assuming batteries to be sole energy storage system for the electric part, minimum key performance indicators have been identified, based on the research and consultations conducted in the preparation work of the Batteries Europe Roadmap

2021, (Batteries Europe 2021). Here the gravimetric energy density of the pack is identified as most important key performance indicator, with a minimum threshold of 350 Wh/kg and above 400 Wh/kg at pack level for the economic viability of full-electric, hybrid electric and more electric aircraft architectures. This constitutes approximately a 30–60% improvement against current state-of-the-art reported in Table 1, and this is expected to be achieved by both improving the cell and the packaging efficiency performance. Desirable values are set at 500 and above 600 Wh/kg in the various categories, identifying these values as achievable based on current Li-ion battery performance development projection by the target year 2035. Likely, airborne batteries will natively belong to Generation-4 solid-state since their market inception, with gravimetric energy density at cell level starting at 400 Wh/kg, and possibly achieving the 700 Wh/kg mark by 2035 (Kühnelt et al. 2023). Minimum cyclability is set at 2,000 cycles (at 1C), to ensure 1-year operation for highly utilized airframes at 6 flights rotations/day (i.e. regional/SMR) and 3-years operation for less utilized airframes at 2 flights rotations/day (commuter/long-range). These durations allow for synchronizing the battery replacement in conjunction with the aircraft scheduled maintenance (i.e. specifically C-checks), hence facilitating the refurbishment of the legacy fleet.

Market projections are therefore derived based on these performance data while assuming fleet penetration of 75% in the commuter, regional and short-medium range market and 50% for the long-range by 2050 and expected aviation industry growth against the reference year 2023. This allows to predict a global market demand for airborne batteries (new deliveries and refurbishment combined) ranging from 55 to 120 GWh/year in the reference year 2050. The market is expected to be segmented in a 10-80-10 fashion, with 10% of the demand driven by smaller aircraft segments (commuter and regional), 10% driven by the long-range aircraft market and 80% driven by the short and medium range market. Impact wise, the fuel consumption reduction, and hence the CO_2 emission reduction benefit, is expected to be equally sized between the lower segments (commuter/regional) and the short and medium range, with a less relevant contribution coming from the long range.

In conclusion, while the hybrid-electric architecture primarily addresses environmental impacts, the more electric architecture is poised to deliver significant economic benefits. The development of both architectures is essential to fully harness the emission reduction potential of aircraft electrification. Additionally, airborne batteries for commercial aircraft are expected to range from 0.5 to 2.0 MWh, with minimal variation in performance requirements across categories. This opens the possibility for a standardized airborne battery, subject to a unified airworthiness framework, simplifying the implementation of both hybrid-electric and more electric architectures across the industry.

However, conclusions regarding the suitability of different electric architectures, as well as market sizing and environmental impacts, could shift dramatically with accelerated advancements in Li-ion battery technology. In the unlikely event that the milestone of 1 kWh/kg is achieved by 2030, hybrid-electric solutions could extend to regional and short- to medium-range aircraft by 2040, significantly influencing the outcomes of this analysis. Nonetheless, such advancements are not expected in the near term, particularly not by 2035, the latest viable entry-into-service date for aircraft architectures that could meaningfully impact the fleet by 2050.

5 Case-Studies for Hybrid Electric Aircraft (Regional Segment)

This section provides a concise overview of viable case studies for hybrid-electric regional aircraft, focusing on the "regional-low" segment, which, consistent with the preceding section, offers the best balance between suitability for hybrid-electric powertrain integration and relevance for passenger transport, thus optimizing environmental and economic impact. Currently, no hybrid-electric passenger aircraft certified under Part-25 exists due to the technology's relatively low maturity. Nonetheless, several high-level aircraft studies have been conducted in recent years, indicating that the clear benefits of hybrid-electric propulsion are challenging to ascertain (Novelli et al. 2023a, b). Hybrid propulsion introduces a complex, multi-dimensional optimization challenge, involving various energy storage, conversion, and propulsion technologies, intertwined with component development, airframe integration, new propulsion strategies, redundancies, and certification procedures; these are challenges that classical aircraft sizing methods struggle to address (Brelje and Martins 2019). Even at preliminary evaluation level, results remain not sufficiently conclusive, with substantial variation in aircraft configurations and estimated performance across different studies. Typically, results are skewed by "optimistic" technology assumptions, particularly regarding the projected energy density capabilities of Li-ion batteries. Despite this, most studies align on the view that the fuel-saving potential of hybrid-electric architecture is limited, and mainly confined in the proximity of the lower bound of the regional-low aircraft segment. Additionally, these studies suggest that retrofitting existing aircraft is not effective, and that hybrid-electric propulsion would require either a clean-sheet aircraft design or substantial modifications to existing airframes.

This section includes three case studies. The first has a research background and draws on the H2020 IMOTHEP project results, alongside preliminary outcomes from the Clean Aviation projects HERA and HECATE. The second and third case studies presents two industrial concepts, i.e. the Heart Aerospace ES-30 and the ATR-EVO, reporting preliminary data and specifications from their ongoing aircraft development programs.

Investigation on the potential of aircraft hybrid electric propulsion: the H2020 Project IMOTHEP and the Clean Aviation projects HERA and HECATE.

The H2020 IMOTHEP project (IMOTHEP—Investigation and Maturation of Technologies for Hybrid Electric Propulsion (GA no. 875006), 2020–2024) has investigated electric technologies for hybrid electric propulsion in close combination with advanced aircraft configurations and innovative propulsion architectures to better substantiate the benefits of this technology with EiS by 2035 in both the regional (40 pax) and SMR (150 pax) segments, (Novelli 2024). Per each segment, both a "conservative" and a "radical" electrification design scenarios have been investigated. Concerning the 150-passenger SMR aircraft, neither the "conservative" (i.e. turbo-electric tube and wing configuration with distributed ducted fans) nor the "radical" turboelectric design (i.e. blended-wing body with distributed propulsion and boundary layer ingestion) could yield major fuel saving gains compared to conventional turbo-fan propulsion (Defoort 2024). Concerning the 40-passenger regional aircraft, the "conservative" parallel hybrid configuration studied a twin turboprop with electric motors assisting the thermal engine by adding power either to the compressor or directly to the rotor shaft with different hybridisation strategies at relatively low degree of hybridisation of up to 15% (Habermann et al. 2023). Accepting a 30% increase in maximum take-off weight due to the introduction of the hybrid powertrain, fuel consumption was predicted to decrease by close to 10% at best for the 200 nm typical mission, but to increase by 6% for the 600 nm design mission, while being highly sensitive to the battery specific energy density. Additionally, the "radical" hybrid-electric regional configuration was conceived as a plug-in hybrid with eight battery-powered electric propellers and a turbo generator to extend range (Atanasov 2022) combined with improved aerodynamics. Here a 55% reduction in block energy is estimated for a full electric 200 nm typical mission, shrinking to 4% for the 600 nm design mission with range extender, resulting in a 20% fuel reduction against the conventional baseline. This gain is materialised despite the significant increase of the MTOW from 15 to 22 tons, mainly due to batteries' weight. However, this configuration tends not to be too sensitive to the battery specific energy, which is a benefit against the "conservative" parallel hybrid. This result indicates the potential of the plug-in hybrid, a paradigm that is currently followed by Heart Aerospace for their novel hybrid aircraft in the regional-low segment and also seems to be scalable to larger aircraft in ultra-short-range operation (DLR).

After IMOTHEP, Clean Aviation has initiated in 2023 the HER programme, developing an ultra-efficient hybrid electric regional aircraft for 50–100 passengers for distances that are typically below 500 km (Clean Aviation 2024). The combination of hybrid electric propulsion that is initially conceived for EiS in 2035 on SAF and batteries and at longer term on H2 fuel cells, electrification of aircraft systems and a highly efficient aircraft configuration aims to reduce CO_2 emission by 30% by 2035 and up to 50% beyond 2035 (excluding the effect of SAF). This objective is being pursued by several interrelated projects, including HERA (HERA—Hybrid-Electric Regional Architecture (GA no. 101102007), 2023–2026) developing the aircraft configuration and systems integration, and HECATE (HECATE—Hybrid ElectriC regional Aircraft distribution TEchnologies (GA no. 101101961), 2023–2026) defining the electrical distribution architecture and all the associated technologies. For this aspect, two aircraft architectures are being evaluated and compared, labelled as use-case A and use-case B (Marco, Neumann De la Cruz, & Reichert), and respectively consisting of a hybrid electric turboprop versus a Distributed Electric Propulsion (DEP), based on an interplay of a thermal engine and electric motor powered by a battery and fuel cells systems. In these projects the energy storage, conversion systems and aircraft electric architecture are also conceptualised. While conclusive results on HERA and HECATE are not yet available, with the projects just turning the eighteenth month of execution as per end of 2024, preliminary data suggest that the target of 30% of CO_2 emission reduction in the 50–100 passenger range by 2035 might be too ambitious to reach, with an indication that the hybrid electric propulsion system alone can be capable of up to 20%. In turn, this seems to be consistent with the results of IMOTHEP, signalling that the emission reduction potential of such technology exists but might be more moderate than expected.

Electrification in the regional-low (30 pax) aircraft segment: the Hearth Aerospace ES-30

The Heart Aerospace ES-30 (Heart Aerospace) represent one of the most concrete development projects for a hybrid electric aircraft available in the industry today, and likely one of the most credible candidates to be the first hybrid electric aircraft certified under Part-25. Developed by Heart Aerospace, a Swedish start-up company whose mission is to electrify aviation, the ES-30 aims to provide an efficient, low-emission solution for regional air travel while serving short-haul routes. At the core of the ES-30's clean sheet design is its hybrid-electric propulsion system, enabling fully electric flights up to a range of 200 km with 30 passengers in a novel and unique 2–1 seating configuration. The ES-30 has a fixed-wing design and is powered by four wing-mounted electric motors/engines, paired with an advanced Li-ion battery system, capable of more than 1 MWh of energy capacity. For extended trips, the aircraft integrates a hybrid energy source, which combines the battery electric with conventional propulsion. At present, it is unclear whether this will happen through a series hybrid configuration, where the four electrically driven propellers will be driven by motors powered by batteries recharged by using an auxiliary generation system based on Jet-A1 or SAF, or through having two propellers driven electrically (inner pair) and two propellers driven by conventional turboprop engines (outer pair, called by Heart Aerospace "independent hybrid system"). The latter configuration clearly offers the advantage of two independent and separated propulsion systems, with clear benefit in redundancy and certification. The hybrid configuration allows reaching distances of up to 400 km flight range with 30 passengers and 800 km flight range with 25 passengers, hence extending the aircraft's operational capability to serve most regional routes in Europe and North America. The battery, initially designed to be equipped with state-of-the-art liquid electrolyte Li-ion battery cells (i.e. generation 3), is planned to incorporate next-generation solid-state batteries (i.e. generation 4), enhancing range and energy efficiency as the battery technology develops over the next years, possibly extending the full electric range up to 400 km, as well as consequently the hybrid electric range. The full aircraft specifications are, at present not published. It is, however, expected to be capable to take-off from runaway of 1,100 m, and fully charge its battery in 30 min, hence allowing an aircraft's turnaround time is comparable to current regional jets, thus aligning well with the high demands of regional airlines. From an environmental perspective, the ES-30 stands out as a low-emission option in the industry. With fully electric operations, it eliminates in-flight carbon emissions entirely for shorter distances, while, when operating in hybrid mode, it still achieves a considerable reduction in emissions compared to conventional fossil-fuel aircraft, estimated at less than half against a regional jet of a similar passenger capacity. As of now, Heart Aerospace plans to type certify the ES-30 by 2029, launching it for commercial use by 2030, marking it as one of the pioneering aircraft in the era of electric aviation. As of end of 2024, the ES-30 has secured 250 orders, with options for additional 120 units and letters of intent signed for additional 191 units (Aerospace Global News), unveiling the first full-scale demonstrator on September 12th, 2024, (Hearth Aerospace).

Electrification in the regional-low (70 Pax) aircraft segment: the ATR-EVO

A second example of the significant efforts that the aeronautic industry is investing in the electrification of air transport is constituted by the ATR EVO (ATR). Designed by ATR, the European leader in regional turboprop manufacturing, the EVO is a next generation "ultra-efficient regional aircraft," engineered to meet the industry's growing demand for sustainable and efficient air travel, specifically catering to regional routes with a focus on minimizing environmental impact, maximizing fuel efficiency, and reducing operational costs. The EVO is targeted to serve regional routes between 500 and 1,500 km (i.e. short-to-medium range), accommodate up to 72–78 passengers with a 2–2 seating configuration. It builds on ATR's established experience with turboprop aircraft but introduces a range of technological innovations. Primarily the EVO combines an ultra-efficient thermal engine re-designed for full compatibility with SAF, and to operate with a battery-powered electrical motor mechanically connected to the propeller gearbox (parallel hybrid), to optimize the engine core size using electrical power. The targeted thermal-electric power split of the 4 MWh drivetrain is not disclosed, however, in normal operations, ATR envisions electrical energy providing up to 20% of the power for the top-of-climb segment and initial cruise, with recharging from the thermal engines potentially occurring later in the cruise and during descent, as well as on the ground between flights. This should result in a DoH between 10 and 20%, requiring a sizing of the battery in the range of few hundreds kWh (two battery packs capable of combined 300–500 kWh, (Aviation Week)), aiming at exploiting the electric powertrain as a booster in a mild-hybrid architecture. As such the EVO targets a 20% fuel and CO_2 reduction against the baseline aircraft, powered with state-of-the-art PW127M engines, possibly achieving up to 30% by implementing additional innovations (e.g. optimized wingtips, redesigned nacelles, lightweight materials, and electric auxiliaries, etc.). ATR targets EiS by 2035 for the EVO, replacing the successful ATR 42/72 product family.

6 Airworthy Structural Batteries: Batteries' Implementation as a Multi-Functional Energy Storage

As elaborated in the previous section, the future of aircraft electrification will hinge on the development of high-performance cells with minimal weight penalties, with energy density being the most influential parameters affecting the viability and benefit derived by the implementation of a hybrid electric propulsion architecture. By 2050, the goal is to achieve system-level energy densities of at least 350–400 Wh/kg (see Table 3). While Li-ion battery technology is currently nearing maturity, it is expected to reach its theoretical limits soon, with usable energy densities projected to reach approximately 550–600 Wh/kg at the cell level.

Although ultra-efficient automotive cell-to-chassis concepts, boasting packaging efficiencies of up to 90% (Battery Design—Cell to Pack 2022) and capable of delivering over 500 Wh/kg at the pack level, represent a significant breakthrough for ground vehicles, these advancements are unlikely to be transferable to aviation. This is largely due to aeronautic thermal runaway and fire safety concerns, as these systems rely on large numbers of cells integrated into a single unit, making them unsuitable for compliance with stringent aviation safety standards.

Multifunctionality in the context of combining electrical energy storage with mechanical load-bearing capabilities, offers an alternative approach to mitigating the weight, and to a lesser extent volume, penalties associated with conventional monofunctional batteries. This can be achieved through two main strategies: either by making batteries structural or by enabling structural components to store electrical energy. Ultimately, the goal is to merge these approaches into a single solution. Over the past few decades, various concepts for structural batteries (SBs) and multifunctional electrochemical energy storage have been explored. These range from adding multifunctional properties to structures or their elements, to innovating at the material or constituent level itself (Kühnelt et al. 2022; Danzi et al. 2021).

For structural batteries to emerge as a viable performance technology in future aircraft, they must not only deliver multifunctional performance but also ensure high operational safety. While high energy density is critical for making structural batteries competitive with conventional battery systems, safety remains a fundamental requirement for airworthiness certification. The mechanical properties of structural batteries must be suitable for specific structural integration scenarios. However, ultimate mechanical strength is not always necessary, as structural designs can accommodate the reduced load-bearing capacity of structural batteries. Furthermore, for widespread adoption in the aerospace industry, structural batteries must be compatible with established aerospace composite materials and manufacturing processes, such as high-strength carbon fibre matrices and autoclave curing. Their processing, manufacturing, and structural integration must also be scalable to meet industry demands.

High-energy semi-solid structural battery technology

As part of a breakthrough research programme funded through the European projects SOLIFLY and MATISSE (SOLIFLY—Semi-SOlid-state LI-ion batteries FunctionalLY integrated in composite structures for next generation hybrid electric airliner (GA no. 101007577), 2021–2023; MATISSE—Multifunctional structures with quasi-solid-state Li-ion battery cells and sensors for the next generation climate neutral aircraft (GA no. 101056674), 2022–2025), the AIT (Austrian Institute of Technology GmbH) has led the development of a multi-electrode-layer semi-solid structural battery technology. It builds upon an energy dense semi-solid Li-ion electrochemistry with in-house developed components: a high-energy NMC811 composite cathode, a graphite composite anode and a non-flammable structural electrolyte film from ionic liquid-thermoplastic mechanically reinforced with filler particles with a final thickness of around 30 μm (Beutl et a. 2024; Krammer et al. n.d). The semi-solid SB cell is assembled in a multi-electrode-layer stack, see Fig. 1a, and has currently an energy density of 96 Wh/kg (271 Wh/l). This is estimated to be 2–3 times higher than state-of-the-art carbon fibre-based SB cells (Siraj et al. 2024; Chaudhary et al. 2024; Bouton et al. 2024). Its structural integration into aeronautic high-strength carbon fibre laminates has been studied for a standard quasi-isotropic unidirectional ply stacking, using the aeronautic AS4/8552 CF-epoxy material (Laurin et al. 2024a, b). As depicted in Fig. 1b, the SB cells are inserted symmetrically into the CF ply stacking where they replace the (three) inner CF plies (± 45°, 90°) that are not aligned with the direction of the main mechanical load (0°). Furthermore, the SB cell must match (with small margin of around 5%) the thickness of the plies it is integrated in to avoid unevenness and material stress. In the present case, the SB cell thickness was adjusted to around 564 μm for integration into 3 plies of the AS4/8552 prepreg material that has a cured ply thickness of around 188 μm. SB cells were not integrated into single outer CF plies, i.e. nr. 2 and 15 oriented at 90°, for efficiency (energy density of single-layer lower than of multi-layer SB cells, additional cell wiring needed) and to maintain the outer CF plies' mechanical protection of the inner CF plies and the SB cells, e.g. against impact.

A first *aeronautic-grade multifunctional structure with energy storage capability* has been demonstrated in SOLIFLY based on a composite stiffened panel (Laurin et al. 2024a, b), see Fig. 1c, which is a standard aeronautic component, e.g. found in the wing box. The multifunctional panel integrates 20 multilayer SB cells of 7 × 8 cm surface area and around 0.6 mm thickness, storing in total around 8 Wh with minimum weight increase (+2.6% against the monofunctional baseline panel). It has been manufactured according to aeronautic standards, i.e. cured in the autoclave using an adapted process that allows full structural curing at 130 °C maximum temperature. Its global mechanical rigidity, assessed in compression up to 18 tons, has been found equivalent to the one of the monofunctional reference. This demonstrator proved the feasibility of aviation-grade multifunctional electrical energy storage and helped to identify challenges concerning the electrical/mechanical integration of the SB cells into CFRP composites.

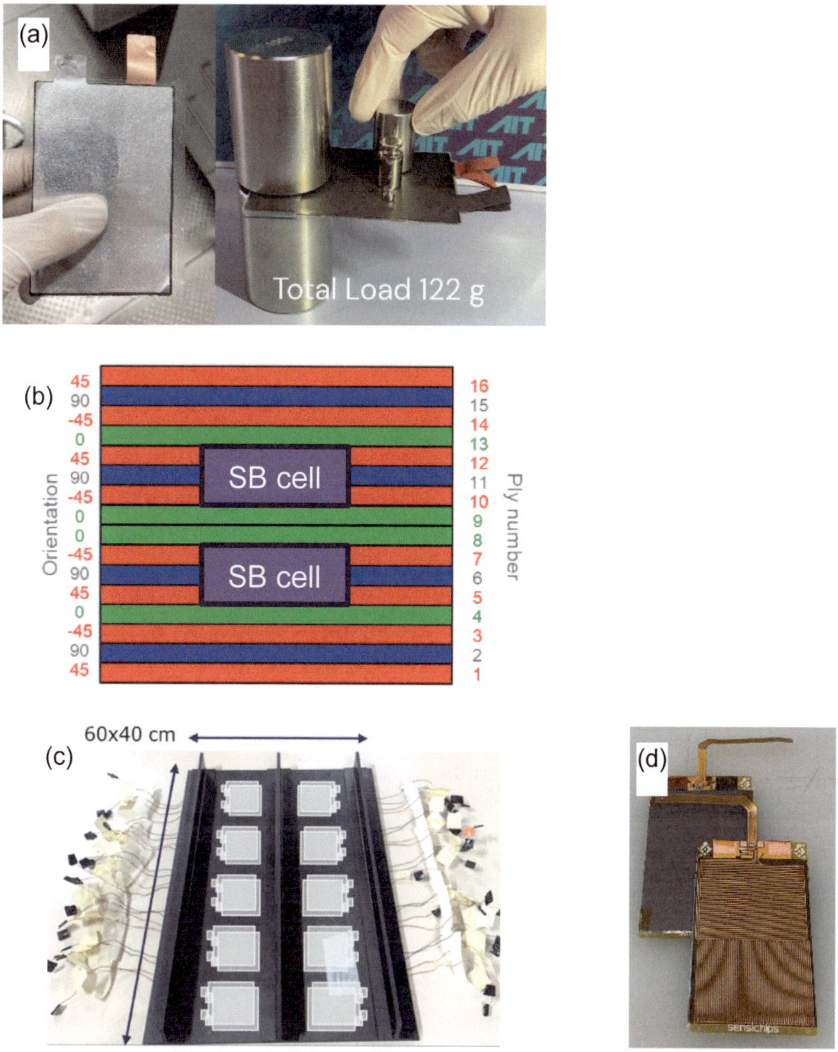

Fig. 1 **a** AIT semi-solid SB cell © AIT/Kühnelt, Beutl; **b** SB cell integration concept into solid unidirectional carbon fibres with quasi-isotropic ply stacking sequence © ONERA; **c** first aero-grade multifunctional energy storage composite panel manufactured and tested in the SOLIFLY project © AIT/Kühnelt (based on photo of ONERA/Laurin); **d** smart SB cell prototype, integrating a micro-chip-based measurement system with the SB cell stack © AIT/Kühnelt, Beutl

The manufacturing of this SB cell technology is fully scalable. The composite electrodes and the semi-solid electrolyte are manufacturable in roll-to-roll processes following conventional manufacturing approach requiring only low (electrodes) to medium (electrolyte) degree of adjustments (Steinke 2024). Best practices for the cell finishing (formation, degassing if required) and structural integration at the composite manufacturer need to be established. However, the fabrication of the SOLIFLY multifunctional demonstrator did not reveal major handling issues at the composite manufacturer.

Table 4 compares AIT's multi-electrode-layer semi-solid SB with state-of-the-art carbon-fibre-based SB and indicates the potential of further improvement steps on the cell KPIs. The multi-layer semi-solid SB achieves significantly higher energy density compared to its carbon-fibre-based counterpart. This advantage is primarily due to greater utilization of the cathode's active material, increasing the full-cell capacity from 120 (Krammer et al. n.d) to 180 mAh/g_{NMC811} (at 0.1C) for the former, compared to 25 mAh/g_{LFP} (at 0.025C) (Chaudhary et al. 2024) for the latter, at comparable active material loading of around 1 mAh/cm^2. Furthermore, the energy density benefits from the higher voltage window of NMC (2.5–4.2 V) compared to LFP (2–3.6 V). The weight penalty introduced by the metallic current collector foils used in the cell stack (cathode: 16 μm Al, anode: 10 μm Cu) could be significantly reduced by replacing them with novel light-weight current collectors, e.g. carbon-based or metal-on-polymer, which are also thinner than the metal foils. This allows to increase the cell energy density to 190 Wh/kg (345 Wh/l), also increasing the active material loading to around 1.3 mAh/cm^2. Adding a low amount of Si (e.g. 10% wt) to the graphite anode would result in 250 Wh/kg (500 Wh/l) at a loading of around 2 mAh/cm^2, and ultimately using Li-metal anode let's project 375 Wh/kg (960 Wh/l) for a loading of around 4 mAh/cm^2. However, the latter case would require the structural electrolyte to be able to withstand dendrite growth.

Integration benefits of structural batteries into composite aerostructures. The benefit of SB integration can be quantified by evaluating the amount of energy stored in the SB against the weight added to the structure, resulting in an *integration energy density* $\Delta GED \Delta GED$:

$$\Delta GED = \frac{E}{\Delta m} = \frac{VED}{\rho_{SB} - \rho_{CF}} \Delta GED = \frac{E}{\Delta m} = \frac{VED}{\rho_{SB} - \rho_{CF}}$$

Especially for *SB integration in solid composite material*, the $\Delta GED \Delta GED$ becomes significantly higher than the *GED GED*, more than tripled for the first-generation AIT semi-solid SB and reaching or surpassing 1 kWh/Δkg for the improved SB technology—compared to the *GED GED* of maximum 500 + Wh/kg for ultra-light monofunctional battery packs that might become achievable when equipped with last-generation (monofunctional) Li-ion ASSB cells with expected *GED GED* of maximum 600 Wh/kg. Tuning the SB density to that of the structural material, could make, at the expense of absolute energy storage capacity, SB cells even massless.

For *SB integration into sandwich composite structures*, the weight benefit becomes apparently less pronounced as the SB replaces the lightweight sandwich core material (e.g. foam, Al or paper hex core). Nevertheless, the integration energy density is still larger than the cell's *GED GED* maintaining zero-volume overhead, in contrast to monofunctional encased batteries. In such integration scenarios, other factors and

Table 4 Cell level KPIs of current and improved SB technology

<table>
<tr><th colspan="2">SB cell</th><th>Composition (cathode | electrolyte | anode)</th><th>GED [Wh/kg]</th><th>VED [Wh/l]</th><th>ΔGED (SB-solid CFRP) [Wh/Δkg]</th><th>Max. C rate</th><th>E [GPa]</th></tr>
<tr><td colspan="2">Carbon-fiber-based SB cell (Siraj et al. 2024; Chaudhary et al. 2024; Bouton et al. 2024)</td><td>LFP on Al foil or coated CF | bicontinuous epoxy-organic liquid (flammable) electrolyte with glass fibre separator | pristine CF</td><td>30–41</td><td>n.a</td><td>n.a</td><td>0.2</td><td>33–76</td></tr>
<tr><td colspan="2">Multi-layer semi-solid SB cell (Beutl et al. 2024; Krammer et al. n.d)</td><td rowspan="3">NMC811 composite | semi-solid, non-flammable ionic liquid-thermoplast with filler particles | graphite composite</td><td colspan="5">(Active material loading increasing from 1 to 4 mAh/cm², maintaining SB cell thickness)</td></tr>
<tr><td rowspan="2">SotA</td><td>As of year 2023</td><td>48</td><td>107</td><td>170</td><td>0.2</td><td rowspan="2">10</td></tr>
<tr><td>As of year 2024</td><td>96</td><td>213</td><td>334</td><td>0.5</td></tr>
<tr><td rowspan="3">Improved</td><td rowspan="3">Reduced weight of passive components; increased anode energy density</td><td>low density current collectors</td><td>~ 190</td><td>~ 345</td><td>~ 1500</td><td rowspan="3">≥ 1</td><td rowspan="3">Improved</td></tr>
<tr><td>+ Si(10%)-gr anode</td><td>~ 250</td><td>~ 495</td><td>~ 1360</td></tr>
<tr><td>+ Li-metal anode</td><td>~ 375</td><td>~ 960</td><td>~ 980</td></tr>
</table>

secondary effects may be more important, such as enabling decentralized energy storage and removing extra cabling that is only installed for redundancy reasons.

SB thermal management. SBs are typically thought to be integrated in a two-dimensional way into solid composite skins or sandwich cores, with one or two rather thin SB cells through the thickness, like the concept depicted in Fig. 1c. As the SB cells are not in thermal contact to each other (unlike many conventional battery modules using pouch or prismatic cells), the heat dissipation properties of the composite materials could enable purely passive thermal management of the SB and avoiding the complexity and weight penalty of active cooling. However, this in turn would require locating SB cells in a temperature-controlled or even heated environment (i.e. inside the aircraft fuselage, or hypothetically in external surfaces of future aircraft acting as high power/low temperature heat exchangers e.g. for fuel cell waste heat rejection) and would rule out many scenarios of integrating SBs into external aircraft structures, at least for aircraft operating at medium to high altitudes.

SB safety, aircraft integration and certification challenges. The high level of fire safety provided by the highly thermally stable (non-flammable) ionic liquid used for the semi-solid structural electrolyte could make it possible to integrate SB inside the aircraft fuselage, even close to the passengers. In fact, several hundreds of square meters (around 500 m^2 in SMR to 1500 m^2 in long-range aircraft) of interior panels for floors, side walls and ceiling are available in large passenger aircraft as well as other scalable components such as seat backwalls or overhead bins, which are non-safety–critical secondary structures that are located in the temperature- and pressure-controlled area. Assuming that 50% of the panels could be equipped with last generation SB (Li-metal anode, as per Table 4, with a quarter (2.5 mm) of the thickness of a typical interior panel), approximately 0.6–1.8 MWh of electrical energy could be stored in these panels, or 3.6–5.5 kWh per passenger, replacing at zero-volume impact a monofunctional battery pack that would need cells with, at minimum, 15–20% higher GED, i.e. 430–450 Wh/kg, not considering the extra weight impact of thermal management and cabling.

The integration of safe SB cells in non-safety–critical secondary structures seems more feasible than in primary aerostructures also for reasons of certification and economic efficiency. For *airworthiness certification* of multifunctional structures, aviation safety authorities consider any type of insert (sensors, antennas, or SBs) in composites as a potential source of delamination. This would place very high durability requirements for SBs. Fire safety would be required in any case.

Economic efficiency of multifunctional energy storage. Apparently, there is still a huge *gap in the lifetime* of aerostructures that are designed to last for decades and modern battery technology (including SB). Aeronautic (structural) batteries need to withstand daily operation over a major maintenance interval that is mandatory every 18–36 months, or 6,000–12,000 flight hours, or 3,000–8,000 flight cycles, typically performing around 220 (330) flight (hours) per month in a commercial aircraft. This requires, on the one hand, a massive improvement in the practical service life of (structural) battery technology. However, achieving tens of thousands of cycles seems achievable for NMC|gr cells under precisely controlled operating conditions (Harlow et al. 2019; Aiken et al. 2022). On the other hand, (multifunctional) electrical energy storage needs to be installed in a scalable and efficiently maintainable and replaceable way, which favours SB integration in non-safety–critical interior components.

Structural batteries enabling multifunctional health monitoring. The increasing introduction of composite materials in aircraft poses challenges for Structural Health Monitoring (SHM) as composite materials age and fail differently and less predictably compared to the well-known aeronautic metal alloys (e.g. Al, steel, Ti). Internal failure cannot always be detected with visual inspection, and non-destructive testing is typically performed only at major maintenance intervals. Furthermore, some structural components are hardly or not accessible for inspection. The smart SB cell currently developed in the MATISSE project (see Fig. 1d) integrates the electrode stack of a semi-solid SB with a micro-chip-based cell and structural monitoring unit (CSMU) (SENISCHIPS 2024; SENSICHIPS 2024) that includes an electrochemical impedance spectrometer with lock-in amplifier, measuring 1st, 2nd and 3rd harmonics, a potentiostat/galvanostat along with on-chip sensors including temperature and mechanical pressure or strain, and outputs digital data instead raw sensor signals. As the CSMU is operational once the SB cell is first charged, it can monitor the SB cell and the structure in which it is embedded over the full lifecycle of the multifunctional part, from manufacturing and curing over operation to 2nd life or dismantling. The self-contained smart SB cell could be integrated as a network of sensors in composite aerostructures, making them smart. At aircraft and operational level, SHM based on this this self-powered, miniaturized system could enable monitoring of very large composite aerostructures (e.g. fuselage, wings or empennage) unlocking substantial benefits in terms of system weight (– 98%, or around 500 kg less than for SMR) and cost (in the order of hundreds of thousands of euros, one order of magnitude less than the figures reported in (Cusati et al. 2021)) compared to conventional MEMS-based SHM systems, as well as time (– 50%) and cost (between 600 k€ and 1.5 M€ per SMR aircraft and year) needed for inspection and maintenance.

7 Conclusions

The chapter on battery systems for air transport climate neutrality provides a comprehensive analysis of the challenges and advancements in developing battery technologies for the aviation industry. As the industry seeks to align with the European Green Deal's target of achieving carbon neutrality by 2050, battery systems play a critical role, especially in the electrification of aircraft. However, the road to achieving this presents significant technological and regulatory hurdles.

Aviation is a notoriously energy-intensive sector, and fully electrifying aircraft remains challenging, particularly for large commercial planes. The chapter highlights the current state-of-the-art in airborne battery systems, focusing on safety, performance, and airworthiness certification specifications. Certification is central to ensuring that batteries meet strict safety standards, particularly in preventing catastrophic events like thermal runaway, requiring that battery modules prevent the propagation of thermal runaway from one cell to another, with compliance set at 20% of the battery cells. These rigorous safety demands impose significant design constraints on battery systems, which can result in a weight penalty, that is a key challenge for future battery development.

Despite these challenges, advancements are being made. Current battery systems deliver gravimetric energy densities between 230 and 290 Wh/kg at the cell level and 170–235 Wh/kg at the module level. However, by 2035, the target for full-electric and hybrid-electric aircraft systems is to achieve gravimetric energy densities of 350–400 Wh/kg at the pack level, a 30–60% improvement over today's capabilities. This will be essential to make full-electric and hybrid-electric aircraft viable for commuter and regional segments.

Current technological projections indicate that hybrid-electric and more-electric aircraft architectures can be deployed at fleet scale in the commuter, regional, and short-medium range segments by 2050. This shift toward electrification is expected to result in a reduction of aviation emissions by approximately 3–3.5% by 2050, combined with a market demand for airborne batteries ranging from 55 to 120 GWh/year in the same reference year. The market is expected to be segmented in a 10-80-10 fashion, with 10% of the demand driven by smaller aircraft segments (commuter and regional), 10% driven by the long-range aircraft market and 80% driven by the short and medium range market. Impact wise, the environmental benefits are expected to be equally sized between the lower segments (commuter/regional) and the short and medium range, while the economic benefits are expected to be mostly concentrated in the short and medium range segment. Moreover, an airborne battery system sized in the range from 0.5 to 2.0 MWh is expected to fit all categories of commercial aircraft, and this simplifies the roll-out of hybrid electric and more electric architectures.

Moreover, two research case studies for hybrid electric aircraft based on the research projects IMOTHEP, HERA and HECATE are presented, complemented by two additional on-going case studies from two aircraft development programmes, i.e. the Heart Aerospace ES-30 and the ATR-EVO. The results converge on the view that the fuel-saving potential of hybrid-electric architecture is limited, and mainly confined in the proximity of the lower bound of the regional-low aircraft segment.

Additionally, the concept of structural batteries is presented, based on the results of the research projects SOLIFLY and MATISSE. These show that such batteries are capable of a gravimetric energy density that can be tripled by the multifunctional integration of the battery cell into the structure achieving the mark of 300 + Wh/kg as per 20,204 state-of-the-art, with the possibility of surpassing the 1 kWh/kg mark in the future. Such concept, extended to non-safety–critical secondary structures, such as interior panels, could add 0.6–1.8 MWh of energy storage capacity to aircraft without increasing volume, serving more electric applications in the regional-high, short and medium range and long range segment.

In conclusion, while full electrification will primarily impact smaller aircraft, hybrid-electric and more-electric solutions are expected to revolutionize larger aircraft categories. By 2050, up to 75% of Europe's commuter and regional fleet could be hybrid-electric, with the potential to reduce CO_2 emissions by up to 40% in certain aircraft segments. The chapter underscores the importance of continued investment in battery technology and regulatory frameworks to achieve climate-neutral air transport in the coming decades.

Acknowledgements The authors are grateful to the European Commission for supporting the present work. The data and the opinions expressed by the authors in this manuscript are derived by the work performed in the European projects IMOTHEP (Grant Agreement no. 875006), ORCHESTRA (Grant Agreement no. 101006771), HIGHSPIN (Grant Agreement no. 101069508), HELENA (Grant Agreement no. 101069681), SOLIFLY (Grant Agreement no. 101007577) and MATISSE (Grant Agreement no. 101056674). Note the views expressed in this publication solely reflects the authors' opinions and neither the European Union, nor the funding Agencies involved in the management of the aforementioned projects can be held responsible for the information it contains.

References

Aerospace Global News (n.d.) ES-30: getting to the heart of the matter. Retrieved October 2024, from https://aerospaceglobalnews.com/news/es-30-getting-to-the-heart-of-the-matter/

Aiken CP, Logan ER, Eldesoky A, Hebecker H, Oxner JM, Harlow JE, Dahn JR (2022) Li[Ni0.5Mn0.3Co0.2]O2 as a superior alternative to LiFePO4 for long-lived low voltage li-ion cells. J Electrochem Soc 169(5):050512. https://doi.org/10.1149/1945-7111/ac67b5

Airbus (2025) Soaring towards future aircraft, Airbus' next-generation single aisle will incorporate novel technologies and cutting edge engineering. Retrieved April 2025, from https://www.airbus.com/en/newsroom/stories/2025-03-soaring-towards-future-aircraft

ATAG (2021) Waypoint 2050: balancing growth in a connectivity with a comprehensive global air transport response to the climate emergency. Air Tranrport Action Group

Atanasov G (2022) Plug in hybrid electric regional aircraft concept for IMOTHEP. Presentation at the 12th EASN international conference on innovation in aviation & space for opening new horizons. Barcelona, Spain. https://doi.org/10.5281/zenodo.7785306

ATR (n.d.) ATR EVO concept. Retrieved October 2024, from https://www.atr-aircraft.com/innovation/atr-evo-concept/

Aviation Week (n.d.) ATR details hybrid-electric study option for EVO. Retrieved October 2024, from https://aviationweek.com/aerospace/emerging-technologies/atr-details-hybrid-electric-study-option-evo

Batteries Europe (2021) Roadmap—application and integration: mobile. Retrieved September 2024, from https://energy.ec.europa.eu/publications/roadmap-application-and-integration-mobile_en

Battery Design—Cell to Pack (2022) Retrieved 9 26, 2024, from https://www.batterydesign.net/cell-to-pack/

Beutl A, Jiang Q, Kühnelt H, Bismarck A (2024) On the feasibility of thermoplastic materials for multifunctional energy storage solutions. In: Proceedings of the twenty-third international conference on composite materials (ICCM23), (p 198). Belfast, UK. Retrieved from https://www.iccm-central.org/Proceedings/ICCM23proceedings/papers/ICCM23_Full_Paper_198.pdf

Bouton K, Schneider L, Zenkert D, Lindbergh G (2024) A structural battery with carbon fibre electrodes balancing multifunctional performance. Compos Sci Technol 256:110728. https://doi.org/10.1016/j.compscitech.2024.110728

Brelje BJ, Martins JR (2019) Electric, hybrid, and turboelectric fixed-wing aircraft: a review of concepts, models, and design approaches. Prog Aerosp Sci 104:1–19. https://doi.org/10.1016/j.paerosci.2018.06.004

Chaudhary R, Xu J, Xia Z, Asp LE (2024) Unveiling the multifunctional carbon fiber structural battery. Adv Mater. https://doi.org/10.1002/adma.202409725

Clean Aviation (2021) Strategic research and innovation agenda

Clean Aviation (2024) Towards disruptive technologies for new generation aircraft by 2035. Retrieved September 2024, from https://clean-aviation.eu/sites/default/files/2024-09/2024-Clean-Aviation-SRIA.pdf

Cuberg (n.d.) Cuberg aviation battery module. Retrieved September 2024, from https://www.datocms-assets.com/115911/1715703379-cuberg_module_validation-2405.pdf

Cusati V, Corcione S, Memmolo V (2021) Impact of structural health monitoring on aircraft operating costs by multidisciplinary analysis. Sensors 21(20):6938. https://doi.org/10.3390/s21206938

Danzi F, Salgado R, Oliveira J, Arteiro A, Camanho P, Braga M (2021) Structural batteries: a review. Molecules 26:2203. https://doi.org/10.3390/molecules26082203

Defoort S (2024) IMOTHEP public deliverable D1.9—synthesis of benefits of HEP for regional and SMR aircraft

DLR (n.d.) DLR showcases aircraft configurations of the future. Retrieved September 2024, from https://www.dlr.de/en/latest/news/2024/dlr-showcases-aircraft-configurations-of-the-future

EASA (2022) Aviation environmental report

EASA (n.d.) Special condition for VTOL and means of compliance. Retrieved 2024, from https://www.easa.europa.eu/en/document-library/product-certification-consultations/special-condition-vtol

Electrical Power Systems (n.d.) EPiC propulsion battery. Retrieved September 2024, from https://epsenergy.com/products-services/epic-propulsion-battery-2/

Elysian Aircraft (n.d.) E9X home page. Retrieved September 2024, from https://www.elysianaircraft.com/

European Commission (n.d.) Community friendly miniliner. Retrieved September 2024, from https://cordis.europa.eu/project/id/864901

Evolito Ltd (n.d.) Evolito homepage. Retrieved September 2024, from https://evolito.aero/battery-solutions/

Habermann A, Kolb M, Maas P, Kellermann H, Rischmüller C, Peter F, Seitz A (2023) Study of a regional turboprop aircraft with electrically assisted turboshaft. Aerospace 10:529. https://doi.org/10.3390/aerospace10060529

Harlow JE, Ma X, Li J, Logan E, Liu Y, Zhang N, Dahn JR (2019) A wide range of testing results on an excellent lithium-ion cell chemistry to be used as benchmarks for new battery technologies. J Electrochem Soc 166(3):A3031. https://doi.org/10.1149/2.0981913jes

Heart Aerospace (n.d.) Introducing the ES-30. Retrieved September 2024, from https://heartaerospace.com/es-30/

Hearth Aerospace (n.d.) Heart aerospace unveils first full-scale demonstrator for 30-seat hybrid-electric airplane. Retrieved October 2024, from https://heartaerospace.com/newsroom/heart-aerospace-unveils-first-full-scale-demonstrator-for-30-seat-hybrid-electric-airplane/

HECATE—Hybrid ElectriC regional Aircraft distribution TEchnologies (GA 101101961). (2023–2026). Retrieved from https://cordis.europa.eu/project/id/101101961

HERA—Hybrid-Electric Regional Architecture (GA 101102007). (2023–2026). Retrieved from https://cordis.europa.eu/project/id/101102007

H55 (n.d.) H55 Homepage. Retrieved September 2024, from https://h55.ch/en/

IMOTHEP—Investigation and Maturation of Technologies for Hybrid Electric Propulsion (GA 875006). (2020–2024). https://doi.org/10.3030/875006

Joby Aviation (n.d.) Joby Aviation Homepage. Retrieved Semptember 2024, from https://www.jobyaviation.com/

Krammer M, Montes S, Kühnelt H, Jiang Q, Lager D, Bismarck A, Beutl A (n.d.) Multifunctionality and processability of a thermoplastic based gel electrolyte cell for the realization of structural batteries. Under Rev

Kühnelt H, Beutl A, Mastropierro F, Laurin F, Willrodt S, Bismarck A, Romano F (2022) Structural batteries for aeronautic applications—state of the art, research gaps and technology development needs. Aerospace 9(1):7. https://doi.org/10.3390/aerospace9010007

Kühnelt H, Mastropierro F, Zhang N, Toghyani S, Krewer U (2023) Are batteries fit for hybrid-electric regional aircraft? J Phys Conf Ser 2526:012026. https://doi.org/10.1088/1742-6596/2526/1/012026

Laurin F, Mavel A, Saffar F, Beutl A, Kühnelt H (2024) Experimental and numerical evaluation of residual mechanical performance of carbon/epoxy laminated coupons after integration of solid battery cells for aeronautical applications. Compos Sci Technol 247(1):110384. https://doi.org/10.1016/j.compscitech.2023.110384

Laurin F, Luco J, Garcia JM, Saffar F, Beutl A, Kühnelt H (2024) Experimental and numerical strategy for the integration of solid-state battery cells into high-performance composite structures for aeronautics. In: Binetruy C, Jacquemin F (ed), Proceedings of the 21st european conference on composite materials, 6, pp 313–320. Nantes, FR. https://doi.org/10.60691/yj56-np80

Lee DS, Allen MR, Nicholas C, Bethan O, Shine KP, Agnieszka S (2023) Uncertainties in mitigating aviation non-CO_2 emissions for climate and air quality using hydrcarbon fuels. Environ Sci Atmos

Marco F, Neumann De la Cruz D, Reichert T (n.d.) Organizational activities in preparation of LCA and LCC estimation in the framework of HERA Project. Retrieved November 2024, from https://project-hera.eu/sites/default/files/2024-10/HERA_T8_2%20presentation.pdf

MATISSE—Multifunctional structures with quasi-solid-state Li-ion battery cells and sensors for the next generation climate neutral aircraft (GA 101056674). (2022–2025). https://doi.org/10.3030/101056674

Novelli P, Defoort S, Tantot N, Zimmer D, Varchetta D, Viguier C, Iodice P (2023a) IMOTHEP European project: an investigation of hybrid electric propulsion for commercial aircraft. AIAA 2023 Aviation Forum, pp AIAA 2023–4131. San Diego, CA. https://doi.org/10.2514/6.2023-4131

Novelli P, Gerada C, Kühnelt H, Lebey T, Makoschitz M, Mezani S, Viguier C (2023b) IMOTHEP public deliverable D6.1—gap analysis and preliminary roadmap on HEP development

Novelli P (2024) Perspectives of hybridization for commercial aircraft: the lessons learned from the IMOTHEP project: Keynote. In: International conference on more electric aircraft towards greener aviation (MEA2024). Toulouse, France. https://doi.org/10.5281/zenodo.10695996

SENISCHIPS (2024) Battery cell management unit. Retrieved 9 26, 2024, from https://sensichips.com/battery-cell-management-unit/

SENSICHIPS (2024) Battery design kit. Retrieved from https://sensichips.com/design-kit/batteries-kit/

Siraj MT, Carlstedt D, Duan S, Johansen M, Larsson C, Xu J, Asp L (2024) Advancing structural battery composites: robust manufacturing for enhanced and consistent multifunctional performance. Adv Energy Sustain Res 4:2300109. https://doi.org/10.1002/aesr.202300109

SOLIFLY—Semi-SOlid-state LI-ion batteries FunctionalLY integrated in composite structures for next generation hybrid electric airliner (GA 101007577) (n.d.) (2021–2023). Retrieved from https://cordis.europa.eu/project/id/101007577: https://cordis.europa.eu/project/id/101007577

Steinke F (2024) SOLIFLY deliverable D4.2—manufacturability, TRL assessment, TRL step-up and exploitation plan for structural batteries. https://doi.org/10.5281/zenodo.13748195

Textron Pipistrel (n.d.) Velis electro homepage. Retrieved September 2024, from https://www.pipistrel-aircraft.com/products/velis-electro/

Transport and Enviroment (2024) Non-CO_2 MRV in EU ETS, a no-regret step to mitigate aviation's full climate impact

On-Board Integration of Hybrid Energy Storage Systems in Heavy Duty Vehicles: The Electric Buses Use Case

Dario Pelosi, Andrea Rampini, Alberto Vazzola, Laura Andaloro, Ilenia Belviso, and Linda Barelli

Abstract Road transportation contributes to 79% of the total transport emissions. To achieve European emissions targets by 2050, a sustainable mobility based on electric vehicles (EVs) powered by batteries or fuel cells is required. Albeit Li-ion batteries represent the best choice for light duty EVs, green transition for heavy duty transport is more challenging because of the greater mileage and the lack of fast-charging infrastructure. To extend mileage range and energy storage lifespan, hybridization of different energy storage technologies represents a promising solution towards a green transition based on zero-emissions heavy duty vehicles. In this Chapter, different systems are presented for a bus application. Three innovative hybrid propulsion systems, consisting of: (i) LiFePO4 battery (LFP)/Proton Exchange Membrane Fuel Cell (PEMFC), (ii) LFP/Vanadium Redox Flow Battery (VRFB) and (iii) Ni–NaCl battery coupled with PEMFC are analysed with reference to a real case study (i.e., urban electric bus) and compared to an electric bus powered by only LFP. With respect to the case study, it is demonstrated that there are benefits of hybridization in terms of mileage range and battery lifespan extension (when LFP is considered), paving the way for an efficient and effective green transition.

Keywords Hybrid energy storage system · Electric mobility · Battery · Fuel cell · Sustainable transportation

D. Pelosi · L. Barelli
Department of Engineering, University of Perugia, Perugia, Italy

A. Rampini (✉) · A. Vazzola
Carlo Rampini S.P.A., Passignano Sul Trasimeno (PG), Italy
e-mail: a.rampini@rampini.it

L. Andaloro · I. Belviso
Istituto Di Tecnologie Avanzate Per L'Energia "Nicola Giordano", Consiglio Nazionale Delle Ricerche, Messina, Italy

R. Scipioni et al. (eds.), *Hybrid Energy Storage*, Lecture Notes in Energy 47,
https://doi.org/10.1007/978-3-031-97755-8_14

1 State-of-the-Art

The move towards a net-zero carbon society by 2050 includes avoiding greenhouse gas (GHG) generation (Saldarini et al. 2022a, b). The transport sector accounts for 15.8% of the global emissions in 2023 with the road transportation contributing to 79% of the total transport GHG emissions.

Sustainable road transport is being achieved through the transition from fossil fuel-powered vehicles to electric vehicles powered by batteries or fuel cells. Although a massive electrification of the light-duty vehicles has begun, heavy-duty vehicles are still based on fossil fuels to travel long distances. This is because of the lack of development and deployment of fast charging or hydrogen refuelling infrastructures all over Europe (Pelosi et al. 2023a). Full electric transportation competitiveness is also strictly connected to the technological progress of batteries. Li-ion batteries (LIBs) represent the most promising solution due to high specific energy, efficiency and expected life. Nevertheless, issues as safety and availability of the raw materials should be addressed (Zubi et al. 2018).

As regards road transport electrification based on LIBs propulsion system, aspects related to range anxiety, and few charging points should be primary addressed (Pelosi et al. 2023b; Vermeer et al. 2022). Such concerns, however, can be well managed in the framework of urban transport (e.g., buses) by optimizing the planning of the daily route and the bus charging (Offer et al. 2010; Wei et al. 2018). Other promising innovative technologies, as open batteries (i.e., vanadium redox flow battery (VRFB)) have the potential to be applied in transport applications supporting conventional batteries in hybrid energy storage sections, especially in the heavy-duty road or marine transport where energy density requirements are less severe, as presented in (Barelli et al. 2024) and (Barelli et al. 2019a, b).

On the other hand, electric vehicles powered by fuel cells present higher mileage range and faster refuelling times than battery vehicles, even if a recent study, by analysing experimental data of battery and fuel cell electric buses operated in the same area, proves that the tank-to-wheel efficiency results of fuel cell electric buses are lower than that of battery electric buses (Estrada Poggio et al. 2023).

However, huge issues to be faced are still related on hydrogen production, distribution and transport. For instance, only few points of hydrogen refuelling stations are placed in Europe, mainly in Germany. Moreover, high costs and reliability of low temperature fuel cells, such as proton exchange membrane fuel cells (PEMFC), still hinder the diffusion of hydrogen-based electric transportation.

As widely proposed for stationary applications, hybridization of energy storage systems (ESSs) allows a wider operating range over multiple time scales, by implementing ESSs with complimentary features (e.g. power intensive with energy intensive technologies). Hybridization of different and complementary (in terms of key performance indexes) technologies could play a key role to date and in the next years. Specifically, hybridization of energy storage systems for on-board road transport extends the vehicle mileage range, while increasing system efficiency. Therefore, hybridizing LIBs with complimentary energy storage as supercapacitors and flywheels results in a LIB life extension due to reduced power solicitations over the driving cycles.

This allows to achieve a higher sustainability degree for battery road transportation, reducing the need for battery replacements. As matter of fact, batteries for road transport must be generally replaced when their State of Health (SoH) goes below the 80% threshold (Pelosi et al. 2023c).

2 Drivers for Installing the Hybrid System

Transportation will face a great transition toward sustainability to achieve the imposed climate targets by 2050 and 2100 (European Commission 2017). Current energy storage technologies for light and heavy-duty vehicles require higher energy and power densities regarding the technical requirements for propulsion to achieve similar performance in terms of mileage range and charging times with respect to internal combustion engine (ICE) vehicles. In this framework, the on-board implementation of complimentary hybrid energy storage systems (HESSs), as supercapacitors or other fast responsive technologies and batteries, or fuel cells and batteries, contributes to extending the vehicles' range and performances. For instance, (Barelli et al. 2019a) analyse different solutions of HESS for a real bus propulsion application, also highlighting the positive effect on Li-ion battery lifespan, which represents a key factor in the view of economic and environmental sustainability of the vehicles. As matter of fact, it is widely demonstrated that operating the Li-ion battery without any harmful power spikes allows to increase its lifespan, while reducing replacements (Barelli et al. 2019b). It is experimentally demonstrated that combining Li-ion battery pack with a flywheel peak shaving system extends the battery lifespan of more than three times compared with a Li-ion battery with a capacity equal to the total HESS capacity in a renewable-based residential microgrid (Barelli et al. 2019b). Another driver for installing on-board hybrid energy storage systems for transportation applications is the extension of vehicle mileage, at the same time reducing the overall time needed for recharging. For instance, as detailed in the following, (Barelli et al. 2012) proposes the hybridization of a battery electric bus with a hydrogen system of suitable power and capacity to allow a 12 h operation under a specific duty cycle.

3 Technology Neutral System Requirements for the Specific Use Case

The electric bus here considered as the specific use case is an 8 m length vehicle for the transportation of 14 seating and up to 30 standing-up passengers in addition to the driver. The vehicle weight achieves about 11 tons when fully loaded. The powertrain includes an electric motor with nominal and maximum power of 70 kW

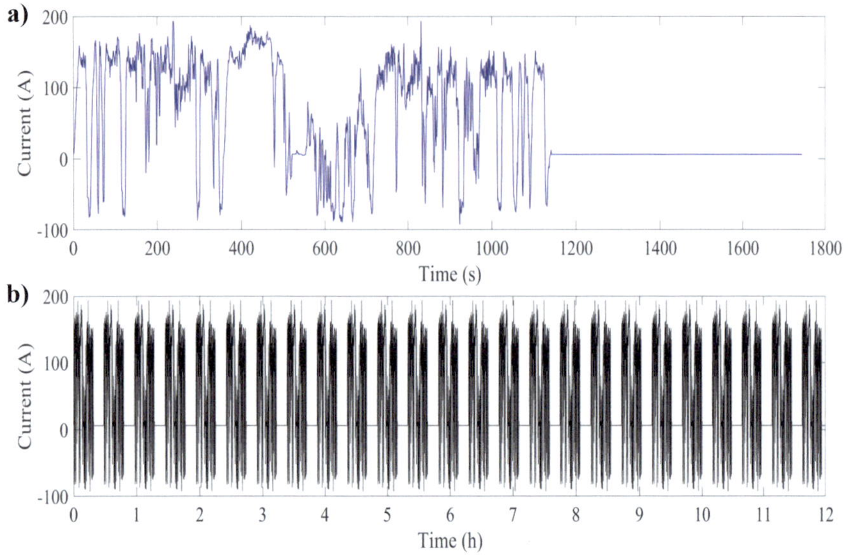

Fig. 1 **a** Measured bus load profile over 29 min; **b** Daily bus load profile (Barelli et al. 2012)

and 140 kW, respectively. The bus has to be operated for 12 h service under the duty cycle depicted in Fig. 1b. The duty cycle is determined by repeating the 29 min current profiles measured during field tests on the electric bus under study. Positive current represents the load evolution required by the electric motor, while negative currents indicate recovered power during braking phases and used to recharge the battery pack.

Key performance indexes for HESS integration are based on the LIB pack operation over the duty daily cycle of Fig. 1 with a single charge, and reduced depth of discharge (DoD), never above 90%. Therefore, mileage extension, reduction of charging time and battery degradation, both contributing to reducing investment and replacement costs, are targeted. The required background on HESSs involves competences in dynamic modelling of powertrain systems including energy storage devices (e.g., closed/open batteries, supercapacitors, and fuel cell systems and related balance of plant).

To provide accurate results, it is also suggested to develop semi-empirical models implementing the performance characteristics of the specific devices to be integrated, e.g., open circuit voltage (V_{ocv}) and internal resistance (R_{bat}^{int}) evolution with the state of charge (SoC) both in charging and discharging for open and closed batteries as in (Barelli et al. 2019a, b), as well as polarization and efficiency curves for fuel cells, as detailed in (Barelli et al. 2012). For batteries, required data can be determined through galvanostatic partial charge and discharge tests, interspersed with open circuit voltage and EIS measurements to determine V_{ocv} and R_{bat}^{int} at different SoCs. To perform such tests, galvanostat/potentiostat instrumentation is generally used.

The cited works provide theoretical background for dynamic modelling of the energy storage technologies cited above.

Dynamic models must include also high-level power management and low-level control strategies. Specifically, a high level power management strategy has to be developed to target in real-time specific objectives, usually according to multi-objective criteria for instantaneous power split among the energy storage devices. Criteria for management optimization are defined according to specific characteristics of integrated devices (e.g., SoC range, maximum discharge C-rates, limitations in C-rates variation over the considered time step), as well as according to expected performance at system level (e.g., cycle efficiency maximization of the overall HESS, or other conditions related to the investigated use case). Provided results have to fully characterize the evolution of devices' operating conditions over the imposed vehicle's duty cycle. Thus, corresponding SoC and efficiency evolutions have to be assessed for each implemented energy storage device.

Moreover, these results have to allow the off-line assessment of the expected devices' lifespan. To this regard, it is highlighted as power management becomes crucial for the real time optimization of HESS to maximize potential benefits, including devices' duration. This is relevant in the case of closed battery integration in HESS. Hence, (Barelli et al. 2019a, b) also developed a methodology to estimate ESS lifetime. Specifically, the Rainflow-counting algorithm, coupled to the cycle-to-failure curve of the considered chemistry, is suggested to assess battery expected lifespan basing on the yearly SoC trends provided by dynamic simulations. It doesn't provide accurate assessments, but it is suitable specifically for comparative analysis among different HESS architectures or power management strategies of the energy storage section. On the other hand, also the aging of the other systems components (e.g. open batteries, fuel cell) have to be assessed according to their features and the expected operating mode.

Due to the potential extension of battery duration, costs for operation and devices' replacement generally can be reduced for HESS, with respect to the case of implementation of single battery packs. On the other hand, installation costs generally increase for HESS considering the integration need of two energy storage devices at least and more complex and smart converters. Therefore, the profitability of HESS integration has to be assessed in consideration of both CAPEX and O&M costs including devices' replacement in a certain timeframe. To this aim the levelized cost of storage (LCOS) metric could be applied for a comparative analysis among two or more solutions.

4 How Can Hybrid Energy Storage Meet the Identified Needs?

Hybrid energy storage systems for transport applications allow mileage range extension and recharging time reduction, as well as lifespan extension of single devices. Benefits provided have to be assessed, according to the criteria indicated in the previous paragraph, by comparing one or more HESS solutions with the case of a single energy storage device on-board installation.

With reference to the specific use case i.e., the 8 m length electric bus, three different HESSs (Systems A, C, and D) are investigated with respect to the case of a bus equipped with Li-ion battery only (System B).

In Systems A and C the Li-ion battery technology (the LFP as implemented in System B) is hybridized with a different technology, while in the case of System D the LFP battery is substituted by a Ni–NaCl battery pack. Investigated ESSs are:

- System A: Hybrid storage system composed by LFP battery pack with PEMFC stack (Proton Exchange Membrane Fuel Cell)
- System B: LFP battery pack (not hybrid ESS).
- System C: Hybrid storage system composed by LFP and VRFB battery packs
- System D: Hybrid storage system composed of Ni–NaCl battery pack (Sodium-Nickel Chloride batteries, also known as Zebra batteries) with PEMFC stack.

This section reports the main technical features of each ESS solution, while the performance assessment of each HESS with reference to the non-hybrid ESS (System B) is discussed in the next section.

Systems A, B, C and D were implemented by means of dynamic models to determine the size of the storage devices, according to the real driving power profile, also considering a suitable SoC control strategy (Barelli et al. 2019a, b). Moreover, PEMFC-LFP, LFP, VRFB-LFP, and Zebra/PEMFC energy storage systems were modelled in Matlab/Simulink environment considering as design constraint the weight of System A (i.e., 1620 kg as detailed in the following). This is highlighted as all investigated ESSs were sized at parity of on-board mass of the storage section. Related performance is compared in reference to the daily working cycle depicted in Fig. 1.

LFP batteries modelling was based on experimental data of a LFP battery module (Evlithium 2024). According to the methodology indicated in the previous section, performance characteristics of the specific devices to be integrated were determined through a dedicated experimental campaign and implemented in the ESS models. Specifically, experimental data of open circuit voltage (V_{ocv}) and internal resistance (R_{bat}^{int}), both in charge and discharge processes, at varying state of charge were measured and implemented in the model. For details, refer to (Barelli et al. 2012).

As regards to VRFB, data required to develop the semi-empirical model can be gathered by mean of measurements on a small single cell, anyway representative for cell voltage and internal resistance assessments. It has to be noted that if a single flow cell, instead of a flow battery system, is tested, pumping losses and system round trip efficiency are not taken into account. As example in (Barelli et al. 2012) the tested cell has an active area of 12 cm^2 and it is composed by a sandwich of several components (aluminium plates for cell compression, copper plates as anodic and cathodic current collectors, graphite plates, Teflon gaskets, carbon felts and the Nafion 115 membrane). BioLogic SP-240 equipment is used to perform the characterization tests. The cell is operated under 50 mA cm^{-2} current density and 432 ml min^{-1} of catholyte and anolyte feeding flow rates. Anolyte and catholyte volumes are 250 ml each one, with 1.5 M vanadium and 5 M sulfuric acid composition.

System D was fully implemented in a city bus following the development of a preliminary numerical simulation model in the MATLAB-Simulink environment. The simulation was used to support the design and dimensioning of the hybrid system (PEMFC and Ni–NaCl batteries), adhering to a weight constraint of 1620 kg and considering specific routes (real-world city duty cycle).

The system was built and integrated into the bus, with both the fuel cell stack and the batteries having been previously tested experimentally. These components were also modelled in Simulink to fine-tune their performance (De Lorenzo et al. 2014). The equivalent circuit parameters for the fuel cell system were derived from the polarization curve provided in the manufacturer's datasheet. Meanwhile, the open circuit voltage of the batteries was calculated using a nonlinear equation based on the state of charge analysis (De Lorenzo et al. 2014).

Both the battery and fuel cell models were validated against real-world operational data acquired during field tests, confirming the accuracy and reliability of the implemented system.

System A. LFP-PEMFC hybrid energy storage section

A hybrid storage system consisting of a proton exchange membrane fuel cell (PEMFC) installed in parallel with an LFP battery pack was modelled and analysed by the authors in (Barelli et al. 2012). The developed dynamic model of the LFP-PEMFC propulsion system, implemented in Simulink environment, allows to characterize the fuel cell dynamic behaviour together with the battery SoC during the daily operation of the bus. For a deep description of the fuel cell dynamic model and the LFP stack, including implemented V_{ocv} and R_{bat}^{int} experimental trends vs. SoC, refer to (Barelli et al. 2012).

The ESS size was determined according to the real driving load demand of Fig. 1b, giving a particular attention to the fuel cell and the hydrogen tank (weight) together with the SoC control strategy. The resulting ESS consists of 140 series connected cells (200 Ah each) (Evlithium 2024) together with a 16.5 kW PEMFC fed by a hydrogen tank of 11.2 kg_{H2} of capacity. Battery pack technical data are summarised below:

- Number and capacity of cells (series connected): 140/200 Ah
- Stack capacity: 95.2 kWh
- Maximum discharge current: 600 A (3C A)
- Standard charge current: 100 A (0.5C A)
- Nominal and peak voltage: 476 V, 560 V

This configuration allowed, with only one initial battery pack charge and H_2 refuelling, 12 h bus operation (over the duty cycle of Fig. 1) with a 10% residual SoC. The system weight was estimated in about 1,620 kg, considering the battery pack, fuel cell, hydrogen cylinders and related piping. This value was considered in the following as the weight constraint for the sizing of all other ESSs.

System B. LFP battery energy storage section

Starting from the battery datasheet (Evlithium 2024), the battery pack was sized to guarantee the nominal (70 kW) and maximum (140 kW) required power and increasing the on-board installed capacity up to the achievement of the maximum weight of 1,620 kg (the weight of both battery modules and auxiliaries is considered). Features of the battery pack, constituted by LFP series connected cells of suitable capacity, are:

- Number and capacity of cells (series connected): 140/260 Ah
- Stack capacity: 123.7 kWh
- Maximum discharge current: 780 A (3C A)
- Standard charge current: 130 A (0.5C A)
- Nominal and peak voltage: 476 V, 560 V

As already indicated above, experimental V_{ocv} and R_{bat}^{int} data, measured on a single battery module, varying the SoC in both charging and discharging processes were implemented in the bus propulsion system dynamic model. The dynamic model implements both the driving and the charging process (0.5 C C-rate is considered for the CC charge), the latter activated when the battery pack SoC achieves the minimum imposed value or at the end of the daily duty cycle.

System C. VRFB-LFP hybrid energy storage section

Batteries capacity Q (expressed in Ah), V_{ocv} and R_{bat}^{int} trends vs. SoC during charge and discharge processes of both LFP battery and VRFB were implemented in the model using look-up tables based on experimental data.

LFP-VRFB hybrid system was sized to cover the nominal load (about 70 kW) by means of the VRFB pack, while a small lithium-ion battery pack provides peak power demand over the nominal value. This strategy was chosen to limit the weight of the VRFB stack (excluding the electrolyte), considering the imposed maximum HESS weight (1620 kg). Consequently, considering 170 kg for the 70 kW VRFB stack (according to an innovative lightweight design including the implementation of composite polymer bipolar plates (Nam et al. 2017)) and auxiliary components (e.g., pumps and piping), about 1,350 kg and 100 kg were considered respectively for the VRFB electrolyte and the auxiliary LFP pack. For what above, regarding the VRFB pack, design constraints were the nominal power and voltage of 70 kW and 476 V (consistent with the values of the conventional system). Under these assumptions, VRFB battery pack was characterized as follow:

- Number of cells: 352 (squared cells with 0.3025 m^2, membrane active area: 55 cm side; single cell nominal and peak voltage: 1.35 V, 1.5 V)

- Nominal voltage: 476 V
- Nominal current density: 50 mA/cm^2
- Nominal power: 72 kW
- Installed capacity: 31 kWh
- Round trip efficiency: 0.87
- Electrolyte energy density: 23 Wh kg^{-1}

Moreover, the size of the LFP auxiliary battery pack was targeted to completely cover the demand peaks during the daily working cycle (12 h). Obtained features are:

- Number and capacity of cells (series connected): 140/16.6 Ah
- Installed capacity: ~ 8 kWh
- Maximum discharge current: 50 A (3C A)
- Standard charge current: 8.3 A (0.5C A)
- Nominal and maximum voltage: 476 V, 560 V.

The auxiliary pack can deliver a maximum constant power of about 24 kW (50 A) and an impulse power up to about 80 kW. It is consistent with the peak power profile of Fig. 1, always under 200 A, considering that the VRFB nominal current is about 151 A. With regards to the HESS power management, the total load demand is split between the two battery packs. A "Switch" block implements such a management strategy by comparing the actual current with the nominal current that the VRFB can supply. The surplus is managed by the LFP additional battery pack. In this way, the latter can compensate the load peaks, while VRFB can cover the base load.

Due to the low on-board installed capacity, the dynamic model includes the VRFB charging phase, i.e., when the VRFB achieves 100% depth of discharge, with a scheduled charge time equals 10 min. This was considered sufficient to fill the tanks with charged anolyte/catholyte fluids. LFP charging was not implemented since the LFP battery pack was sized to completely cover the demand peaks over the daily duty cycle.

The VRFB section of the dynamic model was experimentally validated. In order to validate the model, experimental and simulation results were compared. Specifically, a single VRFB cell was tested setting a constant discharge/charge current equal to 0.6 A. Simulated SoC evolution resulted very close to the experimental one. In particular, in the full discharge test, a gap of about 2% resulted in reference to the total discharge time, while a gap of 6.9% was assessed for the full charge test. Both results can be considered acceptable, even taking into account tests had a long duration over 9 h.

System D. Zebra/PEMFC hybrid energy storage section

System D represents a hybrid solution combining a Zebra (Ni–NaCl) battery with a PEMFC stack.

The PEMFC stack, operating at constant power, recharges the batteries and provides additional energy during periods of high demand, such as accelerations or uphill routes, following a customized control architecture of energy sources (Master

Controller) (Sergi et al. 2014). Thanks to this configuration, the system can extend the bus's operational range by approximately 40% (Andaloro et al. 2013) compared to a not hybrid battery electric system (assuming an equivalent SoC of 20%), while keeping the total vehicle weight below the set limit.

The system architecture includes a battery pack, consisting of six Ni–NaCl batteries, connected in parallel with the DC bus and a 5 kW FC system. The FC system operates at constant power and is supported by two hydrogen tanks, each storing 4.89 kg of H_2 at 200 bar. The current limit of the DC/DC converter has been set to 170 A, corresponding to 29.85 V_{DC} for the 5 kW FC system.

The FC stack consists of 40 series-connected cells with bipolar steel plates, each with an active area of 500 cm^2 and a maximum current density of 400 mA/cm^2 (Ferraro et al. 2009). The cells operating in "dead-end" mode (the anode is closed by an electric valve that allows water purging). The energy produced by the FC is utilized throughout the entire operational period. The FC power recharges the battery and contributes to vehicle traction when the requested power exceeds the battery maximum power.

The features of the Zebra battery are:

- Operating temperature: 310–350 °C
- Number of Cells: 216
- Energy density: 119 Wh/kg
- Weight: 182 kg
- Capacity: 38 Ah
- Rated energy: 21.2 kWh
- Open circuit voltage (0–15% DoD): 557 V
- Max regenerative voltage: 670 V
- Minimum operating voltage: 372 V
- Maximum discharge current: 112 A

With a battery pack of 6 Zebra batteries, the stored energy is estimated at approximately 101.7 kWh (considering a minimum SoC of 20%). The total energy available from the FC stack (hydrogen tanks filled at 200 bar) and the battery pack is 181.7 kWh.

The total weight of the System D is approximately 1611 kg, distributed as it follows:

- Battery pack: 1092 kg
- FC System: 130 kg
- Two hydrogen tanks (H_2 included): 199.78 kg
- DC/DC converter: 50 kg
- Hydrogen tank housing: 120 kg
- Piping, wiring, etc.: approximately 20 kg

5 Time of Deployment, or Implementation (and Challenges to Implementation)

In (Barelli et al. 2019a, b), three different HESSs suitable sized and implemented in the propulsion system of an 8 m-long electric bus (i.e., System A, C, and D) are compared to the case of implementation of a non-hybrid ESS constituted by a LFP battery pack (i.e., System B). All investigated energy storage architectures were sized at parity of on-board installed weight, equal to the total weight of the LFP battery pack hybridised with PEMFC stack (System A) as major design constraint. Main characteristics of the investigated energy storage sections are summarized in Table 1. The comparison was carried out on the same daily profile (i.e., cycle and load duration as depicted in Fig. 1b), considering an operation of 365 days per year. Batteries SoC trends and the consequent required charging time, during the daily bus operation, were estimated by means of the developed dynamic models.

In addition to the daily dynamic simulations, an assessment of the expected LFP battery life was carried out for each investigated ESS architecture. LFP aging was investigated through the analysis of the SoC oscillation on yearly base (annual SoC trend was obtained by repeating the obtained daily operating profile) by applying the Rainflow algorithm method coupled with the LFP cycle to failure curve. Such methodology can estimate battery lifetime when subjected to complex cycles. Obtained results are reported in Table 1 and discussed in the following.

In System A, composed by LFP battery and 16.5 kW PEMFC fed by a 11.2 kg_{H2} tank, thanks to the PEMFC support in continuously providing energy to the battery, SoC never drops below 10%. Thus, only the initial LFP battery charge and hydrogen tank refuelling are required.

Regarding System B, (LFP battery pack only), the SoC trend goes below the minimum threshold before half the day. Hence, a recharge is needed (taking more than 2 h) during operation to terminate the daily cycle.

Table 1 Summary of technical features and the estimated lifespan of the systems

	System A	System B	System C	System D
Technology	LFP/PEMFC	LFP	LFP/VRFB	Zebra/PEMFC
LFP cells' series	140	140	140	–
LFP capacity (kWh)	70	123.7	8	–
VRFB cell series	–	–	352	–
VRFB capacity (kWh)	–	–	31	–
Ni-NaCl cell series	–	–	–	216
Ni-NaCl capacity(kWh)	–	–	–	127.2
On-board H_2 (kg) (kg_{H2})	11.2	–	–	9.78
Lifespan (years)	2.7/4	1.8	9.5/7	10/3
Battery pack recharging/refilling during the daily cycle	1 (LFP)	2 (LFP)	8(VRFB)	1 (Zebra)

On the other hand, the proposed VRFB-LFP HESS (System C) can satisfy the load peaks (about 43 A) and the base load (151 A) by means of LFP and VRFB devices, respectively, as detailed before. Moreover, the VRFB stack can absorb the regenerative power which results in Fig. 14.1 limited by a maximum value of – 93 A.

In detail, System C allows a non-total discharge of the LFP battery, registering a residual SoC slightly lower than 40% at the end of the daily route. However, eight bus stops of 10 min each are needed to refill VRFB tanks. This is in accordance with the decrease in the ESSs mass energy density (about 25 Wh/kg for VRFB) with respect to the other implemented technologies. If System C is compared to System B, the resulting overall energy density corresponds to about one third.

However, System B requires a higher daily recharge time (slightly over 2 h), while System C takes a total recharge time of 80 min. Hence, System C guarantees a longer net driving time of the bus during the considered daily route.

As result, the non-hybrid system based on LFP batteries needed a daily recharging time greater than 70% if compared to VRFB-LFP HESS, although the latter required a high number of stops for anolyte/catholyte refuelling. It is also highlighted as System C requires the handling of aggressive chemicals.

Furthermore, the lifespan of the HESS components was estimated through Rainflow cycle counting algorithm, applying it to the SoC trends of the LFP battery of each system. A bus operation of 365 days per year is considered. Concerning the life expectancy analysis of LFP batteries, the VRFB-LFP hybrid propulsion system shows the highest lifespan. It is remarked as the method based on Rainflow cycle counting algorithm doesn't provide an accurate battery's lifespan estimation. Therefore, provided outcomes can be used mainly to perform comparative analysis among different solutions, rather than as absolute assessments. In the specific case, LFP battery was demanded to satisfy load peaks over the nominal value of 70 kW with a single discharge per day at 60% DoD. Therefore, the LFP pack was greatly downscaled, with only 7% of the nominal capacity of the propulsion system based on the only LFP pack, and life expectancy results in 9.5 years, 3.5 and 5 times the lifespan obtained for LFP battery implemented in the PEMFC-LFP (2.7 years) and LFP (1.8 year) ESS, respectively. Such an outcome directly reflects in lower recurrent costs in relation to LFP battery replacement, with economic advantages in terms of total costs of the bus over its lifespan, though the higher capital cost.

To evaluate the expected life of the whole ESS system, only preliminary assessment was also made on the PEM and VRFB lifespan resulting in an overall greater duration for System C.

PEM lifespan (System A) was estimated in about 4 years, considering 18,000 h of total operation, as suggested for bus fleet application in (Frank de Bruijn 2011; Tom Madden 2010). VRFB lifespan (System C) was assessed equal to 7 years, considering 2920 cycles/year and more than 20,000 working cycles (@100% depth of discharge) indicated for commercial VRFBs (CellCube 2022). ESS systems expected lifespan is reported in Table 1.

Regarding System D, a notable benefit of the PEMFC stack is the ability to maintain a stable state of charge (SoC) in the Ni-NaCl batteries. Thereby, it reduces stress from deep discharges and prolongs the overall lifespan of the batteries themselves.

Zebra battery technology has shown in laboratory tests a lifespan of over 10 years and a cycle life ranging from 1000 to 2000 full cycles. This performance is supported by real-world applications, where batteries are still operational after exceeding 1000 full cycles.

In conjunction with this, the PEMFC stack has been capable of operating for approximately 1000 h without declines in performance (Ferraro et al. 2009). However, the lifespan has reached up to 3 years with non-continuous use of the bus. Additionally, the use of this small fuel cell provides various advantages, including lower costs, reduced hydrogen storage requirements in the onboard tanks, extended range (compared to the equivalent electric vehicle), and shorter recharge times since the batteries are not fully discharged at the end of the journey. These features result in a competitive product in terms of costs, facilitating its introduction into the heavy-duty transport market.

In conclusion, preliminary results obtained in reference to the hybrid systems demonstrated potential advantages in the adoption of HESS with respect to a non-hybrid LFP battery installation. Obtained operating and lifespan data should be reflected in a suitably economic analysis to be performed in consideration of CAPEX and OPEX including replacement costs.

References

Andaloro L, Napoli G, Sergi F, Dispenza G, Antonucci V (2013) Design of a hybrid electric fuel cell power train for an urban bus. Int J Hydrogen Energy 38(18):7725–7732. https://doi.org/10.1016/j.ijhydene.2012.08.116

Barelli L, Bidini G, Ottaviano PA, Pelosi D (2019a) Vanadium redox flow batteries application to electric buses propulsion: performance analysis of hybrid energy storage system. J Energy Storage 24(February):100770. https://doi.org/10.1016/j.est.2019.100770

Barelli L, Bidini G, Bonucci F, Castellini L, Fratini A, Gallorini F, Zuccari A (2019b) Flywheel hybridization to improve battery life in energy storage systems coupled to RES plants. Energy 173:937–950. https://doi.org/10.1016/j.energy.2019.02.143

Barelli L, Longo M, Ottaviano PA, Pelosi D, Zaninelli D, Gallorini F (2024) Vanadium redox flow battery integration in on-board electric systems for hybrid marine applications. IEEE Trans Ind Appl 60(4):6539–6546. https://doi.org/10.1109/TIA.2024.3397786

Barelli L, Bidini G, Ottaviano A (2012) Optimization of a PEMFC/battery pack power system for a bus application. Appl Energy 97. https://doi.org/10.1016/j.apenergy.2011.11.043

Frank de Bruijn (2011) PEMFC lifetime and durability an overview.

CellCube (2022) Datasheet FB250/FB 500 series. https://www.cellcube.com/wp-content/uploads/2021/08/Cellcube_Datenblatt_allgemein_en_01.pdf

European Commission (2017) 2050 long-term strategy | Climate action https://ec.europa.eu/clima/policies/strategies/2050_en

De Lorenzo G, Andaloro L, Sergi F, Napoli G, Ferraro M, Antonucci V (2014) Numerical simulation model for the preliminary design of hybrid electric city bus power train with polymer electrolyte fuel cell. Int J Hydrogen Energy 39(24):12934–12947. https://doi.org/10.1016/J.IJHYDENE.2014.05.135

Estrada Poggio A, Balest J, Zubaryeva A, Sparber W (2023) Monitored data and social perceptions analysis of battery electric and hydrogen fuelled buses in urban and suburban areas. J Energy Storage 72. https://doi.org/10.1016/j.est.2023.108411

Evlithium (2024) 260 Ah Thunder Sky Winston LiFePO4 Battery. https://www.evlithium.com/thunder-sky-winston-battery/lifepo4-260ah.html

Ferraro M, Sergi F, Brunaccini G, Dispenza G, Andaloro L, Antonucci V (2009) Demonstration and development of a polymer electrolyte fuel cell system for residential use. J Power Sources 193(1):342–348. https://doi.org/10.1016/J.JPOWSOUR.2009.02.064

Tom Madden (2010) Progress and challenges for PEM transit fleet applications. UTC Power

Nam S, Lee D, Lee DG, Kim J (2017) Nano carbon/fluoroelastomer composite bipolar plate for a vanadium redox flow battery (VRFB). Compos Struct. https://doi.org/10.1016/j.compstruct.2016.09.063

Offer GJ, Howey D, Contestabile M, Clague R, Brandon NP (2010) Comparative analysis of battery electric, hydrogen fuel cell and hybrid vehicles in a future sustainable road transport system. Energy Policy 38(1):24–29. https://doi.org/10.1016/j.enpol.2009.08.040

Pelosi D, Longo M, Zaninelli D, Barelli L (2023b) Experimental investigation of fast−charging effect on aging of electric vehicle Li−ion batteries. Energies 16(18):6673. https://doi.org/10.3390/en16186673

Pelosi D, Longo M, Bidini G, Zaninelli D, Barelli L (2023a) A new concept of highways infrastructure integrating energy storage devices for e-mobility transition. J Energy Storage 65. https://doi.org/10.1016/j.est.2023.107364

Saldarini A, Barelli L, Pelosi D, Miraftabzadeh S, Longo M, Yaici W (2022a) Different demand for charging infrastructure along a stretch of highway: Italian case study. In: 2022 IEEE international conference on environment and electrical engineering and 2022 IEEE industrial and commercial power systems Europe, EEEIC/I and CPS Europe 2022

Saldarini A, Barelli L, Pelosi D, Miraftabzadeh S, Longo M, Yaic (2022b). Different demand for charging infrastructure along a stretch of highway: Italian case study. In: 2022 IEEE international conference on environment and electrical engineering and 2022 IEEE industrial and commercial power systems Europe, EEEIC/I and CPS Europe 2022. https://doi.org/10.1109/EEEIC/ICPSEurope54979.2022.9854643

Sergi F, Andaloro L, Napoli G, Randazzo N, Antonucci V (2014) Development and realization of a hydrogen range extender hybrid city bus. J Power Sources 250:286–295. https://doi.org/10.1016/J.JPOWSOUR.2013.11.006

Vermeer W, Chandra Mouli GR, Bauer P (2022) A comprehensive review on the characteristics and modeling of lithium-ion battery aging. IEEE Trans Transp Electrification 8(2):2205–2232. https://doi.org/10.1109/TTE.2021.3138357

Wei R, Liu X, Ou Y, Kiavash Fayyaz S (2018) Optimizing the spatio-temporal deployment of battery electric bus system. J Transp Geogr 68:160–168. https://doi.org/10.1016/j.jtrangeo.2018.03.013

Zubi G, Dufo-López R, Carvalho M, Pasaoglu G (2018) The lithium-ion battery: state of the art and future perspectives. Renew Sustain Energy Rev 89(March):292–308. https://doi.org/10.1016/j.rser.2018.03.002

Sustainable and Smart Buildings

Preface

The continuous increase of renewable-based energy systems, both for heating and power generation, at building level requires the development of innovative compact hybrid energy storage. These technologies are able to support the flexible operation of such complex systems, increasing the exploitability of renewables and the overall system efficiency. Part Five of this book focuses on the role of Hybrid Energy Storage Systems (HESS) in behind-the-meter applications and how such kind of hybrid systems could be implemented to meet the dynamic energy needs of modern buildings to support the decarbonization of building stock. Chapter "Buildings (< 50 kWh/day). Integrated Batteries with Phase Change Materials (PCM) for Peak Shaving and Load Management: The HYBUILD Example" gives a brief overview of different energy storage technologies at building level focusing on their integration with onsite renewable energy generation and smart grids. The chapter describes the overall concept, integrating electric batteries, latent, and thermochemical storages, with highly efficient reversible heat pumps for peak shaving and load management and shows the different options developed for continental and Mediterranean climates. Chapter "Hybrid Thermal and Electrical Energy Storage in Office Buildings" investigates the most cost-efficient energy storage solution for a net-zero office in a medium-sized office building in Trondheim, Norway. It determines the optimal combination of latent thermal energy storage and electrical storage for each month during 3 years by solving a convex optimization problem aimed at finding the most cost-efficient energy storage solution. Various factors affecting the performance and cost-efficiency of these technologies, focusing on energy demand, seasonal variations, energy prices, and system response characteristics were investigated. Chapter "Behind-The-Meter. Combination of Li-Ion Batteries and Organic Flow Redox Batteries for BTM Applications" focus on the development of a Hybrid Energy Storage System (HESS), constituted by the integration of a Lithium Titanate and Aqueous Organic Redox Flow battery systems to provide behind-the-meter grid services. In particular, the system has been demonstrated in the Messina community, Italy.

Buildings (< 50 kWh/day). Integrated Batteries with Phase Change Materials (PCM) for Peak Shaving and Load Management: The HYBUILD Example

Andrea Frazzica, Valeria Palomba, Davide Aloisio, Gabriel Zsembinszki, Marco Ferraro, Francesco Sergi, and Luisa F. Cabeza

Abstract The continuous increasing of renewable-based energy systems, both for heating and power generation, at building level requires the development of innovative compact hybrid energy storages. These technologies are able to support the flexible operation of such complex systems, increasing the exploitability of renewables and the overall system efficiency. In this chapter, a brief overview of different storage technologies, such as thermal, electric and hybrid, at building level is provided, mostly focusing on their integration with onsite renewable generation and smart grids. The analysis highlights the relevant role of flexible energy storages at building level as well as the lack of innovative components able to provide multiple services (e.g. heating, cooling, domestic hot water and power) to buildings. In such a background, the experience carried out in the framework of the EU-funded HYBUILD project is described. The overall concept, integrating electric batteries, latent, and thermochemical storages, with highly efficient reversible heat pumps is described, showing the different options developed for continental and mediterranean climates. The fully integrated Mediterranean system, validated both at lab-scale and in a demo building, demonstrated the ability of increasing the share of renewables in buildings, maximizing the self-consumption and increasing the overall energy efficiency of the system.

Keywords Energy storage · Buildings · Batteries · Thermal energy storage · Peak shaving · Load management

---

A. Frazzica (✉) · V. Palomba · D. Aloisio · M. Ferraro · F. Sergi
Istituto di Tecnologie Avanzate per L'Energia "Nicola Giordano", Consiglio Nazionale delle Ricerche, Messina, Italy
e-mail: andrea.frazzica@cnr.it

G. Zsembinszki · L. F. Cabeza
GREiA Research Group, University of Lleida, Lleida, Spain

R. Scipioni et al. (eds.), *Hybrid Energy Storage*, Lecture Notes in Energy 47,
https://doi.org/10.1007/978-3-031-97755-8_15

1 Introduction

The increase of the share or renewable energy into the existing power and heating and cooling infrastructure requires overcoming new challenges (Cabeza and Palomba 2020). Energy storage systems contribute to the stabilization of the variable output of renewable energy (Liu et al. 2024). By storing surplus electricity during high-generation periods and discharging it during low-generation periods, energy storage technologies maintain a balance between supply and demand. Another advantage is the enhancement of the scheduling flexibility of renewable energy sources. Therefore, energy storage brings technical, economic and social opportunities in this new energy system. Energy storage technologies include pumped hydro energy storage, compressed air energy storage, flywheels, supercapacitors, thermal energy storage (TES), batteries, and hydrogen storage.

Batteries serve as a prevalent energy storage medium and are typically classified as short-duration storage systems (Liu et al. 2024). TES is often integrated into concentrated solar power (CSP) systems for short-duration storage due with its large capacity and cost-effectiveness, but recently, its potential for seasonal-storage has been demonstrated (Prieto et al. 2024). At building level, batteries are mainly used for short-term electrical storage, while TES technologies can be either used for short-term (sensible and latent TES) or for long-term storage (thermochemical TES). Therefore, hybrid systems using both batteries and TES can give advantages such as flexibility, short- and long-term storage, energy efficiency, etc.

In the present chapter, a brief overview of the current investigations focused on storage integration in buildings is presented. Then, a case study related to the development and demonstration of innovative hybrid energy storage solutions in buildings, carried out in the framework of the Horizon 2020 project HYBUILD (G. A. 768,824) is described.

2 Integrated Solutions for Hybrid Storage and Power-To-X Systems in Building

In the following sections, a brief state-of-the-art related to different energy storage technologies, i.e., thermal, electric and hybrid, employed in buildings to support heating, cooling and power demand is reported. The main advantages and challenges linked to the use of these technologies are discussed to highlight the relevance of the development of flexible and highly efficient storage solutions to support the future energy systems.

2.1 Thermal Energy Storage + Power-To-X

An overview of the potential inclusion of thermal energy storage (TES) and electrical-thermal energy storage (ETES) in the overall energy system is shown in Fig. 1. Excess electricity from renewable energy sources or for grid balancing can be stored in ETES systems for its later use for heating and cooling purposes (Cisek and Taler 2019). It should be highlighted that both TES and ETES can be installed in the generation side of the energy system (for example, coupled to the energy generation system with molten salts storage), in the transmission and distribution section of the energy system (for example, including heating and cooling networks), and in the demand side to directly cover the load of the user.

There are many reasons to add a TES or ETES device in the energy system. One of the key drivers is given by the need to increase the integration of new renewable energy sources (RES) in the grid (especially the intermittent ones such as wind power and solar energy, to reach the target of up to 100% penetration. Research on this topic has been carried out during the last decade at a global level (Jacobson et al. 2015; Child et al. 2018, 2019) and the importance of a proper mix of energy generation technologies and storage solutions was highlighted (Solomon et al. 2017, 2019), especially in view of a trans-national connection of the energy system (Rasmussen et al. 2012). Other reasons are the reduction of the levelized cost of energy (LCoE) via economic optimization with the adequate balance between production and utilization of energy considering power and time (Kiptoo et al. 2020; Timmons et al. 2020) and peak shifting, to avoid curtailment of the energy produced by RES (Beaudin et al. 2010; Bao et al. 2016; Talluri et al. 2019). On the demand side TES and ETES are integrated two-fold. First, storage solutions help matching the generation

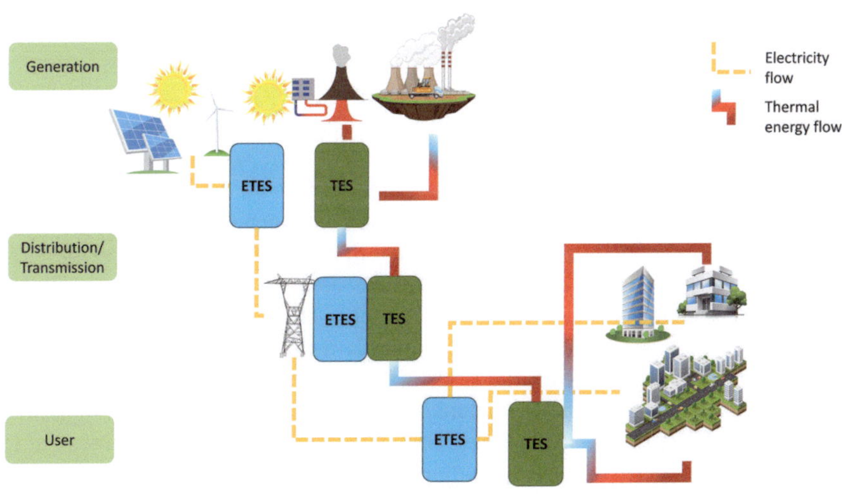

Fig. 1 Possible storage interaction in the energy system. TES, thermal energy storage; ETES, electrical thermal energy storage

capacity with load demand under extremely variable operating conditions to give response to the significant push towards an increased self-consumption and energy balance at district/micro-grid level present today (McKenna et al. 2019). Second, at single building level, TES and ETES are utilized to increase the flexibility of the energy system and achieve contemporarily both human comfort conditions and high efficiency of the system (Palomba and Frazzica 2019; Cisek and Taler 2019).

The need for a combination of solutions for coupling the electricity sector with the heat one has been highlighted in different roadmaps for energy storage technologies (Durand et al. 2013; International Energy Agency 2014; ICEF–Innovation for Cool Earch Forum 2017), but most applications are still linked to the use of electricity storage at the different scales mentioned above.

2.2 *Electric Energy Storage + Power-To-X*

Power-to-X system architectures need the integration of several technologies able to provide the necessary transport and storage of the energy carriers used. This integration requires a combined and optimized use of the energy resources through the management of their production and utilization which normally does not match in time. To do this, energy storage systems are crucial elements to act as buffers between production and consumption, as represented in Fig. 2. This is particularly true if the electricity carrier is considered, given that the presence of renewables introduces an important factor of uncertainty.

Several technologies were and are developed in recent years to realize storage systems more and more efficient, reliable, affordable and generally more performing. In particular, electrochemical storage systems (batteries) have significantly increased their presence on the market, driven by various sectors such as portable devices, transport and stationary services to the electricity grid. The performances currently offered in terms energy and power density, round-trip efficiency and lifetime have pushed these technologies into common use also in building sector, normally coupled with small renewable energy production systems (solar, wind) (IEA 2024). It is important to underline that, in power to X architectures, batteries are not suitable for seasonal storage because they become uneconomical when the stored energy have to be released in weeks or months (Sterner and Specht 2021). On the contrary, they are superior systems when short-term storage is considered, allowing high flexibility and efficiency in power-to-power systems, as schematically reported in Fig. 3. The presence of the storage can have several impact: it maximize the load and generation match; it can reduce stress on the grid, making it more reliable and potentially delaying the need for costly infrastructure upgrades; it can generate revenue or money saving through optimized management of the energy carriers cost and prices; it is able to face unexpected events, maintaining systems stability and allowing superior user comfort (Airò Farulla et al. 2021).

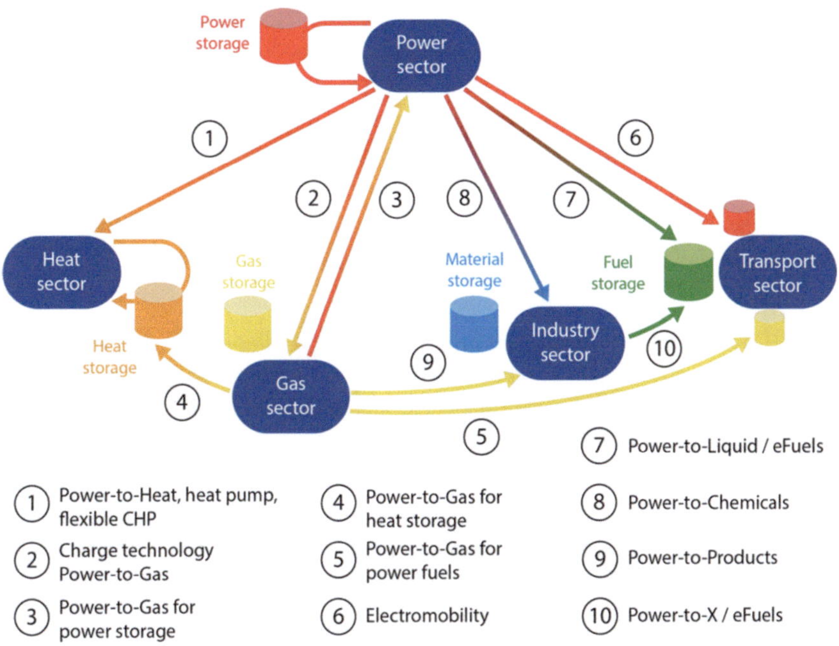

Fig. 2 Sector coupling resulted from power-to-gas and power-to-X (Sterner and Specht 2021)

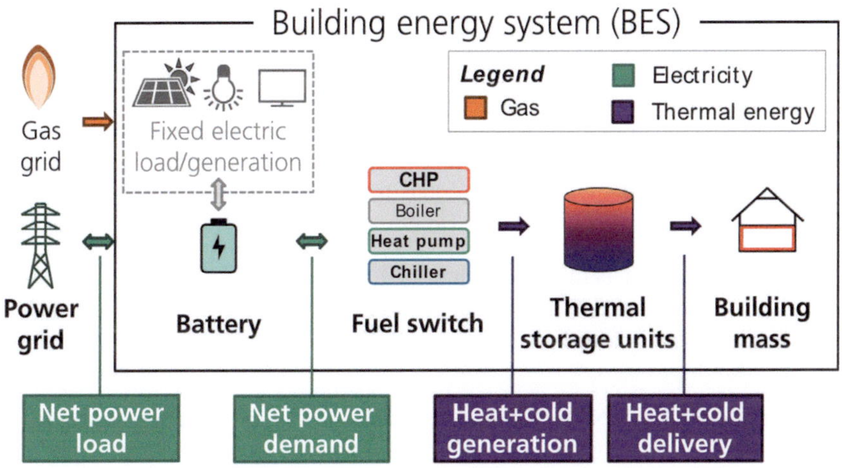

Fig. 3 Building energy system with different flexibility and storage options possible (Klein et al. 2017)

As described in (Rey et al. 2023) the main technology among batteries is currently the Lithium-ion cell (Li-ion). They are nowadays widely used in several application sectors due to high energy and power density, very high round-trip efficiency, good lifetime and rapid charge and discharge times. Different types of lithium battery exist according to active materials used for their realization. Among them, the most important are:

NCA ($Li(Ni,Co,Al)O_2$): the presence of aluminum reduces cobalt use (critical) and also volumetric changes. They ensure long life and overall good performances but lower safety due to cathode instability.

NMC ($Li(Ni,Co,Mn)O_2$): the current most used technology with several different percentages of constitutive components. They offer customizable properties, high capacity, good C-rate and performance. Better safety than NCA but it is still a concerning issue. Stationary market is exploring their use after the first life (second life application).

LFP ($LiFePO_4$): well known for their superior stability, wider temperature window and high lifetime. However, they have a lower nominal voltage and consequent lower energy density. For their characteristics they constitute the dominant technology in Behind the Meter (BTM) applications.

LTO ($Li_4Ti_5O_{12}$): take the name of the different anode material (LTO vs graphite), they offer an exceptional lifetime compared to other lithium batteries and good safety. On the contrary, their cost is the main drawback.

The current evolution stage is called GEN3 with goal of optimized NMC811 and high voltage cell with 3D oxide-structured spinel. The GEN4 (2025–2030) will focus to commercial deploy of solid state batteries.

A schematization usually used to compare the different characteristics of lithium batteries in terms of specific energy (capacity), specific power, safety, performance, life span, and cost is realized via radar diagrams (higher value is better) and shown in Fig. 4 (Miao et al. 2019).

Nickel batteries are generally less energy efficient than Li-ion batteries, but they perform well at high current rates. They are also safer and more robust, which makes their control systems cheaper (i.e. they don't need balancing). The main technology currently developed is:

Nickel-metal-hydride (Ni-MH): they use hydrogen alloys and are common in portable devices. They offer good energy density and C-rate. However, they are heavier and bulkier than Li-ion batteries. Their main advantage is that they don't form dendrites, reducing the risk of overheating and internal short circuits.

Sodium Batteries base their popularity on the abundance and low cost of their main material. Several technologies, mainly divided into low temperature and high temperature sodium batteries, are currently developed. The main technologies with potential use in building are:

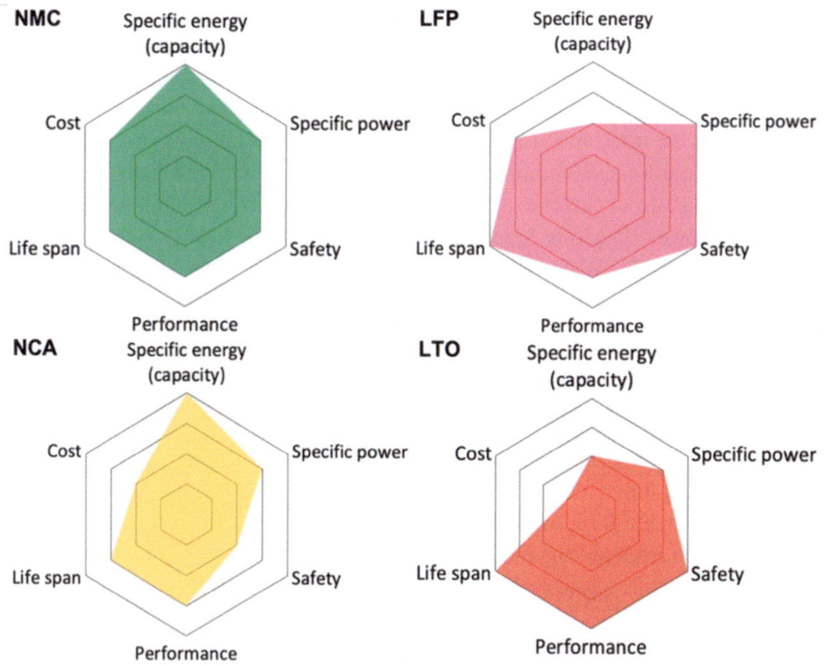

Fig. 4 Radar diagrams to compare different lithium batteries for building applications (Miao et al. 2019)

Sodium ion (low-temp): considered the heirs of lithium-ion, are appearing on the market with different manufacturing process. They feature a lower energy density (probably reaching LFP one) but high interests in research sector could fill the gap. They are safer and cost-effective, offering high C-rate and good lifetime.

Sodium-nickel-chloride (high-temp, $NaNiCl_2$): they operate at high temperatures (270–350 °C) and involve a charging process where nickel and salt transform into molten sodium and nickel chloride. They have a long life cycle and high energy density but require significant thermal management.

ZEBRA cells: they use iron chloride, nickel chloride, or a mix of both with molten sodium tetra chloroaluminate as electrode material. They are safer, less corrosive, and operate over a wider temperature range than other high temperature batteries such as sodium-sulfur batteries. They have a long life cycle (2500–4500 cycles) (Solyali et al. 2022).

Some technologies ($Na–O_2$, $Na–CO_2$) show good performance at the small-scale level, nevertheless, further development activities are needed to improve sodium anodes technology. Sodium-ion based batteries will become widely diffused in next years and also solid state device can improve performance and safety (Wu et al. 2024).

All the technologies previously explained are suitable for small size systems (i.e. below 50 kW). A family that is arousing interest in stationary applications are the Flow Batteries. They are particularly attractive due their capacity to be designed separating the energy content, which depends on the tanks' size, from the power released, which depends on stack size. Electro-active materials (held in tanks) constitute two redox couple solutions. By pumping through the stack, energy exchange can occur. They offer long life cycle, low-maintenance and deep discharge. However overall efficiency is lower than other technologies. Normally they are inserted in larger systems (hundreds ok kW and more), given that the cost depends a lot on the various systems needed for operation. Main technologies developed are:

Zinc-based flow batteries (Zn/Cl_2 or Zn/Br_2): they offer high energy and power density, low environmental impact and low price, gaining popularity among batteries. The main drawbacks are efficiency and volume used for minimum power release.

Vanadium redox batteries (VRBs): they store energy through electron transfer between different ionic states of vanadium. They are known for their long-duration energy storage capabilities, high power release and quick time response. Low energy density and high costs are the main limits.

Polysulfide bromide (PSB) flow batteries: they use a reversible electrochemical reaction between sodium bromide and sodium polysulfide. A polymer membrane separates the electrolytes, allowing sodium cations to transfer between electrodes. PSB batteries are known for their quick response time, high energy efficiency and low environmental impact. High preparation cost is the main challenge for diffuse commercialization.

All the technologies briefly explained show different advantages and disadvantages with Li-ion technology leading the sector. Their main role in power to x economy is to make renewable electricity to become the primary energy source (Breyer et al. 2024). In addition, the flexibility offered by fast response, high C-rate and high efficiency storage system can enable optimized management of the electricity used, allowing price-based behaviors and demand response (DR) programs (Xu et al. 2022). Some works in literature show the importance given by the electricity storage system to implement efficient decision-making power-to-x technologies able to manage uncertainties and variability of these systems (Burre et al. 2020). In particular, at building level, the combination and integration of renewables, batteries, heat pump and/or thermal storage is very interesting. In (Baraskar et al. 2024) how smart control strategies can optimize heat pump operation to reach higher and better self-consumption is highlighted. Some management algorithms were developed and validated using some key performance Indicators (KPIs), showing that seasonal performance factor (SPF) is improved. It practically describes the benefit related to a reduced grid electricity need to meet the household heating demands. In REACT project (Freeman and Coakley 2021), heat pump systems and energy storage combined use (properly shifting when heating or cooling energy is used) enables demand-side management or demand-response strategies. Three paths, using electrical or thermal storage, are possible to manage electricity and heat in buildings and different use strategies and challenges were explored inside this project. The work in (Palomba et al. 2021) explores more in detail the realization and control of hybrid thermal/electrical storage system coupled with heat pump. The system was characterized and studied to improve efficiency and lifetime in a specific use case. Hybridized storage will be further explained in the following section.

2.3 Hybrid Energy Storage + Power-To-X

The definition of hybrid storage is wide, covering the integration of different electric storage technologies (e.g., batteries and supercapacitors (Anta et al. 2024)), electric and mechanical storages (e.g., batteries and flywheels (Li et al. 2024)), electric and chemical storages (e.g., batteries and hydrogen (Jacob et al. 2018)). This paragraph focuses on the analysis of the existing examples of hybrid storages dedicated to the electric and thermal provision to buildings. Indeed, this hybridization approach is considered extremely urgent, taking into account the route towards the decarbonization of buildings (European Commission 2022). Actually, overall, the building sector still represents around 40% of energy consumption in Europe with a large portion of it caused by heating and cooling demand (UNEP 2024).

In this context, most of the reported papers in the literature are focusing on the optimization of the joint operation between electric and thermal storages, using a numerical model approach. For example, Mehrjerdi and Rakhshanib (2019) proposed a hybrid model for an energy storage system integrating both thermal and electrical storage within a building, in which both thermal and electrical loads are modeled using Gaussian probability distributions, considering ideal electrochemical and thermal energy storage systems with a certain efficiency and charge/discharge power (i.e. regardless of the technology). The electric energy supply was provided by the electrical grid, and the goal was to minimize daily energy costs through the optimal operation of the hybrid storage system. Scenario-based stochastic modeling was employed to address the uncertainty in load forecasting. The obtained results showed that the electrical storage system can reduce costs by approximately 15%, and the thermal storage system reduced costs by around 17%, and the coordinated hybrid thermal-electrical storage system can cut costs by about 34%, resulting in the most effective way to integrate energy storage in buildings. Dong et al. (2023) investigated the role of hybrid energy storage, including electric, hydrogen and thermal provision, to support rooftop photovoltaic (PV) operation on a building. A generic model for electric energy storage is used, whereas the thermal energy storage model is based on a water tank. The numerical modelling investigation and optimization demonstrated that the overall integrated system can achieve an annual return on investment of 36.37% and a levelized cost of energy of $0.1016 per kWh. Moreover, it can reduce the annual carbon emissions by 25.5 tons compared to a system with only rooftop PV + electrochemical energy storage. This confirmed the potential of hybrid energy storage solutions to mitigate the stress on the local grid by reducing the peak power needed. Another example, by Brandt et al. (2022) focused on the development of a simplified method to optimize the sizing of hybrid battery/TES systems. The analysis highlighted that integrating batteries with a TES system enhances the system's load-shaving capacity by 20% compared to a system with only electrochemical energy storage. At the same time, hybrid energy storage improves the economics of the system, thanks to the lower capital cost of the TES against the battery, allowing for a reduction of annual expenses in the range of 10 000–60 000 $. Moreover, in the case of climatic zones with high cooling demand, the hybrid solution is always more efficient than the standalone battery or TES system since it increases the available discharge power by up to 40%.

The above reported analysis highlights the huge potential of hybrid storage solutions in buildings, especially when coupled with onsite renewables and innovative heating and cooling provision devices (e.g. heat pumps). Nevertheless, most of these analyses are using standard storage technologies, marking a lack of experimental activities dedicated to the development of novel components, able to maximize the integration of hybrid thermal/electric storages both from the hardware and controlling point of view. In the following, an example of innovative solutions for heating, cooling and power storage and provision in single-family buildings is presented, to highlight also the need of innovative technological packages to make hybrid storages competitive for future market uptake.

3 Description of a Fully Integrated Hybrid Solution for Buildings Integration: The HYBUILD Solution

Within the EU-funded project HYBUILD (Cordis web page 2017), an innovative system to produce heating, domestic hot water (DHW), and cooling was developed for Continental (only heating and DHW) and Mediterranean (including cooling) climates. Most important is that the system included sensible, latent, and electrical energy storage, to increase the use of renewable energy on-site (one of the main KPIs targeted). Moreover, this system integrates a three-media refrigerant/phase change material (PCM)/water heat exchanger in a heat pump using a low GWP refrigerant, R32.

3.1 HYBUILD Solution for Continental Climate

The Continental solution was developed for a multi-family house (MFH) located in Stuttgart (Germany), as being representative of the building typology for the building stock in Continental climate regions in Europe. The innovative system for this Continental climate is shown in Fig. 5. The main components of the system are a PV system connected to the heat pump, ten sensible heat storage DHW storage tanks

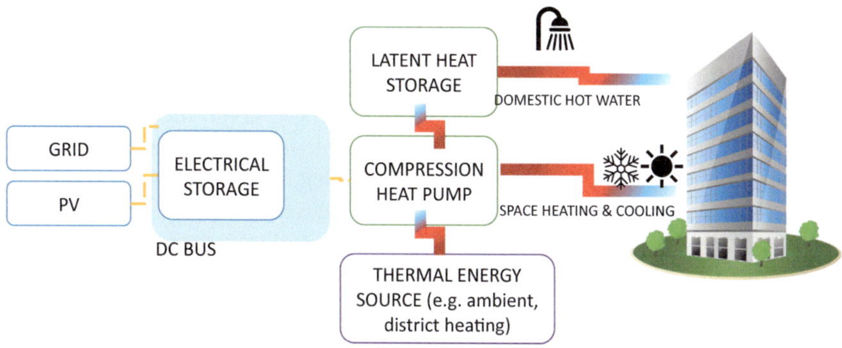

Fig. 5 Schematic diagram of the innovative hybrid system for the continental climate

(one for each dwelling), a high-temperature latent heat storage tank, and a Lithium-Titanate-Oxide electrochemical energy storage system. The latent heat storage, which employs a commercial paraffinic material, is connected to the compressor outlet to store part of the energy contained in the hot refrigerant gas that leaves the compressor, which is used to generate DHW in an efficient way. There are many innovative aspects in the proposed system, such as the direct integration of an innovative three-media refrigerant/PCM/water heat exchanger (RPW-HEX) in the hot superheated section of the heat pump, the use of electric storage combined with both sensible and latent heat storage, and the use of a DC microgrid and innovative control for coupling the electric grid with the thermal distribution. Therefore, the use of PV panels and both the thermal and electrical storage systems help increasing the share of renewable energy.

The sizing of the main components of the system was as follows. The PV system nominal power and the electrical storage capacity were calculated with an iterative process to ensure that 40% yearly self-consumption and self-sufficiency were ensured, therefore 10 kWp of PV peak power and 15 kWh of electrical storage capacity (battery) were used. The water tank (sensible heat storage) was fixed according to the technology provider to 140 L. The heat pump had 30 kW nominal heating power to ensure coverage of building space heating peak demand. Finally, 80 kg of PCM were used to maximize its contribution to the domestic hot water (DHW) production.

The energy consumption of the building was calculated via dynamic simulations with TRNSYS (Klein et al. 1979) simulating each system component using standard or specifically developed types or performance maps provided by the manufacturer or developed experimentally. The annual energy consumption of the different components of the system was 6374 kWh/year for the heat pump, 3654 kWh/year for the electric heater, 658 kWh/year for the fan coil, and 392 kWh/year for the circulation pumps, with a total of 11,078 kWh (Llantoy et al. 2021).

The economic performance of the full system was assessed (Emhofer et al. 2020). The systems showed a payback time of 12.4 year with energy savings of 622 kWh_{el}

per year. That analysis also showed that this system is best suited for low-energy buildings in cold climates.

The environmental performance was also assessed using the life cycle assessment (LCA) methodology using both the ReCiPe and IPCC GWP indicators. For the Continental system, results showed that the overall impact (measured with both indicators) of the innovative system is lower than that of the reference system. The impact for the operational stage (manufacturing stage and disposal stage) is higher than that of the reference system, but the lower impact during the operational stage compensates for it (Llantoy et al. 2021). The analysis of the subsystems considered shows that the sensible heat storage and the PV panels are the subsystems with the higher impact (34% higher and 30% higher, respectively), while the high-temperature latent TES storage subsystem has a contribution of 20%. Finally, the other two subsystems considered have the lowest impact contribution, with 10% the electrical storage and 6% the compression heat pump.

3.2 HYBUILD Solution for the Mediterranean Climate

The Mediterranean solution was developed for a single-family house (SFH) located in Athens (Grece), as being representative of the building typology for the building stock in Mediterranean climate regions in Europe. The innovative system for this Mediterranean climate is shown Fig. 6. The main components of the system are a field of lineal Fresnel solar collectors and a PV system to increase the use of renewable energy sources. To improve the energy efficiency of the heat pump working in cooling mode, a sorption chiller is used in cascade with the heat pump. The heat produced by the Fresnel collectors is used to drive the sorption chiller in summer, but it is also used to contribute to the energy supply for heating and DHW to the building. In addition, the system incorporates a low temperature PCM thermal energy storage tank and an electric battery to help increase the energy efficiency of the system.

The sizing of the main components of the system was as follows. The PV system nominal power and the electrical storage capacity were calculated giving 20.9 m^2 of PV panels and 7.3 kWh of electrical storage capacity (battery). The water tank (sensible heat storage) was fixed according to the technology provider to 800 L fed with 60 m^2 of Fresnel solar collectors. The heat pump had 13.2 kW nominal cooling power to ensure coverage of building space cooling peak demand with a cooling storage capacity of 12 kWh of PCM storage. Finally, the DHW tank had 250 L.

Again, the annual energy consumption of the system was calculated with TRNSYS. The energy consumption of the heat pump was 1564 kWh/year, for the dry cooling was 215 kWh/year, for the adsorption storage was 80 kWh/year, for the DHW electric heater was 552 kWh/year, and for the circulating pumps was 355 kWh/year, with a total annual energy consumption of 2766 kWh/year (Zsembinszki et al. 2021).

The LCA carried out using both the ReCiPe and IPCC GWP indicators showed that the overall impact (measured with both indicators) of the innovative system is

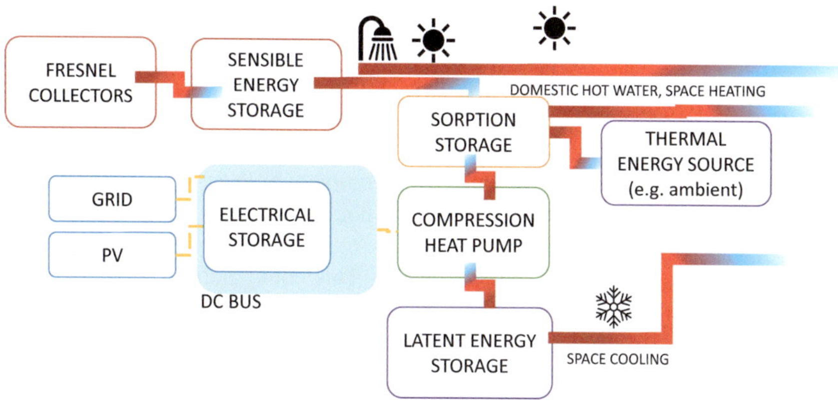

Fig. 6 Schematic diagram of the innovative hybrid system for the Mediterranean climate

higher than for the reference system, mainly due to the higher complexity of the system. Due to the lower use of electricity from the grid compared to the considered reference system, the impact during the operational stage is lower than during the manufacturing and disposal stages. So, in this case, the higher impact in manufacturing due to more components in the system is not compensated by the operational stage. When analysing the subsystems, the higher contribution in the overall impact comes from the TES system, the sorption storage, and the solar field (with 29%, 27%, and 21% contribution respectively). The other subsystems (electrical storage, heat pump, PV panels, and sensible heat storage) have a much lower contribution (14%, 7%, 1%, and 1% contribution respectively) (Zsembinszki et al. 2021).

3.3 *Core Technologies*

The proposed system for the Continental concept includes the following main components:

- Reversible heat pump.
- Latent heat storage.
- Electricity storage.

The reversible heat pump employs R32 refrigerant, a low-GWP refrigerant and its main peculiarity compared to the state-of-art is the use of a DC-driven compressor. Indeed, as shown in the schematics in the previous sections, the HYBUILD system includes a DC bus, for the connection of the PV, the electricity storage and the compression heat pump. Energy from the grid when local production is not available is also transmitted to the various components through the DC bus by using an AC/DC converter. This choice allows a better exploitation of the renewable energy sources on site, since it eliminates the conversion stages between the PV, the electricity storage and the heat pump.

The latent heat storage consists of the desuperheater of the heat pump, as discussed in (Emhofer et al. 2020, 2022). It is an aluminium heat exchanger with passages for three fluids: PCM, heat transfer fluid (HTF), and refrigerant. It is installed in the refrigerant line of the heat pump and uses a commercial PCM with melting point of 64 °C. The PCM is directly charged by the hot refrigerant of the heat pump and releases the heat to the heat transfer fluid, which is, in turn connected to the domestic hot water distribution system of the building. This configuration, with the latent storage embedded inside the heat pump, allows reducing the heat losses that would occur in case of transferring the heat from the refrigerant to a second fluid and then to the PCM.

The proposed system for the Mediterranean concept, instead, includes the following main components:

- Fresnel collectors.
- Reversible heat pump.
- Sorption storage.
- Latent heat storage.
- Electricity storage.

The general idea is to exploit the cascading integration of the sorption system with the vapour compression unit. The energy needed to drive the sorption unit is solar energy, which is harvested by Fresnel collectors. The peculiarity of the Fresnel collectors for the HYBUILD system, is their modularity and the possibility of being used also in small and medium-scale applications.

The sorption unit consists of two modules employing zeolite/water working pair, based on the concept patented by Fahrenheit GmbH, which allows the growth of the zeolite directly on the aluminium heat exchangers. An extensive description of the concept is given in (Velte-Schäfer et al. 2023). The main advantage of the HYBUILD sorption unit compared to other adsorption chillers on the market, is the use of the zeolite, which can operate better at high external ambient temperatures.

The cooling effect to the end-user is provided by the evaporator of the vapour compression unit. The peculiarities in the units are the cascade connection to the sorption unit and the integrated refrigerant-PCM-water heat exchanger. The cascading connection consists in the hydraulic connection of the heat transfer fluid side of the condenser of the compression chiller with the evaporator of the adsorption one. In this way, the heat from the vapour compression chiller is "pumped" to the adsorption chiller, which then discharges it to the ambient by means of a dry cooler. In this way, the condensation temperature of the vapour compression chiller is reduced compared to the ambient temperature. This allows reducing the pressure difference between evaporator and condenser in the vapour compression heat pump, limiting the energy consumption to drive the compressor. The refrigerant-PCM-water heat exchanger is extensively described in (Mselle et al. 2022a, b) and consists of an aluminium heat exchanger of the multi-port extruded tube type, in which there are three separate circuits: for the PCM, for the refrigerant of the vapour compression unit, and for the heat transfer fluid, which is used to actually deliver the cold energy produced to the user. A commercial PCM with melting point of 4 °C was selected for the purpose. It is worth mentioning that the configuration with the three fluids allows the direct integration of the storage inside the vapour compression unit, thus eliminating the needs for extra heat exchangers for their connection.

Finally, the electricity storage for both the Continental and Mediterranean concepts is based on lithium-titanate-oxide (LTO) batteries. This choice was made due to its long lifespan, which aligns with the typical lifespan of PV systems, as well as its exceptional safety—essential for building installations—and high charge and discharge C-rates. This makes it suitable for various services (both energy and power), including operation with on/off heat pumps. Specifically, during the project development, commercial batteries were employed, where the Battery Management System (BMS) was adapted, and communication interfaces were debugged to ensure full control of the storage system by the supervisor. Testing at the cell, module, and pack levels showed strong alignment with the expected specifications. In particular, tests conducted in a climatic chamber at high temperatures demonstrated excellent safety and performance, even at temperatures up to 45 °C. Lab-scale simulations under real operating conditions revealed a high self-consumption rate due to the battery storage (56–62%) and a very low average operating temperature (27 °C), ensuring safe operating conditions.

4 Case Study: HYBUILD System in Mediterranean Climate

In the following, the performed validation and demonstration campaigns for the developed hybrid storage solutions are described. The concept was firstly evaluate in the lab under controlled conditions and then installed in a demo building to analyze the seasonal operation.

4.1 Lab-Scale Validation

The system was first validated at lab-scale at CNR ITAE (Messina, Italy). The test stand with the system connected is shown in Fig. 7.

The validation was done at four different levels. First the operation of the system without the sorption module in terms of typical dynamic evolution was evaluated. Then, the cascade mode in terms of dynamic evolution was tested. Following, an overall energy balance of the system to identify the relative flows and contributions of the components to the overall energy required and supplied by the system was investigated. Finally, the performance maps of the system for the various operating

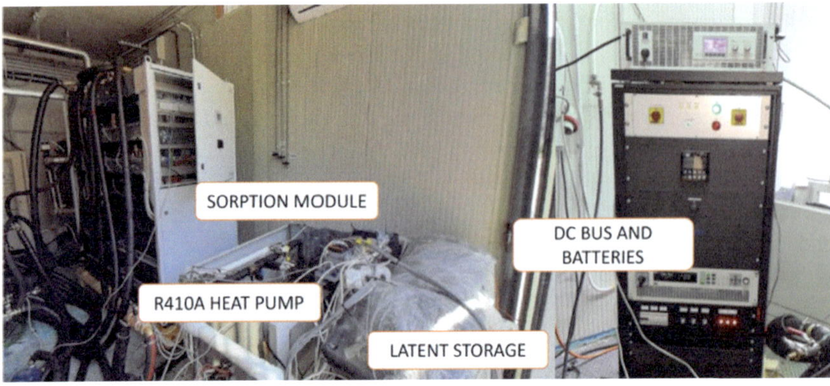

Fig. 7 HYBUILD system installed at CNR ITAE lab (Palomba et al. 2021)

modes as a function of boundary temperatures in the different circuits was calculated. All the results are presented in Palomba et al. (2021); here only a summary is included.

The test without the sorption module were carried out to calculate the electricity consumption of the vapour compression heat pump while charging the latent heat storage at different ambient temperatures, as well as to evaluate the time needed to charge and discharge the storage and the correspondent energy released. The energy that the storage was able to accumulate and release was up to 1.8 kWh, with an average EER (Energy Efficiency Ratio) of the compression heat pump in the range of 3–4. The cascade operation includes, instead, the combination of the sorption storage + compression heat pump + latent heat storage. In this case, the energy efficiency of the heat pump increased and, to achieve the same cooling effect, an EER of 4–7 was measured.

Figure 8 shows the thermal (top) and electric (bottom) powers measured during a test in cascade mode. It is possible to notice the typical cyclic behaviour of the adsorption unit. The behaviour of the compression unit (see as an example the power at the compressor, purple line) is also partly following the oscillations of the sorption unit, due to their cascade coupling. The cooling power at the evaporator of the compressor unit is always in the range of 5–7 kW, with an electricity consumption that is almost constant at 2 kW.

The energy balance for different operating modes is shown in Fig. 9. For consistency in comparison, identical boundary conditions were applied: inlet medium temperature for the sorption module ($MT_{in,sorp}$) or for the compression module ($MT_{in,comp}$) was set at 33 °C, with an inlet high-temperature source (HT_{in}) of 85 °C and an outlet low-temperature target ($LT_{out,comp}$) of 5 °C. A comparison of the first operating mode (charging latent storage) with the parallel charge/discharge mode reveals that approximately one-third of the total evaporation heat is stored in the phase change material (PCM) during parallel charge/discharge operations, while the relative electricity consumption remains unchanged. Similarly, for the charge of the latent storage and the parallel charge/discharge of the latent storage in cascade mode,

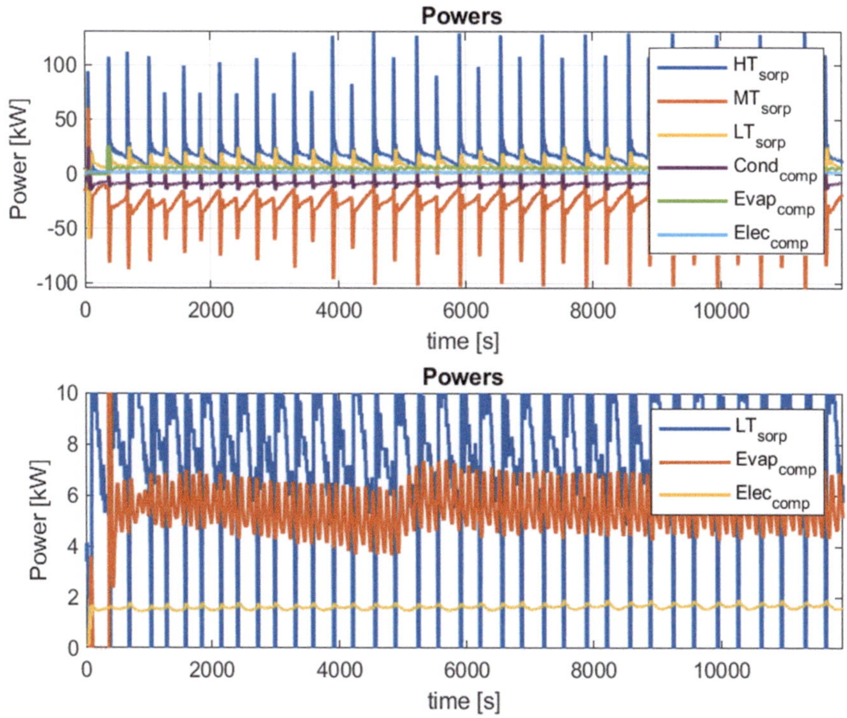

Fig. 8 Example of power in the various components for the HYBUILD Mediterranean concept during lab tests: on top all the powers measured in the prototype, at the bottom a detail of the cooling generation. HT_sorp stands for thermal power driving the adsorption module; MT_sorp stands for thermal power rejected to the ambient by the adsorption module; LT_sorp stands for the cooling power delivered by the evaporator of the adsorption module, cooling down the condenser power of the electric chiller (Cond_comp); Evap_comp stands for the cooling power delivered by the electric chiller; Elec_comp stands for the electric consumption of the electric chiller

the findings indicate that about two-thirds of the overall cooling effect is delivered directly to the user, with the remaining portion of the evaporation heat stored in the PCM. Generally, under these operating conditions, the system achieves a heat input to cooling effect ratio of 3:1, while the compression unit demonstrates an electrical efficiency exceeding 3.

4.2 Demo Implementation

The system described was implemented in a two-floors single-family house, located at Almatret (Lleida, Spain) and built in 1970. Minor renovation was carried out in 2014, replacing windows, blinds and balcony doors to improve their thermal performance. The building serves as medical office and the second floor has a residential

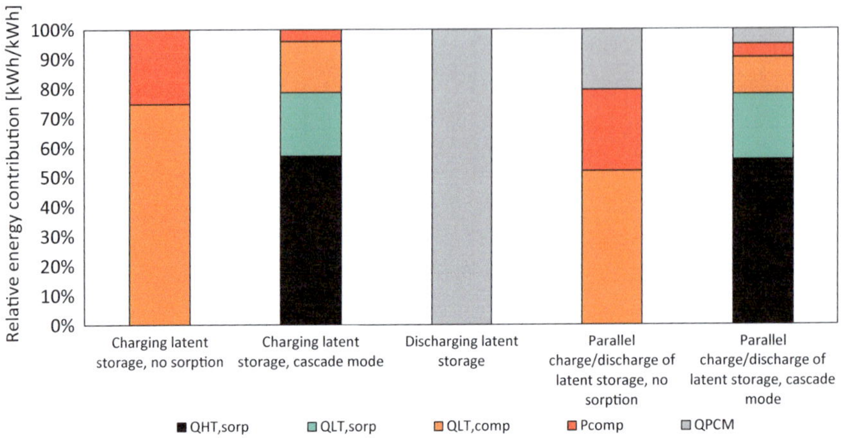

Fig. 9 Energy balance of the HYBUILD Mediterranean concept during lab tests

purpose, where the doctor family lives. The tested system serves the heating and cooling demand of the first floor. This first floor has 8 rooms, with a gross area of 132.5 m^2 and a net/heated area of 112.5 m^2. At the time of the system implementation, the house was heated with a propane gas boiler, and there was no air conditioning system. The main aim of the new system was to provide comfort conditioning during Summer, using the available solar energy. The estimated yearly space cooling demand is around 2600 kWh and the cooling peak power 3 kW (Rossi et al. 2021).

The hybrid multi-energy system tested at Almatret is based in the diagram illustrated in Fig. 5, integrating different components, including 6 modules of Fresnel solar collectors equivalent to 60 m^2 of mirrors surface are, 14 monocrystalline PV panels covering an area of 28 m^2 and providing a total power output of 5.74 kW_p, a sorption chiller connected to a dry cooler, a standard heat pump working with R410A refrigerant, modified to allow the integration of a PCM tank working as evaporator (with RT4 PCM with a nominal melting temperature of 4 °C), a gas boiler working as back-up, a 800 L buffer tank storing the water coming from the Fresnel solar collectors and being used as heat source for the sorption chiller, a DHW tank, an electric battery installed on an electric rack, and connection to the power grid via a DC bus. The system is designed for combined heating, cooling, and DHW production, utilizing renewable energy sources and storage for optimized building energy performance. The components are shown in Fig. 10. The installation was equipped with the required sensors to be able to monitor the performance of the system.

The first evaluated results were the harvested solar energy with the Fresnel solar collectors and the daily accumulated heat flows to the DHW tank, the sorption chiller, and the buffer tank (including ambient heat losses) (Fig. 11). Results show very different behaviour in the different days evaluated, due to sun availability and cooling demand. When the sorption is not active (i.e., the first two days), only about 20% of

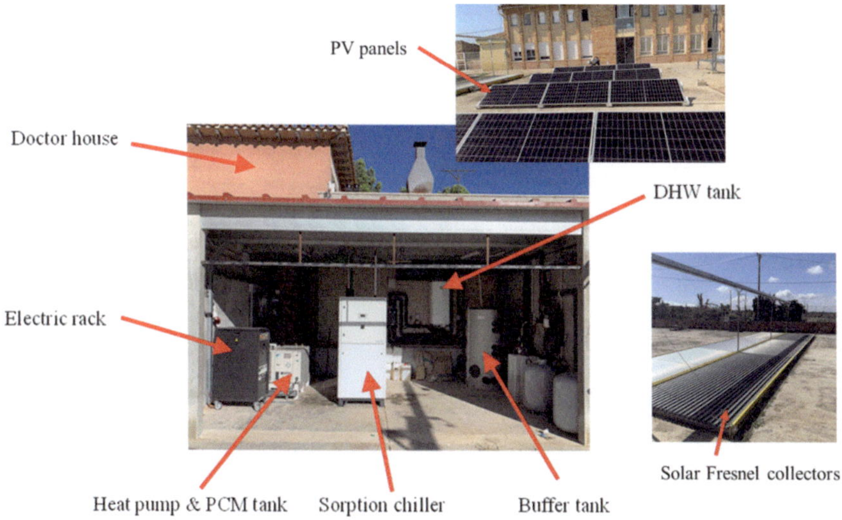

Fig. 10 Demo site located at Almatret (Lleida, Spain). Adapted from (Zsembinszki et al. 2024)

the harvested solar energy is used for DHW. When cooling is needed, around 65% of the energy produced is used by the sorption system (Day 3).

The assessment over a full season was done using the thermal seasonal energy efficiency ratio ($SEER_{th}$) and the thermal seasonal performance factor (SPF_{th}). Experimental results from Almatret (Spain) gave a $SEER_{th}$ of 0.35 while the numerical simulations of a theoretical model in Athens (Greece) gave a $SEER_{th}$ of 0.57. On the other hand, the experimental SPF_{th} was five times higher than the theoretical one.

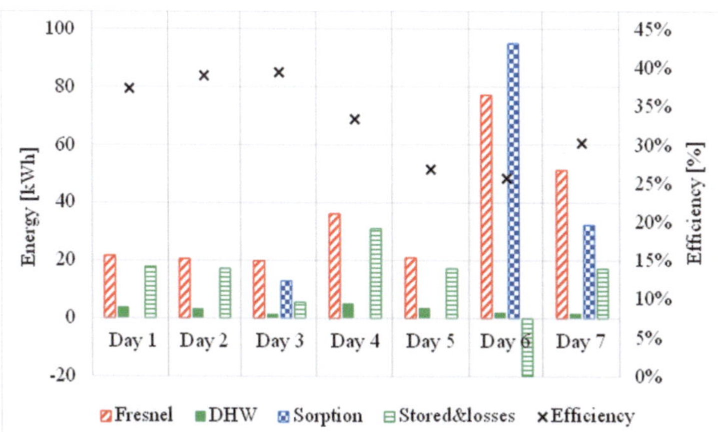

Fig. 11 Harvested thermal solar energy and heat flows in the system up to the buffer tank (Zsembinszki et al. 2024)

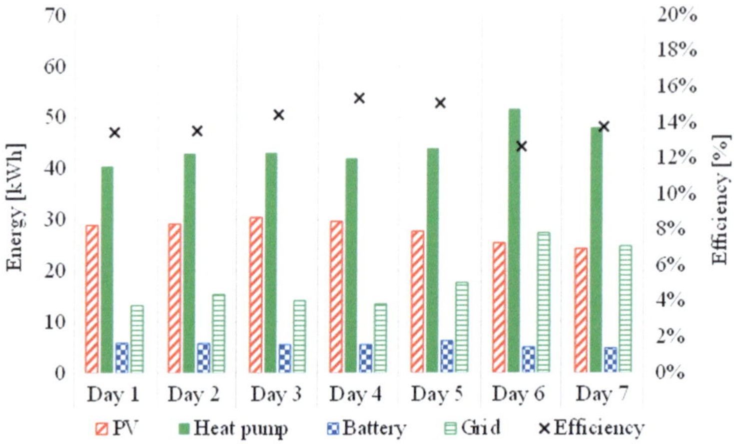

Fig. 12 Harvested electric solar energy and accumulated electricity flows (Zsembinszki et al. 2024)

Looking at the results related to electricity production and use (Fig. 12), first the PV production shows a weekly average of 28 kWh, delivering between 5 and 6 kWh to the battery.

The assessment of the full season electrical performance gives an average daily self-consumption of 0.80, higher than the 0.70 obtained theoretically. The self-sufficiency experimental result was similar to the theoretical one, around 0.50. Finally, the experimental average share of renewables was 0.67. The weekly average of the share of renewables is higher than the self-sufficiency, as expected, due to the contribution of the renewable fraction of the power grid.

5 Lessons Learned and Recommendation for Future Development

The deployment of hybrid energy storage solutions in buildings can help in increasing the exploitation of onsite renewables to support the decarbonization of the building stock. Most of the investigations in this field aimed at analysing the optimal integration of electric and thermal storages through numerical analyses, leveraging on existing standard battery and thermal storage technologies.

The EU-funded HYBUILD project aimed at moving one step further in this sector, by coupling innovative management strategies with new storage components development. The investigation performed both at lab-level and in a real building environment demonstrated that the developed concept is able to promote a PV self-consumption up to 0.8, with self-sufficiency of 0.5 and share of renewables for heating, cooling and power generation up to 0.67. This proved the possibility of significantly reducing the non-renewable primary energy consumption. On the contrary, the integration between concentrating solar collectors and adsorption module showed some criticalities, lowering the overall performance and increasing, under some conditions, the use of dry cooler to properly operate the vapour compression heat pump.

These findings highlight the potential of combining advanced heating and cooling technologies with thermal and electrical storage and renewable energy sources to create more efficient and sustainable building energy systems. On the other hand, it has to be considered that, from the LCA point of view, the complexity of the system increases a lot the impact during the manufacturing phase, which is only partially compensated by the environmental impact during operation.

In general, this first development demonstrates the potentiality of the solution, which requires more studies in different weather conditions to optimize its year-round efficiency. Some design optimizations will be needed to reduce the manufacturing efforts and the system' complexity, to make it more appealing for future commercialization.

Acknowledgements This project has received funding from the European Union's Horizon 2020 research and innovation programme under grant agreement No 768824 (HYBUILD). This work was partially funded by the Ministerio de Ciencia, Innovación y Universidades de España (RTI2018-093849-B-C31–MCIU/AEI/FEDER, UE) and the Ministerio de Ciencia e Innovación (PID2021-123511OB-C31—MCIN/AEI/https://doi.org/10.13039/501100011033/FEDER, UE) and by the Ministerio de Ciencia, Innovación y Universidades–Agencia Estatal de Investigación (AEI) (RED2018-102431-T and RED2022-134219-T). The authors would like to thank the Catalan Government for the quality accreditation given to their research group (2017 SGR 1537 and 2022 SGR 01615). GREiA is certified agent TECNIO in the category of technology developers from the Government of Catalonia. This work is partially supported by ICREA under the ICREA Academia programme.

References

Airò Farulla G, Tumminia G, Sergi F et al (2021) A review of key performance indicators for building flexibility quantification to support the clean energy transition. Energies (Basel) 14. https://doi.org/10.3390/en14185676

Anta A, Gavriluta C, Vettoretti D et al (2024) An optimal power-splitting strategy for hybrid storage systems. In: 2024 IEEE 15th international symposium on power electronics for distributed generation systems (PEDG). IEEE, pp 1–6

Bao H, Ma Z, Roskilly AP (2016) Integrated chemisorption cycles for ultra-low grade heat recovery and thermo-electric energy storage and exploitation. Appl Energy 164:228–236. https://doi.org/10.1016/j.apenergy.2015.11.052

Baraskar S, Günther D, Wapler J, Lämmle M (2024) Analysis of the performance and operation of a photovoltaic-battery heat pump system based on field measurement data. Sol Energy Adv 4. https://doi.org/10.1016/j.seja.2023.100047

Beaudin M, Zareipour H, Schellenberglabe A, Rosehart W (2010) Energy storage for mitigating the variability of renewable electricity sources: an updated review. Energy Sustain Dev 14:302–314. https://doi.org/10.1016/j.esd.2010.09.007

Brandt M, Woods J, Tabares-Velasco PC (2022) An analytical method for identifying synergies between behind-the-meter battery and thermal energy storage. J Energy Storage 50. https://doi.org/10.1016/j.est.2022.104216
Breyer C, Lopez G, Bogdanov D, Laaksonen P (2024) The role of electricity-based hydrogen in the emerging power-to-X economy. Int J Hydrogen Energy 49:351–359. https://doi.org/10.1016/j.ijhydene.2023.08.170
Burre J, Bongartz D, Brée L et al (2020) Power-to-X: between electricity storage, e-production, and demand side management. Chem Ing Tech 92:74–84
Cabeza LF, Palomba V (2020) The role of thermal energy storage in the energy system. In: Reference module in earth systems and environmental sciences. Elsevier
Child M, Bogdanov D, Breyer C (2018) The role of storage technologies for the transition to a 100% renewable energy system in Europe. Energy Procedia 155:44–60. https://doi.org/10.1016/j.egypro.2018.11.067
Child M, Kemfert C, Bogdanov D, Breyer C (2019) Flexible electricity generation, grid exchange and storage for the transition to a 100% renewable energy system in Europe. Renew Energy 139:80–101. https://doi.org/10.1016/j.renene.2019.02.077
Cisek P, Taler D (2019) Numerical analysis and performance assessment of the thermal energy storage unit aimed to be utilized in smart electric thermal storage (SETS). Energy 173:755–771. https://doi.org/10.1016/j.energy.2019.02.096
Cordis web page (2017) HYBUILD project. https://cordis.europa.eu/project/id/768824/it. Accessed 30 Jan 2025
Dong H, Xu C, Chen W (2023) Modeling and configuration optimization of the rooftop photovoltaic with electric-hydrogen-thermal hybrid storage system for zero-energy buildings: Consider a cumulative seasonal effect. Build Simul 16:1799–1819. https://doi.org/10.1007/s12273-023-1066-5
Durand J-M, Duarte MJ, Clerens P (2013) Joint EASE/EERA recommendations for a European energy storage technology development roadmap towards 2030. Brussels
Emhofer J, Marx K, Barz T et al (2020) Techno-economic analysis of a heat pump cycle including a three-media refrigerant/phase change material/water heat exchanger in the hot superheated section for efficient domestic hot water generation. Appl Sci (Switzerland) 10:7873. https://doi.org/10.3390/app10217873
Emhofer J, Marx K, Sporr A et al (2022) Experimental demonstration of an air-source heat pump application using an integrated phase change material storage as a desuperheater for domestic hot water generation. Appl Energy 305:117890. https://doi.org/10.1016/j.apenergy.2021.117890
European Commission (2022) EU renovation wave. In: https://energy.ec.europa.eu/topics/energy-efficiency/energy-efficient-buildings/renovation-wave_en
Freeman J, Coakley D (2021) The role of heat pumps and energy storage in REACT
ICEF—innovation for Cool Earch Forum (2017) Energy storage roadmap—technology and institution
IEA (2024) Batteries and secure energy transitions. Paris
International Energy Agency (2014) Technology roadmap. Energy storage. Paris, France
Jacob AS, Banerjee R, Ghosh PC (2018) Sizing of hybrid energy storage system for a PV based microgrid through design space approach. Appl Energy 212:640–653. https://doi.org/10.1016/j.apenergy.2017.12.040
Jacobson MZ, Delucchi MA, Bazouin G et al (2015) 100% clean and renewable wind, water, and sunlight (WWS) all-sector energy roadmaps for the 50 United States. Energy Environ Sci 8:2093–2117. https://doi.org/10.1039/C5EE01283J
Kiptoo MK, Lotfy ME, Adewuyi OB et al (2020) Integrated approach for optimal techno-economic planning for high renewable energy-based isolated microgrid considering cost of energy storage and demand response strategies. Energy Convers Manag 215:112917. https://doi.org/10.1016/j.enconman.2020.112917
Klein SA, Beckman WA, Mitchell JW, Duffie NA et al (1979) TRNSYS 17, transient system simulation program. University of Wisconsin, Madison, WI, USA

Klein K, Herkel S, Henning HM, Felsmann C (2017) Load shifting using the heating and cooling system of an office building: quantitative potential evaluation for different flexibility and storage options. Appl Energy 203:917–937. https://doi.org/10.1016/j.apenergy.2017.06.073

Li H, Yang F, Chen Y et al (2024) Configuration scheme of battery-flywheel hybrid energy storage based on empirical mode decomposition

Liu T, Li J, Yang Z, Duan Y (2024) Evaluation of the short- and long-duration energy storage requirements in solar-wind hybrid systems. Energy Convers Manag 314:118635. https://doi.org/10.1016/j.enconman.2024.118635

Llantoy N, Zsembinszki G, Palomba V et al (2021) Life cycle assessment of an innovative hybrid energy storage system for residential buildings in continental climates. Appl Sci (Switzerland) 11. https://doi.org/10.3390/app11093820

McKenna R, Fehrenbach D, Merkel E (2019) The role of seasonal thermal energy storage in increasing renewable heating shares: a techno-economic analysis for a typical residential district. Energy Build 187:38–49. https://doi.org/10.1016/j.enbuild.2019.01.044

Mehrjerdi H, Rakhshani E (2019) Optimal operation of hybrid electrical and thermal energy storage systems under uncertain loading condition. Appl Therm Eng 160. https://doi.org/10.1016/j.applthermaleng.2019.114094

Miao Y, Hynan P, von Jouanne A, Yokochi A (2019) Current Li-ion battery technologies in electric vehicles and opportunities for advancements. Energies (Basel) 12. https://doi.org/10.3390/en12061074

Mselle BD, Zsembinszki G, Vérez D et al (2022a) A detailed energy analysis of a novel evaporator with latent thermal energy storage ability. Appl Therm Eng 201:117844. https://doi.org/10.1016/j.applthermaleng.2021.117844

Mselle BD, Zsembinszki G, Veréz D et al (2022b) Experimental assessment of the influence of the design on the performance of novel evaporators with latent energy storage ability. Appl Sci 12:1813. https://doi.org/10.3390/app12041813

Palomba V, Frazzica A (2019) Comparative analysis of thermal energy storage technologies through the definition of suitable key performance indicators. Energy Build 185:88–102. https://doi.org/10.1016/j.enbuild.2018.12.019

Palomba V, Bonanno A, Brunaccini G et al (2021) Hybrid cascade heat pump and thermal-electric energy storage system for residential buildings: experimental testing and performance analysis. Energies (Basel) 14:2580. https://doi.org/10.3390/en14092580

Prieto C, Tagle-Salazar PD, Patiño D et al (2024) Use of molten salts tanks for seasonal thermal energy storage for high penetration of renewable energies in the grid. J Energy Storage 86:111203. https://doi.org/10.1016/j.est.2024.111203

Rasmussen MG, Andresen GB, Greiner M (2012) Storage and balancing synergies in a fully or highly renewable pan-European power system. Energy Policy 51:642–651. https://doi.org/10.1016/j.enpol.2012.09.009

Rey SO, Romero JA, Romero LT et al (2023) Powering the future: a comprehensive review of battery energy storage systems. Energies (Basel) 16

Rossi A, Verber M, Raveduto G et al (2021) Deliverable D4.4: report on system performance

Solomon AA, Child M, Caldera U, Breyer C (2017) How much energy storage is needed to incorporate very large intermittent renewables? Energy Procedia 135:283–293. https://doi.org/10.1016/j.egypro.2017.09.520

Solomon AA, Bogdanov D, Breyer C (2019) Curtailment-storage-penetration nexus in the energy transition. Appl Energy 235:1351–1368. https://doi.org/10.1016/j.apenergy.2018.11.069

Solyali D, Safaei B, Zargar O, Aytac G (2022) A comprehensive state-of-the-art review of electrochemical battery storage systems for power grids. Int J Energy Res 46:17786–17812. https://doi.org/10.1002/er.8451

Sterner M, Specht M (2021) Power-to-gas and power-to-x—the history and results of developing a new storage concept. Energies (Basel) 14. https://doi.org/10.3390/en14206594

Talluri L, Manfrida G, Fiaschi D (2019) Thermoelectric energy storage with geothermal heat integration—exergy and exergo-economic analysis. Energy Convers Manag 199:111883. https://doi.org/10.1016/j.enconman.2019.111883

Timmons D, Elahee K, Lin M (2020) Microeconomics of electrical energy storage in a fully renewable electricity system. Sol Energy 206:171–180. https://doi.org/10.1016/j.solener.2020.05.057

UNEP (2024) Energy efficiency for buildings. https://www.renewableinstitute.org/images/unep%20info%20sheet%20-%20ee%20buildings.pdf. Accessed 30 Jan 2025

Velte-Schäfer A, Zhang Y, Nonnen T et al (2023) Numerical modelling and evaluation of a novel sorption module for thermally driven heat pumps and chillers using open-source simulation library. Energy Convers Manag 291:117252. https://doi.org/10.1016/j.enconman.2023.117252

Wu Y, Shuang W, Wang Y et al (2024) Recent progress in sodium-ion batteries: advanced materials, reaction mechanisms and energy applications. Electrochem Energy Rev 7:17. https://doi.org/10.1007/s41918-024-00215-y

Xu X, Fu Y, Luo Y (2022) Building energy flexibility with battery energy storage system: a comprehensive review. Discov Mech Eng 1. https://doi.org/10.1007/s44245-022-00004-1

Zsembinszki G, Llantoy N, Palomba V et al (2021) Life cycle assessment (LCA) of an innovative compact hybrid electrical-thermal storage system for residential buildings in Mediterranean climate. Sustainability 13:5322

Zsembinszki G, Vérez D, Cabeza LF (2024) Experimental evaluation of a hybrid electrical and thermal energy storage system in a pilot residential house during summer conditions in a Mediterranean climate. J Build Eng 98:111179. https://doi.org/10.1016/j.jobe.2024.111179

Hybrid Thermal and Electrical Energy Storage in Office Buildings

Olav Galteland, Davide Tommasini, Ragnhild Kjæstad Sæterli, and Jorge Salgado-Beceiro

Abstract This chapter investigates the most cost-efficient energy storage solution for a net-zero office building in Trondheim, Norway. For each month from December 2021 to August 2024, we determined the optimal combination of latent thermal energy storage and electrical storage by solving a convex optimization problem aimed at finding the most cost-efficient energy storage solution. Our analysis indicates that a combination of electrical and thermal energy storage, specifically 67 kWh and 104 kWh respectively, represents the optimal solution for the whole period. We explored seasonal variations in optimal storage configurations, finding a greater demand for thermal energy storage during winter months. This is primarily driven by higher energy prices, higher energy price fluctuations, and increased thermal energy demand during winter. While electrical energy storage needs also increase in winter, this is mainly attributed to higher energy prices and energy price fluctuations. Ultimately, our findings highlight that high energy prices and significant energy price fluctuations are the primary drivers for implementing energy storage solutions.

Graphical Abstract

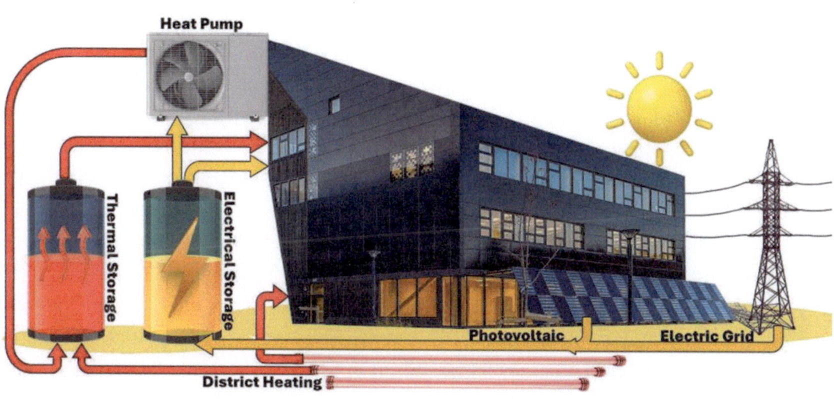

Photo of the ZEB-lab, courtesy of Veidekke

O. Galteland (✉) · D. Tommasini · R. K. Sæterli · J. Salgado-Beceiro
Department of Thermal Energy, SINTEF Energy Research, Trondheim, Norway
e-mail: olav.galteland@sintef.no

R. Scipioni et al. (eds.), *Hybrid Energy Storage*, Lecture Notes in Energy 47,
https://doi.org/10.1007/978-3-031-97755-8_16

Keywords Latent energy storage · Lithium-ion batteries · Net-zero buildings · System optimization · Cost-efficiency

1 Introduction

Consumer-side energy storage in buildings benefits consumers, grid operators, and the general public alike (Medved et al. 2021). For energy consumers, energy storage can provide cost savings through peak shaving and load shifting, and profits through price arbitrage and ancillary services. Energy storage can also supply power during planned or unplanned outages. Consumer-side energy storage enhances grid stability and price stability. For energy systems with intermittent sources, such as photovoltaics (PV), and fluctuating demands, energy storage can help reduce costs (Shaqsi et al. 2020).

Many buildings and households use energy storage. For example, thermal energy storage (TES) in the form of hot water tanks (HWT) is common. However, HWTs are primarily used for hot water, often in combination with waterborne space heating and other applications (Ibrahim et al. 2014). Latent thermal energy storage (LTES), with the use of phase change materials (PCM) (Khudhair and Farid 2021) is an emerging, promising, and market-ready technology (Cartesian 2024; Cowa-TS 2024). LTES offers up to four times the energy density of HWTs, and with decreasing costs, it is expected to achieve commercial breakthroughs soon (Tronchin et al. 2018).

Battery energy storage systems (BESS) are less common in buildings and households due to costs and safety concerns (Conzen et al. 2023). However, with increasing use of consumer-side PV, BESS installations are growing (Sepúlveda-Mora and Hegedus 2021). Various technologies exist, among those the widely employed lithium-ion batteries (LiB) (Campana et al. 2021). One emerging technology with potential for widespread adoption in buildings is vanadium redox-flow batteries (VRFB) (Lourenssen et al. 2019), known for their safety and cost-saving potential.

Both BESS and TES can store energy for a few hours to up to 72 h, making them suitable for mitigating daily energy price fluctuations The building sector already has well-established and highly efficient power-to-heat (P2H) technologies, such as heat pumps (HP), but lacks efficient heat-to-power (H2P) solutions. BESS with P2H is more flexible, as it can meet both electrical and thermal demands. TES, on the other hand, is only efficient for thermal demand. TES on the other hand can only be efficiently used to meet the thermal demand. Smart energy management systems (SEMS) that control both TES and BESS—using factors such as energy costs, building usage, and weather data—could optimize energy use in modern buildings, which are increasingly connected to the grid, district heating networks, and local energy production. However, many variables influencing technology selection, design, control, and sizing remain largely unexplored.

Table 1 compares some of the key performance indicators (KPIs) of BESS (LiB and VRFB) and TES (HWT and LTES). Charge/discharge rate (C-rate) is defined as the charge/discharge power divided by the energy storage capacity of the unit.

The charge and discharge are assumed to be equal for simplicity in this work. A C-rate of 1 indicates that the energy storage can be fully charged or discharged in one hour. BESS generally have higher C-rates than TES, allowing for more rapid charging/discharging. Self-discharge rate is the percentage of stored energy lost over a certain period of time. Regarding energy density, it is generally higher for BESS than TES. The round trip efficiency indicates the energy efficiency of the storage system during charge and discharge cycles, where TES typically exhibit higher round trip efficiencies compared to BESS. In the case of TES, the round-trip efficiency considers only the distribution heat losses of the pipes going towards and from the TES and the energy needed for pumps. Specific energy cost and specific power cost is the turnkey investment cost of the unit divided by the energy and power capacity, respectively. The specific energy cost for BESS is typically higher than TES, while the specific power cost for BESS and TES is at a comparable level. Environmental factors are qualitatively assessed for each technology. TES is generally fire safe, contain no CRMs, and have a low CO2e footprint, as opposed to particularly LiB.

In this study, a zero-emissions building laboratory in Trondheim (Norway) will serve as a case study of how hybrid energy storage systems (HESS) could be implemented to meet the dynamic energy needs of modern office buildings, while keeping costs and environmental impact to a minimum. To minimize energy and investment costs, we will investigate the optimal hybrid energy storage solution by combining LiB and LTES. In this work, we will not specify a LiB chemistry, but rather investigate a general electrical energy storage with charge and discharge rates, self-discharge rates and round trip efficiencies, and investment cost that correspond to state-of-the art BESS systems with lithium iron phosphate (LFP), nickel manganese cobalt

Table 1 KPIs for lithium-ion batteries (LiB) (Le Varlet et al. 2020; Romare and Dahllöf 2017; Kallitsis et al. 2024; Huang and Li 2022), vanadium redox flow batteries (VFRB) (Šimić et al. 2021; AlShafi and Bicer 2021; Bai and Song 2023), hot water tanks (HWT) (International Energy Agency 2024b; Depcik et al. 2020; Høiax 2024; EASE 2024), and latent thermal energy storage (LTES) (International Energy Agency 2024a; Chocontá Bernal et al. 2021; Lamnatou et al. 2018)

KPI	Unit	BESS		TES	
		LiB	VRFB	HWT	LTES
Charge rate	–	0.5–2	0.1–0.5	0.1–0.5	0.02–0.05
Discharge rate	–	0.5–2	0.1–0.5	0.5–2	0.02–0.05
Self-discharge rate	1/day	0.1%	0%	2.5%	0.68%
Round trip efficiency		82–89%	70–80%	87–99%	91–99%
Energy density	kWh/m^3	95–500	150–350	60	80–110
Specific energy cost	€/kWh	350	250–420	40	20–100
Specific power cost	€/kW	350	850–2500	1–15	200–400
Tech. safety	–	Low	Mid	High	High
Use of CRM	–	High	High	Low	Low
Global warming potential	kg_{CO2e}/kWh	48–120	0.04–0.12	0.0005	0.3–0.8
Energy intensive production	MJ/kg	High	Mid	Mid	Low

(NMC), or nickel cobalt aluminum (NCA) LiBs. This approach allows for broader applicability of our findings, as it is not limited to a single LiB chemistry. We will model BESS and TES energy storage in a system with district heating, PV production and a HP, with realistic energy costs and weather data, and then discuss the influence of each of these energy storage solutions. KPIs for TES and BESS corresponding to LTES and LiB, respectively, to highlight the technology advances needed for LTES to compete with HWT.

2 The Zero-Emission Building Laboratory

The zero-emission building laboratory (ZEB-lab) is an office building in Trondheim, designed as a living laboratory for researching topics such as indoor climate, energy systems, and smart energy control systems (Galteland et al. 2023). The building consists of four floors and has a capacity to accommodate approximately 80 office workers. The facility is equipped with a PV system, HP, LTES, and an accumulator tank. It is connected to a local district heating loop, where the thermal energy demand of the building can be met by the HP or by the district heating circuit (Sevault et al. 2019). Importantly, the HPs alone can fully meet the building's peak heating demand, and the peak power of the PV system significantly exceeds the peak electrical load. The building produces a substantial amount of excess energy from the PV system during the months from April to August. The energy system specifications are summarized in Table 2.

The daily mean energy consumption and production from December 2021 to August 2024 is shown in Fig. 1, and in Fig. 2 the energy consumption and production is shown for a week in March 2024. During the summer, the mean daily PV production can double the mean daily electrical consumption, resulting in significant overproduction that is distributed to the neighbouring buildings. In addition, the electricity prices vary through the day, week and seasons. At the Gløshaugen

Table 2 Energy system aspects for the ZEB-lab from December 2021 to August 2024

Aspect	Value
Peak PV system power	185 kW
HP capacity	30 kW
PCM-TES capacity (40–60 °C)	226 kWh
Accumulator tank capacity	800 L
HP Temperature lift	From 35 to 40 °C
District heating temperature	45 °C
Average thermal energy consumption	166 kWh/day
Average electrical energy consumption	319 kWh/day
Average PV production	359 kWh/day

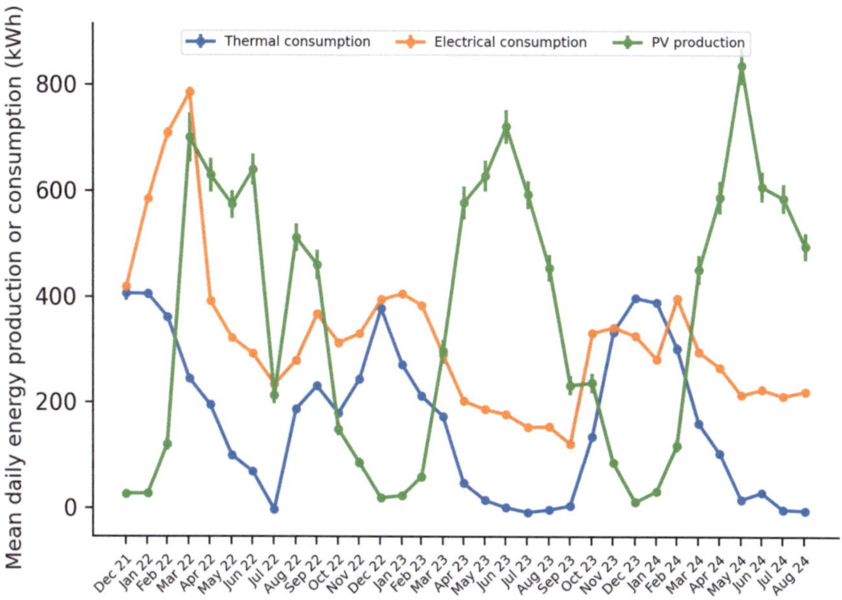

Fig. 1 Mean daily energy consumption and production in the ZEB-lab from December 2021 to August 2024

campus, where the ZEB-lab is located, the building is connected to the local heating loop (Blindheim et al. 2024).

The LTES unit in the ZEB-lab has approximately 3000 kg of organic PCM, and 24 vertical pillow plate heat exchangers with a maximum distance of 40 mm between the plates. The PCM is CrodaTherm 37, with a melting temperature of 37 °C. More thermophysical properties of the material is presented in Table 3. The total energy storage capacity is 226 kWh, it can discharge at an average power of 10.51 kW for 12.2 h and charge at an average power of 13.7 kW for 11 h, or a C-rate of 0.06. The average charge–discharge rate for a full charge/discharge is 3.8 kW, or a C-rate of 0.02. The C-rate is a function of the state-of-charge, but in this work, we will assume it to be constant. The performance of the LTES has been previously reported in detail (Salgado-Beceiro et al. 2022). Water acts as the heat transfer fluid, and the unit can be charged either by the HPs or district heating. The unit can be discharged to preheating of domestic hot water and for spacing heating of the building.

Currently, there is no BESS installed in the ZEB-lab. In this work, we will investigate the hypothetical case of energy cost savings by installing BESS in combination with LTES.. This approach enables us to analyze potential energy cost savings and system performance benefits before making any real-world investments or installations.

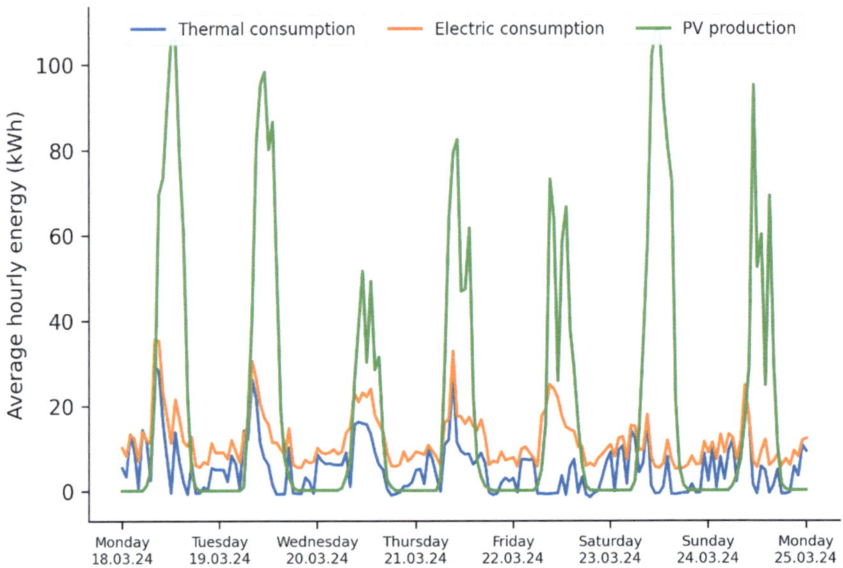

Fig. 2 Mean hourly energy consumption and production from 18 to 25th of March 2024

Table 3 Thermophysical properties CrodaTherm 37

Property	Value
Melting temperature	37 °C
Liquid mass density	819 kg/m^3
Solid mass density	957 kg/m^3
Solid specific heat capacity	2.3 kJ/K/kg
Liquid specific heat capacity	1.4 kJ/K/kg
Thermal conductivity	0.24 W/m/K
Latent heat of fusion	186.6 kJ/kg

3 Objective and Methodology

We will use the energy consumption and production, weather, and energy price data of the ZEB-lab from December 2021 to August 2024 to examine, analyse, and determine the optimal hybrid energy storage solution. The integration of a BESS and the re-dimensioning of the TES system will be simulated, considering scenarios with only TES, only BESS, and combination of both, to determine the most cost-efficient energy storage solution for the ZEB-lab under different energy and weather scenarios. The technologies considered are LiB as the BESS, as the most established technology; and LTES, since it's already installed in the case study.

A convex optimization problem is formulated below and will be solved with the objective of minimizing the electric and thermal energy costs. In summary, the electric

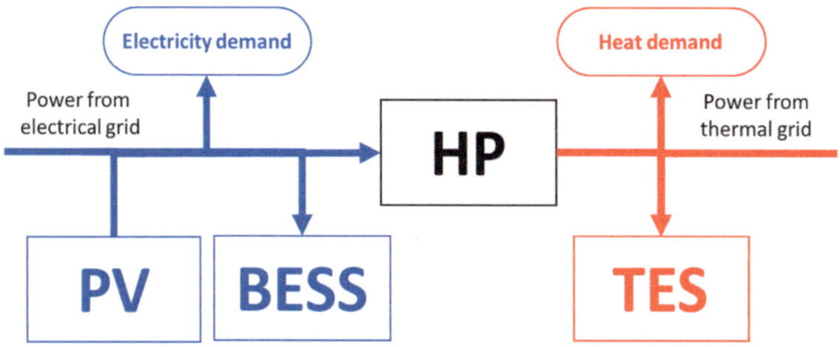

Fig. 3 A diagram of the components in the modelled energy system, with electric energy flows in blue and thermal energy flows in red

and thermal energy demand of the building must be met, and the electric energy production from PV can be used. Hourly data from the ZEB-lab of consumption and production from December 2021 to August 2024 is used and the problem is solved on a monthly basis. The investment costs of storage will be considered after the optimization of charging and discharging schedules for a set of BESS and TES capacities to determine which combination of BESS and TES is most cost-efficient. We will consider specifications of BESS and TES to be representative of LTES, as already installed in the ZEB-lab, and LiB, respectively.

The optimization procedure will be used to analyse the effects of storage capacity and C-rates to determine the most cost-effective storage solution for the energy system. Combinations of BESS and TES will be considered. For the main part of the results, C-rates of 0.05 for TES and 0.5 for BESS will be considered.

In Fig. 3, the components and the electric and thermal energy flows of the system are shown. There are two energy carriers in the energy system, electric and thermal energy. On the electric side, there are two components, PV panels and BESS. On the thermal side there is one component, the TES. The HP connects the two, making it possible to produce thermal energy from electric energy. On each side there is also an energy demand, the electric and thermal demands. In the following section the constrains of these energy flows are defined.

3.1 Constraints Formulation

The constraints of the convex optimization are as follows. The electric energy balance

$$P_{\mathrm{EG}} + P_{\mathrm{PV}} = P_{\mathrm{EC}} + P^{\mathrm{i}}_{\mathrm{HP}} + P_{\mathrm{B}},$$

where the electric power drawn from the grid P_{EG}, and the electric power production from the PV panels P_{PV}, must meet the electric power consumption of the building P_{EC}, the power input to the HP $P^{\mathrm{i}}_{\mathrm{HP}}$, and the power input to the BESS, P_{B}. The power from the electric grid must be positive and the power input to the HP must be positive, $P_{\mathrm{EG}} \geq 0$ and $P^{\mathrm{i}}_{\mathrm{HP}} \geq 0$. We do not consider the possibility of selling electricity back to the grid, e.g. by selling excess PV production or by price arbitrage. PV production that exceeds consumption is curtailed. Similarly for the thermal energy balance,

$$P_{\mathrm{TG}} + P^{\mathrm{o}}_{\mathrm{HP}} = P_{\mathrm{TC}} + P_{\mathrm{T}}$$

where the thermal power drawn from the grid and the output power from the HP meet the thermal power consumption and the thermal power input to the TES. The power from the thermal grid must be positive, $P_{\mathrm{TG}} \geq 0$.

The PV power production, electric and thermal power consumption are constrained to be equal to the actual production and consumption of the ZEB-lab. The HP input and output is

$$P^{\mathrm{o}}_{\mathrm{HP}} = COP \times P^{\mathrm{i}}_{\mathrm{HP}}$$

where $COP = COP(T)$ is the coefficient of performance of the HP. This is an equation of the outside temperature, fitted to a 2nd order polynomial of the data from the HPs in the ZEB-lab,

$$COP(T) = 8.9 \times 10^{-5} T^2 + 2.3 \times 10^{-5} T + 2.4$$

where the temperature is in units of degree Celsius. The HP delivers heat at a constant temperature. The HP electric power capacity is constrained to be $P^{\mathrm{o}}_{\mathrm{HP}} < C_{\mathrm{HP}}$, where C_{HP} equal to 30 kW. For the charging and discharging dynamics of the BESS, we have the following constraint

$$Q^{t+\Delta t}_{\mathrm{B}} = (1 - \epsilon_{\mathrm{B}} \Delta t) Q^{t}_{\mathrm{B}} + \sqrt{\eta_{\mathrm{B}}} P_{\mathrm{B}} \Delta t$$

where $Q^{t+\Delta t}_{\mathrm{B}}$ is the state of charge (in dimension energy) of the BESS in timestep $t + \Delta t$, ϵ_{B} is its self-discharge rate and η_{B} is its round-trip efficiency. We have not used the superscripts for other parameters, as they do not depend on previous timesteps as the states-of-charges do. The self-discharge rate is in dimensions of fraction of energy loss per time. We have a corresponding constraint for the TES,

$$Q^{t+\Delta t}_{\mathrm{T}} = (1 - \epsilon_{\mathrm{T}} \Delta t) Q^{t}_{\mathrm{T}} + \sqrt{\eta_{\mathrm{T}}} P_{\mathrm{T}} \Delta t.$$

The TES and BESS also have constraints on their states of charge and power input and output. The constraints on the states of charge

$$0 \leq Q^{t}_{\mathrm{B}} \leq Q^{\max}_{\mathrm{B}} \quad \text{and} \quad 0 \leq Q^{t}_{\mathrm{T}} \leq Q^{\max}_{\mathrm{T}},$$

where $Q^{\max}_{\mathrm{B}}$ and $Q^{\max}_{\mathrm{T}}$ are the storage capacity of the storages. The constraints on the power input and output are,

Table 4 List of parameters in the constraint formulation

Symbol	Description	Value	Unit
ϵ_{B}	BESS self-discharge rate	3	%/month
ϵ_{T}	TES self-discharge rate	20	%/month
η_{B}	BESS round trip efficiency	0.90	
η_{T}	TES round trip efficiency	0.98	
$\mathrm{Q}_{\mathrm{B}}^{\max}$	BESS storage capacity		kWh
$\mathrm{Q}_{\mathrm{T}}^{\max}$	TES storage capacity		kWh
C_{B}	BESS C-rate	0.5	
C_{T}	TES C-rate	0.05	
COP	HP coefficient of performance. A function of outside temperature	1–4	
C_{HP}	HP capacity	30	kW

$$-\mathrm{C_B P_B} \leq \mathrm{P_B} \leq \mathrm{C_B P_B} \quad \text{and} \quad -\mathrm{C_T P_T} \leq \mathrm{P_T} \leq \mathrm{C_T P_T},$$

where C_{B} and C_{T} are the charge and discharge rates. The initial and end conditions of the storages are set such that they are half-way stored at the start and end of each month.

$$Q_{\mathrm{B}}^{t_s} = Q_{\mathrm{B}}^{t_e} = \frac{Q_{\mathrm{B}}^{\max}}{2} \quad \text{and} \quad Q_{\mathrm{T}}^{t_s} = Q_{\mathrm{T}}^{t_e} = \frac{Q_{\mathrm{T}}^{\max}}{2},$$

where t^{s} and t^{e} are the first and last hours in the month, respectively. A compiled list of the parameters of the constraints are shown in Table 4.

3.2 *Objective Formulation*

The objective of the optimization problem, with the constraints formulated above, is to minimize the energy costs. The energy cost is the sum of the electricity cost and the district heating cost, $C = C_{\mathrm{E}} + C_{\mathrm{T}}$. The investment costs of energy storage will be considered after determining the energy costs for a set of BESS and TES capacities. The energy prices used here are equal to the price model used in Trondheim, Norway and are in the currency of Norwegian kroner (NOK). However, all amounts are presented in EUR for the convenience of the reader. A conversion rate of 1 NOK = 0.085 EUR has been used.

The electricity cost is the sum of the spot price, the grid tariff, and public charges. The grid tariff consists of two parts, the energy and capacity components. The energy component is a fixed amount per kWh during the day and the night. During the day (06-22) it is 0.020 EUR/kWh, and during the night (22-09) it is 0.001 EUR/kWh. The capacity component is an amount per month, calculated from the three hours of the month with the highest power drawn from the grid. These three hours must be on separate days. The capacity component is a staircase function, which is non-convex and has been approximated by a linear function based on the mean of the three maximum powers drawn from the grid. In addition, there are public charges of 68 EUR/year and 0.014 EUR/kWh.

The district heating price has been simplified to be the average value of the electricity price. In the real-world scenario, this cost is calculated after each month, and therefore this cost is not known a priori, but in this optimization procedure it is.

The investment costs of BESS and TES are set to be 325 EUR/kWh and 100 EUR/kWh, respectively, with the specifications given in Table 4.

4 Results and Discussion

4.1 A Single Week

To better understand the results over the entire simulated period, a representative week is shown in detail in Fig. 4. This figure illustrates the time series of the energy prices, energy demand and production, outside temperature, and the stored energy for the 3rd week of March 2024. March 2024 was characterized by a high PV production, while there was still a large electrical and thermal demand. In Fig. 4b, it can be observed that the PV production is much higher than the electrical consumption. In addition, the energy prices are relatively high, as seen in Fig. 4a. Moreover, this figure perfectly describes the behaviour of the prices: the electricity price varies, while the heat price remains constant as previously described in Sect. 2.

Over a year, all these parameters vary. In the winter months the PV production is low and almost zero in December and January, and the thermal demand and the energy prices are relatively high. Conversely, in the summer months, the PV production is high and can be twice the electrical consumption in the time period from April to August, but the energy demand and energy prices are low. The monthly average energy consumption and production, outside temperature, and electric spot price for each month from December 2021 to August 2024 are shown in Fig. 6.

In Fig. 4d the state-of-charge of the TES and BESS is shown throughout this week for storage capacities of 243 and 129 kWh, respectively, which represent the most cost-efficient combination of TES and BESS for this month. The optimization procedure finds this optimal charging and discharging schedule with minimal energy costs. The optimal storage capacities of TES and BESS were determined by evaluating the energy cost savings from the storage system and balancing them against the investment costs associated with the storage. The BESS has a C-rate of 0.5, while the TES has a C-rate of 0.05. For these storage capacities, this corresponds to 64.5 kW and 12.15 kW peak power input and output for BESS and TES, respectively.

The BESS charges rapidly during periods of low electricity prices or high PV production, allowing it to take full advantage of these cost savings. This quick charging is possible due to the BESS's high C-rate. Then, it discharges at a slower rate,

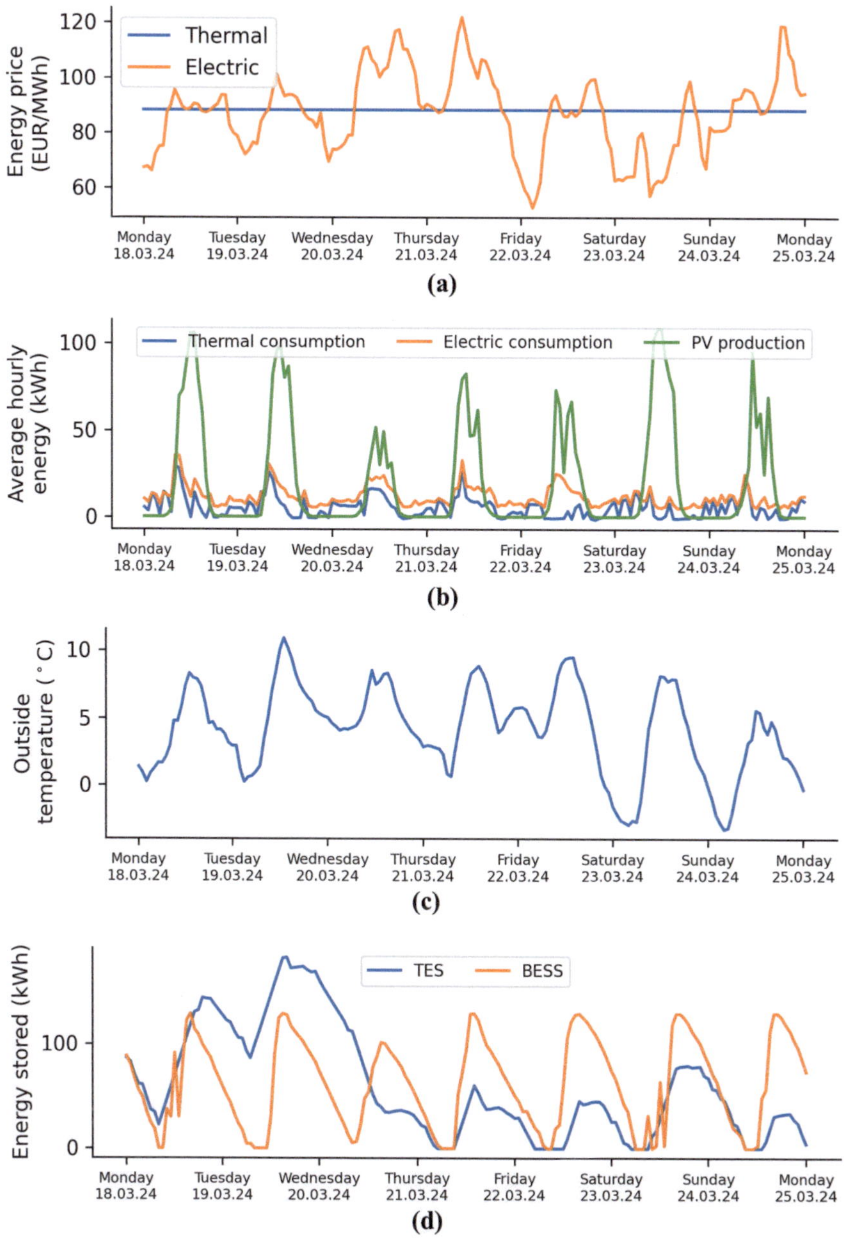

Fig. 4 Timeseries of **a** energy prices, **b** energy demand and production, **c** outside temperature, and **d** energy stored in BESS and TES for a week in March 2024

delivering power over a longer duration. In contrast, the TES charges and discharges at a more consistent rate, regardless of price fluctuations, due to its low C-rate. The BESS has a power capacity to operate with a 23 h-equivalent cycle period. The TES, however, has a much lower C-rate, and operates at an average 78 h-equivalent cycle period.

The equivalent cycle period is the time it takes to complete one equivalent cycle, calculated by dividing the total time by the number of the equivalent cycles. An equivalent cycle represents the average full charge and discharge cycle, determined by converting partial charge and discharge cycles into full cycles.

With this optimal storage, 67% of the PV power is utilized, while without storage for this month, only 40% of the PV power is utilized. It should be noted that in the real building, the PV system is over-dimensioned, and the power is used in neighbouring buildings. In this work, we have not allowed power to be shared with neighbours or sold to the power grid. Any excess PV power is therefore curtailed.

4.2 Optimal Hybrid Energy System

The optimal storage sizes for each month are shown in Fig. 5. Overall, we find that a 67 kWh BESS and 104 kWh TES storage to be optimal solution for the whole period examined, however there are large variations for each month depending on energy costs, weather conditions, and energy demand. For example, in December 2022, the optimal solution is 270 kWh BESS and 740 kWh TES. In this month, the energy price and demand were abnormally high with large fluctuations in energy price, making the energy storage technologies very attractive. On the other hand, in June 2023, the optimal solution is 25 kWh BESS and no TES. In this month, the energy price was low with minimal fluctuations, the energy demand was also low, and the PV production was more than four times the electrical consumption. All these parameters together made both the technologies less convenient in this month. These monthly averages of energy price, outside temperature, energy demand and production are shown in Fig. 6.

With the optimal capacities of the TES and BESS determined for the entire simulation period, the monetary savings in both absolute and relative terms for each month were calculated. The cost savings per month for the optimal hybrid configuration are shown in Fig. 7. The savings are relative to no energy storage installed. Absolute savings are greater during the winter months, particularly when the optimal storage capacity is larger, and decrease during the summer, when smaller storage sizes are optimal. This is because the energy consumption and prices are generally higher in the winter months. In contrast, relative savings follow an inverse pattern, tending to be higher in the summer and lower in the winter. This is because in the summer months there are large amounts of PV production, which is better utilized with energy storage. However, since the energy demand and prices are generally low during the summer, the absolute savings are less significant than the savings during the winter months.

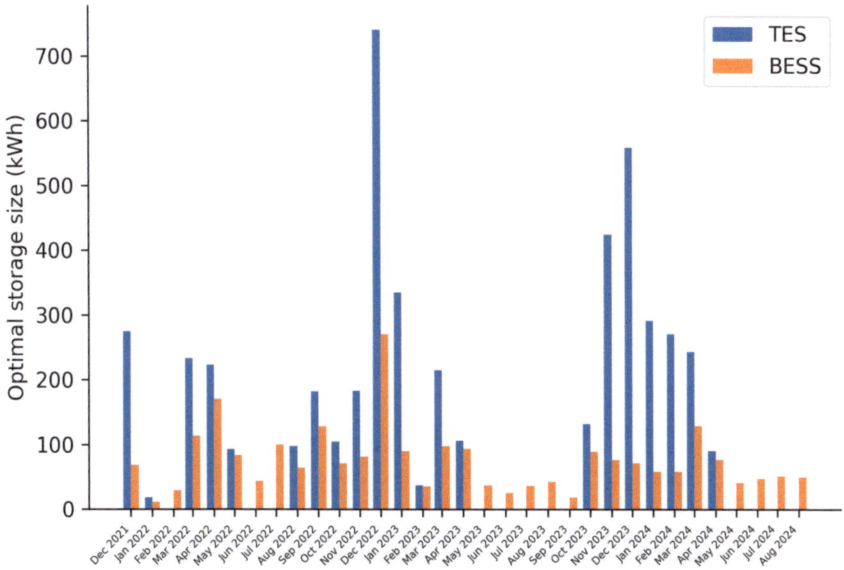

Fig. 5 Optimal combination of BESS and TES storage size for each month from December 2021 to August 2024. BESS C-rate of 0.5 and TES C-rate of 0.05

4.3 Correlations

To quantify what determines if BESS and/or TES are beneficial, we have investigated the correlations between the optimal storage sizes of TES and BESS between various factors. These factors were the monthly mean and standard deviation of thermal load, electrical load, PV production, outside temperature, and electricity spot price. The correlations are normalized and dimensionless numbers between − 1 and 1, representing respectively inverse and positive correlation. These results are presented in Table 5.

There are correlations corresponding to conditions in the winter and summer months. During the winter months, the temperature and PV production are low, and the energy demand and prices are high. In the summer months, the temperature and PV production are high, while the energy demand and prices are low. The two strongest correlations with optimal storage size of TES are the mean and std. of the electricity spot price, and the third strongest correlation is the mean thermal demand. For the BESS, however, the strongest correlations are with the mean and std. electric spot prices. It is clear that TES is much more dependent on having high thermal demand, while BESS is less on high energy demands. As a result, optimal TES size is much more cost-efficient in the winter months than the summer months, while optimal BESS size is less affected by the seasons. A reason for this is that there is very no thermal energy demand in the summer months, but there will always be some

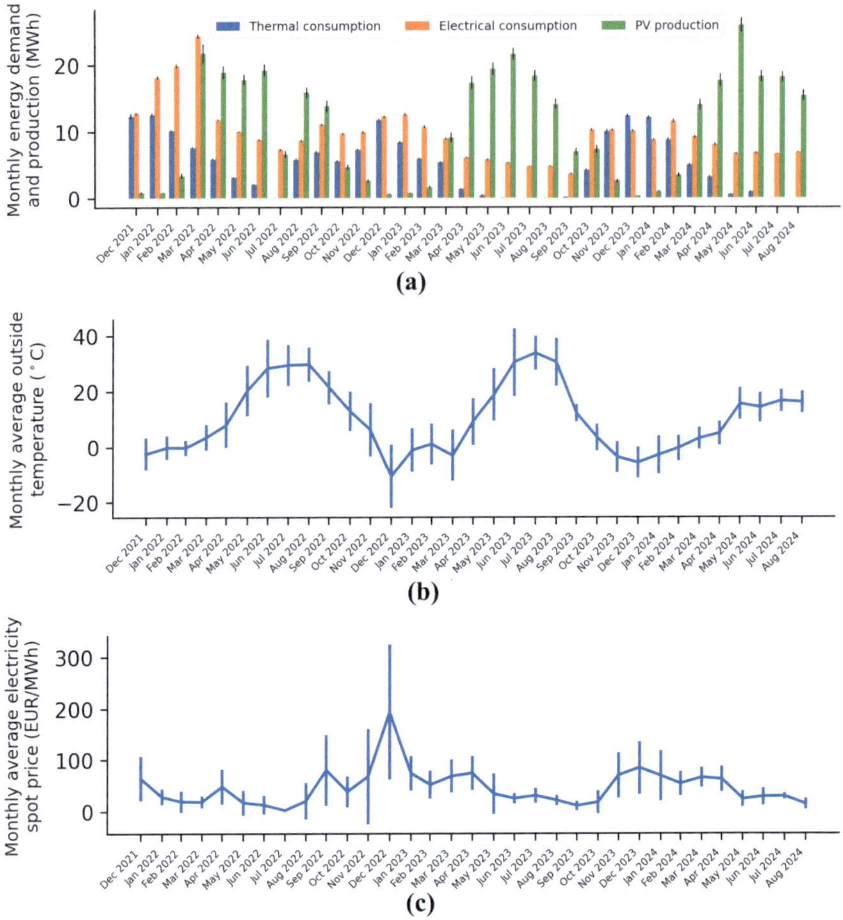

Fig. 6 Monthly average **a** energy consumption and production **b** outside temperature, and **c** electricity spot price

electrical energy demand. In addition, the PV production is higher in the summer months.

4.4 *Geographical Location*

The weather conditions influence the most cost-efficient energy storage solutions. Specifically, the outside temperature influences the heating demand, and the solar irradiation influences the PV production. The geographical location plays a crucial

Fig. 7 Monthly savings with the optimal BESS and TES sizes for the whole simulation period, i.e. 67 and 104 kWh respectively, in absolute (blue) and relative (orange) values

Table 5 Normalized correlations of monthly optimal storage size of BESS and TES to various parameters

		Optimal storage size		Thermal load		Electrical load		PV production		Outside temperature		Electric spot price	
		TES	BESS	Mean	Std.	Mean	Std.	Mean	Std.	Mean	Std.	Mean	Std.
Optimal storage size	TES	1.00	0.70	0.70	0.63	0.29	0.35	-0.52	-0.48	-0.68	0.16	0.84	0.77
	BESS	0.70	1.00	0.30	0.24	0.21	0.14	-0.10	-0.03	-0.35	0.30	0.72	0.68
Thermal load	Mean	0.70	0.30	1.00	0.88	0.69	0.71	-0.73	-0.70	-0.77	-0.10	0.55	0.57
	Std.	0.63	0.24	0.88	1.00	0.54	0.67	-0.59	-0.56	-0.72	-0.19	0.51	0.48
Electrical load	Mean	0.29	0.21	0.69	0.54	1.00	0.83	-0.32	-0.20	-0.53	-0.25	0.11	0.13
	Std.	0.35	0.14	0.71	0.67	0.83	1.00	-0.49	-0.34	-0.63	-0.32	0.19	0.19
PV production	Mean	-0.52	-0.10	-0.73	-0.59	-0.32	-0.49	1.00	0.96	0.65	0.10	-0.42	-0.42
	Std.	-0.48	-0.03	-0.70	-0.56	-0.20	-0.34	0.96	1.00	0.58	0.01	-0.43	-0.44
Outside temperature	Mean	-0.68	-0.35	-0.77	-0.72	-0.53	-0.63	0.65	0.58	1.00	0.21	-0.60	-0.46
	Std.	0.16	0.30	-0.10	-0.19	-0.25	-0.32	0.10	0.01	0.21	1.00	0.30	0.39
Electric spot price	Mean	0.84	0.72	0.55	0.51	0.11	0.19	-0.42	-0.43	-0.60	0.30	1.00	0.89
	Std.	0.77	0.68	0.57	0.48	0.13	0.19	-0.42	-0.44	-0.46	0.39	0.89	1.00

Gold colors indicate positive correlations, and blue colors indicate negative correlations

role in determining the optimal combination of TES and BESS, as it directly influences both energy demand and production. In the case of the ZEB-lab in Trondheim, the extreme seasonal variation in temperature significantly impacts heating and cooling demands, as well as the efficiency of PV generation. Our results show that the optimal configuration of hybrid energy storage requires a higher storage size of TES (in kWh) for the months from October to April (Fig. 5). This is due to the cold climate conditions, which drive higher heating demand, combined with lower PV production during these months. In these conditions, a larger TES system effectively meets the thermal energy needs, while minimizing the reliance on external grid sources. The increased capacity of TES during this period leads to the highest savings over the entire year, as seen in Fig. 7. In contrast, the warmer months with higher PV production can accommodate smaller storage systems, highlighting the seasonal variations that must be considered when designing a hybrid energy storage system. This demonstrates how the cold climate in Trondheim, paired with seasonal fluctuations in PV generation, can optimize the use of a larger TES system, offering substantial energy cost savings.

4.5 Energy Costs

The cost of electricity and grid distribution significantly influences the sizing of BESS, but the integration of TES can further impact this decision. In regions with high electricity costs or unreliable grid infrastructure, larger BESS systems are often necessary to ensure a stable supply and optimize energy usage, especially when combined with renewable sources. However, as we saw in this case study, TES can reduce the required size of BESS by providing an additional means to store and release energy, particularly in sectors with thermal energy needs. By leveraging TES to balance peak demand periods, the overall BESS capacity needed can be reduced, as TES can absorb excess renewable energy during periods of low demand and provide thermal energy when needed, thus enhancing system efficiency and lowering costs. Therefore, the combination of electricity cost, grid distribution needs, and TES capabilities can optimize the sizing of BESS, ensuring the most cost-effective solution (Killer et al. 2020).

In this study, the cost of heat is assumed constant while the cost of electricity is the actual cost seen by the ZEB lab in the relevant time period. District heating costs and policies vary significantly throughout EU (Billerbeck et al. 2023), and connections to free sources of surplus heat separate from the DH network can also be envisioned for other cases. In this study, we have considered energy prices in the region of Trondheim, Norway. The rising overall cost of electricity, combined with large price variations, makes the installation of both BESS and TES more cost-efficient, as indicated by the correlations in Table 5. Table 5 will make a better case for installing both BESS and TES. Higher costs of electricity are also more in line with the costs that are typically seen elsewhere in Europe. For other studies looking at hybrid energy storage solutions, it was found how not only price fluctuations, but grid availability and reliability influenced the selection and sizing of one storage technology or another (Alonso et al. 2024). This dependence shows again how the geographical location plays a key role on optimizing a hybrid energy storage system.

4.6 Charging and Discharging Rates (C-rate)

The charging and discharging rate (C-rate) of LTES is known to be a limiting factor in many applications. This is because of the relatively low thermal conductivity of PCMs. So far in this study, we have considered TES and BESS C-rates of 0.05 and 0.5, respectively. The C-rates are the average rates for the entire charging/discharging process, independent of state-of-charge. In operation, however, we find that the C-rate depends on the state-of-charge (SoC) (Saxena et al. 2019; Suyitno et al. 2023). At the start of charging, the C-rate is at its maximum value and gradually decreases as the SoC rises due to a solid layer that forms on the plates, limiting heat transfer. Similarly, during discharge, the C-rate is highest at the beginning but drops as the SoC decreases. This is because a liquid layer forms between the plates and the solid PCM fraction, acting as an insulating layer and further reducing heat transfer.

To investigate the influence of the TES C-rate on its cost-efficiency, we have calculated the mean energy cost per month as a function of the TES C-rate for five different TES storage capacities, presented in Fig. 8. Only TES as energy storage was considered in this analysis, meaning there was no BESS. All TES storage capacities show a steep incline in energy savings at low C-rates, followed by a plateau at higher C-rates. These results show that the energy cost savings increase of 0.6–1.7% can be expected if the C-rate is doubled from 0.05 to 0.1, while increasing the C-rate from 0.05 to 0.2 we can expect an increased energy cost saving of 1.2–2.8%. The savings depend on the TES storage capacities, more energy cost savings can be expected with larger TES storage capacities. However, high relative savings can be realized for smaller storage sizes.

In this work, we have not considered the investment cost of the LTES. It is expected that the investment cost of the LTES will increase dramatically with increasing C-rates if the increased C-rate is obtained by increased heat exchanger surface area or by introduction of additives.

4.7 Investment Cost

LiBs are a well-established and mature technology in the market, characterized by relatively low CAPEX compared to alternative energy storage solutions. This cost advantage, combined with their high efficiency and adaptability, makes them economically attractive for hybrid installations, particularly when integrated with photovoltaic (PV) systems. However, careful attention to system sizing is essential to avoid over-investment while maximizing operational benefits. LiBs also play a critical role in Demand Response strategies, offering unique energy flexibility that enhances the integration of renewable energy sources. Properly tailored investments in LiBs, considering operational needs and geographical factors, are key to achieving optimal returns while supporting sustainable energy transitions (Chatzigeorgiou et al. 2024).

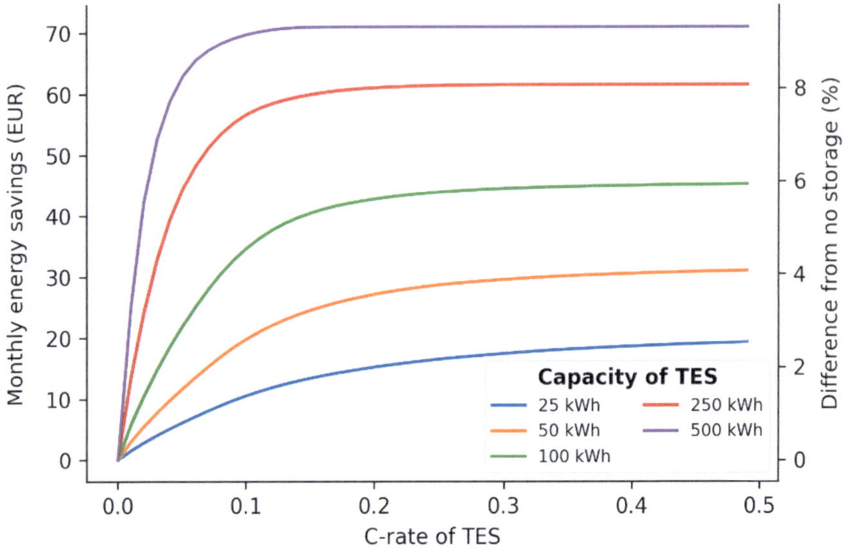

Fig. 8 Average monthly energy cost as a function of the C-rate of the TES for various TES capacities, without any BESS. This does not include investment costs, only electricity spot price, grid tariff, and district heating costs

LTES is a less mature technology than LiBs, and as a consequence it is expected that the price of LTES will reduce at a faster pace than LiBs. To investigate how a reduced price of LTES will impact the optimal storage sizes, we have calculated the overall optimal storage sizes for BESS and TES for investment costs of TES ranging from 20 to 150 EUR/kWh, shown in Fig. 9 The C-rate of TES and BESS were kept constant at 0.05 and 0.5, respectively, and the investment cost of BESS was kept constant at 325 EUR/kWh. So far in the analysis, we have considered a TES investment cost of 100 EUR/kWh, giving an optimal TES storage size of 104 kWh. With lower investment cost of TES, the optimal TES storage size is large, as expected. The optimal storage size decreases approximately linearly until reaching an investment cost of 145 EUR/kWh, beyond which installing TES becomes cost ineffective. Today, the investment cost of LTES is in the price range of 20–100 EUR/kWh and expected to decrease in the coming years. HWTs, on the other hand, are cheaper, with a price range of 10 to 20 EUR/kWh. However, it is important to note that other operational conditions should be considered for HWTs.

The optimal storage size of BESS does increase with increasing investment cost of TES, however to a limited degree, approximately a difference of 15 kWh in this investment cost range (see inset in Fig. 9). The reason for this is that at low investment costs of TES it is more cost efficient to store energy as heat. However, it will always be beneficial to have BESS, as the energy stored as heat cannot be used to meet the electric demand as there is no heat-to-power unit. There will always be a demand to store the energy as electrical energy in the BESS.

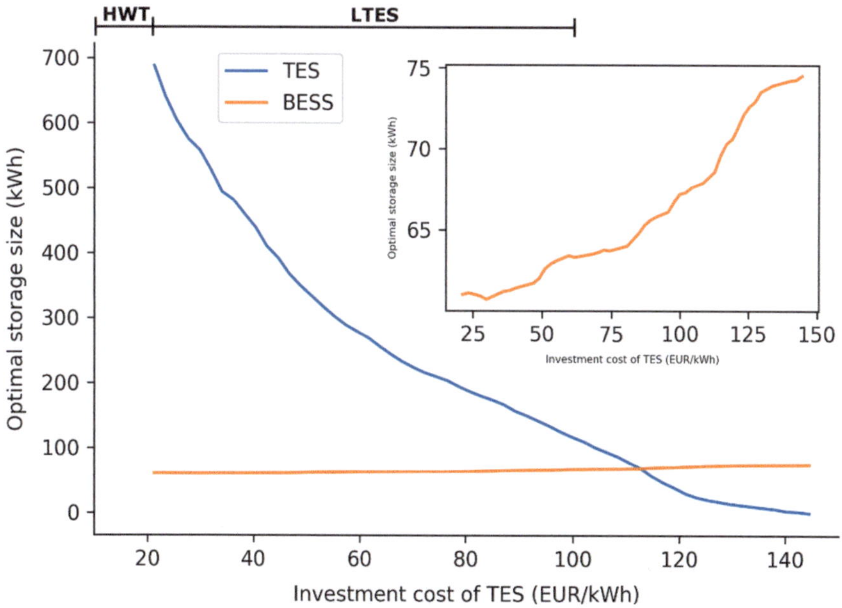

Fig. 9 Optimal capacity of TES (blue) and BESS (orange) as a function of the investment cost of TES. TES C-rate of 0.05 and BESS C-rate of 0.5

4.8 Other Thermal and Electrical Sources in the System

Including a HP would, at first glance, make a good case for installing BESS rather than TES, as a HP will deliver thermal energy at a higher efficiency (electric input times COP). The same argument can be used for PV installations, where storage through BESS initially makes more sense than TES. However, as much as the energy needs of buildings are thermal, it is important to understand whether hybrid energy storage systems with BESS + TES makes sense whenever HPs or PVs are planned or already installed, and what the factors are that influence the size of storages. The potential connection to district heating or other sources of surplus heat, including their costs, as well as the capacity of the HP and also enters into the equation—in cases of high thermal demand, a thermal storage will enable a reduction in HP capacity.

4.9 Environmental, Social, and Safety Aspects

Environmental, social and safety aspects are not easily included in planning and dimensioning tools for energy storages, for a multitude of reasons. Life cycle assessments (LCA) and similar methods exist to address some of these topics, but they are often not implemented in decision making tools (Zakeri and Syri 2015; Kittner et al. 2017). These models are complex and time consuming, while at the same time highly case/boundary specific, making general comparisons of different technologies difficult. The large number of potential parameters to assess, like land and water usage, GHG and particulate matter emissions, is not easily translated into costs and it is often therefore left as a decision for the end-user how to weigh these factors towards each other and towards the technical performance and costs of the technologies.

LCA has also emerged as a valuable tool for evaluating the environmental impact of BESS across all stages of their lifecycle, from material extraction through manufacturing, use, and disposal. Given the building sector's role as a major consumer of energy, reducing greenhouse gas emissions in building applications is essential, particularly with the increasing adoption of net-zero energy buildings, like the ZEB-lab, which integrate renewable energy sources like solar and wind. However, renewable energy's intermittent nature makes storage technologies, such as BESS, essential to ensure a stable and reliable energy supply. LCA studies on BESS for buildings demonstrate both environmental and economic benefits, yet they also highlight challenges, particularly in managing the environmental footprint associated with critical raw materials used in batteries, such as cobalt, lithium, and graphite. Studies indicate that while BESS can significantly reduce greenhouse gas emissions, especially in off-grid or hybrid systems, there remains a need for comprehensive, system-wide assessments that balance environmental impact with costs. Furthermore, fully renewable energy systems can have higher economic and environmental costs, making a mix of on-site generation and grid connectivity a potentially more sustainable option. Moving forward, a balanced approach, informed by LCA, will be crucial to ensuring the sustainability of BESS in building applications (Wrålsen and O'Born 2023).

Here, we have chosen to look at four indicators where BESS and TES are expected to have large environmental impact, to discuss the differences and understand the trade-offs and considerations needed in order to also make hybrid energy storage systems sound from an environmental and social perspective, and that may influence the relative sizes of BESS and TES storages. It is not an exhaustive review of all factors, nor does it review each factor in detail, however it is meant as a guide to the reader on important topics to consider.

4.9.1 Technology Safety

Technology safety is an important aspect when integrated into buildings, whether retrofitted into existing systems or planned as a part of new buildings. In particular, fire safety is an issue with LiB in buildings that require adaptations to fire detection and suppression and containment systems (Chombo and Laoonual 2020). The risk in case of fire for TES depends highly on the choice of material and its properties, but is considered significantly less than for LiB. While HWTs do not pose any particular risk, LTES may in some cases use paraffins or other flammable materials. However, they are not self-sustaining, and usually contained in air-tight, non-flammable containers, and do not release any toxic or flammable vapors, as is the case for LiB.

4.9.2 Use of Critical Raw Materials

LiB are also problematic due to the fact that they use several critical raw materials, namely cobalt, natural graphite, and, to some extent, silicon (Dils 2020). Cobalt is problematic in a number of ways, most notably high global warming potential and eutrophication (Farjana et al. 2019), high possibility for supply chain disruptions (Van den Brink et al. 2020) and human rights violations (Baumann-Pauly 2023). Lithium is not currently on the EU list of CRMs, however that is not to say that it is without controversy. In particular, the water usage for lithium extraction is very high (Vera et al. 2023) an aspect that is not necessarily picked up through LCA studies (Halkes et al. 2024). IEA recognizes the importance of energy storage to reach net zero, and at the same time states the importance of improving the resilience and diversity of supply chains of CRMs (International Energy Agency 2024c). Though VFBs are generally considered more environmentally friendly than LiBs (Weber et al. 2018), they utilize the critical raw material vanadium. Vanadium is solely supplied by four countries, none of which are in EU or associated countries (Plananska 2023). TES storages, be it HWT or LTES, are generally CRM free.

4.9.3 Global Warming Potential

The global warming potential (GWP) is one of the factors regularly assessed through LCA studies. It is, however, as noted earlier, dependent on a number of detailed input factors such as production sites, methods and input factors, including energy mix. The GWP is thus very specific, however a general comparison was given in Table 1. The energy intensity for VRFB and LTES are particularly hard to estimate, as the technologies have yet to reach market maturity, and the manufacturing methodologies are in an early stage of development. LTES is notoriously hard to estimate, since a large portion of the GWP is due to the PCM and thus case specific.

4.9.4 Energy Consumption in Production

The energy consumption is one of the factors included in the GWP assessment, however we also consider it an important factor in itself. The global need for electricity has increased twice as much as the overall energy demand in the past decade, and with the increasing demand for data centres, electrification of industry, increased demands for cooling, and increased electricity use in developing countries, among other factors, the need is expected to increase 6% yearly. The introduction of renewable energy source somewhat alleviates this, with oil, gas and coal set to peak by 2030. However, energy efficiency measures help reach net zero faster. It should also be noted that the intermittency of most renewable energy sources calls for more energy storage solutions. Regarding HWT and LTES, they require a lower energy intensive production than LiB and VRFB due to a simpler assembly process. HWT is the technology with the lowest energy density due to the larger availability of water compared to the other storage mediums. After reviewing data for the four technologies, we found that the energy consumption in production is proportional to the energy density offered by each technology.

5 Summary and Conclusions

This study determined that the most cost-efficient combined capacities for energy storage in a medium-sized office building in Trondheim, Norway, are 67 kWh of battery energy storage system (BESS) and 104 kWh of thermal energy storage (TES), based on the specified system parameters. We investigated various factors affecting the performance and cost-efficiency of these technologies, focusing on energy demand, seasonal variations, energy prices, and system response characteristics.

Key findings highlight that the strongest correlations with optimal storage sizes were observed with energy prices, both for BESS and TES. For this case, combining BESS with HP offered greater flexibility than TES, as BESS could store electrical energy, which, through HP, could be efficiently converted into thermal energy with a COP between 1 and 4. This made BESS + HP the preferred choice for addressing both electrical and thermal demands. However, TES can still play a role, particularly when the cost of TES is low, which is more achievable for HWTs than for LTES. In addition, if the cost of district heating is low, TES will have additional benefits.

Environmental factors, including CO_2 emissions, use of land, water, and critical raw materials, were identified as important, but challenging to quantify comprehensively. Preliminary qualitative assessments suggest that TES generally has a lower environmental impact than BESS. Therefore, hybrid energy storage systems, combining TES and BESS, tend to offer better overall efficiency and cost savings while minimizing environmental harm.

Main Points:

- The optimal storage size is most strongly influenced by energy prices, with high energy prices favouring larger storage systems.
- Combining BESS with HP offers superior flexibility in meeting both electrical and thermal demands.
- TES is more cost-effective when investment costs are low.
- Environmental impacts of TES and BESS are important, with TES generally being less harmful, especially regarding the use of critical raw materials.
- Hybrid systems combining TES and BESS are generally more efficient and cost-effective, but must be tailored to local conditions, including climate and energy prices.

The optimal energy storage configuration depends heavily on location-specific factors, such as energy prices, climate conditions, and energy demand patterns. Key factors influencing the decision-making process include the cost and performance of both TES and BESS technologies, the energy flexibility of the combined system (especially in areas with seasonal variations), and the environmental impact of each storage solution.

In conclusion, for a given location, optimizing hybrid energy storage systems requires a deep understanding of local energy demand, system costs, and environmental impacts. Hybrid energy storage systems, combining TES and BESS, offer strong performance in many cases, but the specific configuration must be tailored to local conditions, with ongoing optimization tools needed for more effective decision-making.

References

Alonso AM, Costa D, Messagie M, Coosemans T (2024) Techno-economic assessment on hybrid energy storage systems comprising hydrogen and batteries: a case study in Belgium. Int J Hydrogen Energy 52:1124–1135

AlShafi M, Bicer Y (2021) Life cycle assessment of compressed air, vanadium redox flow battery, and Molten salt systems for renewable energy storage. Energy Rep 7:7090–7105

Bai H, Song Z (2023) Lithium-ion battery, sodium-ion battery, or redox-flow battery: a comprehensive comparison in renewable energy systems. J Power Sources 580:233426

Baumann-Pauly D (2023) Cobalt mining in the Democratic Republic of the Congo: addressing root causes of human rights abuses. White Paper, NYU Stern Center for Business and Human

Billerbeck A, Breitschopf B, Winkler J, Bürger V, Köhler B, Bacquet A, Popovski E, Fallahnejad M, Kranzl L, Ragwitz M (2023) Policy frameworks for district heating: a comprehensive overview and analysis of regulations and support measures across Europe. Energy Policy 173:113377

Blindheim AD, Zettervall ID, Aamodt LA (2024) Sammenligning Av ulike energilagringsteknologier for Bruk På Gløshaugen. B.S. thesis. NTNU

Campana PE, Cioccolanti L, François B, Jurasz J, Zhang Y, Varini M, Stridh B, Yan J (2021) Li-ion batteries for peak shaving, price arbitrage, and photovoltaic self-consumption in commercial buildings: a Monte Carlo analysis. Energy Convers Manage 234:113889

Cartesian (2024) Cartesian. No. https://cartesian.no/

Chatzigeorgiou NG, Theocharides S, Makrides G, Georghiou GE (2024) A review on battery energy storage systems: applications, developments, and research trends of hybrid installations in the end-user sector. J Energy Storage 86:111192

Chocontá Bernal D, Muñoz E, Manente G, Sciacovelli A, Ameli H, Gallego-Schmid A (2021) Environmental assessment of latent heat thermal energy storage technology system with phase change material for domestic heating applications. Sustainability 13(20):11265

Chombo PV, Laoonual Y (2020) A review of safety strategies of a li-ion battery. J Power Sources 478:228649

Conzen J, Lakshmipathy S, Kapahi A, Kraft S, DiDomizio M (2023) Lithium ion battery energy storage systems (BESS) hazards. J Loss Prev Process Ind 81:104932

Cowa-TS (2024) Cowa-TS.Com. https://www.cowa-ts.com/

Depcik C, Cassady T, Collicott B, Burugupally SP, Li X, Alam SS, Arandia JR, Hobeck J (2020) Comparison of lithium ion batteries, hydrogen fueled combustion engines, and a hydrogen fuel cell in powering a small unmanned aerial vehicle. Energy Convers Manage 207:112514

Dils E (2020) ETC/WMGE Report 5/2020: environmental aspects related to the use of critical raw materials in priority sectors and value chains. European Topic Centre on Waste and Materials in a Green Economy (ETC/WMGE)
EASE (2024) EASE_TD_HotWater.Pdf
Farjana SH, Huda N, Parvez Mahmud MA (2019) Life cycle assessment of cobalt extraction process. J Sustain Min 18(3):150–161
Galteland O, Gouis M, Salgado-Beceiro J, Sevault A (2023) Fourteen months operation of a 200 kWh latent heat storage pilot. In: 2023 8th international conference on smart and sustainable technologies (SpliTech). IEEE, pp 1–5
Halkes RT, Hughes A, Wall F, Petavratzi E, Pell R, Lindsay JJ (2024) Life cycle assessment and water use impacts of lithium production from Salar deposits: challenges and opportunities. Resour Conserv Recycl 207:107554
Høiax (2024) NEPD-5319-4616_Hoiax-Connected-200-Water-Heater.Pdf
Huang Y, Li J (2022) Key challenges for grid-scale lithium-ion battery energy storage. Adv Energy Mater 12(48):2202197
Ibrahim O, Fardoun F, Younes R, Louahlia-Gualous H (2014) Review of water-heating systems: general selection approach based on energy and environmental aspects. Build Environ 72:259–286
International Energy Agency (2024a) Fact sheet thermal latent
International Energy Agency (2024b) Sensible water
International Energy Agency (2024c) World energy outlook 2024
Kallitsis E, Lindsay JJ, Chordia M, Wu B, Offer GJ, Edge JS (2024) Think global act local: the dependency of global lithium-ion battery emissions on production location and material sources. J Clean Product 449:141725
Khudhair AM, Farid M (2021) A review on energy conservation in building applications with thermal storage by latent heat using phase change materials. Therm Energy Storage Phase Change Mater 162–75
Killer M, Farrokhseresht M, Paterakis NG (2020) Implementation of large-scale li-ion battery energy storage systems within the EMEA region. Appl Energy 260:114166
Kittner N, Lill F, Kammen DM (2017) Energy storage deployment and innovation for the clean energy transition. Nat Energy 2(9):1–6
Lamnatou C, Motte F, Notton G, Chemisana D, Cristofari C (2018) Cumulative energy demand and global warming potential of a building-integrated solar thermal system with/without phase change material. J Environ Manage 212:301–310
Le Varlet T, Schmidt O, Gambhir A, Few S, Staffell I (2020) Comparative life cycle assessment of lithium-ion battery chemistries for residential storage. J Energy Storage 28:101230
Lourenssen K, Williams J, Ahmadpour F, Clemmer R, Tasnim S (2019) Vanadium redox flow batteries: a comprehensive review. J Energy Storage 25:100844
Medved S, Domjan S, Arkar C (2021) Contribution of energy storage to the transition from net zero to zero energy buildings. Energy Build 236:110751
Plananska J (2023) Europe's urgent need for primary vanadium extraction
Romare M, Dahllöf L (2017) The life cycle energy consumption and greenhouse gas emissions from lithium-ion batteries. IVL Svenska Miljöinstitutet
Salgado-Beceiro J, Galteland O, Sevault A (2022) Analysis of a latent heat storage unit using a pillow plate heat exchanger during real operations in a zero-emission building
Saxena S, Xing Y, Kwon D, Pecht M (2019) Accelerated degradation model for C-rate loading of lithium-ion batteries. Int J Electr Power Energy Syst 107:438–445
Sepúlveda-Mora SB, Hegedus S (2021) Making the case for time-of-use electric rates to boost the value of battery storage in commercial buildings with grid connected PV systems. Energy 218:119447
Sevault A, Bøhmer F, Næss E, Wang L (2019) Latent heat storage for centralized heating system in a ZEB living laboratory: integration and design. IOP Conf Ser Earth Environ Sci 352, 012042

Šimić Z, Topić D, Knežević G, Pelin D (2021) Battery energy storage technologies overview. Int J Electr Comput Eng Syst 12(1):53–65

Shaqsi AZA, Sopian K, Al-Hinai A (2020) Review of energy storage services, applications, limitations, and benefits. Energy Reports 6:288–306

Suyitno BM, Pane EA, Rahmalina D, Rahman RA (2023) Improving the operation and thermal response of multiphase coexistence latent storage system using stabilized organic phase change material. Res Eng 18:101210

Tronchin L, Manfren M, Nastasi B (2018) Energy efficiency, demand side management and energy storage technologies–a critical analysis of possible paths of integration in the built environment. Renew Sustain Energy Rev 95:341–353

Van den Brink S, Kleijn R, Sprecher B, Tukker A (2020) Identifying supply risks by mapping the cobalt supply chain. Resour Conserv Recycl 156:104743

Vera ML, Torres WR, Galli CI, Chagnes A, Flexer V (2023) Environmental impact of direct lithium extraction from brines. Nat Rev Earth Environ 4(3):149–165

Weber S, Peters JF, Baumann M, Weil M (2018) Life cycle assessment of a vanadium redox flow battery. Environ Sci Technol 52(18):10864–10873

Wrålsen B, O'Born R (2023) Use of life cycle assessment to evaluate circular economy business models in the case of li-ion battery remanufacturing. Int J Life Cycle Assess 28(5):554–565

Zakeri B, Syri S (2015) Electrical energy storage systems: a comparative life cycle cost analysis. Renew SUstain Energy Rev 42:569–596

Behind-the-Meter. Combination of Li-Ion Batteries and Organic Flow Redox Batteries for BTM Applications

Francesco Sergi, Giovanni Brunaccini, Elias Martinez, Sergio Costa, and Davide Aloisio

Abstract This work is focused on the development of a Hybrid Energy Storage System (HESS), constituted by the integration of a Lithium Titanate and Aqueous Organic Redox Flow battery systems. The HESS can provide several Behind-The-Meter services for energy management and efficient use. After an initial section dedicated to BTM services, the system and functionalities offered are deeply described, enhancing system exploitation within some case studies (in particular in an experimental site of Messina) with advantages and benefits offered by the hybrid architecture.

Keywords Hybrid energy storage · BTM services · Organic redox flow batteries · Lithium batteries · Battery management system

F. Sergi · G. Brunaccini · D. Aloisio (✉)
Istituto Di Tecnologie Avanzate per l'Energia "Nicola Giordano", Consiglio Nazionale delle Ricerche, Messina, Italy
e-mail: davide.aloisio@cnr.it

F. Sergi
e-mail: francesco.sergi@cnr.it

G. Brunaccini
e-mail: giovanni.brunaccini@cnr.it

E. Martinez
IREC, Catalonia Institute for Energy Research, Sant Adrià del Besòs, Barcelona, Spain
e-mail: emartinez@irec.cat

S. Costa
Typhoon HIL, Bajci Zilinskog bb, Novi Sad, Serbia
e-mail: sergio.costa@typhoon-hil.com

R. Scipioni et al. (eds.), *Hybrid Energy Storage*, Lecture Notes in Energy 47,
https://doi.org/10.1007/978-3-031-97755-8_17

1 Introduction

Hybrid Energy Storage Systems (HESS) are a rapidly growing technology within the energy storage landscape, offering a versatile and efficient solution for addressing the dynamic challenges posed by modern electrical grids. These systems combine two or more energy storage technologies, typically batteries, to leverage their unique strengths and mitigate their individual limitations (which can appear using single technologies as BESS, Battery Energy Storage System) (Lucà Trombetta et al. 2024; Sergi et al. 2016). The primary purpose of HESS is to enhance grid reliability, efficiency, and flexibility, making them particularly valuable in the context of grid services. The increasing penetration of renewable energy sources like solar and wind, with their inherent intermittency, has amplified the need for energy storage solutions that can provide both short-term power requests and long-term energy storage. Traditional power grids were designed to deliver electricity from centralized power plants to consumers in a predictable, one-directional flow. However, the intermittent nature of renewable energy generation and fluctuating electricity consumption patterns have introduced new challenges to grid management. In this context, storage systems and in particular HESS have emerged as one of the most promising solutions. HESS offers a powerful solution by combining different types of energy storage technologies, optimizing their individual strengths to deliver a flexible, scalable, and cost-effective method for storing and managing electricity.

At the heart of a HESS lies the combination of complementary storage technologies. A typical configuration might integrate high-power density batteries, such as Lithium-Ion (Li-ion) batteries, with large-capacity batteries like Redox Flow Batteries (RFB). Another interesting combination could be realized by the integration of supercapacitors and batteries. In this case, due to the excellent power capability of supercapacitors, the batteries are used as the energy element. Other combinations are also explored (batteries and fuel cells, mechanic storage systems and batteries, etc. (Brunaccini et al. 2017) but are out of scope of this chapter. Hybridization enables a wider range of applications and uses also in the grid services market, providing flexibility in both energy/power delivery and storage. This versatility makes HESS ideal for a variety of applications, from supporting local renewable energy generation to enhancing the stability of large-scale power grids. For example, delivering fast power for short periods makes them ideal for applications like frequency regulation and voltage support, where quick response times are crucial. On the other hand, storing energy over longer durations is essential for services such as energy arbitrage, peak shaving, and load shifting. Therefore, by combining these technologies, a HESS can serve multiple functions within the grid, enhancing its performance across various use cases. Also, the combination of different battery systems can better regulate internally the state of health of both batteries actively and therefore enlarge the useful lifetime of the storing system.

The deployment of HESS is expected to grow significantly in the coming years as grid operators, utilities, and consumers recognize the benefits of hybrid storage technologies. Government policies and incentives aimed at promoting renewable energy and energy storage are also driving the adoption of HESS (Ibrahim et al. 2021]. For example, many countries have implemented feed-in tariffs, tax credits, and subsidies for energy storage systems, making HESS more economically viable for both residential and commercial users. Additionally, regulatory frameworks that support the participation of energy storage in electricity markets are creating new opportunities for HESS to provide grid services and generate revenue (Rezaeimozafar et al. 2022).

In terms of technological advancements, one of the most promising developments in HESS is the integration and development of advanced energy management systems. The use of the new technologies such as big-data access, optimization strategies, AI-based algorithms, can enable the realization of advanced Energy Management Systems (EMS) able to analyze data from the grid, weather patterns, and electricity markets to optimize the operation of HESS (Xiaojuan et al. 2017). For example, they can predict periods of high electricity demand or low renewable energy production, allowing the HESS to charge or discharge at the optimal times. This not only maximizes the economic benefits of energy storage but also improves the efficiency and reliability of the grid. Advanced control systems can also enhance the flexibility of HESS, allowing them to adapt to changing grid conditions in real-time.

In conclusion, Hybrid Energy Storage Systems (HESS) offer a range of advantages for the modern electrical grid, particularly in the context of providing critical grid services. Their ability to enhance grid resilience, defer infrastructure upgrades, and participate in ancillary services markets makes them a valuable tool for grid operators and energy providers. As the energy landscape continues to evolve, with greater emphasis on renewable energy integration and grid flexibility, HESSs are poised to play a central role in the transition to a more sustainable and resilient energy system.

2 Ancillaries Behind-the-Meter Services and HESS as a Viable Player

One of the primary advantages of HESS lies in its ability to offer multiple grid services. These services can be broadly categorized as either "in-front-of-the-meter" (FTM) or "behind-the-meter (BTM)" (Boshell et al. 2019). The main difference between them is related to the beneficiary to which the service is provided: the 'meter' is usually the instrument used to measure the final users' consumption (residential, commercial, industrial) and represent the limit to distinguish FTM and BTM applications. More specifically, FTM services are supplied directly to distribution/transmission grids or power plants, which constitute the bulk of energy generation, since consumers' meters are downstream of the segment served, while, for BTM, the services are provided behind the utility service meter with the main purpose of consumer energy management and electricity bill savings, optimizing energy usage at the consumer level, whether in homes, businesses, or industrial settings. This

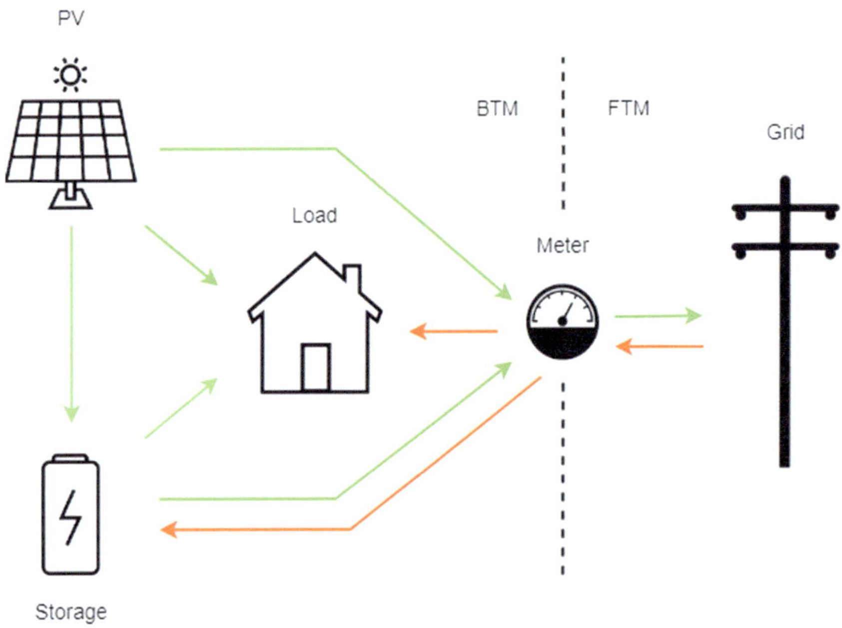

Fig. 1 Behind the meter and Front of the meter electricity exchange with energy storage system (Rearranged from from (Rezaeimozafar et al. 2022)), licensed under CC-BY 4.0

categorization (Rezaeimozafar et al. 2022) does not necessarily identify the point in which the storage system is inserted but rather the service's recipient (Fig. 1).

The classification in (Fitzgerald et al. 2015) identifies services according to stakeholder groups which receive or monetize a major part of the value provided by the service. Thirteen services were defined and linked to beneficiaries, including independent system operators (ISOs), regional transmission organizations (RTOs), utilities, and customers. Some services benefit multiple groups, but the primary beneficiary was considered for classification. The grid is divided into three segments for BESS services: transmission level (bulk generation), distribution level (commercial/industrial customers), and behind-the-meter level (residential, commercial, and industrial end users) (Fig. 2).

Considering the increased integration of energy storage systems into the electric grid, such categorization cannot be univocally defined and some services, usually considered FTM, can be included in BTM. For example, this is the case of service remuneration while considering the BESS/HESS owner point of view.

Therefore, here BTM services that can provide the battery owner with precious benefits will be briefly explained. In (Itani and De Bernardinis 2023), four main categories are identified as business cases, and they are:

1. Self-consumption
2. Energy arbitrage
3. Demand charge management

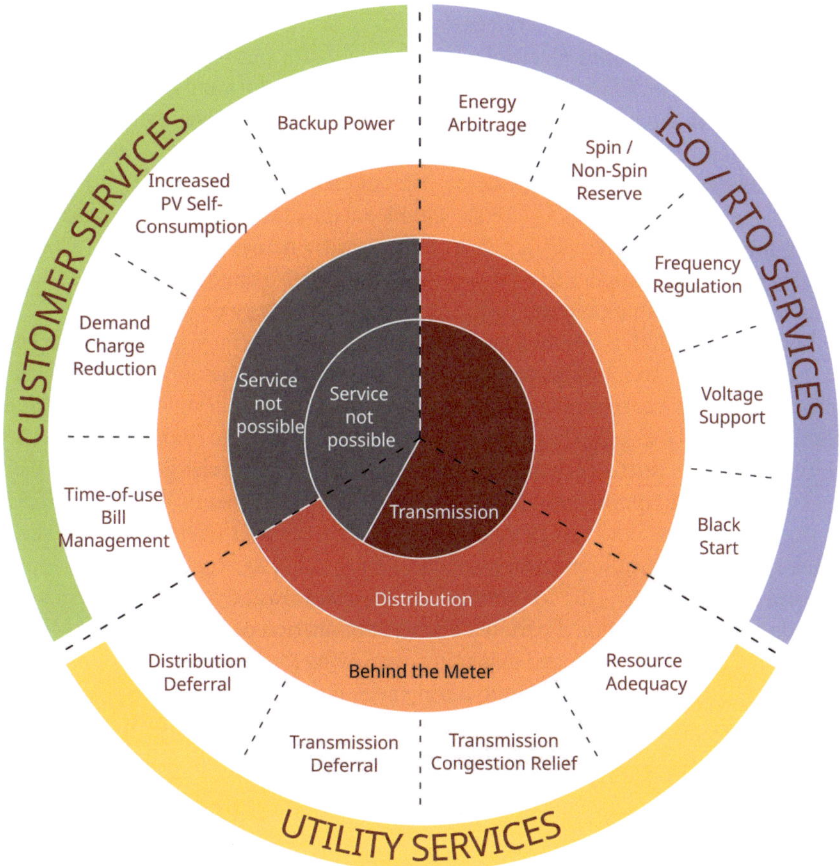

Fig. 2 Grid services and related stakeholders (figure rearranged from (Fitzgerald et al. 2015))

4. Energy and balancing market participation.

Self-consumption of renewable energy is an area where HESSs offer significant advantages. As many homes, businesses, and industrial facilities install renewable energy systems like solar panels, the ability to store energy surplus for later use becomes increasingly important. Without storage, energy surplus generated during periods of high production (e.g., sunny days) is often sent back to the grid, that could not be always able to absorb it efficiently. In some cases, grid operators may even curtail renewable energy production to avoid overloading the grid. HESSs can locally store this energy, allowing consumers to use it later when their own energy demand exceeds production (e.g. at night or during cloudy periods), thus maximizing self-sufficiency. This not only increases the value of renewable energy systems, but also reduces the need for grid electricity, contributing to a more sustainable energy system.

One of the most fruitful services of HESSs is the **energy arbitrage**. Energy arbitrage involves storing electricity during low price periods and discharging it during high price periods. This practice allows grid operators and consumers to take advantage of fluctuations in electricity prices. In many markets, electricity prices can vary significantly throughout the day due to changes in supply and demand. By storing energy during low-price periods and selling it back to the grid during high-price periods, HESS can generate significant cost savings or even extra income, taking advantage of time-varying tariff structures by active management. This not only benefits the grid operator or energy provider but also contributes to overall grid efficiency by smoothing out the peaks and troughs in energy demand.

For end users, both residential and commercial, HESSs offer significant economic benefits through services like **peak shaving** and **demand charge reduction**. They refer to the process of reducing electricity consumption during periods of peak demand, typically to avoid high demand charges or to mitigate grid overload. Many power consumers, such as commercial and industrial users, are charged based on their peak power consumption during a billing period. By using energy stored in a HESS, consumers can smooth their peak demand and reduce their overall energy expenditure. The high-power components of the HESS can discharge quickly to support demand spikes, whereas the long-duration (i.e. with slower dynamics) batteries can provide energy for sustained periods. Such a combination of features makes HESSs ideal solutions for peak demand managing, improving the overall stability of the grid by reducing the strain on the system during peak demand periods.

The role of HESSs in **demand response** programs is also growing. Demand response refers to the practice of adjusting energy consumption in response to price signals or specific grid needs. HESSs can facilitate demand response programs helping to balance the grid, reducing the need for expensive and polluting peaking power plants, and creating a more flexible and efficient energy system. In addition, this service can bring profit to the owner.

Moreover, depending on the various national regulations, HESS can participate in both **energy and balancing market**, by proper management of the stored energy, and improving grid stability and reliability, respectively. One of the most significant services (normally delivered by the power part of the HESS (Sergi et al. 2019) in balancing market is **frequency regulation** (Leonardi et al. 2021). Modern electrical grids require an efficient balancing between electricity supply and demand to maintain a stable frequency. The increasing penetration of renewable energy sources, particularly wind and solar, has led to frequent fluctuations in grid frequency due to their intermittent nature. In such scenarios, high-power batteries within a HESS, such as Li-ion, can respond almost instantaneously to these fluctuations, either by discharging energy to compensate for a generation shortfall or by absorbing excess energy to prevent over-generation. This rapid response, preventing frequency deviations, reduces power outages or damage to electrical equipment.

In addition to frequency regulation, HESS are highly effective in providing **voltage support**. Voltage instability is another challenge of modern grids, especially in regions with a high penetration of renewable energy. As previously mentioned, Solar and wind energy can cause voltage fluctuations due to the variability of the power produced. HESSs can help to maintain voltage stability by absorbing or injecting reactive power into the grid as needed. The high-power components of the HESSs can quickly adjust their output to counteract voltage fluctuations, ensuring that the grid operates within safe voltage limits. This reduces the risk of voltage sags or swells, which can disrupt sensitive equipment and increase wear on electrical infrastructure.

Energy communities, also known as Local Energy Communities (LECs), consist of groups of consumers, producers, and storage systems that work together to optimize energy use and reduce reliance on the grid. These communities typically include a combination of local renewable energy generation, energy storage, and demand response to manage their energy needs. HESSs can play a central role in these communities by storing excess renewable energy and distributing it to members when needed. The Energy Management System (EMS) within the HESS can optimize energy flows, ensuring that energy is used efficiently within the community.

In summary, Hybrid Energy Storage Systems offer a wide range of benefits to both grid operators and consumers, making them a critical component of the modern energy field. Their ability to provide fast-responding, flexible, and scalable solutions to the challenges posed by renewable energy integration, grid instability, and fluctuating energy demand makes them an invaluable asset in the transition to a more sustainable and resilient energy system. In particular, the presence of different storage systems and wider application range enables HESS to release **multiple services** at once, thus increasing the effective value and total income coming from them. Therefore, the role of HESSs in a continuously evolving energy market, with greater emphasis on decarbonization and digitalization, will become more and more important, helping to create a smarter, cleaner, and more reliable grid for the future.

3 Combining Li-Ion Batteries and Organic Flow Redox Batteries for BTM Applications

The combination of Li-Ion and Organic Flow Redox batteries has been studied in the European research project called HYBRIS (Grant Agreement No. 963652): Hybrid Battery energy stoRage system for advanced grid and beHInd-de-meter Segments.

The HYBRIS project is part of the Horizon 2020 research programme funded by the European Union.

The project is built around optimizing a hybrid energy storage system (HESS) by integrating advanced technologies to create efficient, cost-effective, and environmentally friendly energy storage solutions for microgrid applications. The project envisions the integration of lithium titanate oxide (LTO) battery and aqueous organic redox flow battery (AORFB) into a unified hybrid system to address different energy demands, from power intensive requests to long-term energy release.

The central goal of HYBRIS is to demonstrate the viability of hybrid battery systems in different use cases such as:

- Island microgrids, to release energy services where conventional power grids are not available
- Grid-connected microgrids, where energy storage can release services to ensure stability, reduced operational costs and increased renewable energy use
- Energy communities and private users, where usually storage systems are essential to obtain cost-saving and reliability.

HYBRIS is designed with an industrial focus, aiming to bridge the gap between technical development and real-world application. Enabling the aforementioned services, the developed HESS can provide cost savings for the end user, higher energy utilization efficiency, and long-term sustainability.

The hybrid system integrates two complementary battery technologies:

1. LTO batteries, known among all the lithium batteries for their fast response time, high power density, and very long lifetime;
2. AORFB, an innovative redox-flow battery that does not make use of toxic materials, like aqueous organic electrolyte, providing high and long-term energy storage capacity with lower power output.

This combination ensures the realization of a versatile system, capable of managing both high power demands and sustained energy requirements over time. This hybrid system is particularly suitable for advanced grid applications and "behind-the-meter" systems, where energy storage plays a crucial role in balancing power supply and demand.

The HYBRIS system operates under the principle of "Energy as a Service" (EaaS). It is designed not only to serve the needs of the system owner but also to provide energy services to external stakeholders such as Distribution System Operators (DSOs) or new energy market operators such as Balance Service Providers (BSP) or aggregators. This approach prioritizes interoperability, ensuring that the system can be easily integrated with different energy systems, power grids, and applications. In this hybrid configuration, the LTO battery serves as a power buffer, offering fast response and high-power output to facilitate the optimal use of the AORFB's energy storage capacity. The AORFB, in turn, serves as an energy buffer, working at its most efficient operating points. This synergy ensures that the hybrid system can meet a broad range of energy requirements without overburdening either component.

The development of the HYBRIS system involves several key technical components. More specifically they are:

- **LTO** (Lithium Titanate Oxide) Batteries: they are known for their high-power output (high C-rate provided, both in charge and discharge up to 4C), long life cycle (more than 20,000 cycles) (Sergi et al. 2019), and high safety standards. LTO chemistry uses a lithium titanate oxide anode instead of graphite, which guarantees minimum anode deformation, reduced Solid Electrolyte Interface (SEI) and no lithium metal deposition at low temperatures (Gao et al. 2020), avoiding dendrite formation and internal short-circuits. For these reasons, LTO technology allows for rapid charging and discharging without significant degradation, making it suitable for applications where fast response is crucial. LTO batteries also operate efficiently across a wide temperature range, from − 30 to 55 °C, and are resistant to thermal runaway, an important and critical safety issue of common lithium batteries.

This process does not practically change crystal volume and therefore it does not suffer mechanical stress (Fig. 3). This is the anode used by the SCiB cells and chosen as the power element from hybrid system in HYBRIS project. Having such robust anode material increases the operational currents at which the battery can work without suffering material degradation. This enhances the overall health and cyclability of the LiB system and enables it to operate at high currents for short times (peak power consumption).

- **AORFB** (Aqueous Organic Redox Flow Battery): AORFBs provide large-scale energy storage at a lower cost compared to conventional battery technologies. The technology is based on aqueous electrolytes, which are non-flammable and environmentally friendly. While the energy density is lower compared to LTO batteries, the flow battery technology offers flexibility in terms of power and energy capacity, making it ideal for prolonged energy discharge. AORFBs use organic molecules and recyclable materials, further enhancing the system's sustainability. AORFB stores energy in aqueous based liquid electrolyte solutions which flows through a cell stack during charge and discharge processes. Redox

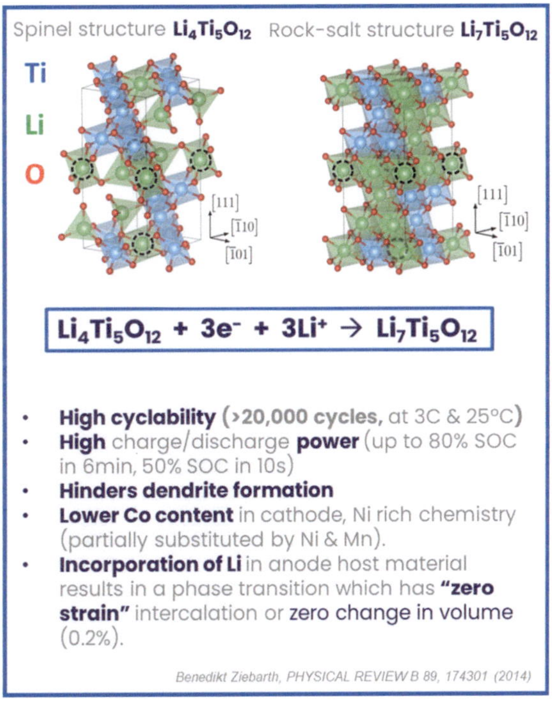

Fig. 3 Lithium Titanate Oxide battery characteristics: LTO unit cell (anode material) and phase transition that undergoes during Li intercalation (Ziebarth et al. 2014), used with permission RNP/25/FEB/088193

reactions occur at the anode and cathode sides that define an electrode compartment (Fig. 4). The electrolyte in each case is called respectively 'anolyte 'and 'catholyte', or 'negolyte' and 'posolyte' as well. Figure 4 shows cyclic voltammograms of the two electrolytes before and after post-treatment compensation, in the reference electrodes.

- **Power Electronics**: The integration of these two battery systems requires advanced power electronics to manage the flow of energy between the batteries and the grid. Studies performed in HYBRIS project demonstrated that the optimum power conversion architecture is the one shown in Fig. 5. This architecture features two stages. The first stage consists of two dc-dc modules that interface the LTO and AORFB batteries with a common dc-link. The second stage consists of a dc-ac module, interfacing the dc-link with the utility grid.

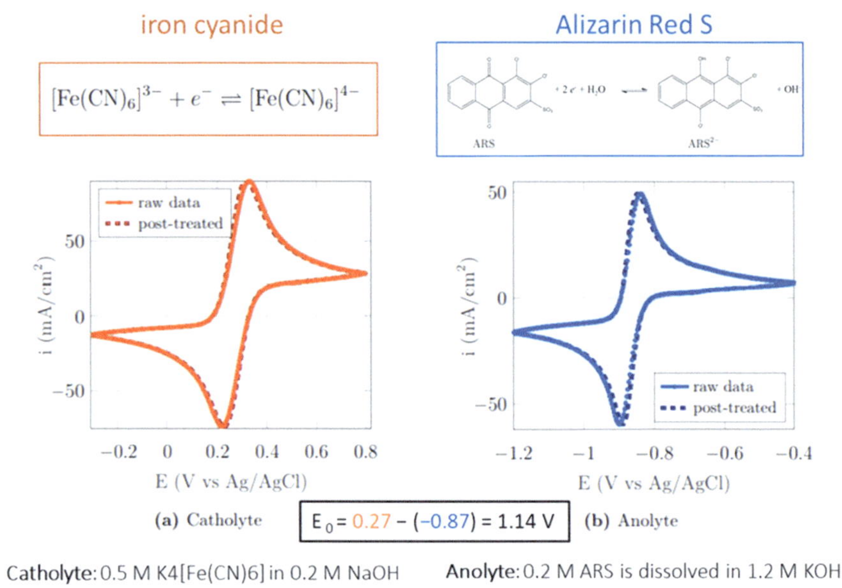

Fig. 4 AORFB: Cyclic voltammograms of the two electrolytes before and after post-treatment compensation ΔRU, v = 100 mV/s (Cazot 2019)

Fig. 5 Optimum power conversion architecture

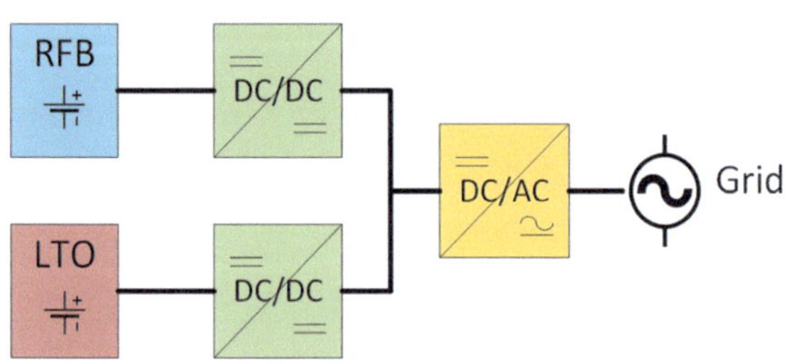

Fig. 6 Power conversion architecture developed in the Hybris project

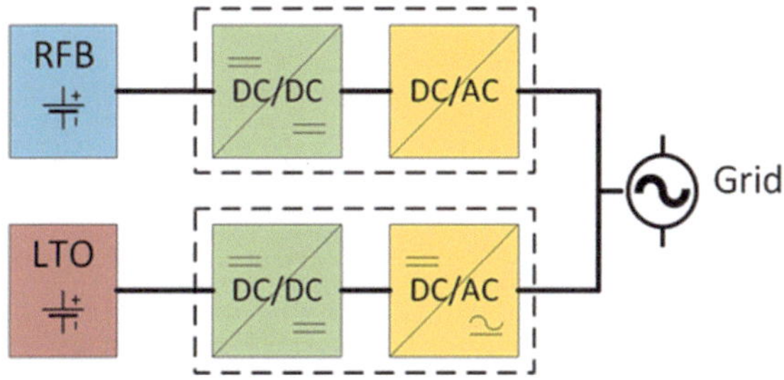

This conversion architecture allows obtaining the best compromise between efficiency, capital cost, and operational cost, compared to other architectures that employ multiple dc-dc modules in parallel, and/or separate dc-ac modules for each battery. The dc-dc modules are implemented with the dual-active-bridge (DAB) converter topology. The DAB converter allows bidirectional power transfer, a high step-up or step-down voltage gain, and provides galvanic isolation, thanks to the use of a high-frequency transformer. Moreover, it features reduced switching losses since all switches feature zero-voltage switching (ZVS) on the turn-on transitions, also known as soft switching. All these features yield on the DAB converter a high efficiency, power density, versatility, and reliability. The dc-ac module is implemented with a three-phase two-level dc-ac converter. The dc-ac converter oversees transferring power from the common dc-link to the utility grid, and vice versa.

Despite the architecture in Fig. 5 can be considered optimal from cost and efficiency point of view, the power conversion architecture reported in Fig. 6 has been chosen for the final development in the Hybris project. The last allows more accurate control of both subcomponents in terms of power exchange with the grid, guaranteeing the provision of power independently to the grid, that increases the reliability of the overall system.

- **Energy Management System** (EMS): The EMS plays a critical role in optimizing the use of both batteries, ensuring that the LTO battery handles short-term power demands while the AORFB manages long-term energy storage. Therefore, this goal is achieved by a proper choice of the set-points dispatched to the two storage subsystems. The EMS also monitors battery health, state of charge (SOC), and other key parameters to ensure efficient operation and longevity of the system.
- **Containerized Design**: to facilitate the deployment of the hybrid system at multiple demonstration sites, the prototype is designed to be portable and housed within a container. This containerized system (Fig. 7) includes all the necessary components, such as battery modules, power converters, control systems, and safety mechanisms. The containerization ensures easy transportation and installation at the different sites.

The use of AORFB technology, which relies on organic molecules and recyclable materials, also helps reduce the environmental impact of energy storage. AORFBs do not rely on rare or toxic materials, making them a greener alternative to conventional

Fig. 7 Hybris containerized hybrid battery energy storage

batteries. The system is also designed to have a long operational life, reducing the need for frequent replacements and minimizing waste.

From an economic perspective, the hybrid system is expected to reduce energy costs by up to 40%, making it a viable solution for both private and commercial applications. The system's flexibility and scalability mean it can be adapted to a wide range of use cases, from small-scale residential energy storage to large-scale industrial applications.

The proposed modules using LTO cells can operate nominally at 1.24 kW (27.6 V @ 45 A) and handle current peaks up to 3 C (135 A). Higher rates up to 5 C (225 A) are possible but require a redesigned cooling system to keep temperatures below 55 °C. Each module has a capacity of 45 Ah, which may be insufficient for some applications. To achieve higher power, more modules need to be stacked. For instance, a 12-module stack can provide 50 kW of power with an energy content of 15 kWh. The AORFB energy capacity is determined by the amount of electrolyte in external tanks, and power depends on the active cell area and number of cells. The simplicity of electrode reactions ensures high lifetime and cyclability, unlike conventional batteries. The system considered uses green, cost-effective technology with organic molecules and recyclable materials. It employs a non-flammable aqueous organic electrolyte, operating safely between 10 and 45 °C, offering advantages like no thermal runaway and greater safety. RFB technology can release energy at constant power for several hours, making it suitable for various energy services. However, it has low power and energy densities due to high volume occupation. Considering the objective of demonstrating the technology, the prototype system was sized with:

1. an AORFB system of 5 kW/15 kWh
2. and an LTO system of 50 kW/15 kWh (Table 1).

The proposed system is the current state of art of respective technologies. Taking a brief outlook of the future of both technologies, lithium batteries are considered mature technology while the AORFBs have currently a lower stage of development. However, both technologies could have important improvements in future years. Lithium batteries are considered at a current stage, called generation 3 (optimized lithium-ion). The close roadmap to 2030 will bring them to generation 4, with the promise of Nichel-rich and low cobalt cathode, higher cell voltage (close to 5 V), but in particular the use of solid-state electrolyte and lithium metal anode. This evolution

Table 1 Hybrid Energy Storage System characteristics

	Units of measurement	Redox Flow KEMIWATT AORFB	Li-ion TOSHIBA SCiB	HESS
Number of cells	–	55	288	–
System nominal capacity	Ah	300	45	–
System nominal voltage	V	50	330	–
System voltage range (min–max)	V	40–70	250–390	–
System max DC current	A	165	200	–
Stored energy	kWh	15	15	30
Peak power	kW	5	50	55

will improve specific energy up to 500 + Wh/kg, + 1000 Wh/L. The following stage (> 2030) will introduce metal air, Lithium-Sulphur and new Ion-based batteries (Itani and De Bernardinis 2023) (Table 2).

On the AORFB side the scenario is less defined and open to important innovations and changes. AORFB constitutes one of the technologies inside the family of Redox Flow Battery (RFB). Both inorganic and organic electrolyte are currently developed with pros and cons for each technology. Only in Europe, several emergent companies (Kemiwatt, involved in the HYBRIS project, Jena Batteries, Green Energy Storage, CMBlu) are developing AORFB at industrial scale using different materials. Major effort is concentrated on the anolyte improvement and, generally, classification is done according to group of anolyte materials. Main technologies are Carbonyl based (anthraquinones, AQS/BQDS all-organic quinone etc.), Quinoxiline based, Viologen based, and Polymer based. Main goals to be reached are reducing materials cost (< 1–2 € kg^{-1}, to meet 0.05 € kW^{-1} h^{-1} $cycle^{-1}$ EU target), improved solubility of redox active species to allow higher energy density (minimum target of 20 Wh L^{-1}, obtained with Vanadium technology), higher stability (value < 0.003%/day for 20 years, able to leave capacity retention > 80%, making them competitive for stationary sector), lifetime (> 10,000 cycles), higher cell potential (> 1 V) and round-trip efficiency

Table 2 Lithium battery evolution, synthetized and adapted from (Itani and De Bernardinis 2023)

	Cell type	Year
Generation 1	Lithium-ion	1991
Generation 2	Lithium-ion	Up to 2005
Generation 3	Optimized lithium-ion	Up to 2025
Generation 4	Solid state lithium-ion and lithium metal	Up to 2030
Generation 5	Metal-air, lithium-Sulphur, new-ion	> 2030

(although not fundamental for RFBs, strongly dependent from the cell type). All these considerations comes from (Brushett et al. 2020; Fontmorin et al. 2022; Sánchez-Díez et al. 2021). Exploration of new materials combination (redox active species with high solubility and stable molecules) will drive the process for performance improvement. In this process, the various mechanisms understanding combined with research on new materials (helped by modern technologies such as computational models, Machine learning etc.) will bring several improvements in coming years to this kind of batteries (Zhu et al. 2024).

4 Control and Communication Architecture

The control architecture of the HYBRIS system is designed to manage the interaction between the two different battery systems and implement the control strategy. This infrastructure enables the system to operate in different modes and deliver several kinds of services such as frequency regulation, peak shaving, energy arbitrage, and increased self-consumption of renewable energy.

The platform operates at two levels (Fig. 8): the former on the cloud and the latter on the HESS system. At cloud level, two elements have been developed: the *EMS (Energy Management System)* and an *ABMS (Advanced Battery Management System)*. The EMS constitutes the logic controller of the system. Through the information coming from the other sub-systems, it implements the optimization algorithm that governs the system behavior, by proper selecting the working setpoints of the two batteries according to the (present and past) state and the selected application/service. The data coming from the physical layer provide the EMS with SoC, battery voltage and current values, temperatures, warnings and alarms, grid status and local requests/production. The building energy data and on field telemetries are also collected by the EMS to feed its control algorithm and give feedback (to allow a close control loop.) In fact, consumption forecast and solar production are used to realise energy optimisation control and real data helps to perform real-time corrections. This EMS computes general strategies depending on external and internal factors, as well as it ensures the transmission of information to the local controller, actuating a separation layer between the IT and OT operations.

The ABMS, at cloud level, uses battery data to feed an internal model of the involved systems. The output of the model feeds the EMS while updating the working points of the two batteries (by considering impacts of the requests on ageing of single batteries). The ABMS periodically sends data to the EMS with yield charts, advanced state of health, power in function of temperature for each battery and technology preference indicator. Then, this information is exploited to optimize long term Hess battery control and maximise its life cycle operations.

At physical level, a Supervisory Control and Data Acquisition (SCADA) system plays a key role in managing the flow of data between the different components of the HESS. The SCADA is basically a device that acquire all the HESS battery data by instance historical data and historical error/warning messages but as well real

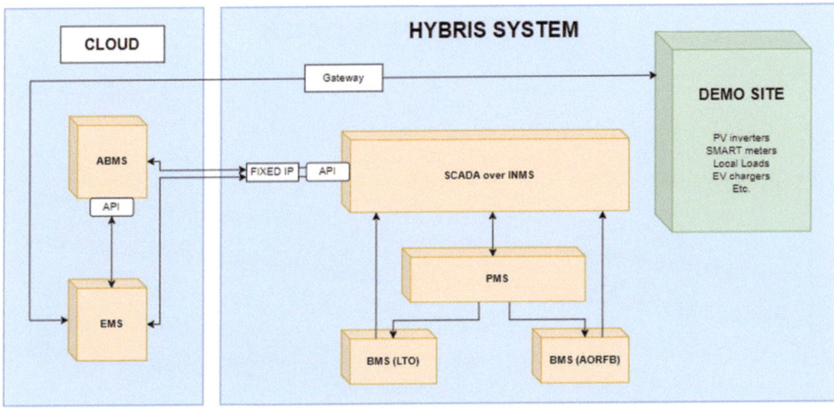

Fig. 8 Overall HYBRIS communication diagram

time data. This device will communicate with several entities. It collects real-time data from two other sub-systems, the *Power Management System (PMS)* and *Battery Management System (BMS)*, including information on the state of charge (SoC) of the batteries, energy consumption, and grid status. This data is then used by the EMS to take informed decisions.

The PMS is responsible for dispatching the operation set points of the LTO and AORFB battery systems with a direct connection with battery inverters and control of the same. Substantially it acts like a passthrough gateway supervisor to manage the two battery inverters (LTO and AORFB). On the other hand, the proprietary BMS monitors the health and status of the batteries, ensuring that they operate within safe and efficient parameters, and informs the higher-level components on the real-time battery status. The SCADA collects all the recovered datapoints and shares them with cloud level.

The realized infrastructure ensures proper communication among each subsystem. In a general view the communication between all the parties could be summarized as follows:

An important element for digitalization of the system and simulation of its behavior, is the HiL device model. This model was built using Typhoon HIL Control Center's freemium software installation and deployed to interact directly with the communication assets on the site. In this way, the real control and communication hardware is connected to the "digital twin"—in this case, a high-fidelity mathematical model of the battery system. These digital twin models are then run in real-time, receiving commands from the control system and returning the resulting measurements just as the real, physical network would. In doing so, the control design and communication infrastructure can be tested and pre-commissioned under real conditions, without the same cost and safety risks as physical tests (Costa and Kavgic 2022).

Three separate levels of HiL system models were developed, with each model building on the previous one, as shown in Fig. 9. At the individual battery level

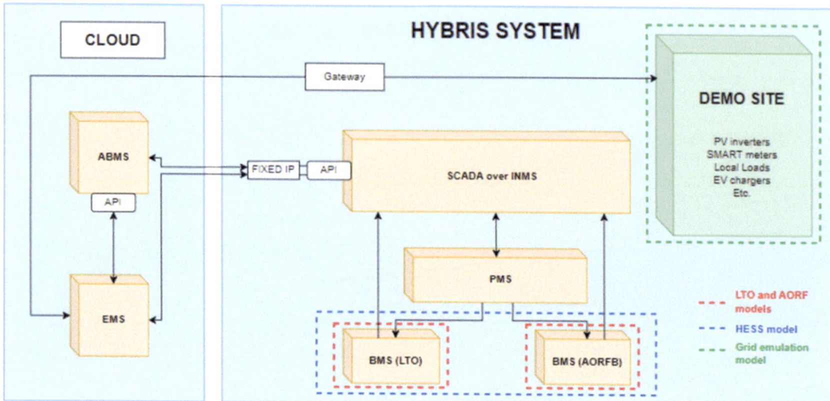

Fig. 9 Representation of the scope of the Digital Twin models in the physical HYBRIS system (Sergi et al. 2022), licensed under CC-BY 4.0

is the model of the LTO and AORFB battery systems, which served the role of supporting the Energy Management System (EMS) development and control of individual batteries. This was later integrated into a HESS-level model, which additionally includes the converters, ancillary loads, and the corresponding local grid balancing and grid forming functionality that they provide. Finally, the HESS system is integrated with a Grid Emulation model to make a site-level model, which allows for control system validation in a real environment.

The individual AORFB and LTO battery models were originally developed in collaboration with IREC and CEA within Matlab/Simulink. This model represented the dynamics of each battery system individually, integrating the design parameters of the final battery systems and reporting on the stack voltage (V), stack current (A), state of charge (SoC) of each cell and the battery stack, and the battery stack temperature. These outputs were considered sufficient for cloud-based estimation of the state of health (SoH), as well as for the cloud-based EMS to properly coordinate and control the individual battery systems.

For real-time testing, the AORFB and LTO models were characterized using an enhanced self-correcting cell model utilizing equivalent circuit representations of the behavior of the original battery system within the Typhoon HIL Control Center toolchain (Typhoon HIL n.d.; Cazot 2019) To validate that the accuracy of these modeling approaches, a 60-min simulation was run involving charging of the battery stacks (LTO and AORFB) with 2 kW of power for 20 min, followed by a 20-min 2 kW discharge, and finally 20 min of 0 W power demand. The simulation was performed with a time resolution of 100 μs in the Typhoon HIL environment and 200 ms for the Simulink environment.

Various measurements were taken once every two seconds, resulting in 1800 data points. Measured and compared variables for the LTO battery stack are displayed in Fig. 10:

Fig. 10 LTO battery stack comparison

1. Vstack/V—voltage of the stack
2. Current/A—current of the stack
3. SOC/%—state of charge of each battery cell and of the entire battery stack
4. Temperature/°C—temperature of the battery stack in Celsius.

Measured and compared variables for AORFB battery stack are displayed in Fig. 11:

1. Vstack/V—voltage of the stack
2. Current/A—current of the stack

Fig. 11 AORFB battery stack comparison

3. SOC/%—state of charge of each battery cell and of the entire battery stack.

As shown in Figs. 10 and 11, validation of AORFB and LTO battery performance within the Matlab/Simulink and Typhoon HIL Control Center models were found to be virtually identical (Sergi et al. 2022). For this reason, the Typhoon HIL Control Center Battery Cell model was used exclusively for the planned real-time tests. This allowed for rapid control development and initial validation of EMS performance in a safe environment, prior to physical battery integration. An example of the battery-level model adapted for use in rapid control development is shown in Fig. 12.

To support communication between the model and the cloud, the HiL model includes a Modbus Device component, which is parameterized based on the specific network in which the Typhoon HIL device is housed. Additionally, local control logic was added to this component as needed by implementing either with code or as logical structures using the built-in library components. Scada Input and output components are added as needed to make the data outputs available to other partners.

To ensure continuous availability of data, the Typhoon HIL device was run in a standalone boot configuration mode. In this mode, the model self-boots upon powering up the device, and runtime interaction with the model takes place via MODBUS commands sent over the Ethernet port of the HIL device. While the functionalities of Standalone boot mode are more limited, it does allow the HiL model to

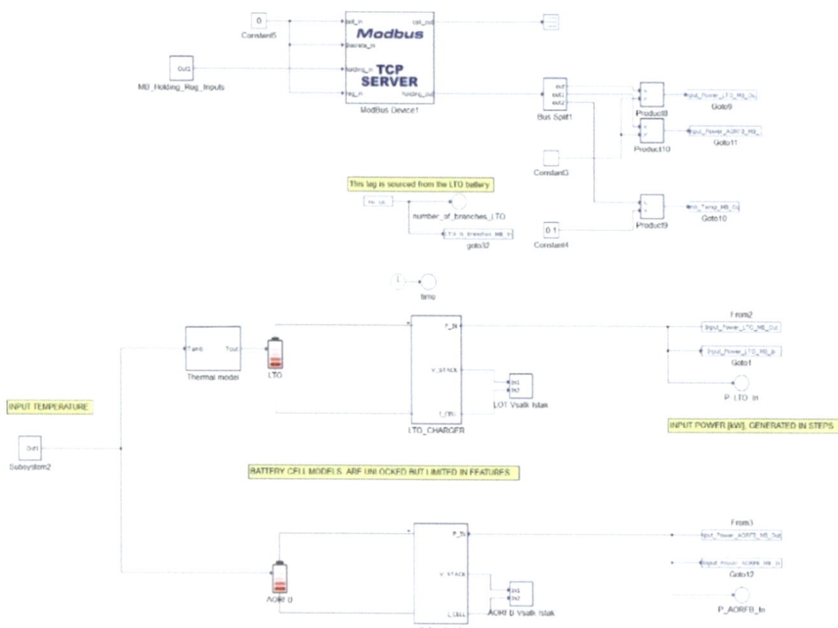

Fig. 12 Example of an LTO and AORFB model with a Modbus Device component and associated local control logic implemented (top)

run in real-time without requiring a dedicated computer available to send or receive data.

This battery level model formed the core of what would later become the integrated HESS model. The HESS-level model allowed EMS developers to validate their control actions in real-time, using the same commands that would be used on the physical battery. This required integrating a bi-directional inverter that integrated the same grid forming functionalities intended for the physical LTO inverter, as well as the necessary communication interfaces which translated the EMS commands into real-time commands to stimulate the model. This is shown on the left side of Fig. 13.

Once these tests were implemented, the HYBRIS HESS model was then connected to a real-time representation of one of the physical sites, utilizing the same parameters as the existing on-site assets, gathered from electrical diagrams of the site itself and nameplates of the actual devices on site. A simple example of this is shown in the scheme in Fig. 13. With this site-level model, full EMS control functionality could be validated under a variety of conditions.

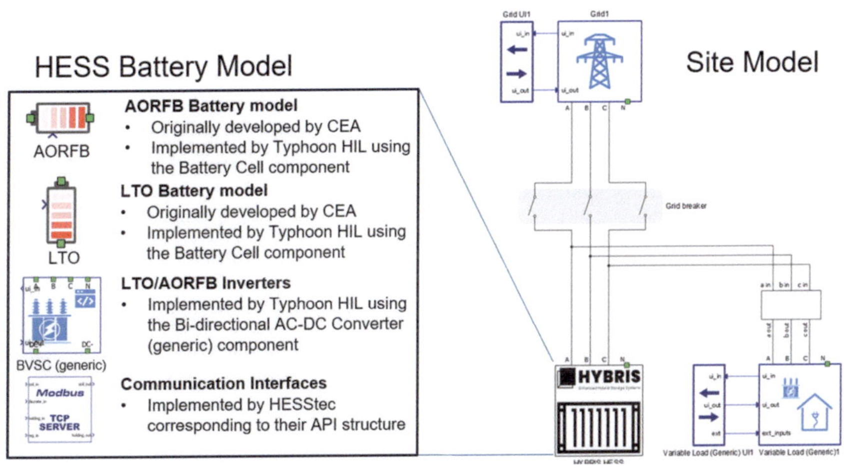

Fig. 13 Representation of the components of the virtual HESS model (left) and its placement within a physical site model (right)

5 Preliminary Parallel Testing of the Batteries Through Common Switch and Controlled by SCADA and Validation of the HESS Performance with Real Case Studies and HIL Simulation

Once defined the systems and communicating architecture (see Sect. 4) preliminary tests have been applied in parallel to the containerized systems. Previously to any service application, it is necessary to characterize the systems working in parallel and test the communication and lecture of commands and data, respectively, through the SCADA controlling commands. Therefore, a simple test of discharge and charge cycles at low power has been launched in parallel at the Li-ion (TOSHIBA-SCiB) battery and the AORFB (KEMIWATT) battery systems. Voltage, state of charge, current, power setpoints and temperature have been registered among other relevant information of the systems given by the respective BMS. The first approach shows that several preconfigured limitations of the systems must be considered and plausibly modified to provide services to customers. For example, de-ratings setpoints modify automatically power setpoints given a certain SoC at each system. These behaviors must be considered when the automatized response wants to be implemented. The preliminary measurements done on the hybrid system via a router with fixed IP and controlled by SCADA are shown in Fig. 14. The router derives the commands to a switch following MODBUS protocol. The switch in turn derives the commands to the two different battery systems. These measurements are done using both batteries at the same time. Data has been recorded directly from respective converters. On the left side of the figure a discharge–charge curve for two power setpoints on the Li-battery system is shown. Voltage, SoC, current and power as functions of time measured for the LiB system are represented. Red dotted lines indicate the change in power setpoints. On the right-side similar measurements carried out on AORFB system are reported. In both cases we observe the derating adjustments present on the batteries.

To demonstrate the performance of the HESS under real systems, both physical and virtual prototypes were developed within the HYBRIS project. These prototypes were deployed at several demonstration sites to validate performance for several of the behind-the-meter use cases described previously.

The containerized design of the HYBRIS HESS yields certain advantages when testing the prototype for multiple demonstration sites. Despite the AORFB's low energy density, placing both batteries inside of a containerized unit allows for much easier logistical deployment at different sites. This is in large part due to the capability of pre-assembly and commissioning of the entire battery unit ahead of first deployment. For subsequent re-deployments of the battery the liquid components of the AORFB need only to be drained and transported separately before being reincorporated into the system at the new site. Once all proper connections to site equipment are made and validated, the battery is ready to provide energy services.

Within the HYBRIS project, for instance, this allowed a single battery to be tested at multiple case study sites, saving on total system costs for multiple battery

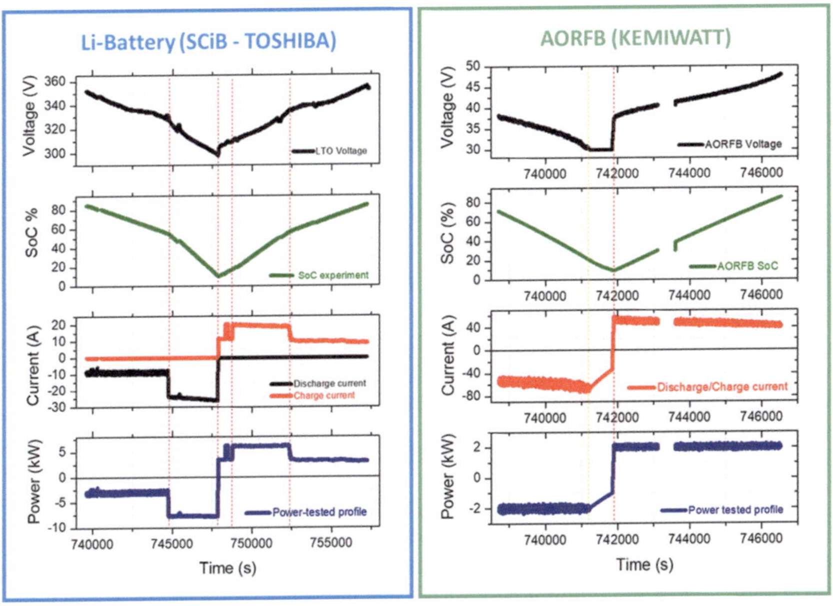

Fig. 14 Preliminary test on hybrid system. On the left side, LIB discharge–charge curve for two power setpoints. From top to bottom: voltage, SoC, current and power as a function of time. On the right-side same measurement carried out on AORFB system

prototype systems. While this method of re-deployment is not recommended for extremely short periods, due to the need for ground infrastructure to support the container's weight and the expertise required for ensuring proper air-tight cycling of the AORFB fluids, repurposing a single battery container to multiple locations can be useful in cases where a static battery is needed to provide regular energy and power services over months at a time.

The HYBRIS project demonstrated battery system performance at a total of three sites. The first of these case studies covers a small energy community in Messina managed by Solidarity and Energy, with several PV panels and an ESS already installed. The primary intention of this case study is to validate the capability of the HESS storage to augment the energy storage capabilities on site, increasing the self-consumption of PV energy and avoiding power outages (SAIFI, load shedding events, etc.). In these cases, the HESS can provide island operation for a limited period until power is restored.

The second case study involves a business park of 20+ mixed business companies in Belgium, managed by Quares. At this site, the focus is on demonstrating the capabilities of using the HESS to provide energy services for the grid in the form of peak shaving, enhanced self-consumption, and load shifting/energy arbitrage. For these cases, the advantages of the long cycle life and flexible storage capacity of the redox flow battery is expected to provide consistent energy bill savings that couldn't normally be achieved solely via a single, lithium battery solution.

The final case study involves a glass factory in Tholen, Netherlands. This site is equipped with PV, and will integrate a cloud-controllable EV charger, with the goal of developing a local energy community in the near future. The primary purpose of the HESS in this case is to improve energy bill savings—like the Belgium case.

In particular, the demo site in Messina was the first one tested and the only one in which the system was physically installed. For the other two sites, virtual implementation was considered. Historical data and data coming from the Messina site were used for real model development and virtual implementation. Data includes information such as the load demand on site and grid frequency from which the battery control should react to. For the purpose of this chapter, the Messina site will be considered with the various services provided and the benefits brought to the small community.

To resume the main services considered for the use case Table 3 is provided.

To show the effective functionality of the system the services described were tested both in field test and simulated way. Some of them are briefly reported below.

In Figs. 15 and 16 the commutation from grid-following **to island mode** of the HESS is demonstrated through the HIL for a small grid, by using the site model shown in Fig. 13. In Fig. 15, the grid breaker is closed, with the battery operating in grid following mode. The EMS sends an Active power reference of 30 kW to the LTO battery and 5 kW to the AORFB, forcing the charge using excess power from the grid. The resulting battery status and battery state of charge are all visible.

Following a fault in the grid, the grid breaker tripped. This forces the batteries to switch to grid-forming configuration led by the LTO inverter operating in droop mode. To meet the load demand, the EMS ensures that 35 kW from the LTO battery and 5 kW from the AORFB. The resulting conditions are shown in Fig. 16.

Since the HESS model supports real-time interaction, utilizing the developed communication infrastructure and load and generation profiles reproduce the site itself. This type of integration represents a realistic representation (like a digital twin) of the demonstration site. The EMS can send commands to this digital twin in real-time, precisely as it would control the physical system. Then it receives back the resulting dynamics within the grid, including all critical measurements. Afterwards, it needs to estimate the proper controls for the next time window, such as peak voltage and current, state of charge of the battery, and other parameters.

Once aggregated data coming from simulation, the total energy savings achieved over a set period can be calculated. This allows maximum-fidelity comparisons of system performance under "what-if" scenarios. Such HiL approach demonstrates that an AORFB system and its integration within a HESS can be modeled in a real-time framework.

In some countries, such as Belgium and the Netherlands, a "capacity fee" is charged, which bills consumers not only for the energy consumed but also for the maximum 15-min average monthly peak power. The goal of the **peak shaving** functionality is to reduce these monthly peaks and, consequently, the yearly average. Therefore, in this case, the HESS is controlled to keep the grid power below a predefined peak target. If the shaving of the peak fails, e.g., the portion of the peak is too high or the peak energy exceeds the available amount, the target is reset. To

Table 3 BTM services explored with technical issues and provided benefits

Service	Implementation	Target result and/or expected benefit (KPI metric)	Other details
Island operation and SAIFI reduction	The HESS inverter switches the use mode, normally grid-connected, to provide energy to loads connected to privileged line	No grid interruption/ fault in island operation. The target for the site is to avoid 40 interruptions per year (practically 100%)	Average interruption < 5 min. Time response of few seconds
Peak shaving	HESS intervention to reduce peak power used inside 15 min gate time over a certain threshold	Reduce the peak buildings capacity more than 10 kW monthly. Results are validated at the end of a month	A peak target for the current month is set at the beginning, which we shouldn't exceed. All 50-kW provided can be used. LTO prevailing
Energy arbitrage	HESS management to buy and sell energy, according to the energy prices to generate profit	Evaluate cost revenue (€) considering monthly use of arbitrage (in terms of hours in which the service is active)	Total energy usable set at a % of the total kWh. Service activated on command. LTO and AORFB combined use
Increase to self-consumed energy	HESS maintains the grid as close as possible to 0 kW, storing the excess of PV production, and releasing it when necessary	Increasing the self-consumption by > 75%. Self-sufficiency is increased using HESS even at night using about up to 15KWh	Primary use of solar energy, HESS charged only with excess. HESS working during day and night
Demand/response	HESS follows D&R profiles. Simulated D&R service from the TSO is considered to set the overall power profile for a specific period (not known a priori)	Benefit in terms of revenue of the services (€) to the community and more in general benefits to the whole electric grid	To release this service, a certain amount of energy must be considered unusable for other services. This can be done setting a low threshold of SOC for other services
Energy community	HESS share energy with the whole community	A minimum energy sharing of 50 kWh/ day to provide cost savings and social benefits to disadvantaged users	PV production and load consumption typical/historical profile can be used to manage the system and provide benefits to the whole community

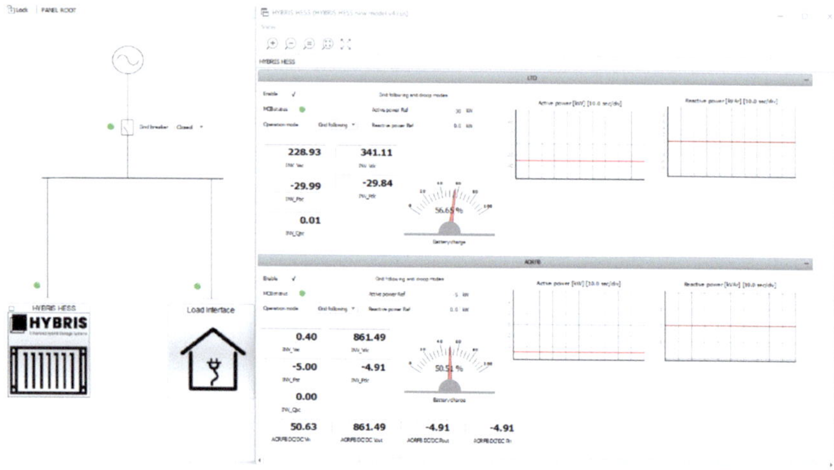

Fig. 15 HiL simulation of HESS operation in grid following mode (battery charging)

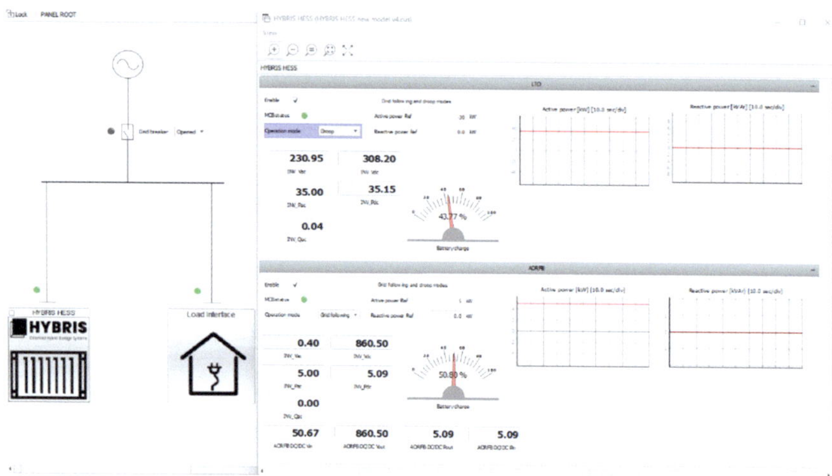

Fig. 16 HiL simulation of HESS operation in Island mode (battery discharging)

achieve this, the EMS uses historical, forecasts, and real-time measurements. If a peak is soon expected, the EMS uses a faster (one-minute basis) loop to control the battery discharge. The LTO will provide most of the value here, but the AORFB could provide the flexibility to remove a portion of the base load. By updating the peak target as the month progresses, the system can avoid shaving peaks lower than the monthly average. The principle of the functionality is described in the Fig. 17.

From historical data the highest peak of the last months can be retrieved. To reduce the peak, the target is set at a fraction of the historical peak, e.g. 60% (5 kW in this case, picture (a)). If the grid power (on a one-minute basis) is observed to exceed the

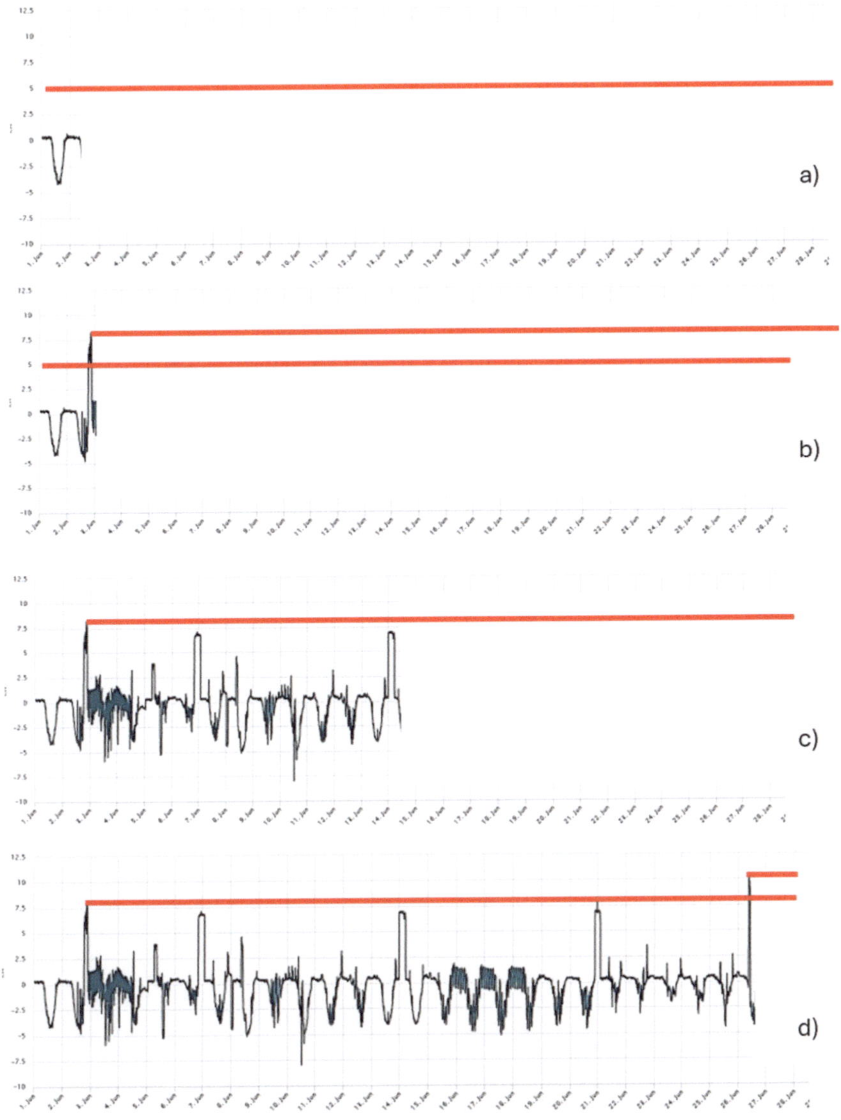

Fig. 17 Peak shaving behaviour during one month of operation

peak target, the algorithm will try to shave it down to the target limit. If the target (on a 15 min basis) is exceeded, then the target will be recalculated (e.g. 8 kW). This operation is done daily (picture (b)). All peaks which remain below the new target are not shaved (picture (c)). At the end of the month, a new consumption peak is encountered and, in this case, also successfully shaved (picture d) and target was not reset. Overall, without this functionality a peak of more than 10 kW would not have been managed, with a consequent impact on the user. The next month's target is based on last month's final target, where again a percentage is applied (50%) with the aim of achieving a better result.

The principle of **arbitrage** relies on the knowledge of the published day-ahead electricity prices. By strategically scheduling the purchase and selling of energy through the battery's charge–discharge cycles, the user can exploit price fluctuations to generate profit. The potential gains are influenced by both the price variability and the difference between the purchase and selling prices. This difference can be impacted by national taxes and transmission charges. For instance (as illustrated in Fig. 18), this gap can be relatively small (as in Netherlands) because there is only a 21% VAT and no transmission fees. Therefore, since the primary goal is to buy at low prices and sell at high prices, this case is an advantageous application case of arbitrage.

In other instances (e.g., Belgium), the selling price is generally much lower than the purchase price, as illustrated in Fig. 19. This discrepancy arises from the combination of transmission and distribution charges (which account for about two-thirds of the total cost per kWh, only while the prosumer drains power from the grid), and other taxes. However, when energy is sold back to the grid, the prosumer is only compensated for the energy itself (without any reimbursement to compensate for transmission and distribution charges). Consequently, arbitrage is less profitable in this scenario.

Finally, **self-consumption** involves using locally produced energy to balance the building's energy consumption. When production exceeds demand, the energy surplus is stored. This stored energy can be later used to reduce grid electricity consumption and generate profit. This concept is like arbitrage, since the surplus energy is stored and used later (at no cost). The goal is to use stored energy when the

Fig. 18 DAM price without transmission fees

Fig. 19 DAM price with transmission fees

Fig. 20 DAM price self-consumption

purchase price is highest. As illustrated in Fig. 20, this is equivalent to selling energy to the grid and receiving the purchase price (orange plot). On the contrary, sending the excess energy back to the grid would only yield the sell price (green plot). Therefore, once again, in countries where there is a significant gap between the purchase and sell prices, there is a strong incentive to maximize self-consumption.

As shown in these few examples of applications, the benefits coming from the HESS system and its corresponding release of energy services have a direct benefit on the end user energy exploitation. In addition, indirect advantages to the main electric grid support an overall better management of the energy flows, thanks to a local optimized use of the energy.

6 Conclusions

A first of a kind hybrid energy storage system made of the combination of an Aqueous Organic Redox Flow battery and a Lithium Titanate battery has been demonstrated in a European project called Hybris through preliminary parallel testing and hardware-in-the-loop tests, designed for behind-the-meter grid services applications. The results showed that both digital twin models and HiL representations can reproduce real case studies. The system has been designed to provide both energy and power intensive services. Control and communication architecture, using both cloud and local solutions have been integrated together with batteries and power electronics in a unique containerized energy storage system, easy to transport and install to the end user premises.

To achieve more details on the real operation of the HEES and to evaluate the ability to provide behind-the-meter grid services, the system has been demonstrated in the Messina (Italy) community, where the different use cases have been validated in real environment.

Features offered by the combined use of two different systems extend the range of possible applications and enable the possibility to perform multiple services at once. In the case of a small community as in the Messina case study, the benefits are evident and involve several aspects, from a better use of the green energy locally produced, to balancing of social differences, through a fairer use of energy within the community.

Acknowledgements The authors acknowledge Aug-E company that developed EMS and performed test campaign during Hybris project.

Funding This work has been developed in the framework of a European Project called HYBRIS—*Hybrid Battery energy stoRage system for advanced grid and beHInd-de-meter Segments*, funded by European Commission (Grant Agreement No. 963652) in the Horizon 2020 program.

References

Boshell AS, Kamath F, Kanani S, Mehrotra H (2019) Utility-scale batteries innovation landscape brief

Brunaccini G, Sergi F, Aloisio D, Ferraro M, Blesznowski M, Kupecki J, Motylinski K, Antonucci V (2017) Modeling of a SOFC-HT battery hybrid system for optimal design of off-grid base transceiver station. Int J Hydrogen Energy 42:27962–27978. https://doi.org/10.1016/j.ijhydene.2017.09.062

Brushett FR, Aziz MJ, Rodby KE (2020) On lifetime and cost of redox-active organics for aqueous flow batteries. ACS Energy Lett 5:879–884. https://doi.org/10.1021/acsenergylett.0c00140

Cazot M (2019) Development of analytical techniques for the investigation of an organic redox flow battery using a segmented cell

Costa S, Kavgic A (2022) How digital twins accelerate the energy transition [WWW Document]. https://www.typhoon-hil.com/blog/digital-twins-accelerate-energy-transition/

Fitzgerald G, Mandel J, Morris J (2015) The economics of battery energy storage: how multi-use, customer-sited batteries deliver the most services and value to customers and the grid (technical appendix)

Fontmorin J-M, Guiheneuf S, Godet-Bar T, Floner D, Geneste F (2022) How anthraquinones can enable aqueous organic redox flow batteries to meet the needs of industrialization. Curr Opin Colloid Interface Sci 61:101624. https://doi.org/10.1016/j.cocis.2022.101624

Gao X, Zhou Y-N, Han D, Zhou J, Zhou D, Tang W, Goodenough JB (2020) Thermodynamic understanding of Li-dendrite formation. Joule 4:1864–1879. https://doi.org/10.1016/j.joule.2020.06.016

Ibrahim H, Rezkallah M, Ilinca A, Ghandour M (2021) 10—Hybrid energy storage systems. In: Kabalci E (ed) Hybrid renewable energy systems and microgrids. Academic Press, pp 351–372. https://doi.org/10.1016/B978-0-12-821724-5.00004-0

Itani K, De Bernardinis A (2023) Review on new-generation batteries technologies: trends and future directions. Energies (Basel) 16. https://doi.org/10.3390/en16227530

Leonardi SG, Aloisio D, Brunaccini G, Stassi A, Ferraro M, Antonucci V, Sergi F (2021) Investigation on the ageing mechanism for a lithium-ion cell under accelerated tests: the case of primary frequency regulation service. J Energy Storage 41:102904. https://doi.org/10.1016/j.est.2021.102904

Lucà Trombetta G, Leonardi SG, Aloisio D, Andaloro L, Sergi F (2024) Lithium-ion batteries on board: a review on their integration for enabling the energy transition in shipping industry. Energies (Basel) 17. https://doi.org/10.3390/en17051019

Rezaeimozafar M, Monaghan RFD, Barrett E, Duffy M (2022) A review of behind-the-meter energy storage systems in smart grids. Renew Sustain Energy Rev 164:112573. https://doi.org/10.1016/j.rser.2022.112573

Sánchez-Díez E, Ventosa E, Guarnieri M, Trovò A, Flox C, Marcilla R, Soavi F, Mazur P, Aranzabe E, Ferret R (2021) Redox flow batteries: status and perspective towards sustainable stationary energy storage. J Power Sources 481:228804. https://doi.org/10.1016/j.jpowsour.2020.228804

Sergi F, Brunaccini G, Aloisio D, Randazzo N, Polito RM, Pietrucci M, Fadda MG, Ferraro M, Antonucci V (2019) Evaluation of a Li-Titanate battery module in primary frequency control ancillary service conditions. J Energy Storage 24:100805. https://doi.org/10.1016/J.EST.2019.100805

Sergi F, Arista A, Agnello G, Ferraro M, Andaloro L, Antonucci V (2016) Characterization and comparison between lithium iron p hosphate and lithium-polymers batteries. J Energy Storage 8:235–243. https://doi.org/10.1016/j.est.2016.08.012

Sergi F, Aloisio D, Lucà Trombetta G (2022) Hybrid storage optimal sizing and control

Typhoon HIL (n.d.) Description of the battery cell component in schematic editor [WWW Document]. https://www.typhoon-hil.com/documentation/typhoon-hil-software-manual/References/battery_cell.html

Xiaojuan L, Qi G, Haiying D (2017) Multi objective optimization of hybrid energy storage micro grid based on CMOPSO algorithm. Acta Energiae Solaris Sinica 279–286

Zhu F, Chen Q, Fu Y (2024) Perspectives on aqueous organic redox flow batteries. Green Energy Environ 9:1641–1649. https://doi.org/10.1016/j.gee.2024.08.003

Ziebarth B, Klinsmann M, Eckl T, Elsässer C (2014) Lithium diffusion in the spinel phase $Li_4Ti_5O_{12}$ and in the rocksalt phase $Li_7Ti_5O_{12}$ of lithium titanate from first principles. Phys Rev B 89:174301. https://doi.org/10.1103/PhysRevB.89.174301